Clean and Low Carbon Energy

Clean and Low Carbon Energy

Editors

Dong Li
Fuqiang Wang
Zhonghao Rao
Chao Shen

MDPI • Basel • Beijing • Wuhan • Barcelona • Belgrade • Novi Sad • Cluj • Manchester

Editors

Dong Li
School of Civil Engineering and Architecture, Northeast Petroleum University
Daqing
China

Fuqiang Wang
School of New Energy, Harbin Institute of Technology
Weihai
China

Zhonghao Rao
School of Energy and Environmental Engineering, Hebei University of Technology
Tianjin
China

Chao Shen
Department of Building Thermal Energy Engineering, Harbin Institute of Technology
Harbin
China

Editorial Office
MDPI
St. Alban-Anlage 66
4052 Basel, Switzerland

This is a reprint of articles from the Topic published online in the open access journals *Energies* (ISSN 1996-1073), *Nanomaterials* (ISSN 2079-4991), *Clean Technologies* (ISSN 2571-8797), and *Sustainability* (ISSN 2071-1050) (available at: https://www.mdpi.com/topics/clean_low_carbon_energy).

For citation purposes, cite each article independently as indicated on the article page online and as indicated below:

Lastname, A.A.; Lastname, B.B. Article Title. *Journal Name* **Year**, *Volume Number*, Page Range.

ISBN 978-3-7258-1347-6 (Hbk)
ISBN 978-3-7258-1348-3 (PDF)
doi.org/10.3390/books978-3-7258-1348-3

Contents

Ruitong Yang, Fuqiang Wang, Zhonghao Rao, Chao Shen and Dong Li
Advancing Sustainable Energy Solutions: Innovations in Clean Energy Applications and Conventional Energy Efficiency Upgrade
Reprinted from: *Energies* **2024**, *17*, 2441, doi:10.3390/en17102441 . . . . **1**

Yafen Han, Shuai Li, Haidong Liu and Yucong Li
Lattice Boltzmann Simulation of Coupling Heat Transfer between Solid and Gas Phases of Nanoporous Materials
Reprinted from: *Nanomaterials* **2022**, *12*, 3424, doi:10.3390/nano12193424 . . . . **5**

Bo Ning, Yanjiang Yu, Jing Zeng and Qiuping Lu
A Review of Stimulation Technologies for Weakly-Consolidated Natural Gas Hydrate Reservoirs
Reprinted from: *Energies* **2022**, *15*, 7195, doi:10.3390/en15197195 . . . . **18**

Kirsi Spoof-Tuomi, Hans Arvidsson, Olav Nilsson and Seppo Niemi
Real-Driving Emissions of an Aging Biogas-Fueled City Bus
Reprinted from: *Clean Technologies* **2022**, *4*, 59, doi:10.3390/cleantechnol4040059 . . . . **35**

Anukriti Pokhriyal, José Luis Domínguez-García and Pedro Gómez-Romero
Impact of Battery Energy System Integration in Frequency Control of an Electrical Grid with Wind Power
Reprinted from: *Clean Technologies* **2022**, *4*, 60, doi:10.3390/cleantechnol4040060 . . . . **53**

Xianrong Zhuang and Lingying Pan
Study on the Impact of Clean Power Investment on Regional High-Quality Economic Development in China
Reprinted from: *Energies* **2022**, *15*, 8364, doi:10.3390/en15228364 . . . . **68**

Guilherme Jesus, Martim L. Aguiar and Pedro D. Gaspar
Computational Tool to Support the Decision in the Selection of Alternative and/or Sustainable Refrigerants
Reprinted from: *Energies* **2022**, *15*, 8497, doi:10.3390/en15228497 . . . . **91**

Muhammad Hussain, Hussain Al-Aqrabi and Richard Hill
PV-CrackNet Architecture for Filter Induced Augmentation and Micro-Cracks Detection within a Photovoltaic Manufacturing Facility
Reprinted from: *Energies* **2022**, *15*, 8667, doi:10.3390/en15228667 . . . . **111**

Da-Sheng Zhang, Xiao-Hua Gu, Si-Wen Liu, Yan Liu, Qing-Yun Zhou, Shang-Wen Zhu, et al.
Study on Properties of Regenerated Fluorinated Polyurethane Rigid Foam Prepared by Degrading Waste Polyurethane
Reprinted from: *Sustainability* **2022**, *14*, 15685, doi:10.3390/su142315685 . . . . **127**

Djati Wibowo Djamari, Muhammad Idris, Permana Andi Paristiawan, Muhammad Mujtaba Abbas, Olusegun David Samuel, Manzoore Elahi M. Soudagar, et al.
Diesel Spray: Development of Spray in Diesel Engine
Reprinted from: *Sustainability* **2022**, *14*, 15902, doi:10.3390/su142315902 . . . . **144**

Tao Xing, Shutai Yu, Jingliang Tang, Huiliang Liu, Feng Zhen, Yongming Sun, et al.
Liquid–Liquid Extraction of Volatile Fatty Acids from Anaerobic Acidification Broth Using Ionic Liquids and Cosolvent
Reprinted from: *Energies* **2023**, *16*, 785, doi:10.3390/en16020785 . . . . **166**

Zhao Xin-gang, Wang Wei, Hu Shuran and Liu Xuan
Impacts of Government Policies on the Adoption of Biomass Power: A System Dynamic Perspective
Reprinted from: *Sustainability* **2023**, *15*, 1723, doi:10.3390/su15021723 . . . . **177**

Asrin Seyedzahedi and Salah Bahramara
Facilitating Investment in Photovoltaic Systems in Iran Considering Time-of-Use Feed-in-Tariff and Carbon Market
Reprinted from: *Energies* **2023**, *16*, 1067, doi:10.3390/en16031067 . . . . **188**

Likun Ai, Yiping Lu, Jiade Han and Wenxu Suo
Simulation of the Temperature of a Shielding Induction Motor of the Nuclear Main Pump under Different Turbulence Models
Reprinted from: *Energies* **2023**, *16*, 2792, doi:10.3390/en16062792 . . . . **208**

James Chilver-Stainer, Anas F. A. Elbarghthi, Chuang Wen and Mi Tian
Power Output Optimisation via Arranging Gas Flow Channels for Low-Temperature Polymer Electrolyte Membrane Fuel Cell (PEMFC) for Hydrogen-Powered Vehicles
Reprinted from: *Energies* **2023**, *16*, 3722, doi:10.3390/en16093722 . . . . **223**

Dai Geng, Di Wang, Yushuang Li, Wei Zhou and Hanbing Qi
Detection Stability Improvement of Near-Infrared Laser Telemetry for Methane Emission from Oil/Gas Station Using a Catadioptric Optical Receiver
Reprinted from: *Energies* **2023**, *16*, 3854, doi:10.3390/en16093854 . . . . **241**

Y. Nhu Quach and Ocktaeck Lim
An Investigation of the Effect of Propylene Gas Flame on Emissions and Temperature Distribution of a Preheated Metal Plate
Reprinted from: *Sustainability* **2023**, *15*, 12306, doi:10.3390/su151612306 . . . . **257**

Bin Xiong and Qi Sui
Does Carbon Emissions Trading Policy Improve Inclusive Green Resilience in Cities? Evidence from China
Reprinted from: *Sustainability* **2023**, *15*, 12989, doi:10.3390/su151712989 . . . . **270**

Hongqiang Wang, Wenyi Xu and Yingjie Zhang
Research on Provincial Carbon Emission Reduction Path Based on LMDI-SD-Tapio Decoupling Model: The Case of Guizhou, China
Reprinted from: *Sustainability* **2023**, *15*, 13215, doi:10.3390/su151713215 . . . . **286**

Rong Zhuang, Feng Zhou, Xuwen Tian, Buqing Xu, Shaocong Li and Guoyuan Ma
Performance Study of Booster-Driven Hybrid Cooling Units for Free Cooling in Data Centers
Reprinted from: *Sustainability* **2023**, *15*, 14558, doi:10.3390/su151914558 . . . . **306**

Yubin Lin, Qinhui Wang, Chao Ye, Yao Zhu and Haojie Fan
Experimental Research on the Gas-Solid Flow Characteristics in Large-Scale Dual Fluidized Bed Reactor
Reprinted from: *Energies* **2023**, *16*, 7239, doi:10.3390/en16217239 . . . . **325**

Shangwen Zhu, Xiaohua Gu, Mengyi Duan, Feiyang Yu, Danyi Zhao, Siwen Liu, et al.
Analysis of the Influence Factors of the Crude Oil Temperature Maintenance System of Solar Sewage Heat Pumps in Cold Regions
Reprinted from: *Energies* **2023**, *16*, 8124, doi:10.3390/en16248124 . . . . **344**

Yubin Lin, Qinhui Wang, Guilin Xie, Mengxiang Fang and Zhongyang Luo
Numerical Simulation Study on the Gas–Solid Flow Characteristics of a Large-Scale Dual Fluidized Bed Reactor: Verification and Extension
Reprinted from: *Energies* **2024**, *17*, 1304, doi:10.3390/en17061304 . . . . **364**

energies

MDPI

Editorial

Advancing Sustainable Energy Solutions: Innovations in Clean Energy Applications and Conventional Energy Efficiency Upgrade

Ruitong Yang [1], Fuqiang Wang [2], Zhonghao Rao [3], Chao Shen [4] and Dong Li [1,*]

1 School of Civil and Architectural Engineering, Northeast Petroleum University, Daqing 163318, China; rt_yang@stu.nepu.edu.cn
2 School of New Energy, Harbin Institute of Technology, Weihai 264209, China; wangfuqiang@hitwh.edu.cn
3 School of Energy and Environmental Engineering, Hebei University of Technology, Tianjin 300401, China; 2021101@hebut.edu.cn
4 Department of Building Thermal Energy Engineering, Harbin Institute of Technology, Harbin 150006, China; chaoshen@hit.edu.cn
* Correspondence: lidong@nepu.edu.cn

Citation: Yang, R.; Wang, F.; Rao, Z.; Shen, C.; Li, D. Advancing Sustainable Energy Solutions: Innovations in Clean Energy Applications and Conventional Energy Efficiency Upgrade. *Energies* **2024**, *17*, 2441. https://doi.org/10.3390/en17102441

Received: 13 May 2024
Accepted: 20 May 2024
Published: 20 May 2024

The transition to clean and low-carbon energy sources is an imperative driven by the urgent challenges of climate change, energy security, and environmental sustainability. As we face these challenges, it becomes essential to foster technologies and strategies that not only support sustainable growth but also align with environmental conservation. This editorial presents a curated collection of innovative research studies that contribute significantly to advancing our understanding and implementation of sustainable energy solutions.

Innovations in clean energy and its applications are pivotal for sustainable growth. The integration of solar energy with advanced materials is further exemplified by Hussain et al. [1], who introduce a novel architecture for detecting micro-cracks in photovoltaic cells, thereby enhancing the durability and efficiency of solar panels. Zhu et al. [2] investigate the performance enhancements of solar collectors in heating systems, offering insights that can optimize the use of solar thermal systems in industrial applications like crude oil heating. Spoof-Tuomi et al. [3] assess the real driving emissions of biogas-fueled buses, supporting the use of biogas as a clean alternative in public transport systems. Moreover, given the significant potential of natural gas hydrates (NGHs) as a clean alternative energy source, enhancing their production through effective reservoir stimulation is crucial. Ning et al. [4] provides a comprehensive overview of current and potential methods for enhancing the production of NGHs, focusing on weakly consolidated reservoirs.

Energy efficiency and storage methods not only contribute to our understanding of clean energy technologies but also provide practical approaches to reducing carbon emissions and enhancing energy efficiency. Energy efficiency is furthered through technological innovations that address both industrial needs and environmental concerns. Pokhriyal et al. [5] focus on the role of battery energy storage systems in stabilizing grid frequency, a key aspect in the integration of fluctuating renewable energy sources such as wind power. Energy conservation is another critical area of focus. Zhuang et al. [6] explore the application of rotary boosters in hybrid cooling units to improve the energy efficiency of the information technology industry, where energy-efficient cooling solutions are critical for sustainable operations.

In the realm of sustainable practical applications, Lin et al. [7,8] investigate the optimization of dual fluidized bed reactors for clean energy processes like biomass gasification, providing valuable data to enhance the efficiency and sustainability of industrial energy use. Jesus et al. [9] develop a computational tool to assist in selecting sustainable refrigerants, aligning with environmental regulations and reducing the ecological impact of refrigeration systems. Geng et al. [10] address methane emission detection stability, offering innovative optical designs that enhance the accuracy and reliability of environmental monitoring. Han et al. [11] research the thermal conductivity of nanoporous aerogels, aiming to improve insulation materials that are essential for energy conservation in buildings and industrial processes. The integration of renewable energy into power grids is crucial for maintaining energy reliability. Xing et al. [12] explore an environmentally friendly liquid–liquid extraction method for recovering volatile fatty acids (VFAs) from anaerobic acidification broth using ionic liquids (ILs) and cosolvents. This breakthrough could lead to more sustainable practices in the recovery of valuable compounds from organic waste, with potential applications in biofuel production and other industries. Chilver-Stainer et al. [13] delve into the optimization of gas flow channels in low-temperature Polymer Electrolyte Membrane Fuel Cells (PEMFCs), crucial for enhancing the power output of hydrogen-powered vehicles. Ai et al. [14] focus on improving the operational efficiency of shielding induction motors in nuclear applications, emphasizing the importance of accurate temperature management for safety and reliability.

Urban and industrial clean energy solutions reflect the integration of policy and technology. Effective policy instruments and regulations can significantly influence the adoption and success of clean energy technologies. Xiong and Sui [15] analyze the impact of carbon emissions trading policies on urban green resilience, highlighting the socio-economic benefits of these policies in fostering sustainable urban development. Seyedzahedi and Bahramara [16] evaluate the impact of policy instruments on solar investments in Iran, showing how specific incentives can boost the economic viability of solar energy. Furthermore, the role of bioenergy in a sustainable energy future is underscored by studies on biomass power. Xin-gang et al. [17] evaluate the influence of policy instruments on the adoption of biomass technologies, highlighting how strategic policy frameworks can accelerate technology uptake and efficiency. Wang et al. [18] assess various strategies for reducing carbon emissions in Guizhou Province, China, highlighting the necessity of integrated policies for sustainable development and carbon reduction. Zhuang and Pan [19] investigate the impact of clean power investment on regional, high-quality economic development in China, utilizing a spatial Durbin model to analyze panel data from thirty provinces from 2010 to 2019, offering valuable insights for policy-makers aiming to align economic growth with environmental sustainability goals.

Emerging challenges in clean energy require innovative solutions that also offer significant opportunities for technological advancement and environmental sustainability. Zhang et al. [20] propose a method for regenerating waste polyurethane into high-performance materials, contributing to waste management and resource efficiency. Djamari et al. [21] explore developments in diesel engine technology, focusing on the optimization of spray patterns to reduce emissions and improve fuel efficiency. Quach and Lim [22] explore the impact of propylene gas flames on metal plate emissions and temperature distribution during preheating. The study discovered that adjusting the equivalence ratio significantly affects emissions and temperature distribution. These studies contribute to optimizing industrial processes by enhancing our understanding of environmental and operational effects.

These studies not only reflect the broad spectrum of clean energy technologies but also highlight the necessity of interdisciplinary approaches that integrate engineering, economics, and policy-making to fully harness the potential of these innovations. The transition to clean and low-carbon energy is not merely a technological challenge; it is also a socio-economic and political journey that requires holistic strategies and collaborative efforts. It requires a comprehensive approach that considers the complex interplay between energy systems, environmental policies, market dynamics, and societal needs. Therefore, interdisciplinary research that bridges the gap between science, engineering, economics, and policy is crucial for developing effective solutions and strategies. In conclusion, the topic of "Clean and Low Carbon Energy" is at the forefront of addressing one of the most pressing issues of our time—climate change. The research in this area not only advances scientific knowledge and technological innovation but also shapes a sustainable and resilient energy future for generations to come.

As we venture further into this critical field, it becomes a shared obligation to ensure that the evolution towards cleaner energy models is inclusive, fair, and congruent with the overarching objectives of sustainable development. We extend our heartfelt gratitude to the authors, peer reviewers, and the editorial team whose invaluable contributions have enriched this publication. It is our sincere hope that the scholarly works contained within this issue will spark interest and serve as a catalyst for further research and innovation in the years ahead.

Funding: This research received no external funding.

Conflicts of Interest: The authors declare no conflicts of interest.

References

1. Hussain, M.; Al-Aqrabi, H.; Hill, R. PV-Crack Net Architecture for Filter Induced Augmentation and Micro-Cracks Detection within a Photovoltaic Manufacturing Facility. *Energies* **2022**, *15*, 8667. [CrossRef]
2. Zhu, S.; Gu, X.; Duan, M.; Yu, F.; Zhao, D.; Liu, S.; Liu, Y. Analysis of the Influence Factors of the Crude Oil Temperature Maintenance System of Solar Sewage Heat Pumps in Cold Regions. *Energies* **2023**, *16*, 8124. [CrossRef]
3. Spoof-Tuomi, K.; Arvidsson, H.; Nilsson, O.; Niemi, S. Real-Driving Emissions of an Aging Biogas-Fueled City Bus. *Clean Technol.* **2022**, *4*, 954–971. [CrossRef]
4. Ning, B.; Yu, Y.; Zeng, J.; Lu, Q. A Review of Stimulation Technologies for Weakly-Consolidated Natural Gas Hydrate Reservoirs. *Energies* **2022**, *15*, 7195. [CrossRef]
5. Pokhriyal, A.; Domínguez-García, J.; Gómez-Romero, P. Impact of Battery Energy System Integration in Frequency Control of an Electrical Grid with Wind Power. *Clean Technol.* **2022**, *4*, 972–986. [CrossRef]
6. Zhuang, R.; Zhou, F.; Tian, X.; Xu, B.; Li, S.; Ma, G. Performance Study of Booster-Driven Hybrid Cooling Units for Free Cooling in Data Centers. *Sustainability* **2023**, *15*, 14558. [CrossRef]
7. Lin, Y.; Wang, Q.; Ye, C.; Zhu, Y.; Fan, H. Experimental Research on the Gas-Solid Flow Characteristics in Large-Scale Dual Fluidized Bed Reactor. *Energies* **2023**, *16*, 7239. [CrossRef]
8. Lin, Y.; Wang, Q.; Xie, G.; Fang, M.; Luo, Z. Numerical Simulation Study on the Gas–Solid Flow Characteristics of a Large-Scale Dual Fluidized Bed Reactor: Verification and Extension. *Energies* **2024**, *17*, 1304. [CrossRef]
9. Jesus, G.; Aguiar, M.; Gaspar, P. Computational Tool to Support the Decision in the Selection of Alternative and/or Sustainable Refrigerants. *Energies* **2022**, *15*, 8497. [CrossRef]
10. Geng, D.; Wang, D.; Li, Y.; Zhou, W.; Qi, H. Detection Stability Improvement of Near-Infrared Laser Telemetry for Methane Emission from Oil/Gas Station Using a Catadioptric Optical Receiver. *Energies* **2023**, *16*, 3854. [CrossRef]
11. Han, Y.; Li, S.; Liu, H.; Li, Y. Lattice Boltzmann Simulation of Coupling Heat Transfer between Solid and Gas Phases of Nanoporous Materials. *Nanomaterials* **2022**, *12*, 3424. [CrossRef] [PubMed]
12. Xing, T.; Yu, S.; Tang, J.; Liu, H.; Zhen, F.; Sun, Y.; Kong, X. Liquid–Liquid Extraction of Volatile Fatty Acids from Anaerobic Acidification Broth Using Ionic Liquids and Cosolvent. *Energies* **2023**, *16*, 785. [CrossRef]
13. Chilver-Stainer, J.; Elbarghthi, A.; Wen, C.; Tian, M. Power Output Optimisation via Arranging Gas Flow Channels for Low-Temperature Polymer Electrolyte Membrane Fuel Cell (PEMFC) for Hydrogen-Powered Vehicles. *Energies* **2023**, *16*, 3722. [CrossRef]
14. Ai, L.; Lu, Y.; Han, J.; Suo, W. Simulation of the Temperature of a Shielding Induction Motor of the Nuclear Main Pump under Different Turbulence Models. *Energies* **2023**, *16*, 2792. [CrossRef]
15. Xiong, B.; Sui, Q. Does Carbon Emissions Trading Policy Improve Inclusive Green Resilience in Cities? Evidence from China. *Sustainability* **2023**, *15*, 12989. [CrossRef]
16. Seyedzahedi, A.; Bahramara, S. Facilitating Investment in Photovoltaic Systems in Iran Considering Time-of-Use Feed-in-Tariff and Carbon Market. *Energies* **2023**, *16*, 1067. [CrossRef]
17. Xin-gang, Z.; Wei, W.; Shuran, H.; Xuan, L. Impacts of Government Policies on the Adoption of Biomass Power: A System Dynamic Perspective. *Sustainability* **2023**, *15*, 1723. [CrossRef]
18. Wang, H.; Xu, W.; Zhang, Y. Research on Provincial Carbon Emission Reduction Path Based on LMDI-SD-Tapio Decoupling Model: The Case of Guizhou, China. *Sustainability* **2023**, *15*, 13215. [CrossRef]
19. Zhuang, X.; Pan, L. Study on the Impact of Clean Power Investment on Regional High-Quality Economic Development in China. *Energies* **2022**, *15*, 8364. [CrossRef]
20. Zhang, D.; Gu, X.; Liu, S.; Liu, Y.; Zhou, Q.; Zhu, S.; Zhu, Y. Study on Properties of Regenerated Fluorinated Polyurethane Rigid Foam Prepared by Degrading Waste Polyurethane. *Sustainability* **2022**, *14*, 15685. [CrossRef]
21. Djamari, D.; Idris, M.; Paristiawan, P.; Abbas, M.; Samuel, O.; Soudagar, M.; Herawan, S.; Chandran, D.; Yusuf, A.; Panchal, H.; et al. Diesel Spray: Development of Spray in Diesel Engine. *Sustainability* **2022**, *14*, 15902. [CrossRef]

22. Quach, Y.; Lim, O. An Investigation of the Effect of Propylene Gas Flame on Emissions and Temperature Distribution of Preheated Metal Plate. *Sustainability* **2023**, *15*, 12306. [CrossRef]

nanomaterials

Article

Lattice Boltzmann Simulation of Coupling Heat Transfer between Solid and Gas Phases of Nanoporous Materials

Yafen Han *, Shuai Li, Haidong Liu and Yucong Li

School of Energy and Power Engineering, Northeast Electric Power University, Jilin 132013, China
* Correspondence: hanyafen@126.com

Abstract: In order to deeply study the heat conduction of nanoporous aerogel, a model of gas-solid heat conduction was established based on the microstructure of aerogel. The model was divided into two subdomains with uniform mesh because of the different gas-solid characteristics, and simulation was performed on each domain using the lattice Boltzmann method. The value of temperature on the boundaries of subdomains was determined by interpolation. Finally, the temperature distribution and the thermal conductivity were maintained. It can be concluded that when the gas-phase scale was fixed, the temperature distribution of the solid phase became more uniform when the scale increased; when the solid-phase scale was fixed, the temperature jump on the gas-solid interface decreased with the increase in the gas-phase scale; and the thermal conductivity of gas-solid coupling varied with the scale of the gas phase or solid phase, showing a scale effect in varying degrees.

Keywords: heat transfer; nanomaterials; aerogel; temperature distribution; thermal conductivity

1. Introduction

Aerogels, as a super insulation material, have a unique structure that is an open nanoporous and continuous three-dimensional skeleton. Therefore, they exhibit outstanding properties such as high porosity, high specific surface area and low thermal conductivity [1]. They have been applied in many fields such as aerospace, industrial insulation and military technology [2]. In the thermal transport of aerogel, the heat needs to continuously pass through the porous network composed of gas and solid. The complex microscopic characteristics and the scale effect of gas and solid make it difficult to study the thermal conductivity. With the depletion of natural resources, all countries in the world are committed to the conservation and efficient use of energy. The insulation requirements of aerogel insulation materials are becoming more and more urgent. Accordingly, it is significant to further study the gas-solid coupling thermal conductivity of aerogels in order to optimize and improve their adiabatic performance.

At present, the researches on the gas-solid coupling heat transfer of aerogel mainly focus on theoretical and numerical calculation methods. In terms of theoretical studies, Swimm et al. proposed a calculation model of coupling the thermal conductivity of aerogels and studied the coupling effect of their effective thermal conductivity [3]. Based on the microstructure of aerogels and the model, Zhao J et al. improved the prediction accuracy of the thermal conductivity of aerogels [4]. Given that the above two models are based on the thermal bridge effect, the calculation process is extremely difficult. In addition, Wei et al. used a cubic array composed of small spheres to establish a gas-solid coupling heat conduction model for aerogels [5]. The calculation results showed that the nanostructure and the nanoscale effect of the solid grains were the main reasons for the extremely low thermal conductivity of the materials. In terms of numerical methods, Coquard R. et al. established a model for the overall heat transfer inside aerogels with cellulose [6]. Under the assumption of "uniform approximation", the heat-transfer law of gas-solid two phases was studied. Zhao J. et al. established a three-dimensional numerical model of silica aerogel based on the random

Citation: Han, Y.; Li, S.; Liu, H.; Li, Y. Lattice Boltzmann Simulation of Coupling Heat Transfer between Solid and Gas Phases of Nanoporous Materials. *Nanomaterials* **2022**, *12*, 3424. https://doi.org/10.3390/nano12193424

Academic Editor: Seung Hwan Ko

Received: 7 August 2022
Accepted: 27 September 2022
Published: 29 September 2022

aggregation structure of porous secondary nanoparticles and studied the overall thermal conductivity of aerogel by considering the coupling thermal conductivity [7].

In recent years, the lattice Boltzmann method (LBM) has been widely used in the field of micro-nanoscale heat transfer [8]. It can not only reveal the heat transfer and the intrinsic mechanism of porous insulating materials, but it can also make up for the shortcomings of the macro methods. In addition, the LBM is high efficiency and stable. It has become a powerful tool for the analysis of micro-scale heat transfer. Song proposed a local-effective-viscosity multirelaxation-time lattice Boltzmann pore-network coupling model (LEV-LBM-PNM) to simulate gas transport in nanoporous media [9]. The proposed LEV-LBM-PNM accurately predicts gas-apparent permeability by accounting for gas slip in irregular pore shapes and surface roughness. Qu established a modified lattice LBM to predict effective thermal conductivity of aerogel materials [10]. The modified LBM scheme introduces an additional coefficient to regulate significant differences in inherent thermal conductivity between solid and gas phases. Kan proposed a random internal morphology and structure generation-growth method [11], The model was imported into the lattice Boltzmann algorithm as a fully resolved geometry and used to investigate the effects on heat transfer in nanoporous materials.

In this paper, the gas-solid coupling heat conduction model is established, and the temperature distribution and thermal conductivity of silica-aerogel gas-solid coupling are simulated by LBM based on the region splitting. The thermal transfer of gas-solid coupling heat conduction is analyzed and discussed in depth.

2. Physical Model

Figure 1 displays the model of gas-solid coupling heat conduction of aerogels. Figure 1a,b are a three-dimensional model and a two-dimensional gas-solid center section, respectively. The model consists of gas-solid two phases, which are air and silica, respectively, and the characteristic scales along the x direction are expressed by d_g and d_s. The length, width and height of the model were expressed as L_x, L_y, L_z. The red cross section is the interface of the gas-solid two phases. The front and back walls are kept at a constant temperature; the temperature of the front wall is T_1 = 300 K, and the temperature of the back wall is T_2 = 301 K; the other walls were adiabatic. The heat flow transferred from the high temperature wall (front wall) to the low temperature wall (back wall) along the x-axis.

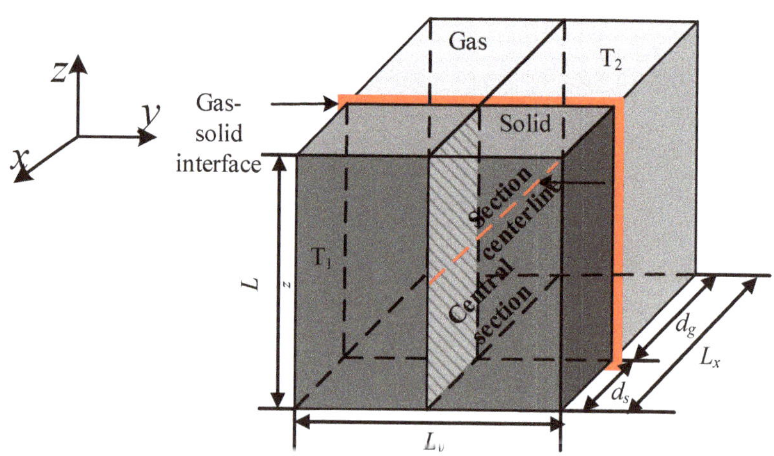

(a) Three-dimensional stereo model

Figure 1. *Cont.*

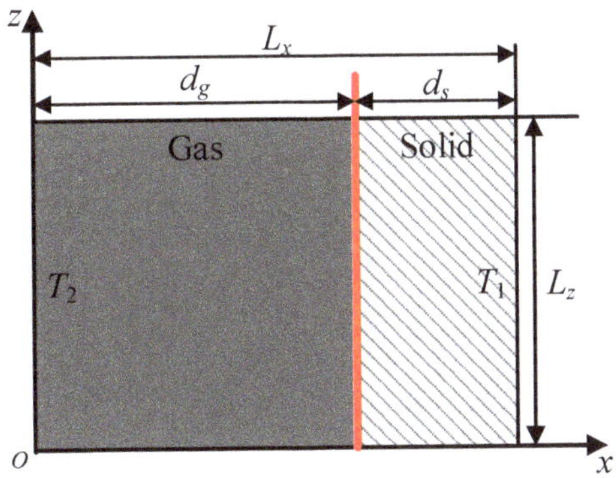

(**b**) Two-dimensional gas-solid center section

Figure 1. Gas-solid coupled thermal conduction model.

3. Mathematical Model

3.1. Basic Algorithm

3.1.1. Gas-Phase Basic Algorithm

The three-dimensional fifteen-speed (D3Q15) lattice Boltzmann model was used to solve the gas steady heat conduction of aerogel. The evolution equation was as follows:

$$f_k(\boldsymbol{r}_g + \boldsymbol{c}_k \Delta t_g, t_g + \Delta t_g) - f_k(\boldsymbol{r}_g, t_g) = -\frac{1}{\tau}[f_k(\boldsymbol{r}_g, t_g) - f_k^{eq}(\boldsymbol{r}_g, t_g)] \quad (1)$$

where subscripts g, k were the gas phase and the discrete velocity directions of the lattice point, respectively; $\boldsymbol{r}_g$ was the space position vector, $\boldsymbol{c}_k$ was the discrete lattice velocity (as shown in Figure 2); t_g was the time, Δt_g was the time step; τ was the dimensionless relaxation time; f_k with f_k^{eq} were, respectively, the internal energy distribution function and the corresponding equilibrium distribution function of the discrete velocity directions of the different lattice points in the gas phase.

$$f_k^{eq} = \begin{cases} (2/9)\rho c_p T & k = 0 \\ (1/9)\rho c_p T & k = 1 \sim 6 \\ (1/72)\rho c_p T & k = 7 \sim 14 \end{cases} \quad (2)$$

where T was temperature; ρ was the air density; c_p was the specific heat capacity of the air.

$$\boldsymbol{c}_k = \begin{cases} (0,0,0)c, k = 0 \\ (\pm 1,1,0)c, (0,\pm 1,0)c, (0,0,\pm 1)c, k = 1 \sim 6 \\ (\pm 1,\pm 1,\pm 1)c, k = 7 \sim 14 \end{cases} \quad (3)$$

$$\tau = \frac{9}{5}\frac{\lambda}{\rho c_p c^2 \delta t} + 0.5 \quad (4)$$

where λ was the thermal conductivity of the air; c was the lattice speed that theoretically could take any positive value only to insure the τ value within (0.5,2); $c = \delta x/\delta t$ was the space step [12,13]. Temperature T and heat flux q could be obtained, respectively, according to the following formula [14]:

$$T = \frac{1}{\rho c_p}\sum_k f_k \quad (5)$$

$$q_g = \sum_{k=0}^{Q} (c_k f)\frac{\tau - 0.5}{\tau} \quad (6)$$

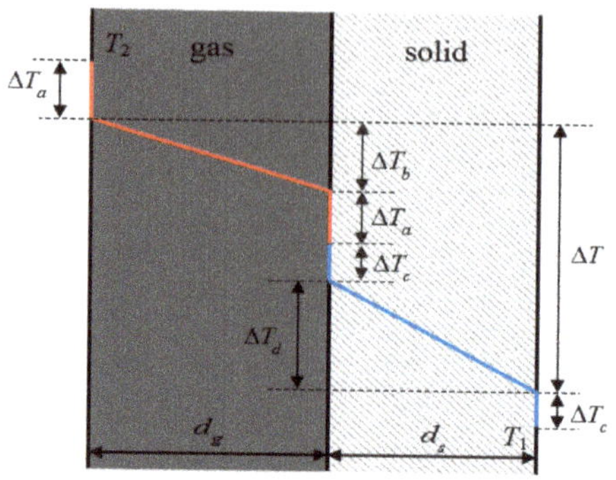

Figure 2. Calculating model of effective thermal conductivity of gas-solid coupling thermal conductivity.

When the temperature reached equilibrium state, the effective thermal conductivity could be calculated on the temperature T and heat flux q.

3.1.2. Solid-Phase Basic Algorithm

The three-dimensional fifteen-speed (D3Q15) lattice Boltzmann model was used to solve the solid steady heat conduction of aerogel. The evolution equation was as follow:

$$e_k(\boldsymbol{r}_s + \boldsymbol{v}_k \Delta t_s, t_s + \Delta t_s) = e_k(\boldsymbol{r}_s, t_s)[1 - \beta] + \beta e_k^{eq}(\boldsymbol{r}_s, t_s) \quad (7)$$

where subscripts s, k were the solid phase and the discrete velocity directions of the lattice point, respectively; $\boldsymbol{r}_s$ was the space position vector, $\boldsymbol{v}_k$ was the discrete lattice velocity (as shown in Figure 2); t_s was the time, Δt_s was the time step; β was a weight function e_k with e_k^{eq} were, respectively, the phonon energy density distribution function and the equilibrium phonon energy density distribution function of the discrete velocity directions of the different lattice points in the solid phase.

The total phonon energy density e is calculated in reference [15]. The total energy density e of each lattice point is discretized into the energy density e_k of each discrete direction

$$e = \sum_{k=1}^{Q} e_k \quad (8)$$

Since the phonon energy density e and the equilibrium phonon energy density e_k^{eq} of all lattice points need to be redefined at each time step, assuming that the probability of phonon scattering in all directions is equal, i.e. isotropic, the equilibrium phonon energy density

$$e_k^{eq} = \frac{1}{Q} \sum_{k=1}^{Q} e \quad (9)$$

$$v_k = \begin{cases} (0,0,0)v, k = 0 \\ (\pm 1, 1, 0)v, (0, \pm 1, 0)v, (0, 0, \pm 1)v, k = 1 \sim 6 \\ (\pm 1, \pm 1, \pm 1)v, k = 7 \sim 14 \end{cases} \quad (10)$$

where the lattice speed $v = \Delta r_s / \Delta t_s$, Δr_s was the solid-phase space step.

Weight function $\beta = \Delta t_s / \tau_0$, τ_0 was the particle relaxation time.

The normal heat-transfer rate of the yoz plane within the model is

$$q_s = \sum_{k=0}^{Q} \boldsymbol{v}_k \cdot e_k(r, t) \quad (11)$$

3.2. Boundary Conditions

In the LBM, for calculation accuracy, stability and computational efficiency, the boundary conditions should be reasonably selected according to the actual situation. Since the influences of solid skeleton were not taken into account in the gas heat conduction, the mirror bounce format was adopted for no friction losses. Moreover, the energy distribution function could free transfer in this boundary condition. That was equivalent to removing the boundary, and the results would not be affected by the width L_y and the height L_z. The L_x direction is the constant temperature boundaries T_1 and T_2. The specific forms were implemented as follow:

$$f_{i'}(r_b, t) = f_i\left(r_f, t\right) \tag{12}$$

where r_b was the boundary lattice point, $r_f = r_b - e_i\delta t$ were internal lattice point; $f_{i'}$ was a mirror-symmetric distribution function of f_i. The f_i could be obtained by the inner lattice points that were adjacent the boundary wall.

3.3. Calculation Method of Effective Thermal Conductivity

Owing to both the gas-phase region and the solid-phase region being stable and the heat flow in the two sub-regions being equal, q is used to represent the heat flow in the whole region, that is $q = q_s = q_g$. Because the characteristic scales of the whole system region are smaller than or close to the average free path of the gas molecules and phonons, obvious scale effects appear in both sub-regions, and the temperature distribution of materials is no longer continuous. The effective thermal conductivity can be obtained according to Fourier's law and the geometric relationship in Figure 3.

$$k_{\text{eff}} = \frac{q \cdot (d_g + d_s)}{\Delta T_a + \Delta T + \Delta T_c} \tag{13}$$

where q is the system heat flow; ΔT_a is the jumping for the gas-phase boundary temperature; ΔT_c is the jumping for the solid-phase boundary temperature.

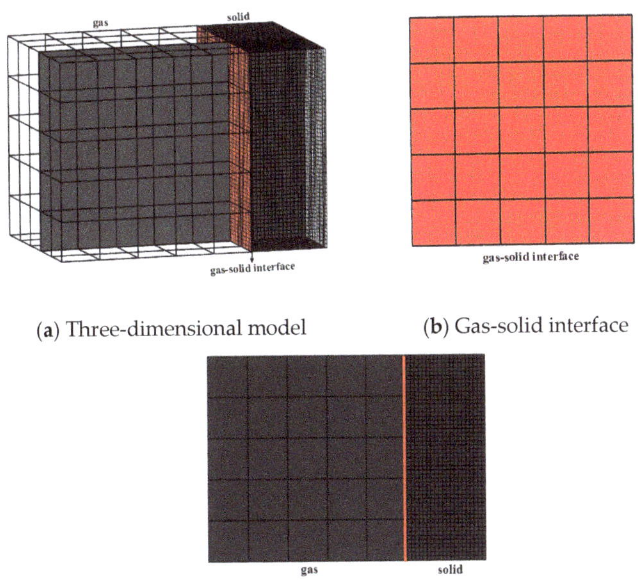

(**a**) Three-dimensional model (**b**) Gas-solid interface

(c) Gas-solid Center Section

Figure 3. Gas-solid coupled thermal conduction model with lattice partition.

3.4. Region Splitting

Although the hybrid LBM has been proven to provide great advantages over the structured LB in terms of geometry flexibility and the treatment of interfaces [16,17], the focus of this paper is to study the performance of gas-solid coupling heat transfer at the interface. Hence, considering the accuracy and simplicity, the "region splitting" is selected.

The basic idea of region splitting is to divide the whole system region into different sub-regions, and then using the corresponding LBM model to calculate each sub-region independently [18].

According to the gas-solid coupled thermal conduction model, the gas-solid coupled thermal conduction model is divided into two sub-regions: the gas phase and the solid phase, which are represented by Ω_g and Ω_s, $\Gamma_{as} = \Omega_a \cap \Omega_s$, because the average free path of a gas molecule (about 70 nm) is much larger than that of a phonon (about 0.6 nm) of silica solid. To simulate the heat transfer at the solid boundary, we set a virtual boundary node at the gas region XA, as shown in Figure 4. Therefore, different size meshes are used to subdivide the sub-regions, which are set up as $\Delta x_a = \Delta y_a = \Delta z_a$, $\Delta x_s = \Delta y_s = \Delta z_s$. Generally, the time step and space step in the gas-phase and solid-phase regions are different, so the physical quantities involved in the virtual boundary need to be calculated by special methods.

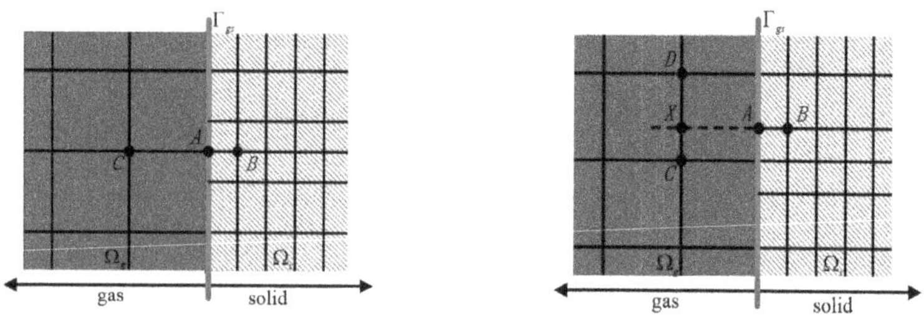

(a) The Lattice point A belong to two-phase region. (b) The Lattice point A belongs to the solid phase region only

Figure 4. Boundary lattice interpolation of two-dimensional gas-solid center section.

Set $n = \Delta x_g / \Delta x_s$ as an integer, that is, the lattice size of the gas phase is several times larger than that of solid phase. Because the lattice size of gas phase and solid phase is different, some lattices will coincide on the virtual boundary, and some lattices will disperse on the virtual boundary, as shown in Figure 5. Because the lattice size of the gas-solid two-phase is different, the time step of the gas-solid two-phase also has a special relationship. If the gas-phase region evolves from time t_0 to time $t_g = t_0 + \Delta t_g$, the solid-phase region needs to evolve n time steps: $t_s = t_0 + \Delta t_s$, $t_s = t_0 + 2\Delta t_s, \cdots, t_s = t_0 + n\Delta t_s$. The physical quantities on lattice A at the virtual boundary time t are calculated by the second-order interpolation method to ensure the second-order accuracy of the calculation results. Since there is no temperature gradient in the y and z directions, and the boundary is slip free adiabatic, the specific implementation steps are as follows:

(1) If $A \in \Omega_g \cap \Omega_s$, as shown in Figure 5a, then

$$T(A) = \frac{1}{n+1}[T(B) + T(C)] \tag{14}$$

Here $C \in \Omega_g$ and $B \in \Omega_s$ are the nearest lattice points in the vertical direction of the virtual boundary.

(2) If $A \notin \Omega_g, A \in \Omega_s$, as shown in Figure 5b, then

$$T(A) = \frac{1}{n+1}[\alpha T(C) + (1-\alpha)T(D) + nT(B)] \tag{15}$$

Here $D \in \Omega_g$, $\alpha = |XC|/|DC|$.

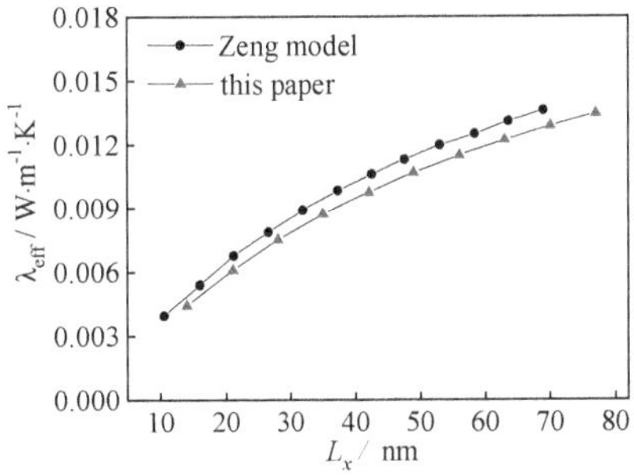

Figure 5. Verification of numerical simulation results.

The gas-solid two-phase coupling is performed by the second-order interpolation method. According to the region splitting, the gas-solid two-phase sub-regions are calculated by their respective LBM models, and the coupling calculation of the two regions is made by counting the macroscopic physical parameters, T, at the interface. Finally, the heat transport process of the whole gas-solid is obtained.

4. Program Verification

4.1. Gas-Phase Sub-Area Verification

Figure 5 shows the numerical simulation results of the thermal conductivity of the gas-phase region at the pore scale of 10–80 nm, which is compared with the Zeng model [19] under the same conditions.

It can be seen from the figure that the simulation results of this paper are basically consistent with the variation of the theoretical analysis results of the Zeng model, which is slightly smaller than the Zeng model. This is because the potential was weak, and it only considered pairwise interaction between atoms. As a result, the simulation results deviated from this paper in 60~70 nm, although they have the same change trend. Hence, the reliability of the LBM program could be proved.

4.2. Solid-Phase Sub-Area Verification

Figure 6 shows a comparison between the simulation results of thermal conductivity of the solid-phase region and the results of reference [15] under the same conditions. It can be seen from the figure that the simulation results of the solid-phase region in this paper are basically consistent with the variation trend of reference [15].

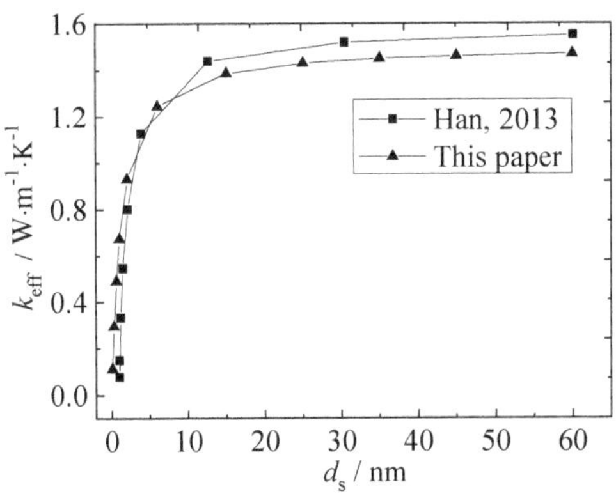

Figure 6. Validation of solid-state numerical simulation.

When the characteristic scale of the solid-phase region is between 0 and 10 nm, the thermal conductivity shows a significant jump, showing the scale-effect characteristics while the thermal conductivity is close to 1.47 $W{\cdot}m^{-1}{\cdot}K^{-1}$ of silica in the macroscopic state at 10–60 nm.

5. Simulation Results and Analysis

Figure 7 is a steady-state temperature distribution cloud diagram of a two-dimensional gas-solid center section, in which the gas-phase region d_g is 60 nm, and the solid-phase region d_s is 2 nm, 3 nm, 4 nm, and 5 nm, respectively. Because the solid-phase scale is much smaller than the gas-phase scale, the solid-phase region is presented in the form of enlarged views. From the temperature distribution profiles in Figure 7a–d, since the gas-phase scale is 60 nm, it can be seen that the gas-phase characteristic scale is smaller than the average free path of the gas molecule and the discontinuous temperature diffusion mainly concentrated at the boundary of x = 0 nm and x = 60 nm, while the temperature distribution in the central region is relatively uniform. From the corresponding temperature distribution diagram of the solid-phase region, it can be seen that the scattering between phonons is dominant because the characteristic scale of the solid phase is much larger than the average free path of the silica phonons at 0.6 nm. At the same time, the characteristics of the concentrated distribution of the solid-phase boundary temperature are weakened. With the increase in the characteristic scale of the solid phase, the transition between the different parts of the temperature distribution becomes more uniform and tends to the temperature diffusion phenomenon under the macroscopic state.

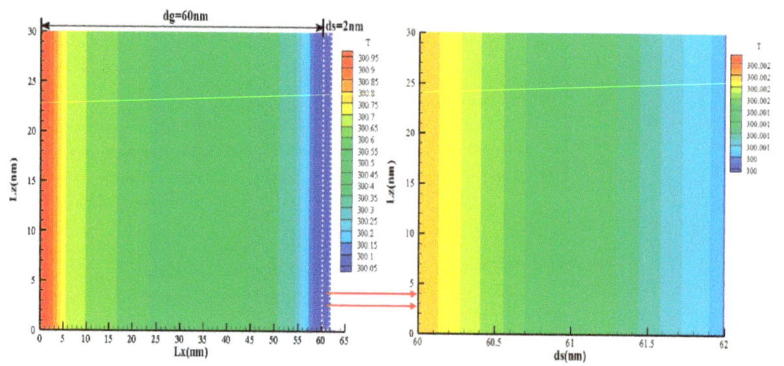

(a) $d_g + d_s$ = 62 nm

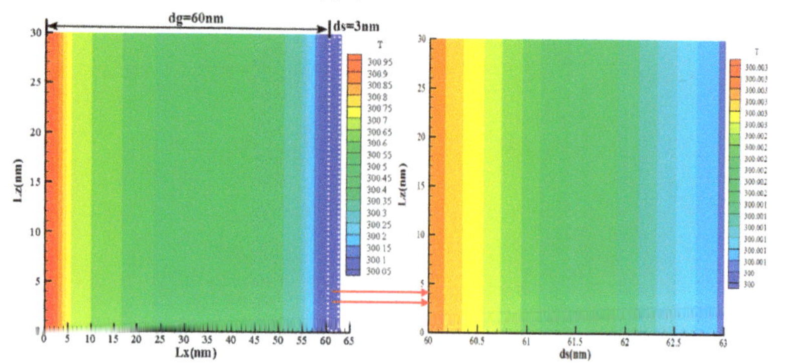

(b) $d_g + d_s$ = 63 nm

Figure 7. *Cont.*

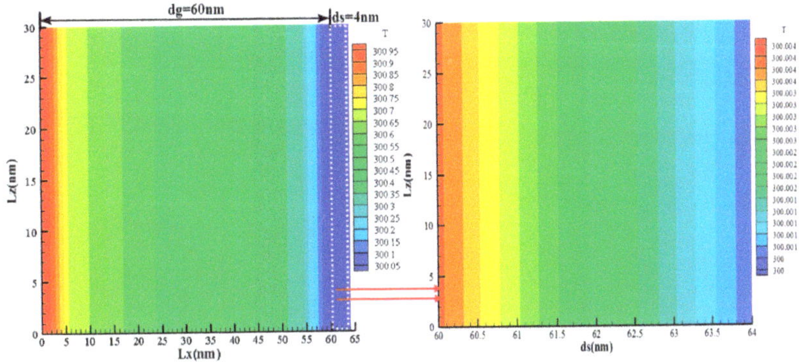

(**c**) $d_g + d_s = 64$ nm

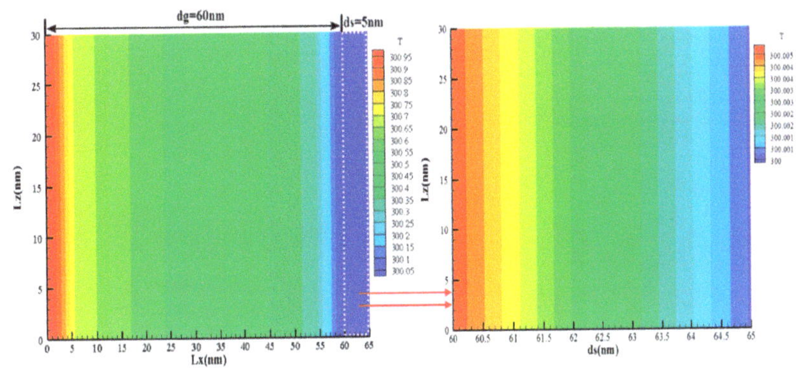

(**d**) $d_g + d_s = 65$ nm

Figure 7. Cloud map of steady temperature distribution of gas-solid coupling.

Figure 8 shows the two-dimensional steady-state temperature distribution of the gas-solid central section of the gas-solid coupled heat conduction model. It can be seen from the graph that the gas-phase region d_g is 60 nm, and since the gas-phase characteristic scale is small, the free movement of the gas molecules is severely restricted, which hinders the thermal transport of the heat molecules to the cold molecules, and a strong boundary scattering effect occurs. Therefore, the temperature jump is obvious at the boundary $x = 0$ nm and $x = 60$ nm. It can be seen from the large graph of the respective solid-phase region that there is a significant temperature jump at the interface of gas-solid coupling $x = 60$ nm, and The temperature jump of the gas-phase region is significantly higher than the temperature jump of the solid-phase region. This is because the characteristic scale of the solid-phase region is much larger than the average free path of the silica phonons, at which time the thermal transport between the phonons plays a dominant role, and the scattering between the phonons and the boundaries is relatively weakened. Hence, the temperature jump between the two boundaries is weakening.

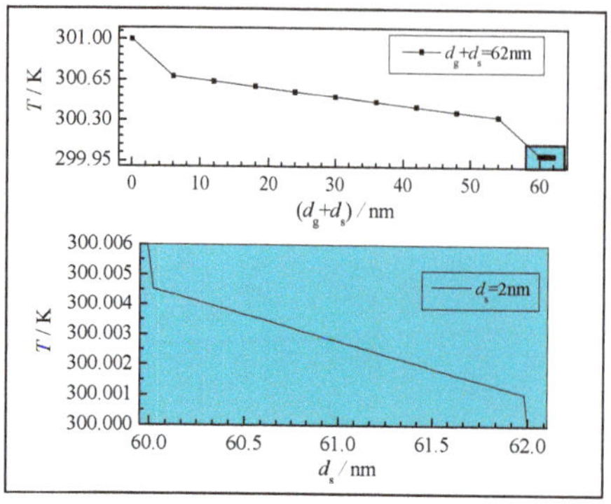

(a) $d_g + d_s = 62$ nm

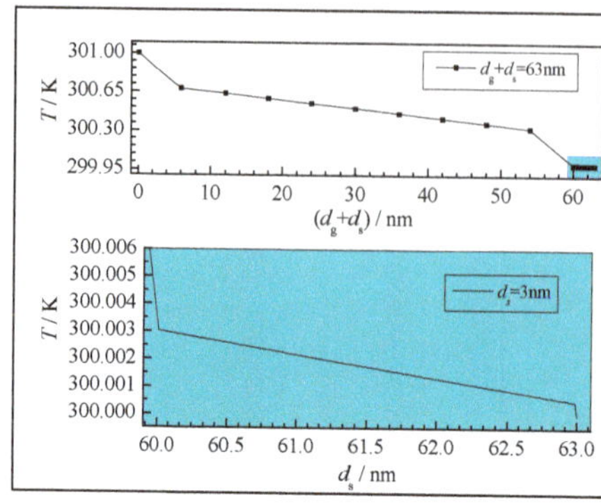

(b) $d_g + d_s = 63$ nm

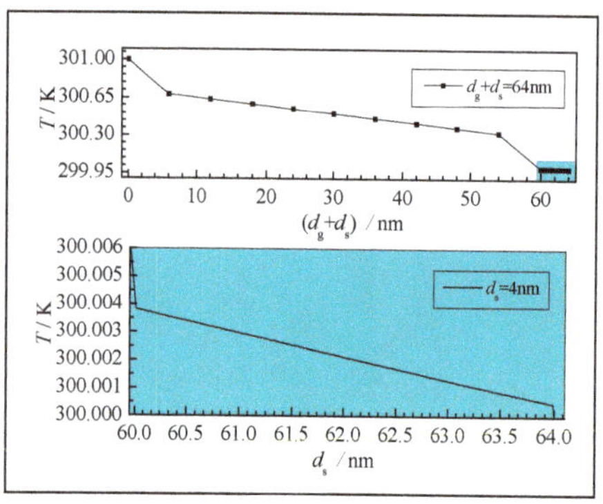

(c) $d_g + d_s = 64$ nm

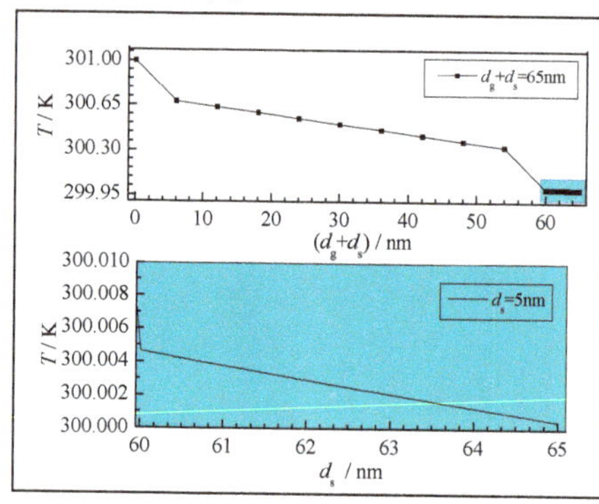

(d) $d_g + d_s = 65$ nm

Figure 8. Steady temperature distribution curve of two-dimensional gas-solid center section.

Figure 9 is a temperature jump diagram at the gas-solid interface when the solid-phase region is 5 nm. It can be seen from the graph that the temperature jump at the interface of the gas-solid coupling decreases with the increase in the gas-phase scale. The temperature jump at 10–70 nm decreases sharply when it reaches 70 nm, and the decrease amplitude begins to be gentle.

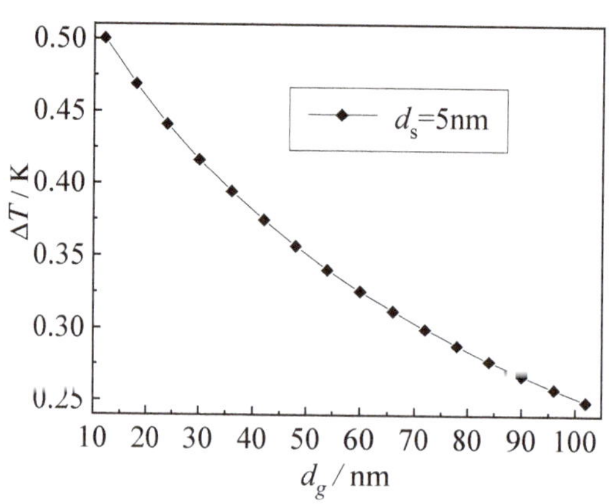

Figure 9. Temperature jump at gas-solid interface.

This is because when the gas-phase scale is smaller than the mean free path of the gas molecules, the boundary scattering effect is obvious, and the temperature jump at the boundary will be more violent. With the increase in the gas-phase scale, the discontinuous temperature distribution at the boundary caused by the boundary scattering effect will gradually weaken at same time, leading to the decrease in the boundary temperature jump value.

Figure 10 is a schematic diagram of the dimensionless temperature distribution of the gas-solid two-phase system. As shown in Figure 10a, since the gas-phase characteristic scale is 60 nm, the temperature distributions in the dimensionless gas-phase regions in the four cases are in good agreement, and the boundary temperature jumps are consistent; as shown in Figure 10b, the temperature jumps in the solid-phase region decrease with the increase in the solid-phase scale, and the temperature diffusion is more and more close to the macroscopic heat transport characteristics. This is because the solid-phase scale is larger than the average free path of phonon 0.6 nm. With the further increase in the solid-phase scale, the scattering effect between the phonon and the boundary will gradually disappear, the scattering between phonon and phonon will be enhanced, and the energy transfer will be smoother. Hence, the macroscopic thermal transport characteristics will gradually appear.

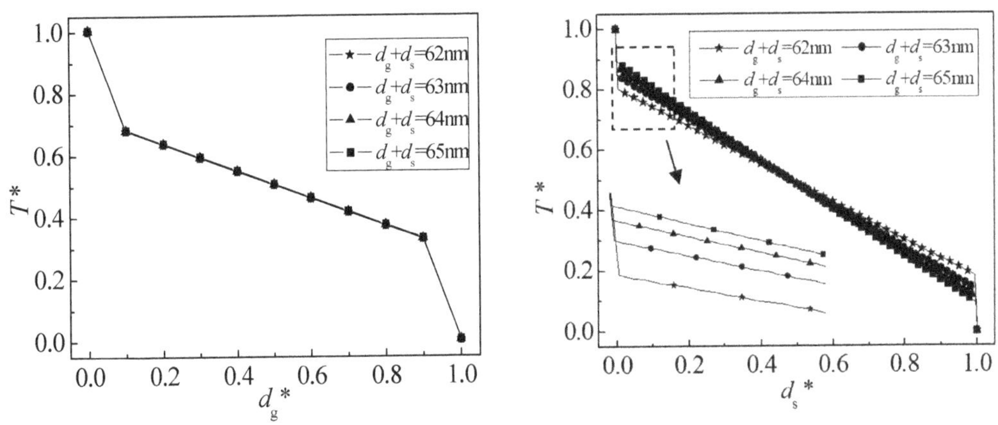

(**a**) Gaseous region

(**b**) Solid region

Figure 10. Dimensionless gas-solid temperature distribution. * dimensionless parameters.

Figure 11 shows the relationship between the gas-solid coupling thermal conductivity and the solid-gas scale. It can be seen from Figure 11a that when the solid-phase scale is less than 1 nm, the increase in thermal conductivity is small, and when the solid-phase scale is greater than 1 nm, the thermal conductivity begins to increase rapidly. As can be seen from Figure 11b, when the gas-phase scale is between 10 and 70 nm, the increase in thermal conductivity is larger, and when the gas-phase scale reaches about 70 nm, the change trend of thermal conductivity begins to be stable.

This is because the characteristic scales of the gas-solid two phases are smaller or closer to the mean free path of their molecules and phonons, and boundary scattering will severely restrict the heat transport between the molecules and phonons, showing the obvious scale effect. With the increase in scale, scattering will rapidly weaken and thermal conductivity will correspondingly increase.

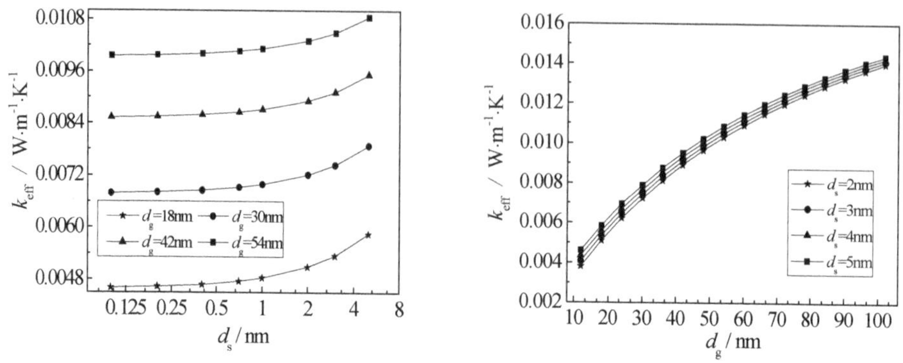

(**a**) Variation law with solid-phase size (**b**) Variation law with gas-phase scale

Figure 11. Dependence of Coupled Thermal Conductivity on Solid-Gas Scale.

6. Conclusions

In this paper, the gas-solid coupling heat conduction model of nanoporous aerogels is established. Based on the theory of regional splitting, the gas-solid two-phase simulation is carried out by using the respective LBM. The second-order interpolation is used to transfer temperature at the gas-solid interface. Finally, the internal temperature distribution and the variation of thermal conductivity of gas-solid coupling are systematically studied. The main conclusions are as follows:

(1) When the gas-phase scale is 60 nm, the boundary of the gas-phase region shows obvious discontinuous temperature distribution characteristics, and the temperature jump is obvious; when the transition of the temperature distribution in the solid-phase region is gradually uniform, the solid-phase scale changes from 2 nm to 5 nm.

(2) When the solid-phase scale is 5 nm, the temperature jump at the gas-solid interface decreases with the increase in the gas-phase scale. The temperature jump decreases sharply when the gas-phase scale is 10–70 nm and begins to change gently when the gas-phase scale reaches 70 nm.

(3) The variation of thermal conductivity of gas-solid coupling with solid-phase scale is basically the same. The variation range of thermal conductivity is relatively small when the solid-phase scale is less than 1 nm. When the solid-phase scale is larger than 1 nm, the variation range of thermal conductivity begins to increase rapidly.

(4) The thermal conductivity of gas-solid coupling increases gradually with the increase in the gas-phase scale. The thermal conductivity increases greatly between 10 and 70 nm. When the thermal conductivity reaches about 70 nm, the increase in thermal conductivity begins to slow down gradually.

Author Contributions: Conceptualization, Y.H. and S.L.; methodology, Y.H. and S.L.; software, S.L.; validation, H.L. and Y.L.; writing—original draft preparation, S.L.; writing—review and editing, H.L. All authors have read and agreed to the published version of the manuscript.

Funding: This research received no external funding.

Data Availability Statement: The study did not report any data.

Conflicts of Interest: The authors declare no conflict of interest.

References

1. Han, Y.F.; Xia, X.L.; Liu, H.D. Characteristic of heat conduction in nano-insulation material. *J. Funct. Mater.* **2014**, *45*, 03017–03019.
2. Shu, X.; Liu, Z.; Ding, Y.; Yang, H.; Luo, P. Preparation and application of nano-silica aerogels in thermal insulation: An overview. *Mater. Rev.* **2018**, *32*, 788–795.
3. Swimm, K.; Reichenauer, G.; Vidi, S.; Ebert, H.P. Gas Pressure Dependence of the Heat Transport in Porous Solids with Pores Smaller than 10μm. *Int. J. Thermophys.* **2009**, *30*, 1329–1342. [CrossRef]

4. Zhao, J.J.; Duan, Y.Y.; Wang, X.D.; Wang, B.X. Effects of solid–gas coupling and pore and particle microstructures on the effective gaseous thermal conductivity in aerogels. *J. Nanopart. Res.* **2012**, *14*, 1024. [CrossRef]
5. Wei, G.; Zhang, X.; Yu, F. nano-porous structures and effective thermal conductivity of aerogel super insulator. *J. Therm. Sci. Technol.* **2005**, *4*, 107–112.
6. Coquard, R.; Baillis, D.; Grigorova, V.; Enguehard, F.; Quenard, D.; Levitz, P. Modelling of the conductive heat transfer through nano-structured porous silica materials. *J. Non-Cryst. Solids* **2013**, *363*, 103–115. [CrossRef]
7. Zhao, J.J.; Duan, Y.Y.; Wang, X.D.; Wang, B.X. A 3-D numerical heat transfer model for silica aerogels based on the porous secondary nanoparticle aggregate structure. *J. Non-Cryst. Solids* **2012**, *358*, 1287–1297. [CrossRef]
8. Ge, C. *Investigation of Gas Thermal Conductivity in Nanopore by Molecular Dynamics*; Harbin Institure of Technology: Shenzhen, China, 2010.
9. Song, W.; Yin, Y.; Landry, C.J.; Prodanovic, M.; Qu, Z.; Yao, J. A Local-Effective-Viscosity Multirelaxation-Time Lattice Boltzmann Pore-Network Coupling Model for Gas Transport in Complex Nanoporous Media. *SPE J.* **2020**, *26*, 461–481. [CrossRef]
10. Qu, Z.G.; Fu, Y.D.; Liu, Y.; Zhou, L. Approach for predicting effective thermal conductivity of aerogel materials through a modified lattice Boltzmann method. *Appl. Therm. Eng.* **2018**, *132*, 730–739. [CrossRef]
11. Kan, A.; Mao, S.; Wang, N.; Shi, B. Simulation and Experimental Study on Thermal Conductivity of Nano-Granule Porous Material Based on Lattice-Boltzmann Method. *J. Therm. Sci.* **2019**, *30*, 248–256. [CrossRef]
12. Wang, J.; Wang, M.; Li, Z. A lattice Boltzmann algorithm for fluid–solid conjugate heat transfer. *Int. J. Therm. Sci.* **2007**, *46*, 228–234. [CrossRef]
13. Wang, M.; Pan, N. Predictions of effective physical properties of complex multiphase materials. *Mater. Sci. Eng. R* **2009**, *63*, 1–30. [CrossRef]
14. Cai, J.; Huai, X. Study on fluid–solid coupling heat transfer in fractal porous medium by lattice Boltzmann method. *Appl. Therm. Eng.* **2010**, *30*, 715–723. [CrossRef]
15. Han, Y. *Mechanism and Characteristics of Heat Conduction in Nanocomposites*; Harbin Institure of Technology: Shenzhen, China, 2013.
16. Di Ilio, G.; Chiappini, D.; Ubertini, S.; Bella, G.; Succi, S. Fluid flow around NACA 0012 airfoil at low-Reynolds numbers with hybrid lattice Boltzmann method. *Comput. Fluids* **2018**, *166*, 200–208. [CrossRef]
17. Di Ilio, G.; Chiappini, D.; Ubertini, S.; Bella, G.; Succi, S. Hybrid lattice Boltzmann method on overlapping grids. *Phys. Rev.* **2017**, *95*, 013309. [CrossRef] [PubMed]
18. Guo, Z.; Shi, B.; Wang, N. A nonuniform lattice Boltzmann method based on domain decomposition. *Chin. J. Comput. Phys.* **2001**, *18*, 181–184.
19. Zeng, S.; Hunt, A.; Greif, R. Transport properties of gas in silica aerogel. *J. Non-Cryst. Solids* **1995**, *186*, 264–270. [CrossRef]

energies

MDPI

Systematic Review

A Review of Stimulation Technologies for Weakly-Consolidated Natural Gas Hydrate Reservoirs

Bo Ning [1,2], Yanjiang Yu [1,2,*], Jing Zeng [1,2] and Qiuping Lu [1,2]

1 National Engineering Research Center of Gas Hydrate Exploration and Development, Guangzhou 511458, China
2 Guangzhou Marine Geological Survey, China Geological Survey, Ministry of Natural Resources, Guangzhou 510075, China
* Correspondence: yuyanjiang@mail.sysu.edu.cn

Abstract: As an unconventional clean resource with huge reserves and a wide distribution, natural gas hydrates (NGHs) have good application prospects. However, due to limited understanding and available production technology for NGHs, there is still a large gap between current production tests and commercial exploitation. A breakthrough in reservoir stimulation technologies is key to realizing the industrialization of NGHs in the future. Through a comprehensive summary of hydrate production test cases in Japan and China, this paper highlights the difficulties in the transformation of weakly-consolidated reservoirs. By systematically reviewing the theory and technology of hydrate reservoir transformation and engineering applications, this paper elucidates in detail the technical principles and mechanisms of several available stimulation technologies for weakly-consolidated reservoirs, and assesses the feasibility of their application to increase the production of NGHs. Existing problems and challenges are summarized and future prospects are discussed. Finally, suggestions are put forward for research and development of transformation technology for weakly-consolidated NGHs reservoirs in the future.

Keywords: NGHs; weakly-consolidated; reservoir stimulation; hydraulic fracturing; hydrojet; hydraulic dilatancy; hydroelectric effect; electromagnetic wave; dipole moment

Citation: Ning, B.; Yu, Y.; Zeng, J.; Lu, Q. A Review of Stimulation Technologies for Weakly-Consolidated Natural Gas Hydrate Reservoirs. *Energies* **2022**, *15*, 7195. https://doi.org/10.3390/en15197195

Academic Editors: Reza Rezaee and Dameng Liu

Received: 16 August 2022
Accepted: 27 September 2022
Published: 30 September 2022

1. Introduction

As an inclusion compound formed by water molecules and guest molecules associated with them, natural gas hydrates (NGHs) exist in the form of crystals under low temperature and high pressure. NGHs in nature mainly refer to methane hydrate occurring in permafrost areas and deep-sea sediments under low temperature and high pressure. The decomposition of methane hydrate per unit volume can produce 150–180 standard volumes of methane gas and 0.8 standard volumes of water. NGHs have the advantages of huge reserves, wide distribution, shallow burial depth and high energy density. It is estimated that the organic carbon stored in the form of NGHs on the Earth accounts for 53% of total global organic carbon [1,2], which is twice the total carbon content of coal, oil and natural gas. Therefore, NGHs are considered an ideal clean alternative resource for the 21st century and have good application prospects.

Since 2002, production testing and research into NGHs at Mallik Station in the Arctic region of Canada have confirmed that NGHs can be decomposed into gaseous and liquid states by heating and depressurization, representing a prelude to the efficient exploitation of NGHs. A combination of carbon dioxide injection and depressurization was completed in the Alaskan tundra in 2012, and production tests were conducted. In 2012, Japan took the lead in exploration of the offshore South China Sea Trough, aiming to extract NGHs in sandy sediments that had been discovered, and, using depressurization as the main method, carried out the world's first production test of NGHs in this area of the sea. In 2017, a second production test was carried out in this sea area, which determined that there were good prospects for the exploitation of sandy NGHs with high permeability by depressurization [3,4].

However, the vast majority of the world's NGHs occur in clayey silt or silty sediments on the seabed. In 2017, China carried out a production test of NGHs in the Shenhu sea area on the northern slope of the South China Sea, and successfully obtained natural gas from muddy siltstone NGH deposits 203 to 277 m below the seabed of 1266 m. Continuous and stable gas production lasted for 60 days, with cumulative gas production of 309,000 m^3. In 2020, China took the lead in using horizontal well drilling and production technology to carry out the second round of production tests of NGHs in the Shenhu sea area of the South China Sea. After one month of production tests, the total gas production was 861,200 m^3, with a daily gas production 28,700 m^3, which was 5.57 times the daily gas production of the first vertical well production test [5]. The tests demonstrated that muddy siltstone NGHs have good exploitation potential. They also demonstrated that the application of new technology can play an important role in improving the single well productivity of NGHs.

Many hydrate production tests have been carried out throughout the world, and a series of breakthroughs have been achieved, resulting in a degree of consensus on the approach to NGH exploitation. Although the reserves of NGHs are huge, there are complex changes in solid, liquid and gas phases of NGHs and the average daily production of NGHs is still far below the threshold for commercial exploitation. Wu Nengyou et al. [6] compared the gap between the average daily production capacity of NGHs and the threshold production capacity for commercial exploitation, pointing out that there was still a gap of two to three orders of magnitude between current production capacity and the threshold for industrial production (see Figure 1). Rapid increase in daily average production using reservoir stimulation technologies is key to realizing the industrialization of NGHs.

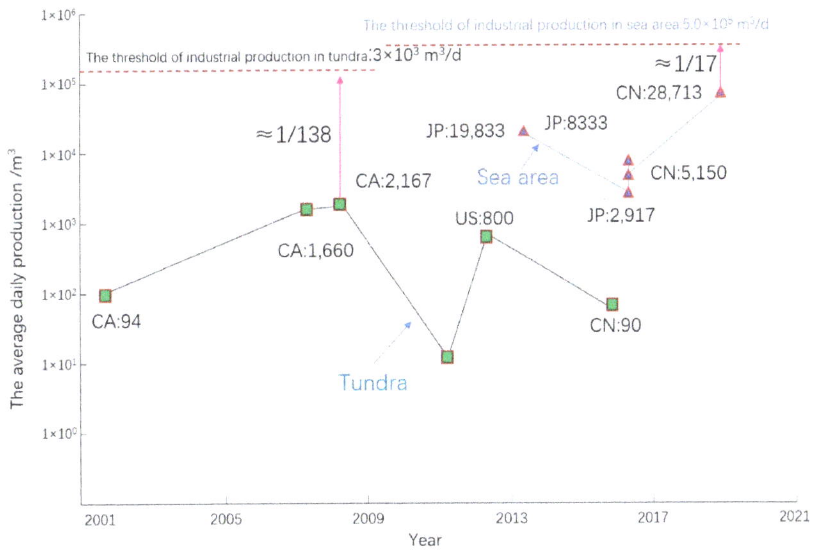

Figure 1. Relationship between average daily production test capacity and threshold of industrial production for NGHs (Reprinted with permission from Ref. [6]. 2020, Wu et al.).

Most research on reservoir stimulation technologies is still at the stage of model construction and numerical simulation, with a lack of corresponding experimental verification and basic support for field implementation, and understanding of how to increase production and reservoir stimulation remains uncertain. Considering the complex phase transitions of NGHs, they are susceptible to the effects of temperature-pressure multi-field coupling factors.

By systematically examining the theory and technology of hydrate reservoir transformation, combined with the assessment of the outcomes of two rounds of NGH reservoir transformation in the South China Sea, this paper highlights the difficulties in the transformation of weakly-consolidated reservoirs, analyzes the methods and basic principles of weakly-consolidated reservoir transformation of NGHs, considers the optimization of several types of transformation technologies for weakly-consolidated reservoirs, and elaborates the principles and mechanisms for each technology. The paper provides a comparative analysis of the feasibility of their practical application, summarizes existing problems and challenges, and discusses the prospects for these technologies. The analysis is anticipated to provide valuable information for the development of efficient and safe NGH mining technology.

2. Difficulties in Stimulation for Weakly-Consolidated Reservoirs

NGHs are mainly distributed in sediments along the continental slope of the deep sea continental shelf with a shallow burial depth. The reservoir lithology is dominated by weakly-consolidated muddy siltstones with a low temperature and loose structure. Presently, the internationally recognized exploitation methods mainly include depressurization, thermal shock, carbon dioxide replacement and the combined application of these methods [7]. However, through field practice and experimental simulation, it has been confirmed that the depressurization method and the scheme based on depressurization are the best ways to realize the exploitation of NGHs.

NGHs are decomposed by reducing the pressure of their reservoirs below the thermodynamic equilibrium line. Because NGH particles participate in the construction of cementation of reservoirs in solid form, the hydrate crystal skeleton in reservoirs disappears and the cementation capacity among particles decreases along with the decomposition of NGHs. The loose structure of weakly-consolidated NGHs reservoirs makes it difficult to maintain a large production pressure difference, resulting in stress deformation or even the collapse of reservoirs, which block the pressure conduction channel, such that the pressure is unable to conduct, and decomposition of NGHs is terminated. Simultaneously a great deal of heat is absorbed in the decomposition process, which leads to decrease in the reservoir temperature. In turn, when the temperature decreases to a certain extent this affects the phase equilibrium of NGHs and promotes the secondary formation of NGHs. To break this thermodynamic equilibrium, it is necessary to continue reducing the reservoir pressure so that the hydrate can continue being decomposed, thus entering a cycle of pressure drop—temperature drop—continuous pressure drop—corresponding temperature drop (see Figure 2).

Based on test results for hydrate-bearing cores, KONO et al. [8] suggested that the existence of hydrate and the change in reservoir pore structure caused by the hydrate decomposition process would significantly affect the permeability of reservoirs. Ruan Xuke, Li Xiaosen et al. [9] simulated the secondary formation of NGHs in the process of depressurization and exploitation in the laboratory. They suggested that the secondary formation of hydrates was mainly limited near the depressurized gas outlet, causing significant changes in the local hydrate saturation, temperature and pressure, and leading to a significant reduction in the gas production rate. Li Yanlong et al. [10] observed that there was a "memory effect" in NGHs, i.e., under certain conditions, the water decomposed from NGHs was more easily able to reform hydrates than water without a hydrate formation history, which led to increased risk of secondary formation blockage in the process of NGH exploitation, as well as decreased efficiency of reservoir seepage.

The decomposition of NGHs is a complex phase transition and seepage process involving interactions among multi-phase, multi-component and multi-physical fields [11]. Both complex morphological changes to gas, liquid and solid phases in the decomposition process and the secondary formation of hydrates lead to changes in pore structure, porosity, permeability, saturation and seepage channels of weakly-consolidated reservoirs. The mutual influence and restriction among these complex phenomena make the application of reservoir stimulation technologies very challenging.

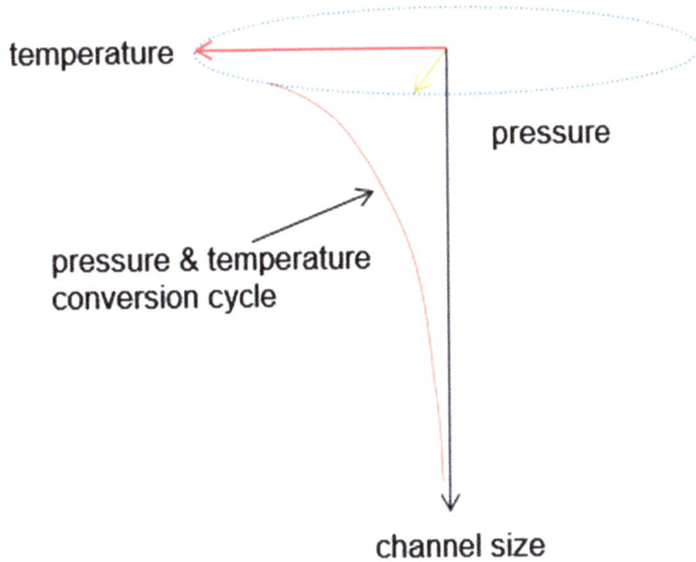

Figure 2. Schematic diagram of pressure and temperature conversion cycle.

3. Stimulation Technologies for Weakly-Consolidated Reservoirs

On the basis that the gas and water produced by the decomposition of NGHs follow the seepage mechanisms of porous media when flowing in reservoirs, methods to achieve increased production of NGHs are mainly divided into three categories: expanding the hydrate decomposition front, enhancing the hydrate decomposition rate and improving the hydrate seepage condition [12]. Based on the first round of production tests, the second round in the Shenhu sea area of the South China Sea in 2020 greatly increased the daily gas production to 28,700 cubic meters using complex structure wells and reservoir stimulation technologies [5]. Therefore, the stimulation of weakly-consolidated hydrate reservoirs should be based on the creation of an artificial fracture network to increase the permeability of reservoirs near the wellhole, enlarging the contact area between the wellhole and NGH reservoirs, accelerating the speed of pressure conduction and improving the seepage capacity of reservoirs. In addition, it is also necessary to remove blockages in pores or fractures, to maintain the geometric size of effective fractures, and to enhance the flow efficiency of reservoir fluids.

3.1. Challenges of Hydraulic Fracturing Technology

In recent years, hydraulic fracturing technology has been well applied, especially in the field of shale gas exploitation in North America. With the increasing development of research and the application of horizontal well multi-stage fracturing technology [13], the technology has been widely used in the fields of unconventional resources, such as ultra-low permeability shale oil and gas reservoirs, low-permeability and tight sandstones, and deep and coalbed methane reservoirs, and has achieved remarkable results. Hydraulic fracturing technology has the advantages of improving the complexity of fractures, expanding the distribution of fractures, greatly increasing the seepage area and enhancing the overall efficiency of exploitation of reserves.

Although it has been recognized that NGH exploitation efficiency by hydraulic fracturing has improved, the mechanism of fracturing technology is still in its infancy, few related monographs have been published, and there is still a lack of in-depth reports on the characteristics of NGHs and their impact on the fracturing effect. Additionally, the feasibility of NGH fracturing is also controversial [14,15]. Through experiments, Too et al. [14] confirmed that sandy reservoirs with an NGH saturation of 50–75% were compressible. Fractures of muddy/sandy interbedded NGH reservoirs tended to extend along the interface of sand-mud interbeds [16]. However, clayey silty NGH reservoirs with low saturation were easily compacted and difficult to fracture [17]. Liu Lele et al. [18] analyzed the permeability of NGH strata and pointed out that the stress sensitivity of NGH strata was strong, and that the strength of NGH reservoirs composed of fine particles, such as fine silt and clay was much weaker than that of the low-permeability oil and gas reservoir rocks, with the strength of strata after complete hydrate decomposition being reduced to as little as one tenth of the original.

Because of the special physical properties of NGHs, the hydrate crystalline solids not only participate in the construction and cementation of rock skeleton, but are also formed in the pores between reservoir particles to occupy the space of reservoir pore fluid, resulting in the permeability of hydrate storage decreasing to a very low level. However, once NGHs are decomposed, the porosity and permeability changes greatly—the porosity decreases gradually, but the permeability increases sharply. These factors may affect NGH hydraulic fracturing. As a result of the stress change caused by the decomposition/formation of NGHs and the formation and extension of fracturing fractures, the permeability of reservoirs will change greatly. Through numerical simulation, Sun et al. [19] demonstrated that to increase NGH production, hydraulic fracturing has an obvious effect in the early stages of exploitation, but has a limited effect in the middle and late stages of exploitation, and the overall effect of fracturing might change.

The results of two production tests of NGHs in the China Sea area showed that horizontal wells can provide improved exploitation efficiency, with the production of a single well found to be 5.57 times that of a vertical well [5]. In the field of unconventional resources, the multi-stage fracturing technology of horizontal wells has the advantages of greatly increasing the seepage area and improving overall exploitation efficiency. Using horizontal wells, the original single fracture is transformed into complex fractures, and the complex fractures of different scales are further transformed into effective fracture transformation volumes. As a result, the complexity of each fracture is further enhanced (see Figure 3). In this way, the interaction between natural fractures and artificial fractures is made full use of to maximize fracture transformation volumes. Thus, it becomes more and more critical in field applications to make use of horizontal well multi-stage fracturing technology to enhance the hydrate production of a single well.

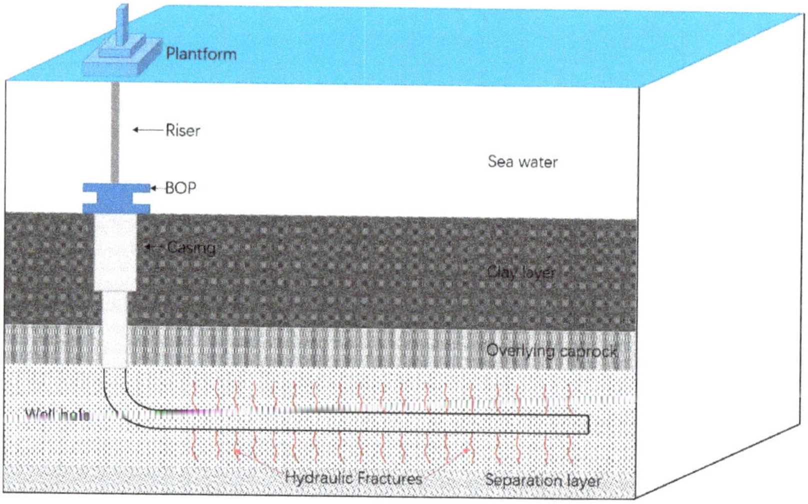

Figure 3. Schematic diagram of physical model for horizontal well fracturing.

For hydraulic multi-stage fracturing in weakly-consolidated reservoirs, it is necessary to determine the feasibility boundary of NGH fracturing, analyze the influence of reservoir type, geomechanical properties and conditions, and stratum sensitivity, and consider fracture initiation and propagation laws. It is also important to carry out research on fracture parameter optimization methods, special perforation technology, multi-stage fracturing tools and methods, multi-scale fracturing process parameter optimization methods, low-damage fracturing fluid and fracture effectiveness maintenance technology. In addition, due to the shallow burial of NGHs, the decomposition and phase transition of hydrates are complex, and the deformation amplitude of strata is large. Therefore, how best to achieve effective multi-stage modification is also one of the technical challenges of hydraulic fracturing.

3.2. Hydraulic Self-Sealing Advantages of Hydraulic Jet Fracturing Technology

Hydraulic jet fracture (HJF) is a new stimulation technology, developed from high pressure abrasive water jet technology based on the transformation relationship among kinetic energy, potential energy and pressure in the Bernoulli equation, which combines hydraulic jet perforation with hydraulic fracturing technology. In 1998, Surjaatmadja of the Halliburton Company in the United States proposed the hydraulic jet fracturing method [20]. For this method, a high-speed fluid generated by a special jet tool with a propping agent is utilized to penetrate the casing and rock to form a hole, and then the fluid generates a pressure higher than the fracture pressure at the bottom of the hole to create a main fracture with a single direction. This increases the seepage area, decreases the reservoir filtration resistance, and provides a good migration path for stratum fluids to enter the wellhole (see Figure 4).

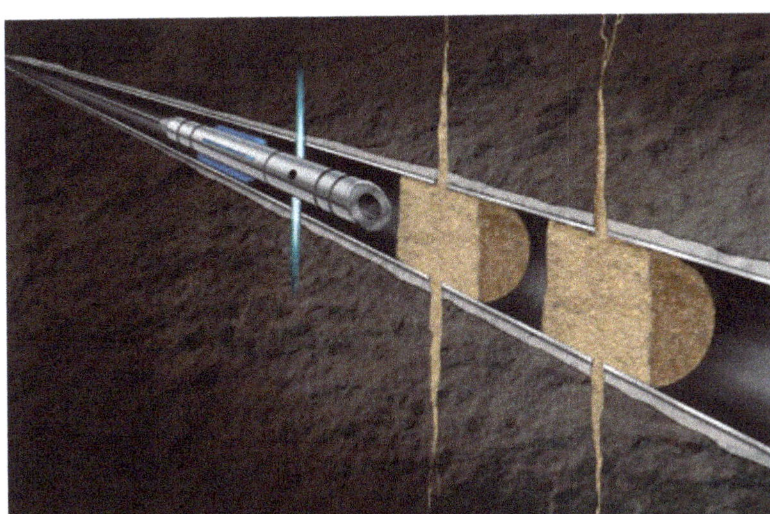

Figure 4. Schematic diagram of hydraulic jet fracturing.

In the late 1990s, the Halliburton Company first carried out field tests and hydraulic jet technology operations in Texas, New Mexico, as well as other oil wells in the United States, and achieved remarkable effects. Subsequently, the technology was successfully implemented in hundreds of wells in North America and the Gulf of Mexico, and, in 2003, it was successfully tested in the 1-RJS. 512HA well in Campus Bay, Brazil [21]. This was the first time that hydraulic jet fracturing technology had been applied to marine stimulation operations. In December 2005, the Halliburton Company cooperated with the Changqing Oilfield Company to successfully complete stimulation operations in the Zhuangping 3 and Jingping 1 wells of the Jing'an Oilfield. This was the first test of this technology in China [22], and it was then widely applied in the Changqing oilfield, the southwest oil and gas field, the Xinjiang oilfield and other oilfields in China. Hydraulic jet fracturing technology is now used in thousands of horizontal wells and tens of thousands of vertical wells all over the world.

For hydraulic jet fracturing, the stratum rock is eroded by a high-speed abrasive jet to form holes, and the low pressure area formed near the high-speed jet entrains the fluid around the holes into the holes, resulting in the pressure in the holes being higher than the stratum fracture pressure, which induces the stratum to crack and achieve hydraulic self-sealing (see Figure 5). Thus, this process does not need to use an additional mechanical sealing device and can carry out the fracture stimulation on the producing layer fast and accurately. This has solved the difficult problem of implementing multi-stage stimulation because the effective sealing cannot be operated in a long, open-hole horizontal well. Therefore, this technology has the advantages of higher construction safety, of a shorter operation cycle, and a wider range of application.

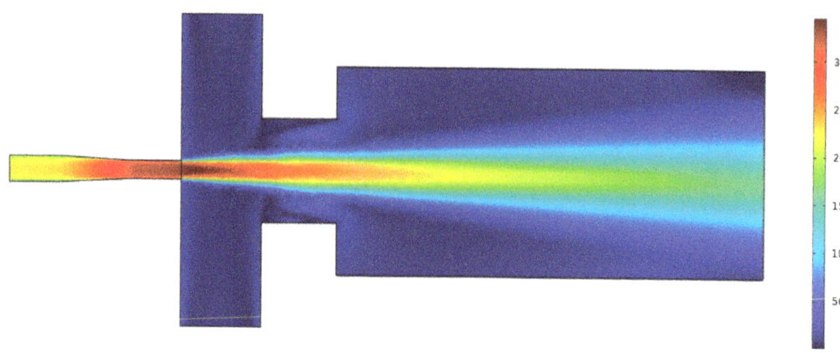

Figure 5. Flow-field analysis of hydraulic jet.

Because of the shallow burial depth and weak rock cementation strength of weakly-consolidated NGHs reservoirs, the conventional mechanical sealing device cannot support the borehole wall stably, and the development of the micro-fracture and the extension direction of the fracture is uncontrollable, which creates significant challenges to the implementation of large-scale reservoir stimulation. On the basis of the fine reservoir characterization and the reservoir stress distribution, and using the hydraulic self-sealing characteristics formed by the high-speed jet, the dominant target reservoir is closely cut, the resistance at the end of the hole is reduced, and the fracturing fluid is forced to follow the principle of minimum resistance and move forward along the dominant channel direction made by the abrasive jet, so as to induce the development of fractures along the set direction, thus achieving control of the fracture direction. This technology provides technical support for the multi-stage stimulation of weakly-consolidated NGH reservoirs, and, currently, is one of the most effective reservoir stimulation technologies. Future research should focus on how to improve the efficiency of stimulation construction, strengthen the intervention and treatment of secondary hydrate formation and sand production in the later exploitation process, and maintain the seepage capacity of dominant channels.

3.3. *Fracture Complexities in Weakly-Consolidated Reservoirs Enhanced Using Rock Dilatancy*

Dilatancy is an important rock property, referring to the rock deformation phenomenon whereby the total volume of the rock medium increases under the action of shear stress or pore fluid pressure increase when the total stress is still compressive stress. As early as the 1950s and 1960s, Brigeman and Handlin [23,24] studied rock dilatancy. Subsequently, scientists and scholars pursued more systematic studies on this issue, and the process of micro-fissure before macroscopic rupture was also elucidated. In general, when a certain stress combination in the rock medium exceeds a certain limit (the initial dilatancy yielding pressure), part of its deformation is recoverable and the other part is unrecoverable [25]. The former is called reversible elastic deformation and the latter is called irreversible dilatant deformation. Analyzed in terms of the internal structure of rock deformation, dilatancy deformation is an irreversible deformation caused by the sliding of particles and particle interfaces in rocks and the static propagation of micro-fissures, which, from a micro-perspective, can be regarded as the rotation, rolling and rearrangement of rock particles.

The application of rock dilatancy deformation to the field of oil and gas originated from an accidental discovery by the Imperial Oil Company in the treatment of SAGD wells. The production of oil sand wells has been greatly improved following high-pressure injection operations, which has led to the systematic study of the dilatancy of weakly-consolidated sandstones. Dilatancy in oil sands is a combination of shear damage and tensile microfracture. When shear dilatancy occurs, the shear stress on rock particles exceeds the friction strength, and the original structure of rock particles changes due to the disturbance of the particle sliding and rolling, but the particles still contact each other, and, with the continuous injection of high-pressure fluids, tensile fracture occurs. Thus, the particles are separated from each other, which not only increases the total volume of pores in rocks, but also increases the permeability of rocks [26] (see Figure 6a,b).

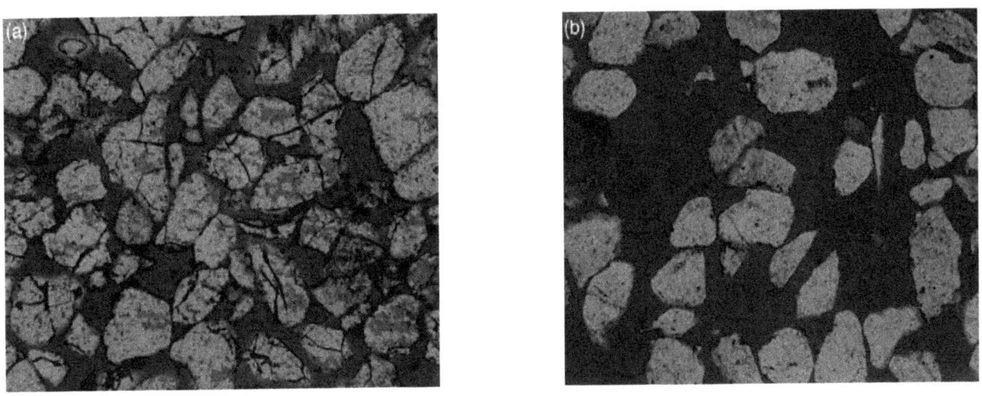

Figure 6. (**a**) Rock micrograph before dilatancy. (**b**) Rock micrograph after dilatancy.

Through rock mass dilatancy, shear expansion zones and independent tensile microfracture areas can be formed around and between wells, which can greatly improve the fluid injection capacity and contact area of injected fluids. Therefore, dilatancy stimulation technology has been tested when transforming water injection wells in the oilfield. Yan Xinjiang et al. [27] carried out cyclic loading triaxial experiments on the sandstone reservoir rocks of the Dongying Formation in the Bohai Oilfield. The mechanical properties and deformation parameters of Dongying Formation sandstones under low confining pressure were tested, the dilatancy angle was measured, and dilatancy simulation of loose sandstones under different conditions was carried out. The test results showed that the loose sandstones were able to produce strong dilatancy under low effective stress, and the dilatancy technology of high-pressure injection was able to form a high-permeability dilatancy fracture network and remove reservoir pollution blockage. Hence, this technology provides a feasible method for the stimulation of water injection wells.

The burial depth of NGHs is shallow, and hydraulic dilatancy technology, as a high-pressure reservoir stimulation process, has a destructive impact on the integrity of the caprock. There is also a risk of local communication of the dominant horizon, which occurred in the early SAGD project. In pilot testing during the Christina Lake SAGD project by the Canadian Cenovus Resource Company, because the pretreatment time was too short, premature dilatancy with large displacement led to the formation of tensile fractures between wells and the communication of bottom water layers, and the results were poor after production. However, this technology was later used successfully after process improvement.

Much experience has been gained in the use of hydraulic dilatancy technology in the treatment of weakly-consolidated loose sandstones, such as oil sands [28]. Through dynamic analysis of the elastic mechanics and thermoelastic mechanics mechanism of porous media, the stress, pore pressure state and water saturation of strata around the well can be adjusted in time, and the rock mass pre-dilated in advance, to form a large volume dilatancy area in the vertical direction. After completion of the dilatancy, the injection pressure and displacement are gradually increased [29], thus achieving dilatancy treatment of the reservoir rock mass (see Figure 7).

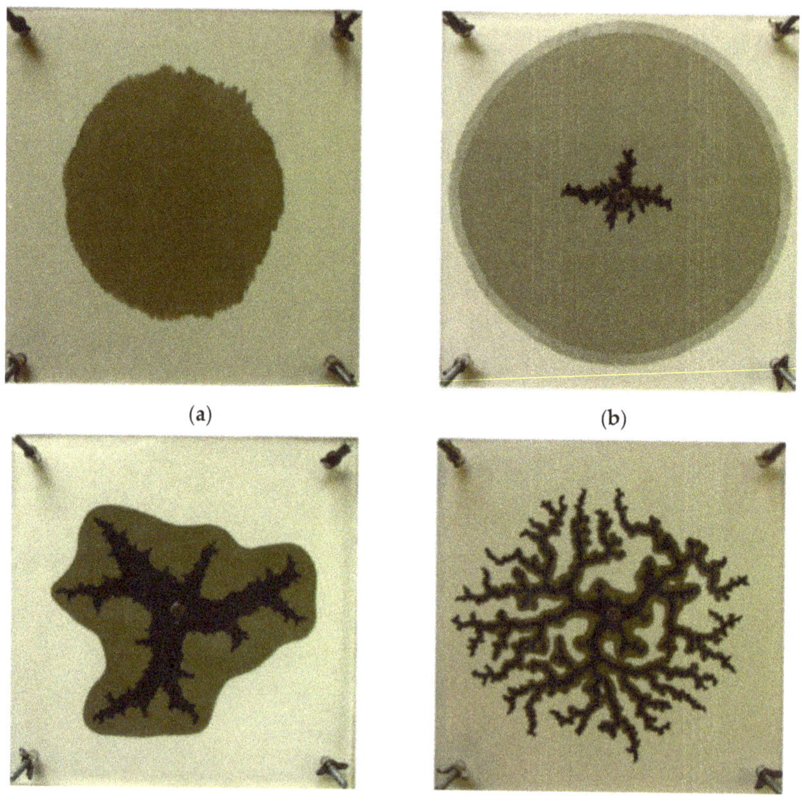

(a) (b) (c) (d)

Figure 7. Schematic diagram of hydraulic dilatancy process. (**a**) Treatment of pre-dilatancy. (**b**) Formation of dilatancy area. (**c**) Extension of dilatancy area. (**d**) Dilatancy of large volume.

Although NGHs have the same rock mass conditions as oil sands, and are also weakly-consolidated undiagenetic reservoirs, NGH reservoirs, compared with oil sand reservoirs, have fine particles, low permeability, a high mud content and strong adsorption capacity among particles, especially in the Shenhu Sea area of the South China Sea. Because hydrate crystals are hidden in the pores of particles, the pressure in the pores is increased, the permeability of the pores is reduced, and the injected high-pressure fluids have difficulty entering the pores of particles. The mutual dislocation and separation among rock particles are realized by increasing the pressure in the pores and creating a large number of shear micro-fractures for dilatancy. Thus, for diffusion-type hydrates, use of this technology may lead directly to the tensile fracture of the rock mass, and the dilatancy of the rock mass cannot be fulfilled. However, for some leakage-type hydrate reserves, because hydrate crystals participate in the construction of the reservoir skeleton and have larger crystal structures, it is possible to use rock dilatancy deformation to form shear dilatancy zones and independent tensile micro-fractures to increase the porosity and permeability of reservoirs. In this way, the pressure conduction area and the gas production capacity of hydrates are improved. Further investigation and development are needed so that the stimulation of different types of NGHs resources can be achieved using the mechanism of hydraulic dilatancy.

3.4. Stimulation Technology Based on Electro-Hydraulic Effect

When pulse high voltage is applied to a group of electrodes in a liquid, the gaps between electrodes are instantaneously broken down to generate a strong arc spark discharge. High-density electric energy is converted into high-strength light energy, sound energy, electromagnetic energy, mechanical energy and chemical energy in a very short time with the help of high-temperature strong plasmas formed in a discharge channel in the liquid. Many experts and scholars all over the world have carried out theoretical and experimental research on this special discharge phenomenon. In 1955, Yutkin, an engineer in the former Soviet Union, found a way to generate huge levels of mechanical energy by discharging in the liquid. Thus, the mechanical energy produced by the electro-hydraulic effect can be used in the field of engineering applications and is referred to as the "electro-hydraulic effect" [30].

By exciting the arc discharge in a sealed cavity filled with liquids, some of the liquids in the discharge channel are instantaneously vaporized, decomposed and ionized into high-temperature plasmas. These plasmas expand quickly and generate a kind of rapid radial mechanical energy to produce strong shock waves and associated bubbles, which affect the surrounding reservoir rock masses, providing a path for the stimulation of reservoirs [31]. In the 1980s, several tools and technologies developed using the hydro-electric effect were introduced into the field of oil and gas exploitation and were widely used in some countries of the former Soviet Union [32]. As one of the conventional stimulation measures for oil and gas fields, in the 1990s, a number of downhole electric pulse detonation instruments were introduced in China and widely popularized in Xinjiang, Daqing, Ansai and other oilfields, achieving good application results [33]. The Institute of Electrical Engineering of the Chinese Academy of Sciences, PetroChina Exploration, the Development Research Institute, Xi'an Jiaotong University and some oilfields have carried out research for theory development as well as product development and application [34].

Presently, research on the mechanism of the hydro-electric effect have resulted in more systematic understanding. Through discharge in the water medium, high temperature gas and high temperature plasma are produced, resulting in strong shock waves acting on the reservoir rocks and fluids, with acceleration as high as 3000 times the acceleration of gravity [35]. Under the action of the shock wave, the respective point particles of discontinuous media, such as rock and liquid, vibrate violently. Moreover, under the action of the highly-accelerated impact, when the fatigue strength of the rock is exceeded, new micro-fractures or macro-fractures are caused (see Figure 8). At the same time, under the influence of the shock wave, the clay cement on the surface of rock particles is shaken off, and the clay particles filled in the pore throats between the rock particles are loosened or migrated [36]. This results in removal of the blockage of the pore throats, expansion of the pore throat radius and pore connectivity, and improvement in the permeability of reservoirs [37].

Mostly stored in muddy silt deep-sea sediments, NGHs are associated with a weakly-consolidated rock structure with a high mud content. Following the impact of an elastic wave, the continuity of the reservoir media is destroyed, and tensile fractures are easily formed. However, after the shock wave disappears, the fractures close rapidly and exhibit obvious plastic deformation. According to feedback from field applications, the use of the strong blasting ability of the hydro-electric effect is better for dealing with dense rocks with brittle failure characteristics, such as calcareous dolomite, siltstone, etc., and is slightly worse for rocks with relatively strong plasticity, such as mudstone [38]. On the other hand, due to the low cementation strength of hydrate reservoirs, under the action of a shock wave, the fractures formed in rocks expand forward at high speed and become large in scale with a long extension distance, which can be used to achieve deep reservoir stimulation. How to balance the above advantages and disadvantages to provide useful guidance for NGH reservoir stimulation requires further study.

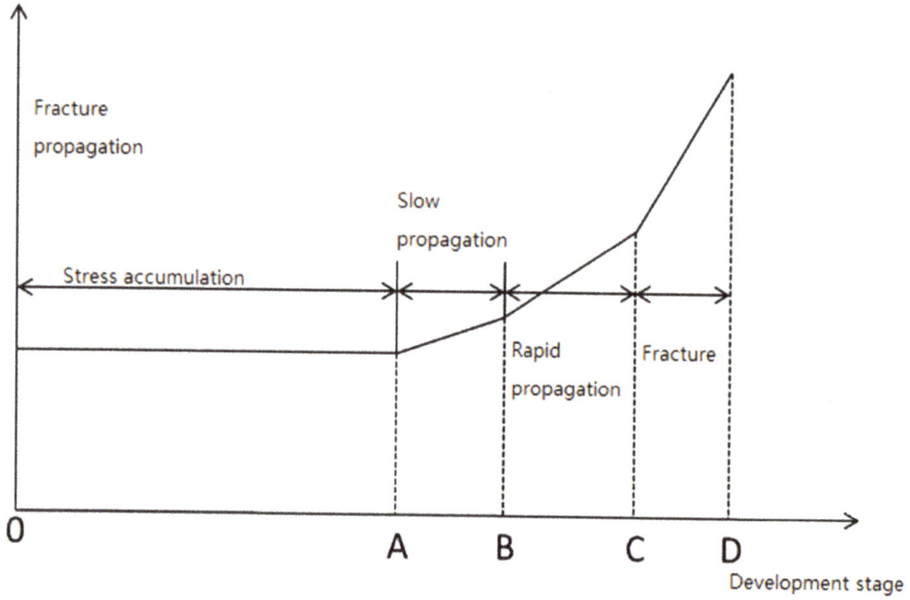

Figure 8. Relationship between fracture propagation and discharge.

In addition to stimulating the reservoir before exploitation, the high-pressure electro-hydraulic pulse stimulation device can be installed in the well for a long time as an auxiliary completion tool (see Figure 9), which will not interfere with the exploitation of NGHs, and can continuously discharge and shock the production layer during the subsequent exploitation process and produce conductive fractures. Additionally, blockage of seepage channels can be removed in time, the smoothness of the high-permeability channel can be maintained, and continuous production of NGHs can be guaranteed.

3.5. Selective Action of Electromagnetic Waves on Polar Molecules

In recent years, when evaluating NGH exploitation methods, a variety of heating modes have been applied to improve the efficiency of thermal stimulation methods. Liang et al. [39] used electric heating to heat hydrate reservoirs and compared the energy efficiency with hot water injection. They concluded that the efficiency of electric heating in vertical wells was higher than that of hot water injection. Rahim I et al. [40] analyzed the role of microwave and plasma radio frequencies in hydrate decomposition. Wang Bin, Zhao Jiafei et al. [41,42] carried out studies on heating hydrates by microwave. M R Davletshina et al. [43] of the Ufa State Petroleum Technological University and Li, D.L. and Liang, D.Q. of the Guangzhou Institute of Energy Conversion Chinese Academy of Sciences carried out experiments on heating hydrates by microwave [44]. It was found that microwaves have a very marked effect on NGH decomposition, and that hydrates can be decomposed in a very short time using low-power microwave radiation. In addition to the obvious "heating effect", electromagnetic waves also have a "non-thermal effect". Jin Youhuang et al. [45] conducted a microwave test on heavy oil cores in the Liaohe Oilfield and the "non-thermal effect" of microwaves on breaking molecular bonds was confirmed. Jeambey et al. [46] used alternating electromagnetic fields with different frequencies, intensities and waveforms acting on oil shale, finding that the hydrocarbons in the samples changed greatly, the molecular bonds of heavy hydrocarbons were broken, and a large number of small molecular hydrocarbons were formed.

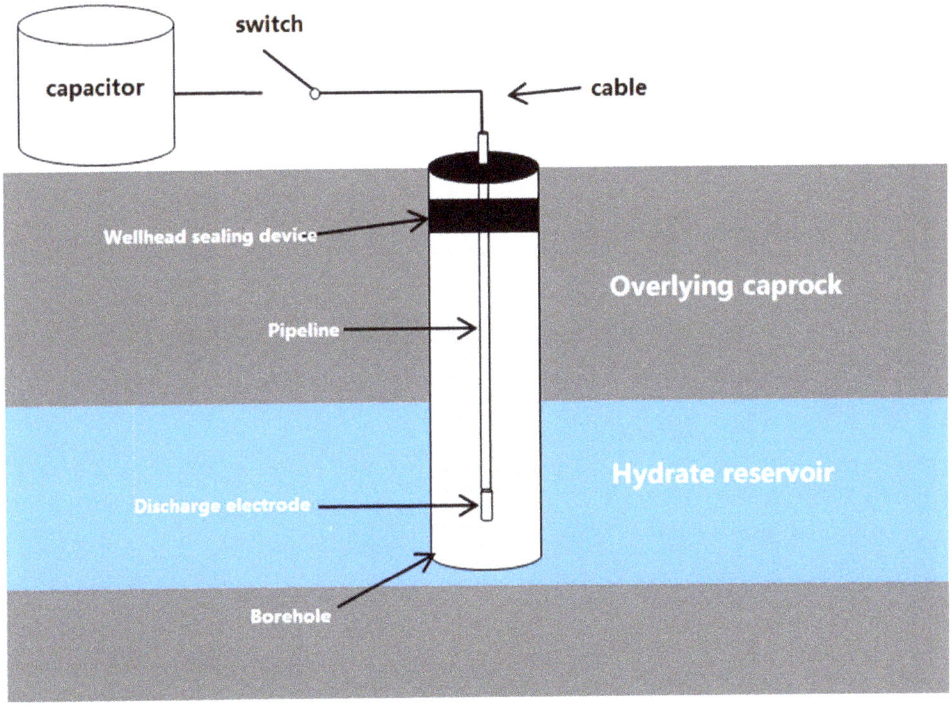

Figure 9. Schematic diagram of high-pressure electro-hydraulic pulse stimulation operation.

Due to the complexity of underground porous media conditions and the differences in the electrical parameters between different interfaces, the waveform path, and the amplitude and intensity of electromagnetic waves, will change in the process of propagation in underground porous media [47]. The frequency change of electromagnetic waves and the electrical characteristics of rocks determine the skin depth. The higher the frequency, the more obvious the skin effect is, and the smaller the penetration distance of electromagnetic waves [48]. The microwave frequency commonly used for heating in industry is 2450 MHz, and its penetration depth is about 0. 5~3 cm [49]. The decomposition effect of electromagnetic waves on hydrates is mainly due to the heating of the surface layer and the transfer of heat. Therefore, it is difficult to use electromagnetic waves to heat large-scale hydrate ore bodies.

NGHs in nature mainly refer to methane hydrate, which is mainly composed of water molecules and methane molecules. Water is a polar molecule in which the positive and negative charge centers do not coincide, so there is a certain dipole moment [50]. Methane is a tetrahedral nonpolar molecule with a dipole moment of 0. At low temperature, the positive and negative charge centers of polar molecules rotate under the action of the electric field to follow the direction of the external electric field [51]. When a large number of disorderly polar molecules are placed in an alternating electric field, the orientation of these polar molecules will change with the polarity of the electric field (see Figures 10 and 11).

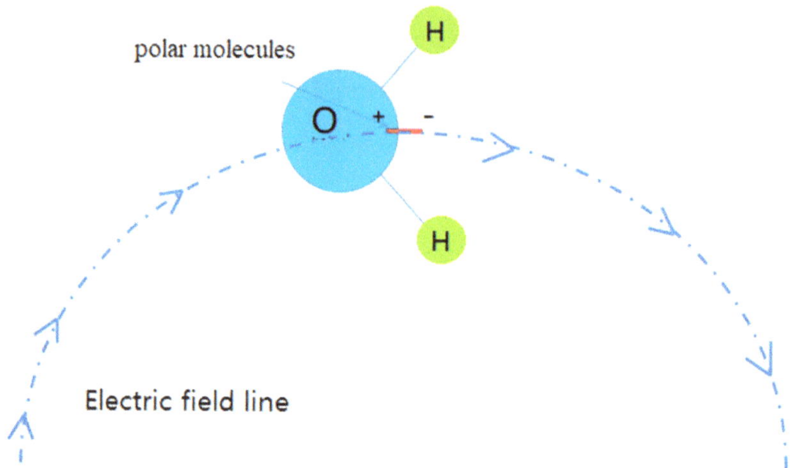

Figure 10. Schematic diagram of dipole moment affected by electric field.

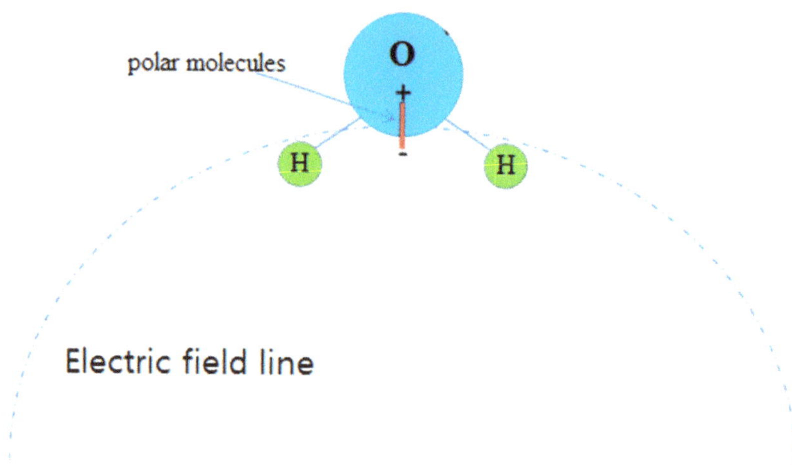

Figure 11. Illustration of deflection of dipole moment under action of electric field.

The application of technologies related to electromagnetic waves may become one of the methods used to more efficiently exploit NGH resources in the future. The transmission characteristics of electromagnetic waves in different media follow Maxwell's equations in the theory of electromagnetic wave propagation [52]. At the same time, the alternating change of magnetic and electromagnetic fields produces radiation of electromagnetic waves in the medium. The lower the electromagnetic wave frequency is, the smaller the influence of the medium, and the greater the distance it travels. In future, we can study and utilize the natural property of the dipole moment of polar molecules, which can strongly absorb electromagnetic wave energy, and the characteristics of non-polar molecules, which do not interact with electromagnetic wave field because of their zero dipole moment, to separate polar molecules from non-polar molecules and apply low-frequency electromagnetic fields directionally to hydrate ore bodies. The stronger the external electric field is, the faster the polarity changes after interacting with the dipole moment of polar molecules. Therefore, the bonds between water molecules are broken, gas molecules can escape from the holes between water molecules, and the gas in NGHs is separated through the non-heating electromagnetic effect.

From the perspective of engineering applications, combined with the geological conditions of NGHs in the South China Sea, this paper compares and analyzes several stimulation technologies suitable for weakly-consolidated reservoirs, and discusses the technical principles and mechanisms, as well as providing an adaptability analysis for the exploitation of NGHs, as shown in Table 1.

Table 1. Comparative analysis of several stimulation technologies.

Technology	Principle	Restrictions	Adaptability Analysis
Hydraulic fracturing	High-pressure pump units are used to inject the fracturing fluid into the wellhole at a rate that exceeds the absorption capacity of the stratum, which is forced to fracture, and then proppant is squeezed in to support the fractured fracture.	1. The operation cost is high, and the construction equipment occupies a large area. 2. There are certain requirements for reservoir thickness and sealing integrity. 3. The fracturing fluid shall be compatible with the physical properties of reservoirs.	The technology is mature, and the support facilities are complete, but the geological requirements are high, and the applicability of unconsolidated plastic rock masses is poor.
Hydraulic jet fracturing	High-pressure abrasive water jet is used to penetrate the casing and stratum to form spindle-shaped holes to relax the stress near the wellhole and increase the seepage area.	1. There is a long construction period. 2. It belongs to the near-wellhole transformation, and the effective distance is limited. 3. It is easy to induce sand production of the stratum.	The construction safety is high, and the application range is wide, but the construction efficiency needs to be improved and the control of sand production needs to be strengthened.
Rock dilatancy	By injecting the high-pressure fluid, the pore pressure is increased, the pore volume of rock masses is increased, and the complex large volume micro-tension-shear fracture area is formed.	1. The physical property of the reservoir is selective. 2. There is a long operation time. 3. It is necessary to carefully control the volume of stimulation.	The action distance is long, the volume of the affected ore bodies is large, no proppant is needed, and the application to weakly-consolidated loose sandstones is mature.
Electrical pulse detonation	The shock wave produced by electrode high voltage discharge is used to shock the reservoir and increase the complexity of wellhole fractures.	1. Accurate stimulation can be carried out for the quasi-reservoir. 2. It can be placed in the well as a completion production string, and the production will not be affected during the stimulation.	It is mainly used near the wellhole to dredge the diversion channel and increase the permeability and has great potential for production and application.
Electro-magnetic wave resonance	The high-frequency resonance of the electromagnetic wave is utilized to drive water molecules to generate heat by friction to increase the heat, and the thermal radiation is utilized to promote the decomposition of hydrates, so that the thermal efficiency is higher.	1. It is effective for open hole wells and cannot be used in casing and screen wells. 2. Electromagnetic waves have a limited penetration distance in the stratum and has a heating effect on the shallow surface of the borehole wall, so it is difficult to heat the hydrate ore body on a large scale.	The reservoir is accurately heated to improve the hydrate decomposition rate; it can be placed in the well as a completion production string, and heating does not affect production.

4. Discussion and Recommendation

In recent years, NGHs, as an alternative resource with huge potential, have attracted wide attention. Several countries have invested huge amounts of money to carry out exploration, production testing and research into NGHs. The breakthrough in the production test projects of NGHs in China and Japan has inspired confidence in exploiting this clean natural gas resource.

However, weak-consolidated NGHs reservoirs have the characteristics of loose structure, weak bearing capacity and easy collapse, which greatly limit the scope for application of reservoir stimulation technologies. With decomposition of the hydrate, physical and mechanical properties of the rock stratum change accordingly. The gas and water generated by hydrate decomposition increase the pore pressure sharply, and the degree of cementation between the clay particles is weakened due to increase in the local stress concentration, and the geomechanical stability of the reservoir reduces significantly [53]. How reservoir transformation and stimulation technology affect the mechanical and physical properties of the formation will be key areas of research in the future.

At present, the exploitation of NGHs has just begun, and no commonly accepted method of exploitation has yet been developed. Most methods focus on the use of stimulation measures in the initial stage of NGH exploitation. Due to restrictions caused by the lack of understanding and research on NGHs, the seepage capacity of NGH reservoirs needs to be enhanced, the migration problems of gas, liquid and solid in reservoirs after the decomposition of NGHs need to be solved, and the mechanism of multi-phase flow in muddy silt reservoirs remains unclear. At present, we cannot effectively improve reservoir seepage capacity, which greatly restricts NGH exploitation efficiency.

Although there is a complex situation of multi-phase transition and multi-field coupling in NGHs, from the macro-perspective, the important factor that determines the decomposition rate of NGHs is the seepage capacity of the stratum. The development of horizontal wells and hydraulic fracturing technology has greatly improved the rate of extraction of shale gas, and has also increased the final recovery of a single well. The level of recovery of production wells has increased to between 15% and 35%, reducing the production cost of shale gas, leading to the industrialization of shale gas, providing inspiration for the industrialization of hydrates. As a result of NGH production testing in the China Sea area, the contact area between the wellhole and reservoir has been increased using complex structure wells. Together with stimulation reconstruction of the reservoir near the wellhole, the increased contact area can greatly increase production. Based on the seepage mechanism in porous media, the permeability of the stratum is increased by expanding the size and total area of the fracture, which, on the one hand, provides a channel for low pressure conduction in the wellhole to promote the continuous decomposition of NGHs, and, on the other, reduces resistance to the migration of decomposed water to the wellhole. These measures are beneficial in improving the decomposition efficiency of NGHs and increase gas production of a single well by improving and maintaining the seepage capacity of the stratum. However, the long-term effect of reservoir stimulation cannot be determined due to the short production test cycle.

The first two rounds of production testing of NGHs in the sea area mainly focused on near-wellhole stimulation, and it is urgently required to carry out more extensive and large-scale controllable stimulation technologies to address the problem of the release of deep NGH resources, and to further improve the effect of increasing and stabilizing the production of a single well. Therefore, it is suggested that, in subsequent production test projects of NGHs in the South China Sea, in light of the above-mentioned reservoir stimulation technologies, that systematic research should be carried out, giving priority to hydraulic dilatancy stimulation technology which can be supplemented by electromagnetic detonation and other technical means to dredge solid particle blockages near the wellhole, focusing on the phase transition dynamic treatment technology of NGHs, and studying well control safety and fracture monitoring technologies in the process of reservoir stimulation. Adaptability evaluation of weakly-consolidated reservoir stimulation technologies should be carried out, and related scientific research and experimental simulation should be conducted to ensure smooth progress of NGH industrialization.

5. Conclusions

(i) The key issues for increasing NGH production are improvement in the decomposition rate of hydrates, expansion of the size of seepage channels and maintenance of the long-term effectiveness of seepage capacity. The breakthrough of the development of reservoir stimulation technology is key to achieving the industrialization of NGHs in the future.

(ii) Several types of weakly-consolidated reservoir stimulation technologies have been optimized, which are suitable for different geological conditions and application scenarios. In the diffusion-based stratum, hydraulic jet fracturing technology has obvious advantages, while in the leakage-based stratum, hydraulic dilatancy technology exhibits better adaptability. Electric pulse detonation technology and electromagnetic wave resonance technology can continuously stimulate the reservoir in the later production process and there is no need to stop production.

(iii) The understanding of, and research into, NGHs is still at a relatively early stage. When exploring different reservoir stimulation technologies, we should pay attention to the combination of geology and engineering, strengthen process quality control, and avoid environmental safety risks.

Author Contributions: Conceptualization, B.N. and Y.Y.; methodology, B.N.; validation, J.Z. and Q.L.; formal analysis, B.N.; resources, B.N. and J.Z.; data curation, J.Z.; writing—original draft preparation, B.N.; writing—review and editing, J.Z.; visualization, B.N.; supervision, Y.Y.; project administration, Y.Y. All authors have read and agreed to the published version of the manuscript.

Funding: This research was funded by High-tech ship research project of the Ministry of Industry and Information technology (CJ05N20).

Conflicts of Interest: The authors declare no conflict of interest.

References

1. Solan, E.D.; Koh, C.A. *Clathrate Hydrate of Natural Gases;* CRC Press: Boca Raton, FL, USA, 2008.
2. Kvenvolden, K.A. Gas hydrates-geological perspective and global change. *Rev. Geophys.* **1993**, *31*, 173–187. [CrossRef]
3. Yamamoto, K.; Terao, Y.; Fujii, T.; Ikawa, T.; Seki, M.; Matsuzawa, M.; Kanno, T. Operational overview of the frist offshore production test of methane hydrates in the eastern Nankai Trough. In Proceedings of the Offshore Techology Conference, Houston, TX, USA, 5–8 May 2014. [CrossRef]
4. Yamamoto, K.; Wang, X.X.; Tamaki, M.; Suzuki, K. The second offshore production of methane hydrate in the Nankai Trough and gas production behavior from a heterogeneous methane hydrate reservoir. *RSC Adv.* **2019**, *9*, 25987–26013. [CrossRef] [PubMed]
5. Ye, J.L.; Qin, X.W.; Xie, W.W.; Lu, H.L.; Ma, B.J.; Qiu, H.J.; Liang, J.Q.; Lu, J.A.; Kuang, Z.G.; Lu, C.; et al. Main progress of the second gas hydrate production test in the South China Sea. *China Geolog.* **2020**, *47*, 557–568.
6. Wu, N.; Li, Y.; Wan, Y.; Sun, J.; Huang, L.; Mao, P. Prospect of marine natural gas hydrate stimulation theory and technology system. *Nat. Gas Ind.* **2020**, *40*, 102. [CrossRef]
7. Moridis, G.J.; Collett, T.S. Strategies for gas production from hydrate accumulations under various geologic conditions. In Proceedings of the TOUGH Symposium, Berkeley, CA, USA, 12–14 May 2003.
8. Konno, Y.; Yoneda, J.; Egawa, K.; Ito, T.; Jin, Y.; Kida, M.; Suzuki, K.; Fujii, T.; Nagao, J. Permeability of sediment cores from methane hydrate deposit in the Eastern Nankai Trough. *Mar. Pet. Geol.* **2015**, *66*, 487–495. [CrossRef]
9. Xuke, R.; Xiaosen, L.; Mingjun, Y.; Feng, Y. Influences of gas hydrate reformation and permeability changes on depressurization recovery. *Acta Petrol. Sin.* **2015**, *36*, 612–618.
10. Yanlong, L.I. Nucleation probability and memory effect of methane-propane mixed gas hydrate. *Fuel* **2021**, *291*, 120103.
11. Shuxia, L.; Shangping, G.; Yueming, C.; Ningtao, Z.; Didi, W. Advances and recommendations for multi-field characteristics and coupling seepage in natural gas hydrate development. *Chin. J. Theor. Appl. Mech.* **2020**, *52*, 831–835.
12. Zhang, W.; Shao, M.; Jiang, C.; Tian, Q. World progress of drilling and production test of natural gas hydrate. *Mar. Geol. Quat. Geol.* **2018**, *38*, 1–13.
13. Konno, Y.; Jin, Y.; Yoneda, J.; Uchiumi, T.; Shinjou, K.; Nagao, J. Hydraulic fracturing in methane-hydrate-bearing sand. *RSC Adv.* **2016**, *6*, 73148–73155. [CrossRef]
14. Too, J.L.; Cheng, A.; Khoo, B.C.; Palmer, A.; Linga, P. Hydraulic fracturing in a penny-shaped crack. Part II: Testing the frackability of methane hydraulic-bearing sand. *J. Nat. Gas Sci. Eng.* **2018**, *52*, 619–628. [CrossRef]
15. Ito, T.; Igarashi, A.; Suzuki, K.; Nagakubo, S.; Matsuzawa, M.; Yamamoto, K. Laboratory Study of Hydraulic Fracturing Behavior in Unconsolidated Sands for Methane Hydrate Production. In Proceedings of the Offshore Technology Conference, Houston, TX, USA, 5–8 May 2008.
16. Shan, L.; Fu, C.; Liu, Y.; Qi, Y. A feasibility study of using frac-packed wells to produce natural gas from subsea gas hydrate resources. *Energy Sci. Eng.* **2020**, *8*, 1247–1259. [CrossRef]
17. Liu, Y.; Fukun, S.; Xuhui, Z.; Xiaobing, L. Experimental studies on the propagation characteristics of hydraulic fracture in clay hydrate sediment. *Chin. J. Theor. Appl. Mech.* **2020**, *52*, 224–234.
18. Liu, L.; Zhang, X.; Lu, X. Review on the Permeability of Hydrate-Bearing Sediments. *Adv. Earth Sci.* **2012**, *27*, 733–746.
19. Sun, J.; Ning, F.; Liu, T.; Liu, C.; Chen, Q.; Li, Y.; Cao, X.; Mao, P.; Zhang, L.; Jiang, G. Gas production from a silty hydrate reservoir in the South China Sea using hydraulic fracturing: A numerical simulation. *Energy Sci. Eng.* **2019**, *7*, 1106–1122. [CrossRef]
20. Surjaatmadja, J.B. Subterranean Formation Fracturing Methods. U.S. Patent No.5765, 16 June 1998.
21. Surjaatmadja, J.B.; Willett, R.; McDaniel, B.W.; Rosolen, M.A.; de AFranco, M.L.; dos Santos, F.C.; Fernandes, P.D.; Carneiro, F.A.; Bezerra de Lima, B.; Cortes, M. Selective Placement of Fractures in S2 Horizontal Wells in Offshore Brazil Demonstrates Effectiveness of Hydra Jet Stimulation Process. *SPE Drill. Complet.* **2007**, *22*, 137–147. [CrossRef]
22. He, P.; Liu, H. Hydraulic Jet Fracturing Technology in Thin Oil Reservoir with Bottom Water Application. *J. Yanan Univ. (Nat. Sci. Ed.)* **2012**, *12*, 31.
23. Brigeman, P.W. Volume changes in the plastic stage of simple compression. *Appl. Phys.* **1949**, *20*, 1241–1251. [CrossRef]
24. Handin, J.; Hager, R.V.; Friedman, M.; Feather, J.N. Experimental deformation of sedimentary rocks under confining pressure: Pore Pressure effects. *Bull. Am. Assoc. Petrol. Geol.* **1963**, *47*, 717–755.

25. Huang, W.; Shen, M.; Zhang, Q. Study of unloading dilatancy property of rock and its constitutive model under high confining pressure. *Chin. J. Rock Mech. Eng.* **2010**, *29*, 3475–3479.
26. Bratli, R.K.; Risnes, R. Stability and failure of sand arches. *SPE J.* **1981**, *21*, 63–67. [CrossRef]
27. Yan, X.; Li, M.; Fan, B.; Yu, J.; Hou, Z.; Yuan, Y. Study on Mechanical Properties of Unconsolidated Sandstone Dilatation. *Technol Superv. Pet. Ind.* **2020**, *10*, 24–26,31.
28. Zhao, R.; Sun, X.; Xu, B.; Luo, C.; Meng, X. Status and prospect of SAGD quick start technology. *Oil Drill. Prod. Technol.* **2020**, *42* 417–424.
29. Xu, B.; Wong, R.C.K. Coupled finite-element simulation of injection well testing in unconsolidated oil sands reservoir. *Int. J. Numer. Anal. Methods Geomech.* **2013**, *37*, 3131–3149. [CrossRef]
30. Li, Y.; Sun, Y.; Liu, Y.; Zhang, L.; Zheng, J.; Huang, Y.; Xu, X.; Sun, Y. Electrohydraulic Effect and Sparker Source: Current Situation and Prospects. *High Volt. Eng.* **2021**, *47*, 753–755.
31. Mok, Y.S.; Ahn, H.T.; Kim, J.T. Treatment of Dyeing Wastewater by Using Positive Pulsed Corona Discharge to Water Surface *Plasma Sci. Technol.* **2007**, *1*, 71–75. [CrossRef]
32. Klotz, J.A.; Krueger, R.F.; Pye, D.S. Effect of Perforation Damage on Well Productivity. *J. Pet. Technol.* **1974**, *26*, 1303–1314. [CrossRef]
33. Wu, W.; Huang, S. Application and development of high power pulsed discharge in water. *Mod. Electron. Technol.* **2003**, *5*, 85–87.
34. Qu, Y.; Wang, X. Electric pulse plugging removal and stimulation equipment. *Nat. Gas Ind.* **1997**, *3*, 88–89.
35. Lu, H.; Nie, B.; Chen, X.; Xu, X. Experimental research on coal crushing by using high-voltage electrical pulse based on electrohydraulic effect. *J. Saf. Sci. Technol.* **2020**, *16*, 83–85.
36. Shi, D.; Wang, D.; Liu, S. Analysis and Application on the Mechanism of Plug Removal with Electric Plus. *Oil Drill. Prod. Technol.* **2002**, *24*, 73–75.
37. Lu, X.; Wang, S.; Sui, M.; Huang, P. Mechanism analysis and application of electric pulse plugging removal and injection enhancement. *Nat. Gas Oil* **2011**, *29*, 61–62.
38. Zhang, X.; Liu, B.; Shen, T. Application of the Electric Detonation Broken Down Technology in Low-permeability Oilfield. *Oil Drill. Prod. Technol.* **2010**, *33*, 68–70.
39. Liang, Y.P.; Liu, S.; Wan, Q.C.; Li, B.; Liu, H.; Han, X. Comparison and optimization of methane hydrate production process using different methods in a single vertical well. *Energies* **2019**, *12*, 124. [CrossRef]
40. Rahim, I.; Nomura, S.; Mukasa, S.; Toyota, H. Decomposition of methane hydrate for hydrogen production using microwave and radio frequency in-liquid plasma methods. *Appl. Therm. Eng.* **2015**, *90*, 120–126. [CrossRef]
41. Wang, B.; Dong, H.; Fan, Z.; Liu, S.; Lv, X.; Li, Q.; Zhao, J. Numerical analysis of microwave stimulation for enhance energy recovery form depressurized methane hydrate sediments. *Appl. Energy* **2020**, *262*, 114559. [CrossRef]
42. Zhao, J.; Fan, Z.; Wang, B.; Dong, H.; Liu, Y.; Song, Y. Stimulation of microwave stimulation for the production of gas form methane hydrate sediment. *Appl. Energy* **2016**, *168*, 25–37. [CrossRef]
43. Davletshina, M.R.; Stolpovsky, M.V.; Chiglintseva, A.S.; Gimaltdinov, I.K. Features of decomposition of gas hydrate when exposed to microwave radiation. In *IOP Conference Series: Materials Science and Engineering, Mathematical Methods in Engineering and Technology*; IOP Publishing Ltd.: Bristol, UK, 2020; Volume 919, pp. 64–71. [CrossRef]
44. Li, D.L.; Liang, D.Q.; Fan, S.S.; Li, X.S.; Tang, L.G.; Huang, N.S. In situ hydrate dissociation using microwave heating: Preliminary study. *Energy Convers. Manag.* **2008**, *49*, 2207–2213. [CrossRef]
45. Ai, Z.; Meng, Z.; Ai, Y.; Fu, B.; Tang, Q. The Effect of Microwave Radiation on Viscosity Reduction and Dehydration of Heavy Oil and Technology Research. *J. Microw.* **2016**, *1*, 92–95.
46. Calhoun, G.J. System for Recovery of Petroleum from Petroleum Impregnated Media. U.S. Patent 4817711, 4 April 1989.
47. Rabl, P.; Zoller, P. Molecular dipolar crystals as high fidelity quantum memory for hybrid quantum computing. *Phys. Phys. Rev. A* **2007**, *76*, 042308. [CrossRef]
48. Wei, W.; Luo, X.Z.; Cai, J.C.; Hu, X.Y.; Li, Y.N. Fractal Study on Skin Depth of Electromagnetic Wave through porous Rocks. *Prog. Geophys.* **2014**, *29*, 2416–2421.
49. Fatykhov, M.A.; Bagautdinov, N.Y. Experimental investigations of decomposition of gas hydrate in a pipe under the impact of a microwave electromagnetic field. *High Temp.* **2005**, *43*, 614–619. [CrossRef]
50. Ma, T.; Gao, J. Thinking and practice of the construction of the new form of physics textbooks in universities. *Coll. Phys.* **2016**, *35*, 17–23.
51. Pupillo, G.; Micheli, A.; Büchler, H.; Zoller, P. *Cold Molecules: Creation and Applications*; CRC Press: Boca Raton, FL, USA, 2009.
52. Zhou, A. *Research of Low-Resolution Pulse-GPR Technology*; Jilin University: Changchun, China, 2011.
53. Zhu, C.Q.; Zhang, M.S.; Liu, X.L.; Wang, Z.; Shen, Z.; Zhang, B.W.; Zhang, X.T.; Jia, Y.G. Gas hydrates: Production, geohazards and monitoring. *J. Catastrophol.* **2017**, *32*, 51–56.

clean technologies

MDPI

Article

Real-Driving Emissions of an Aging Biogas-Fueled City Bus

Kirsi Spoof-Tuomi [1,*], Hans Arvidsson [2], Olav Nilsson [1] and Seppo Niemi [1]

1 School of Technology and Innovations, University of Vaasa, Box 700, FI-65101 Vaasa, Finland
2 RISE Research Institutes of Sweden, Box 5053, SE-90403 Umeå, Sweden
* Correspondence: kirsi.spoof-tuomi@uwasa.fi

Abstract: Transition to low emission transportation and cleaner cities requires a broad introduction of low- and zero-carbon alternatives to conventional petrol- and diesel-powered vehicles. New-generation gas buses are a cost-effective way to reduce local air pollutants from urban transportation. Moreover, major greenhouse gas (GHG) savings may be achieved using biogas as the power source. The main objective of this research was to investigate CH_4 and other gaseous emissions of a biogas-fueled urban bus equipped with a three-way catalyst (TWC) in real-world conditions. The study focused on emissions from a six-year-old gas-powered city bus, supplementing emission data from aging bus fleets. Impaired CH_4 oxidation and NO_x reduction were observed in the catalyst after its service life of 375,000 km–400,000 km. The main reason for low CH_4 and NO_x conversion over the TWC was concluded to be the partial deactivation of the catalyst. Another critical issue was the fluctuating air-to-fuel ratio. The results show that the efficiency of exhaust after-treatment systems should be closely monitored over time, as they are exposed to various aging processes under transient driving conditions, leading to increased real-world emissions. However, the well-to-wheels (WTW) analysis showed that an 80% GHG emission benefit could be achieved by switching from diesel to biomethane, giving a strong environmental argument for biogas use.

Keywords: real-driving emission; portable emission measurement system; Euro VI; urban bus; catalyst deactivation; compressed biogas; well-to-wheels analysis

Citation: Spoof-Tuomi, K.; Arvidsson, H.; Nilsson, O.; Niemi, S. Real-Driving Emissions of an Aging Biogas-Fueled City Bus. *Clean Technol.* **2022**, *4*, 954–971. https://doi.org/10.3390/cleantechnol4040059

Academic Editors: Dong Li, Fuqiang Wang, Zhonghao Rao and Chao Shen

Received: 27 July 2022
Accepted: 22 September 2022
Published: 2 October 2022

1. Introduction

There is a worldwide consensus that significant reductions in greenhouse gas (GHG) emissions are needed to avoid the worst impacts of climate change, and various laws and regulations have already been implemented to combat and respond to global warming. In July 2021, the European Commission adopted an extensive legislative package, Fit for 55, with the goal of reducing the economy-wide GHG emissions by at least 55% by 2030 compared to 1990 levels [1]. This is a substantial increase from the previous 40% target. Achieving the 55% reduction in GHG emissions over the next decade is crucial for Europe to achieve climate neutrality by 2050. Moreover, Finland has set itself the goal of becoming carbon neutral by 2035 [2]. This is one of the most ambitious targets of any country in the industrialized world.

In 2019, GHG emissions from domestic transportation accounted for 21 percent of Finland's total greenhouse gas emissions and about 30 percent of the energy sector's GHG emissions [3]. Road transportation is likely to remain a significant contributor to air pollution in the coming decades, especially in urban areas [4]. Transition to low emission transportation and cleaner cities will undoubtedly require a broad introduction of low- and zero-carbon alternatives to conventional petrol- and diesel-powered vehicles.

New generation gas buses are a cost-effective way to reduce CO_2 and local pollutants from urban transportation. Fueling with gas reduces pollutant emissions, including carbon monoxide (CO), nitrogen oxides (NO_x), and particulate matter (PM), as shown, e.g., by Biernat et al. [5]. Moreover, major GHG savings can be achieved by using biogas as the power source. This is based on the fact that producing biomethane from organic waste material results in fuel that contains only biogenic carbon, and combustion of such fuel releases only biogenic CO_2, which is, unlike CO_2 from fossil fuels, not considered to contribute the climate change [6].

Buses running on biogas are becoming more common in Finland as cities and transportation companies invest in greener alternatives. For example, in the western coastal city of Vaasa, biogas buses have been touring since 2017. Life cycle GHG emissions from biogas vehicles largely depend on the extent of methane (CH_4) leakage throughout the fuel life cycle, and unintended CH_4 emissions from different stages of the fuel chain can narrow their potential climate benefits. Methane is a powerful greenhouse gas with a global warming potential (GWP) 28–34 times that of CO_2 over a 100-year timescale [7]. In addition, due to the strong C–H bonds of methane, it is one of the most difficult hydrocarbons to treat catalytically [8], and insufficient removal rates of exhaust after-treatment systems at low loads and low exhaust temperatures may lead to increased real-world CH_4 emissions [9].

Besides exhaust gas temperature, another critical issue is the effect of rapid changes in exhaust gas composition—typical in real-world driving conditions—on after-treatment devices. This phenomenon is particularly evident when dealing with stoichiometric gas engines using three-way catalytic converters (TWC), requiring a very precise control of air-to-fuel ratio (AFR), as some deviations from the stoichiometric lambda value can interfere with the catalyst efficiency [10]. For example, Rodman Oprešnik et al. [11] reported instantaneous, local rises of THC emissions as a result of occasional inadequate lambda control of a CNG bus during transient regime and, consequently, increased cumulative emissions.

The main objective of this research was to investigate CH_4 and other gaseous emissions plus fuel consumption of a biogas-fueled urban bus in real-world operation. The actual driving emissions were recorded using a portable emissions measurement system (PEMS). The key advantage of on-board measurements is that they can truly demonstrate the emission characteristics of vehicles under various traffic conditions, operating cycles, and ambient conditions, including those that are challenging to replicate in the laboratory, such as varying road gradients [4]. The load on the lines that buses serve and the number of passengers may also affect exhaust emissions under actual traffic conditions [12].

Exhaust emissions under real-world conditions were examined by Lv et al. [13]. The authors showed an underestimation of road emissions of gas- and diesel-powered heavy vehicles; emission factors under real-driving conditions were significantly higher than in previous chassis dynamometer studies, likely caused by frequent accelerations, decelerations, and start-stop operation. In a recent study, Rosero et al. [14] investigated the effects of passenger load, road grade, and congestion level on real-world emissions and fuel consumption of urban Euro VI CNG and Euro V diesel buses. As the road grade and congestion level increased, both buses' fuel consumption and CO_2 emissions increased by 6–55%. Gallus et al. [15] studied the impact of driving style and road grade on gaseous exhaust emissions of Euro V and Euro VI diesel vehicles. CO_2 and NO_x emissions, measured with PEMS, showed a linear increase with road grade. Chen et al. [16] investigated the impact of speed and acceleration on emissions of heavy-duty (HD) vehicles in Shanghai. They found that congestion conditions with low speed and frequent deceleration and acceleration increased THC and CO emissions. Ozener & Ozkan [17] reported that the acceleration effect on both fuel consumption and emission values was significant. They concluded that the real-driving emission data could be effectively used in developing cleaner engine calibrations and more economical operations.

In addition, gaseous emissions are strongly affected by starting conditions. The cold-start emissions challenge has been highlighted, e.g., in [18,19]. During the first minutes of operation, emissions are high because the after-treatment equipment has not reached the appropriate temperature required to efficiently remove gaseous pollutants. Faria et al. [20] also showed a substantial increase in energy consumption for cold-start, leading to increased CO_2 emissions during the cold-start period. The problem of cold-starts is considered more pronounced at low ambient temperatures, as lower ambient temperature increases the cold-start running duration [20,21].

One crucial topic rarely addressed in real-driving emissions (RDE) studies is the catalyst deactivation and deterioration over time. Indeed, the presence of catalyst poisons and other impurities in the feed, the fluctuating exhaust gas composition and flow rate in the converter, as well as high temperatures and temperature gradients, all increase the possibility of catalyst deactivation [22]. Therefore, to ensure a significant reduction of emission levels throughout the vehicle's useful life, EU regulation has adopted dedicated "emission durability" periods, i.e., the minimum mileage or time after which the engine is still expected to comply with applicable emission limits. For example, for category M3 buses, the required emission durability period is six years or 300,000 km, whichever comes first [23]. However, the useful life of urban buses is usually much longer; e.g., the Finnish bus fleet's average age is 12.5 years [24]. Therefore, emission levels after the emission durability period and closer to the service life of the vehicles need to be investigated.

This study focused on emissions from a six years old gas-powered city bus, supplementing emission data from aging bus fleets. PEMS measurements were performed in real-traffic conditions on a regular bus line in Vaasa in collaboration with the University of Vaasa and RISE Research Institutes of Sweden. In addition to methane emissions, gaseous emissions of NO_x, CO, and CO_2 were measured. Both cold-start and warm-engine emissions were recorded. We conducted two measurement campaigns, the first in March 2022 and the second in June 2022. In addition, the total carbon footprint of compressed biogas (CBG) is discussed in terms of its GHG reduction potential, defined as the percentage reduction in life cycle GHG emissions relative to its fossil counterpart natural gas and traditional diesel fuel.

2. Materials and Methods

2.1. Test Vehicle

Exhaust emission tests in real-driving conditions were carried out on a Scania Euro VI bus owned by the City of Vaasa and operated by Wasa Citybus. The CBG-fueled bus was equipped with a spark ignition engine with a displacement of 9.3 dm^3 and a power of 206 kW. The vehicle was equipped with exhaust gas recirculation (EGR) and a three-way catalytic converter. Table 1 presents the characteristics of the test vehicle and Table 2 summarizes the engine technical specifications.

Table 1. Vehicle technical specifications.

Parameter	Value
Model name	Scania Citywide LE
Model year	2016
Gross vehicle weight (kg)	19,100
Curb weight (kg)	12,960
Max passenger number	75
Axle configuration	4×2
Gearbox	6-speed automatic transmission
Accumulated mileage (km)	375,000 (Test 1), 400,000 (Test 2)
After-treatment system	TWC
Other systems	EGR
Exhaust emission norm	Euro VI-C

Table 2. Engine technical specifications.

Parameter	Value
Model	Scania OC09 101
Engine type	Spark ignition engine
Fuel	CNG/CBG

Table 2. *Cont.*

Parameter	Value
Number of cylinders	5
Compression ratio	12.6:1
Total displacement (L)	9.3
Maximum power (kW@rpm)	206 kW@1900 rpm
Engine peak torque (Nm@rpm)	1350 Nm@1000–1400 rpm

2.2. Portable Emissions Measurement System

The real-driving gaseous emissions of CH_4, CO, CO_2, NO, and NO_2 from the tested city bus were measured and recorded using an on-board VARIOplus Industrial device manufactured by MRU Messgeräte für Rauchgase und Umweltschutz GmbH. VARIOplus measures CH_4, CO, and CO_2 concentrations using a non-dispersive infrared (NDIR) sensor, and NO_x concentrations are measured using electrochemical cells. Table 3 shows the technical characteristics of the measurement apparatus used in this work.

Table 3. Technical characteristics of VARIOplus Industrial.

Parameter	Measurement Method	Accuracy
CH_4	NDIR—Non-dispersive infrared, range 0–10,000 ppm	±2%
CO	NDIR—Non-dispersive infrared, range 0–10%	±0.03% or * ±3% reading
CO_2	NDIR—Non-dispersive infrared, range 0–30%	±0.05% or * ±3% reading
NO	electrochemical, range 0–1000 ppm	±5 ppm or * 5% reading
NO_2	electrochemical, range 0–200 ppm	±5 ppm or * 5% reading
O_2	electrochemical, range 0–10%	±0.2 Vol-% abs.
Sampling	1 Hz	

* = whichever is larger.

The engine speed, torque, coolant temperature, air flow, lambda, and the vehicle speed were recorded from the vehicle engine control unit (ECU) via an on-board diagnostics (OBD) system using Scania Diagnosis & Programmer (SDP3) software version 2.50.3 (in Test 1) and version 2.52.1 (in Test 2), copyright Scania CV AB, Scania Suomi Oy, Vaasa, Finland. The vehicle's position in terms of latitude, longitude, and altitude, and the vehicle speed data were registered using an external global positioning system (GPS). A dedicated weather station was used to register the ambient temperature, pressure, and relative humidity. The real-world emission data obtained with PEMS and the GPS and the weather data were collected and stored with the DEWESoft data acquisition system. All data were recorded with a frequency of 1 Hz. Prior to the data processing, the SDP3 and DEWESoft data were synchronized based on the vehicle speed from the ECU and the GPS.

An external power unit supplied the electrical power to the PEMS system. Figure 1 depicts the system set-up.

2.3. Test Route

Emission tests were performed in real-driving conditions on an urban route in Vaasa, i.e., in normal traffic and with normal driving patterns and typical passenger loads. The selected test route was the same route the bus usually travels daily. The measurements started in the morning at the same time and the same driver from Wasa Citybus was used in both measurement campaigns. Figure 2 shows the driving circuit chosen for the tests. The length of one circuit was 25.5 km, and the same circuit was run three times. The total test duration was approx. 3 h. The route included both urban and rural driving. The speed profile of the driving circuit is presented in Figure 3. Table 4 shows the percentages and mean velocities for three different driving speed ranges. The passenger load varied between 5 and 30 percent during the tests.

Figure 1. Measurement system set-up.

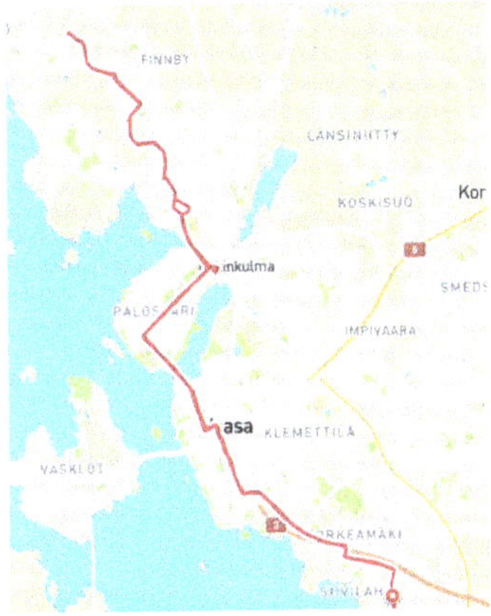

Figure 2. Driving circuit.

The first measurement campaign was performed in March 2022, and the second in June 2022. In June, only warm engine measurements were recorded, while in March, both cold-start and hot-start emissions were investigated.

2.4. Fuel

The fuel used in the test was CBG from a commercial filling station. The methane content of the fuel was 97% by volume. The other main components of the fuel were CO_2 (2.2 vol.-%), nitrogen (0.5 vol.-%), and oxygen (0.3 vol.-%), so the energy content of the fuel was solely related to the methane concentration. The calculated lower heating value (LHV) of the gas was 46.4 MJ/kg.

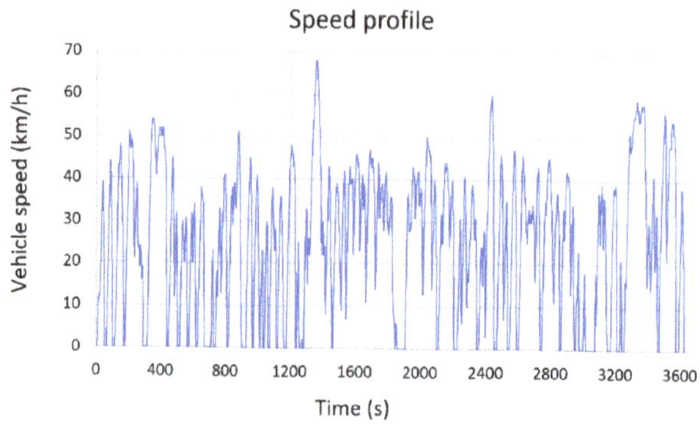

Figure 3. Speed profile of the driving circuit.

Table 4. Shares of driving speed ranges.

	Speed Range	**Time (min)**	**%**	**Mean Velocity**
Urban driving	0–30 km/h	102	56	12
Urban driving	30–50 km/h	65	36	38
Rural driving	50–75 km/h	16	9	57
Total		182		25

2.5. Calculation Procedure

2.5.1. Calculation of Fuel Mass Flow

The instantaneous fuel flow ($\dot{m}_{fuel}$) in kg/s was calculated based on the recorded instantaneous air flow ($\dot{m}_{air}$) and lambda (λ) values and the stoichiometric air-to-fuel ratio (AFR_{stoich}), according to Equation (1).

$$\dot{m}_{fuel} = \frac{\dot{m}_{air}}{AFR_{stoich} \times \lambda} \quad (1)$$

To determine AFR_{stoich}, the stoichiometric oxygen demand ($n_{O2,stoich}$) in moles per kg of fuel was calculated first, based on the chemical composition of the fuel (Equation (2)). In the equation, w_c, w_{H2} and w_{O2} are the fuel mass fractions of carbon, hydrogen and oxygen in the fuel.

$$n_{O2,\,stoich} = \frac{w_c}{0.012011} + \frac{1}{2} \times \frac{w_{H2}}{0.002016} - \frac{w_{O2}}{0.031999} \quad (2)$$

As air contains 20.95% of oxygen, the stoichiometric air demand ($n_{air,stoich}$) in moles per kg fuel could be determined by Equation (3):

$$n_{air,stoich} = \frac{n_{O2,stoich}}{0.2095} \quad (3)$$

Finally, the stoichiometric air demand in kg of air per kg of fuel was calculated by multiplying $n_{air,stoich}$ by the molar mass of air (M_{air}), see Equation (4):

$$AFR_{stoich} = n_{air,stoich} \times M_{air} \quad (4)$$

2.5.2. Calculation of Fuel Consumption

The total fuel mass (ΣFC_i) over the test cycle was calculated based on the instantaneous (second-by-second) fuel mass flows according to Equation (5).

$$\Sigma FC_i = \left(\frac{1}{2}\dot{m}_{fuel,0} + \dot{m}_{fuel,1} + \dot{m}_{fuel,2} + \ldots + \dot{m}_{fuel,n-1} + \frac{1}{2}\dot{m}_{fuel,n}\right) \quad (5)$$

2.5.3. Calculation of Exhaust Mass Flow

The instantaneous exhaust gas mass flow ($\dot{m}_{exh.,i}$) (wet basis) in kg/s was determined based on the recorded air flow and the calculated fuel flow values (Equation (6)):

$$\dot{m}_{exh.,i} = \dot{m}_{air,i} + \dot{m}_{fuel,i} \tag{6}$$

2.5.4. Emissions Dry–Wet Correction

The emission concentrations were measured on a dry basis. Dry concentration (c_{dry}) was converted to a wet basis with the dry–wet conversion factor ($K_{d\text{-}w}$):

$$c_{wet} = K_{d\text{-}w} \times c_{dry} \tag{7}$$

$K_{d\text{-}w}$ was calculated according to the UN/ECE Regulation 49 [25], Equation (8):

$$K_{d-w} = \left(\frac{1}{1 + a \times 0.005 \times (c_{CO2} + c_{CO})} - k_{w1}\right) \times 1.008 \tag{8}$$

where a is the molar hydrogen to carbon ratio of the fuel, and

$$k_{w1} = \frac{1.608 \times H_a}{1000 + (1.608 \times H_a)} \tag{9}$$

where H_a is the intake air humidity in g water per kg dry air.

2.5.5. Calculation of Mass Emissions

Second-by-second mass flow of the pollutant ($\dot{m}_{gas}$) in g/s was calculated using Equation (10):

$$\dot{m}_{gas} = u_{gas} \times c_{gas} \times \dot{m}_{exh.} \tag{10}$$

where u_{gas} is the ratio between the density of pollutant and the density of exhaust gas, and c_{gas} is the instantaneous concentration of the pollutant in raw exhaust in ppm (wet basis). The instantaneous u values were calculated following the UN/ECE Regulation No 49 [25], according to Equations (11)–(14):

$$u_{gas,i} = \frac{\rho_{gas}}{(\rho_{exh.,i} \times 1000)} \tag{11}$$

$$\rho_{gas} = \frac{M_{gas}}{22.414} \tag{12}$$

where M_{gas} is the molar mass of the gas component in g/mol, ρ_{gas} is the density of the gas component in kg/m^3, and $\rho_{exh.,i}$ the instantaneous density of the exhaust gas in kg/m^3, derived from Equation (13):

$$\rho_{exh.,i} = \frac{1000 + H_a + 1000 \times \left(\frac{\dot{m}_{fuel,i}}{\dot{m}_{dry\ air,i}}\right)}{773.4 + 1.2434 \times H_a + k_{fw} \times 1000 \times \left(\frac{\dot{m}_{fuel,i}}{\dot{m}_{dry\ air,i}}\right)} \tag{13}$$

where k_{fw} is the fuel specific factor of wet exhaust, obtained from Equation (14):

$$k_{fw} = 0.055594 \times W_\alpha + 0.0080021 \times W_\Delta + 0.0070046 \times W_\varepsilon \tag{14}$$

where W_α is the hydrogen content (wt%) of the fuel, W_Δ the nitrogen content (wt%), and W_ε the oxygen content (wt%) of the fuel.

The mass of gaseous emissions (m_{gas}) in grams per test cycle was calculated using Equation (15).

$$m_{gas} = \sum_{i=1}^{i=n} u_{gas,i} \times c_{gas,i} \times \dot{m}_{exh.,i} \times \frac{1}{f} \tag{15}$$

where f is the data sampling rate in Hz.

The final results are expressed in g/kWh and in g/km, i.e., the total mass of each pollutant over the test cycle was divided by the engine cycle work or by the distance covered in km.

2.5.6. Calculation of Cycle Work

The engine work (W_i) in kWh over the test cycle was calculated based on the instantaneous (second-by-second) engine power values (P_e), according to Equation (16):

$$W_i = \frac{\left(\frac{1}{2}P_{e,0} + P_{e,1} + P_{e,2} + \ldots + P_{e,n-2} + P_{e,n-1} + \frac{1}{2}P_{e,n}\right)}{3600} \tag{16}$$

2.5.7. Calculation of Effective Power of the Engine

The instantaneous engine power in kW was calculated by using each pair of recorded engine speed and torque values (Equation (17)):

$$P_e = \frac{2 \times \pi \times N \times \tau}{60 \times 1000} \tag{17}$$

where N is the engine speed in rpm and τ is the engine torque in Nm.

3. Results and Discussion

3.1. Ambient Conditions

Table 5 summarizes the average ambient conditions during the tests.

Table 5. Ambient conditions during the tests.

Ambient Condition	Test 1	Test 2
	March 2022	June 2022
Temperature (°C)	−5 °C	+18 °C
Pressure (kPa)	102.5	100.5
Humidity (%)	65.5	54.7

3.2. Gaseous Emissions

In the current legislation, the regulatory in-service conformity (ISC) emission test applies the 20% power threshold as a boundary condition for Euro VI-C bus engines. However, Mendoza Villafuerte et al. [26] showed that a large fraction of urban operation is not considered if the current power threshold boundary for post-processing the PEMS data is applied, and up to 80% of the data may be excluded from the emission analysis. They also showed that cold-start emissions, which are currently also excluded from the analysis, could account for a significant proportion of total emissions. To give a more accurate depiction of real-driving emissions, no power threshold boundaries were applied in this study. In addition, in Test 1, both cold-start and hot-start emissions were recorded. In Test 2, unfortunately, only hot-start emissions were successfully recorded.

3.2.1. Hot-Start Emissions

A test was considered a hot-start once the coolant temperature had reached 70 °C for the first time or stabilized within ±2 °C over a period of 5 min, whichever came first [27]. Specific emissions were calculated in both g/kWh and g/km, and the results are presented separately for the total trip and for urban and rural sections of the circuit (Figure 4). Although the tests performed did not fully reflect the ISC tests in the type-approval procedure regarding boundary conditions and route requirements, the Euro VI standard limits (ISC limit) are also presented for comparative purposes.

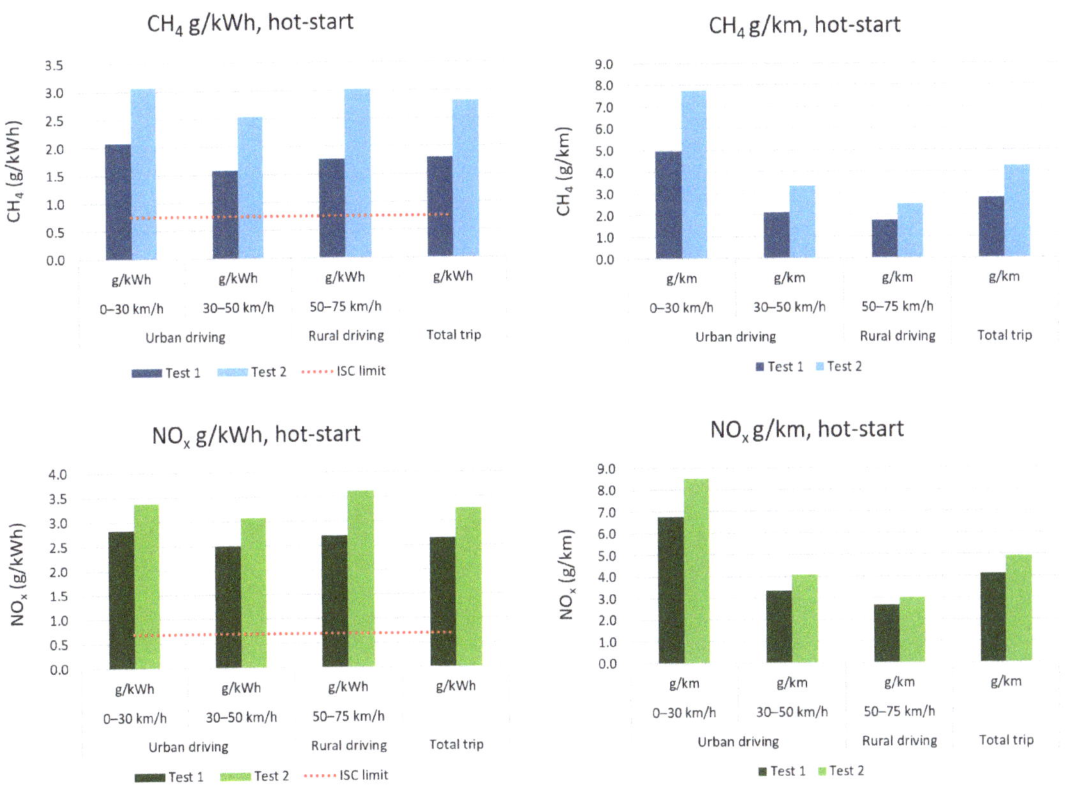

Figure 4. Specific CH_4 and NO_x emissions in g/kWh and g/km in hot-start tests.

CO emission values were low and well below the ISC limit of 6 g/kWh in both tests, indicating efficient oxidation of CO in the catalyst. In contrast, relatively high values were observed for CH_4 and NO_x, indicating impaired CH_4 oxidation and NO_x reduction in the catalyst after its service life of 375,000 km (Test 1). After 400,000 km (Test 2), the catalyst efficiency had further deteriorated. Here, it should be noted that according to EU Regulation EC 595/2009 [23], the minimum mileage or time after which the engine is still expected to comply with applicable emission limits for category M3 buses, is 300,000 km or six years, whichever comes first. Hence, the required "emission durability" period had already been exceeded in our case. Nevertheless, the bus has passed the regular technical inspections valid in Finland, including CO_2 and HC measurements.

The primary reason for relatively high CH_4 and NO_x emissions after the TWC was assumed to be the low CH_4 reactivity due to a partial deactivation of the catalyst. In addition to the low CH_4 oxidation rate, low CH_4 reactivity also means that methane-based reducing agents for NO_x reduction do not work, leading to substantial NO_x breakthrough from the catalyst, also concluded by Van den Brink & McDonald [28].

One of the most important reasons for the deactivation of the TWC in automotive applications is chemical deactivation [29], mainly caused by lubricating oil additives and other impurities in the exhaust gases. For example, Winkler et al. [30] observed a significant increase in hydrocarbon emissions during CNG operation over a relatively short TWC lifetime of 35,000 km. Contaminants originating from the lubricating oil, such as calcium, phosphorus, and magnesium, detected on the catalyst's surface, appeared to affect especially CH_4 oxidation. In addition to lubricating oil, another source of catalyst poisons is the impurities in the fuel. The CBG used in this study contained small traces of commonly encountered catalyst poison sulfur (<2.3 mg/Nm^3) and siloxanes (0.7 mg/Nm^3). Althoug the amounts of these compounds were very low, they could have had a deactivating effec on the emissions control system.

Furthermore, the light-off of a TWC in gas-fueled engine exhaust typically occur at higher temperatures compared to gasoline engines [31]. Indeed, methane is the mos difficult hydrocarbon to oxidize due to its high stability [8]. A typical light-off temperatur for methane is 400 °C [8], but in a deactivated catalyst, significantly higher temperatures up to 500–600 °C [32], may be required to break the strong C–H bonds in methane. At low loads (Figure 5), common in a city bus's driving profile, the exhaust gas temperature was too low to allow the deactivated catalyst to work effectively.

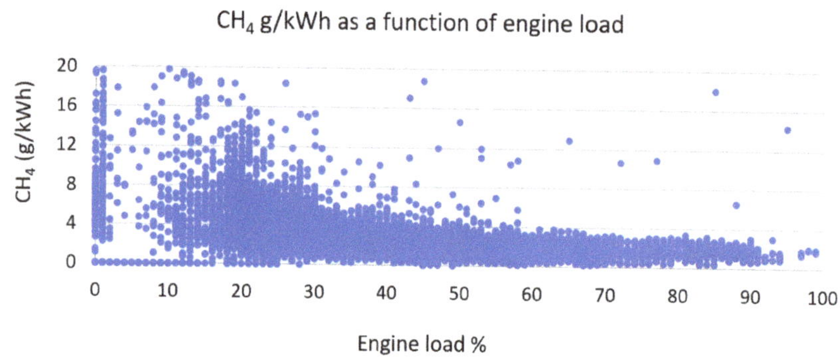

Figure 5. Specific CH_4 emissions as a function of engine load%.

Thus, restoring the catalytic activity of a deactivated TWC is a critical consideration. In some cases, depending on the adsorbed poison, the activity of the poisoned catalyst can be at least partially restored by regeneration [22]. For example, SO_2 can be removed from the catalyst under elevated temperatures and anoxic or very rich conditions, as shown by Auvinen et al. [32]. Careful control of the exhaust gas composition during regeneration could provide significant benefits in terms of CH_4 emissions. However, under real-driving conditions, the rapidly and dramatically varying exhaust gas temperature and composition between oxidizing and reducing environment make the on-board regeneration difficult to control.

Another possible deactivation mechanism for the TWC is thermal degradation. Three-way catalysts are known to lose their activity when exposed to high temperature (>800 °C) oxidizing environments, typically occurring during fuel shut-off phases [33]. Switching off the fuel flow, e.g., during engine braking, is a strategy of the automotive industry to improve fuel economy. Thermal degradation is critical to the catalyst's performance since these changes are typically irreversible.

In addition to the partial deactivation of the catalyst, another probable reason for the relatively high emissions was the fluctuating lambda value. Indeed, close control of the exhaust gas composition is essential for high emission conversion as the composition of the gas entering the TWC significantly affects its catalytic efficiency [34]. For simultaneous conversion of HC, CO, and NO_x species in the TWC, the engine must be operated within a very narrow AFR window—near stoichiometric conditions—due to a rapid drop in NO_x conversion efficiency on the lean side and a non-complete conversion of hydrocarbons both in lean and rich stoichiometry [10]. For example, Lou et al. [34] detected the highest TWC conversion efficiency when AFR was controlled between 0.995 and 1. The narrow AFR range over which significant conversion of natural gas exhaust emissions is possible presents a challenging control problem. As seen in Figure 6, lambda was outside the optimal range for a significant part of the time in our experiments.

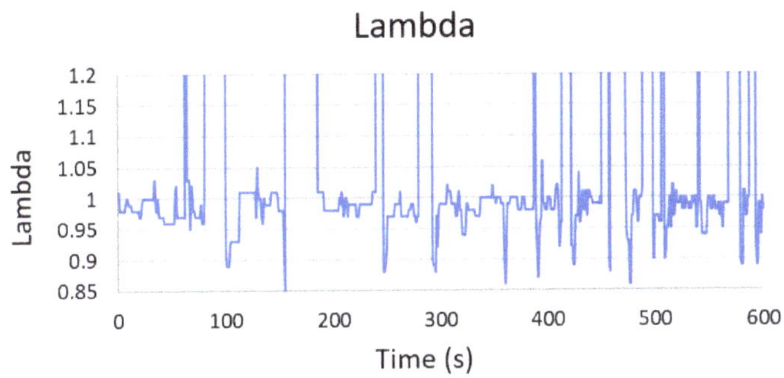

Figure 6. Fluctuating lambda values under real-driving conditions.

In sum, deterioration of the exhaust after-treatment systems over time should be monitored as they are exposed to different aging processes resulting in elevated real-world emissions. Our results indicate a catalyst replacement need after 375,000 km of service life. In addition, a precise lambda control is absolutely necessary to ensure high conversion rates throughout the vehicle's lifetime.

3.2.2. Cold-Start Emissions

Cold-start emissions were recorded from the moment the coolant temperature had reached 30 °C for the first time and continued until the coolant temperature was stabilized within ±2°C over 5 min [27]. In Test 1 (at −5 °C), the cold-start period lasted 14.5 min. The combined cold- and hot-start emissions were calculated according to EU Regulation 1718 [35]: the vehicle was driven over a cold-test cycle followed by nine hot-test cycles, identical to the cold one in a way that the work developed by the engine was the same as the one achieved in the cold cycle.

Figure 7 illustrates CH_4 and NO_x emissions during cold-start versus hot-start. During the cold-start, CH_4 emissions were 2.3 times higher and NO_x emissions 1.4 times higher than those during the hot-start. This highlights the temperature sensitivity of catalytic emission control systems, which is also evidenced in Figure 8.

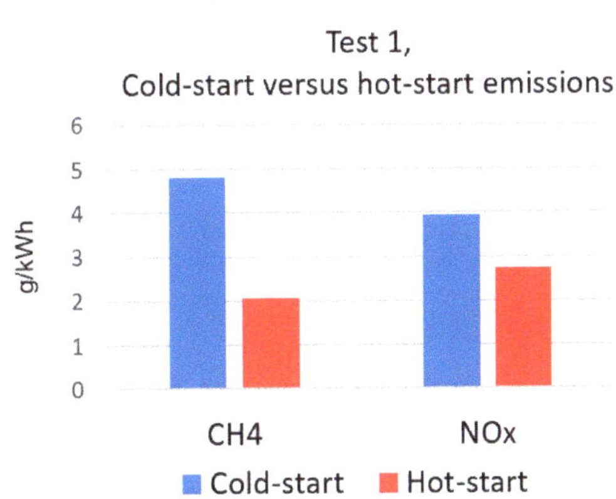

Figure 7. Cold-start versus hot-start emissions.

Over the combined cold- and hot-start cycles, CH_4 emissions increased by 30%, NO_x by 13%, and CO by 33% compared to hot-start-only measurements.

The cold-start emissions challenge is more pronounced at low ambient temperatures because it then takes longer for the TWC to reach effective operating temperature, leading to a prolonged period of high emission rates [18,20,21].

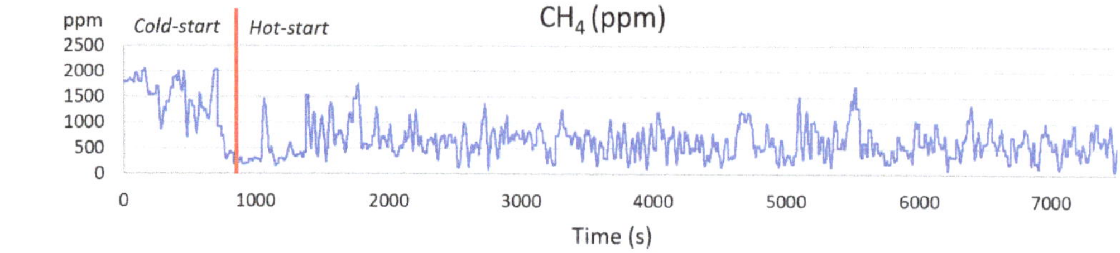

Figure 8. CH_4 emissions during combined cold- and hot-start test.

3.3. *Well-to-Wheels Analysis*

In the transport sector, well-to-wheels (WTW) analysis is a commonly used method for assessing the carbon intensity of a fuel. Carbon intensity refers to the amount of greenhouse gases—including CO_2, nitrous oxide, and methane—released during the production and consumption of a transportation fuel, measured in grams of carbon dioxide equivalent per megajoule of energy (g CO_2-eq./MJ).

3.3.1. Fuel Consumption

The total fuel consumption in the hot-start test at −5 °C was 21.9 MJ/km (6.1 kWh/km), corresponding to 0.306 kg/kWh and 47.1 kg/100 km. In June, at +18 °C, the vehicle showed better fuel economy with fuel consumption of 19.8 MJ/km (5.5 kWh/km), corresponding to 0.283 kg/kWh and 42.7 kg/100 km (Figure 9).

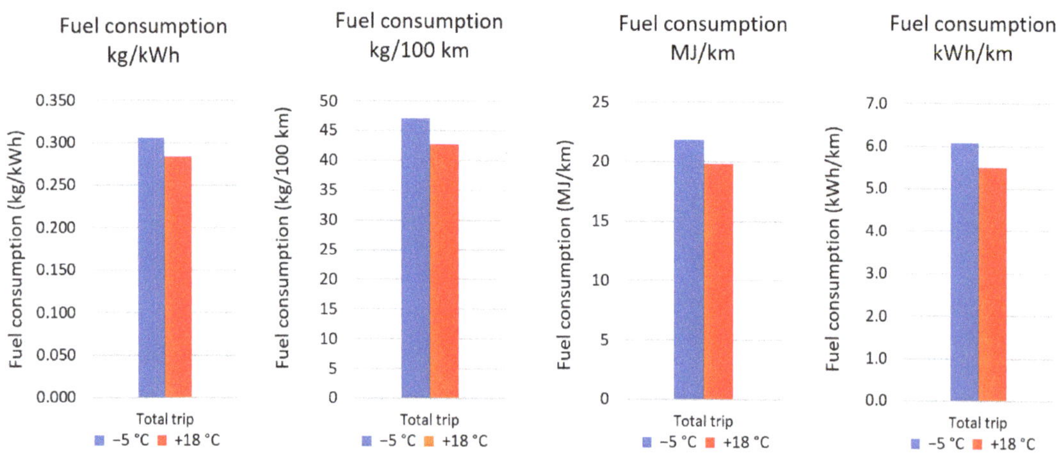

Figure 9. Fuel consumption in hot-start tests at −5 °C and +18 °C.

3.3.2. Biogas Production Process

The life cycle steps for CBG investigated in this study are feedstock collection and transportation, biogas production, biogas processing to biomethane, biomethane compression, and finally, combustion in an engine. The CBG was produced at Stormossen waste treatment plant near Vaasa. The anaerobic digestion process at Stormossen is divided into two separate process lines. Biogas reactor 1 is fed with wastewater sludge, and the raw material used in biogas reactor 2 is municipal biowaste, supplied within a radius of 40 km [36].

In 2020, raw biogas production at Stormossen was 2.7 million Nm^3, of which 52% was upgraded into biomethane, 32% was used for heat and electricity production, and the rest was flared [37]. The methane content of the raw biogas was 62%.

The biogas upgrading is executed by an amine scrubber. The main advantages of chemical absorption with amine solvents are a high methane recovery rate in the upgraded biogas and a low methane slip of <0.1% [38]. In addition, amine solvents are effective at near atmospheric pressure and thus consume a low quantity of electric energy [39]. On the other hand, chemical scrubbing liquids require substantial thermal energy during regeneration, which must be supplied as process heat [39]. After the refining stage, biogas contains 97–98% methane. Finally, the processed biomethane is piped to a gas filling station near the biogas plant. At the refueling station, the gas is pressurized to 300 bar and stored in gas cylinders.

Although the combustion of waste-based biomethane is considered carbon-neutral in Finland's national GHG inventories (CO_2 emissions from biogas combustion are reported as zero), the use of biomethane may still have climate impact from the above-mentioned earlier stages of the fuel chain. For CBG production, the major contributors of GHG emissions are energy consumption and fugitive losses of methane during digestion and upgrading processes [40]. In addition, some GHG emissions form during the collection of wastes and residues.

3.3.3. GHG Inventory

In this study, the calculation of GHG emissions begins with feedstock collection and transportation. GHG emissions from these steps are based on the following assumptions. Transportation distance 40 km and diesel B7 fuel consumption 20 l/100 km. The lower calorific value of diesel B7 fuel is 43 MJ/kg. The biocomponent of diesel fuel was assumed to be hydrotreated vegetable oil made from waste materials, so the calculated well-to-tank emission factor for diesel B7 was 14.7 g CO_2-eq./MJ fuel, based on the JRC [41] data. Tank-to-wheels CO_2 emission factor for diesel B7 was set at 68.4 g CO_2-eq./MJ fuel [42]. The heat and electricity needs of biogas production and upgrading processes are covered internally by the plant's own CHP biogas engine and were, therefore, ignored in the GHG inventory. Methane emissions were calculated assuming a methane loss of 1% during anaerobic digestion [43] and 0.1% during the upgrading process [39]. Methane emissions are converted to CO_2-equivalents using a 100-year time horizon global warming potential (GWP) factor of 28 [7]. The energy demand for biomethane compression to 300 bar is 0.25 kWh/m^3 (NTP) [44], and the electric energy for compression is taken from the public grid. The CO_2 emission factor for electricity generation in Finland in 2020 was 68.6 g CO_2-eq./kWh [45]. Table 6 summarizes the main assumptions and input data used in the calculation.

Table 6. CBG well-to-tank GHG emissions.

Parameter	Value	Unit	g CH_4/MJ$_{bio\text{-}CH4}$	g CO_2-Equivalent /MJ$_{bio\text{-}CH4}$	Source
Feedstock collection and transportation					
Diesel trucks, diesel fuel biocomponent 7%	40	km		1.95	[41,42]
Biogas production and refining					
Total biogas production	2,716,000	Nm^3			[37]
52% of raw gas for upgrading	1,412,320	Nm^3			[37]
Methane content (62%)	875,638	Nm^3			[37]
Total biomethane production	31,522,982	MJ			
Heat demand *					
- Anaerobic digestion	0.19	kWh/Nm$^3_{raw\ gas}$			[43]
- Upgrading	0.110	kWh/kWh$_{bio\text{-}CH4}$			
Electricity demand *					
- Anaerobic digestion	0.14	kWh/Nm$^3_{raw\ gas}$			[43]
- Upgrading	0.0136	kWh/kWh$_{bio\text{-}CH4}$			
Methane losses					
- Anaerobic digestion, 1%	6368	kg	0.202	5.66	[43]
- Upgrading, 0.1%	630	kg	0.020	0.56	[39]

Table 6. *Cont.*

Parameter	Value	Unit	g CH_4/$MJ_{bio-CH4}$	g CO_2-Equivalent /$MJ_{bio-CH4}$	Source
Compression					
Electricity demand	0.25	kWh/m^3 (NTP)		0.48	[44,45]
CBG well-to-tank GHG emissions				**8.65**	

* Covered internally by the plant's own CHP biogas engine.

After the anaerobic digestion, the digestate is dewatered and composted to be used as a soil improvement product or as landscaping soil. The digestate treatment is not included in the above table. Any fertilizer or sludge credits are also not considered in GHG calculations.

The GHG benefits associated with transition from fossil-based natural gas or diesel to biomethane were calculated by comparing well-to-wheels CO_2-Equivalent emissions shown in Table 7. Well-to-tank GHG emission factors for compressed natural gas and diesel fuel were taken from the JRC report [41]. Tank-to-wheel GHG emissions for gas buses are based on the CO_2 and CH_4 emission results recorded in this study, but CO_2 emissions are considered only for fossil CNG. Tank-to-wheels CO_2 emission factor for diesel buses was taken from [42]. The average fuel consumption from Test 1 and 2 in this study was 20.8 MJ/km, and this value is applied to both CBG and CNG bus. It is well known that compression-ignition diesel engines have higher thermal efficiency compared to spark ignition gas engines. Therefore, the fuel consumption of a diesel bus was set at 80% of that of a gas bus, based on the VTT's (Technical Research Centre of Finland) comprehensive report on city bus emissions measurements [46].

Table 7. Well-to-wheels CO_2 Equivalent emissions for CBG, CNG, and diesel B7.

	CBG	CNG	Diesel B7
GHG emissions			
Well-to-tank (g/MJ_{fuel})	8.65	13.0	14.7
Tank-to-wheels			
• CO_2 (g/MJ_{fuel})		46.6	68.4
• CH_4 (g/MJ_{fuel})	0.1708	0.1708	
Total GHG (g CO_2-eq./MJ_{fuel})	13.4	64.4	83.1
Fuel consumption (MJ/km)	20.8	20.8	16.7
Specific GHG (g CO_2-eq./km)	**279**	**1342**	**1385**

Figure 10 shows the percentage changes in life cycle GHGs for the studied fuels. Shifting from conventional diesel to fossil natural gas does not show meaningful GHG benefits, bearing in mind the higher thermal efficiency of compression-ignition engines compared to spark-ignition gas engines. However, for biomethane, the situation is very different; 80% GHG emission benefit is achieved by switching from diesel to biomethane. With more precise methane emission control, GHG emission savings would advance towards 90%.

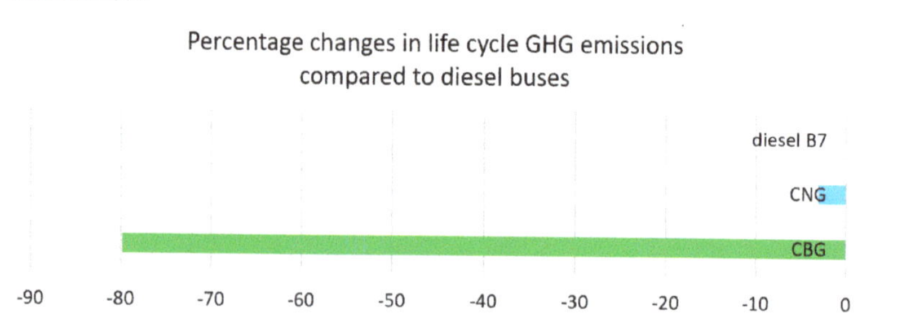

Figure 10. Percentage changes in life cycle GHGs.

This gives a strong environmental argument for biogas use. Increasing biogas use would be a quick and cost-effective way to reduce GHG emissions from urban traffic. Unfortunately, the potential of renewable gas is not acknowledged in the current EU emission standards, which only focus on tank-to-wheels emissions. Changing the measurement method to life cycle-based WTW instead of tailpipe measurement would enable a proper assessment of GHG emissions of future vehicle technology and fuel combinations. However, the results of this study can be utilized in designing strategies for transitioning to sustainable urban transport systems.

4. Conclusions

Transition to low-emission transportation and cleaner cities requires a broad introduction of low- and zero-carbon alternatives to conventional petrol- and diesel-powered vehicles. This paper presents the results of real-driving emission measurements from a Euro VI biogas-powered city bus equipped with a TWC. In addition, the lifetime carbon intensity of CBG was investigated and compared to its fossil counterpart CNG and traditional diesel fuel. The main findings were, first, for the bus:

- The rapid changes in exhaust gas temperature and composition under transient driving conditions seemed to be a critical challenge to an efficient operation of the TWC.
- Unimpressive CH_4 oxidation and NO_x reduction were observed in the catalyst after its service life of 375,000 km–400,000 km. In contrast, CO emissions were low, indicating efficient oxidation of CO in the catalyst.
- The primary reason for deficient CH_4 and NO_x conversion over the TWC was assumed to be the low CH_4 reactivity due to a partial deactivation of the catalyst. At low loads, common in a city bus's driving profile, the exhaust gas temperature was too low to allow efficient CH_4 oxidation. In addition to the low CH_4 oxidation rate, low CH_4 reactivity also means that methane-based reducing agents for NO_x reduction do not work, leading to substantial NO_x breakthrough from the catalyst.
- In addition, during the cold-start, CH_4 emissions were 2.3 times and NO_x emissions 1.4 times as high as those during the hot-start, highlighting the temperature sensitivity of catalytic emission control systems.
- Based on the above, deterioration of the exhaust after-treatment systems over time should be monitored as they are exposed to different aging processes resulting in elevated real-world emissions.
- Another critical issue was the fluctuating air-to-fuel ratio. Lambda was outside the optimal range for a significant part of the time, likely reducing the TWC efficiency. This highlights the need for precise lambda control to ensure high conversion rates throughout the vehicle's lifetime.

  Additionally,

- The WTW analysis showed an 80% GHG emission benefit by switching from diesel to biomethane, giving a strong environmental argument for biogas use. With more precise methane emission control, GHG emission savings would advance towards 90%.

The presented real-driving emission results are of great importance in supplementing the emission data for aging gas-powered HD vehicles, filling the gap of data on emissions closer to the service life of the vehicles. After all, the average age of bus fleets in Finland, for example, is over 12 years. The results of this study can also be utilized in scheduling catalyst maintenance or replacement activities.

In the future, it would be worthwhile to repeat the weather-related comparison with a completely new bus or with a new catalyst on an old bus.

Author Contributions: Conceptualization, K.S.-T. and S.N.; data curation, K.S.-T.; formal analysis, K.S.-T. and H.A.; funding acquisition, K.S.-T. and S.N.; investigation, K.S.-T., H.A. and O.N.; methodology, K.S.-T. and H.A.; project administration, K.S.-T.; supervision, S.N.; validation, K.S.-T. and H.A.; visualization, K.S.-T.; writing—original draft, K.S.-T.; writing—review and editing, K.S.-T., H.A., O.N. and S.N. All authors have read and agreed to the published version of the manuscript.

Funding: This research was funded by The European Regional Development Fund (ERDF) through The Council of Tampere Region, under Sustainable Growth and Jobs 2014–2020—Structural Funds Programme of Finland, grant number A75906. The APC was funded by MDPI/Clean technologies (2020 Best Paper Award).

Data Availability Statement: Not applicable.

Acknowledgments: The authors would like to thank Tuomas Wentin from Scania, Mikko Lähdesmäki from Wasa Citybus, and the City of Vaasa for their support and collaboration. The main author would also like to thank Gasum Oy for awarding a personal grant to support the research.

Conflicts of Interest: The authors declare no conflict of interest.

Abbreviations

The following abbreviations are used in this manuscript:

AFR	air-to-fuel ratio
CAN	controller area network
CH_4	methane
CO	carbon monoxide
CO_2	carbon dioxide
CBG	compressed biogas
CNG	compressed natural gas
ECU	engine control unit
EGR	exhaust gas recirculation
GPS	global positioning system
HC	hydrocarbon
HD	heavy-duty
ISC	in-service conformity
NDIR	non-dispersive infrared
NO	nitrogen monoxide
NO_2	nitrogen dioxide
NO_x	nitrogen oxides
OBD	on-board diagnostics
PEMS	portable emissions measurement system
PM	particulate matter
RDE	real-driving emissions
THC	total hydrocarbons
TWC	three-way catalyst
WTW	well-to-wheels

References

1. Communication from the Commission to the European Parliament, the Council, the European Economic and Social Committee and the Committee of the Regions. Stepping up Europe's 2030 Climate Ambition. Investing in a Climate-Neutral Future for the Benefit of Our People. EU COM/2020/562 Final. Available online: https://eur-lex.europa.eu/legal-content/EN/TXT/?uri=CELEX%3A52020DC0562 (accessed on 14 May 2022).
2. Finnish Government; Marin's Government; Government Programme. Strategic Themes. 3.1 Carbon Neutral Finland that Protects Biodiversity. Available online: https://valtioneuvosto.fi/en/marin/government-programme/carbon-neutral-finland-that-protects-biodiversity (accessed on 2 May 2022).
3. Liikennefakta. Liikenteen Kasvihuonekaasupäästöt ja Energiankulutus. Finnish Transport and Communications Agency Traficom. Available online: https://liikennefakta.fi/fi/ymparisto/liikenteen-kasvihuonekaasupaastot-ja-energiankulutus (accessed on 2 May 2022).
4. Franco, V.; Kousoulidou, M.; Muntean, M.; Ntziachristos, L.; Hausberger, S.; Dilara, P. Road vehicle emission factors development: A review. *Atmos. Environ.* **2013**, *70*, 84–97. [CrossRef]
5. Biernat, K.; Samson-Bręk, I.; Chłopek, Z.; Owczuk, M.; Matuszewska, A. Assessment of the Environmental Impact of Using Methane Fuels to Supply Internal Combustion Engines. *Energies* **2021**, *14*, 3356. [CrossRef]
6. Biogenic Carbon Dioxide. The Finnish Innovation Fund Sitra. Available online: https://www.sitra.fi/en/dictionary/biogenic-carbon-dioxide/ (accessed on 2 May 2022).

7. Myhre, G.; Shindell, D.; Bréon, F.-M.; Collins, W.; Fuglestvedt, J.; Huang, J.; Koch, D.; Lamarque, J.-F.; Lee, D.; Mendoza, B.; et al. Anthropogenic and Natural Radiative Forcing. In *Climate Change 2013: The Physical Science Basis*; Contribution of Working Group I to the Fifth Assessment Report of the Intergovernmental Panel on Climate Change; Stocker, T.F., Qin, D., Plattner, G.-K., Tignor, M., Allen, S.K., Boschung, J., Nauels, A., Xia, Y., Bex, V., Midgley, P.M., Eds.; Cambridge University Press: Cambridge, UK; New York, NY, USA, 2013; pp. 659–740. Available online: https://www.ipcc.ch/site/assets/uploads/2018/02/WG1AR5_Chapter08_FINAL.pdf (accessed on 16 May 2022).
8. Stoian, M.; Rogé, V.; Lazar, L.; Maurer, T.; Védrine, J.C.; Marcu, I.C.; Fechete, I. Total Oxidation of Methane on Oxide and Mixed Oxide Ceria-Containing Catalysts. *Catalysts* **2021**, *11*, 427. [CrossRef]
9. Pan, D.; Tao, L.; Sun, K.; Golston, L.M.; Miller, D.J.; Zhu, T.; Qin, Y.; Zhang, Y.; Mauzerall, D.L.; Zondlo, M.A. Methane emissions from natural gas vehicles in China. *Nat. Commun.* **2020**, *11*, 4588. [CrossRef]
10. Di Maio, D.; Beatrice, C.; Fraioli, V.; Napolitano, P.; Golini, S.; Rutigliano, F.G. Modeling of Three-Way Catalyst Dynamics for a Compressed Natural Gas Engine during Lean–Rich Transitions. *Appl. Sci.* **2019**, *9*, 4610. [CrossRef]
11. Rodman Oprešnik, S.; Seljak, T.; Vihar, R.; Gerbec, M.; Katrašnik, T. Real-World Fuel Consumption, Fuel Cost and Exhaust Emissions of Different Bus Powertrain Technologies. *Energies* **2018**, *11*, 2160. [CrossRef]
12. Gis, M.; Pielecha, J.; Gis, W. Exhaust emissions of buses LNG and Diesel in RDE tests. *Open Eng.* **2021**, *11*, 356–364. [CrossRef]
13. Lv, L.; Ge, Y.; Ji, Z.; Tan, J.; Wang, X.; Hao, L.; Wang, Z.; Zhang, M.; Wang, C.; Liu, H. Regulated emission characteristics of in-use LNG and diesel semi-trailer towing vehicles under real driving conditions using PEMS. *J. Environ. Sci.* **2020**, *88*, 155–164. [CrossRef]
14. Rosero, F.; Fonseca, N.; López, J.M.; Casanova, J. Effects of passenger load, road grade, and congestion level on real-world fuel consumption and emissions from compressed natural gas and diesel urban buses. *Appl. Energy* **2021**, *282*, 116195. [CrossRef]
15. Gallus, J.; Kirchner, U.; Vogt, R.; Benter, T. Impact of driving style and road grade on gaseous exhaust emissions of passenger vehicles measured by a Portable Emission Measurement System (PEMS). *Transp. Res. D Transp. Environ.* **2017**, *52*, 215–226. [CrossRef]
16. Chen, C.; Huang, C.; Jing, Q.; Wang, H.; Pan, H.; Li, L.; Zhao, J.; Dai, Y.; Huang, H.; Schipper, L.; et al. On-road emission characteristics of heavy-duty diesel vehicles in Shanghai. *Atmos. Environ.* **2007**, *41*, 5334–5344. [CrossRef]
17. Ozener, O.; Ozkan, M. Assessment of real driving emissions of a bus operating on a dedicated route. *Therm. Sci.* **2020**, *24*, 66–73. [CrossRef]
18. Giechaskiel, B.; Valverde, V.; Kontses, A.; Suarez-Bertoa, R.; Selleri, T.; Melas, A.; Otura, M.; Ferrarese, C.; Martini, G.; Balazs, A.; et al. Effect of Extreme Temperatures and Driving Conditions on Gaseous Pollutants of a Euro 6d-Temp Gasoline Vehicle. *Atmosphere* **2021**, *12*, 1011. [CrossRef]
19. Reiter, M.S.; Kockelman, K.M. The problem of cold starts: A closer look at mobile source emissions levels. *Transp. Res. D Transp. Environ.* **2016**, *43*, 123–132. [CrossRef]
20. Faria, M.V.; Varella, R.A.; Duarte, G.O.; Farias, T.L.; Baptista, P.C. Engine Cold Start Analysis Using Naturalistic Driving Data: City Level Impacts on Local Pollutants Emissions and Energy Consumption. *Sci. Total Environ.* **2018**, *630*, 544–559. [CrossRef]
21. Yusuf, A.A.; Inambao, F.L. Effect of cold start emissions from gasoline-fueled engines of light-duty vehicles at low and high ambient temperatures: Recent trends. *Case Stud. Therm. Eng.* **2019**, *14*, 100417. [CrossRef]
22. Lassi, U. Deactivation Correlations of Pd/Rh Three-way Catalysts Designed for Euro IV Emission Limits. Effect of Ageing Atmosphere, Temperature and Time. Academic Dissertation, University of Oulu, Department of Process and Environmental Engineering, Oulu, Finland, 2003. Available online: http://jultika.oulu.fi/files/isbn9514269543.pdf (accessed on 8 June 2022).
23. Regulation No 595/2009 of the European Parliament and of the Council of 18 June 2009 on Type-Approval of Motor Vehicles and Engines with Respect to Emissions from Heavy Duty Vehicles (Euro VI) and Amending Regulation (EC) No 715/2007 and Directive 2007/46/EC and Repealing Directives 80/1269/EEC, 2005/55/EC and 2005/78/EC. EC 595/2009. Available online: https://eur-lex.europa.eu/legal-content/EN/TXT/PDF/?uri=CELEX:02009R0595-20200901&from=EN (accessed on 14 May 2022).
24. ACEA. Facts and Figures. Average age of the EU Vehicle Fleet, by Country. Available online: https://www.acea.auto/figure/average-age-of-eu-vehicle-fleet-by-country/ (accessed on 20 June 2022).
25. Regulation No 49 of the Economic Commission for Europe of the United Nations (UN/ECE)—Uniform Provisions Concerning the Measures to be Taken against the Emission of Gaseous and Particulate Pollutants from Compression-Ignition Engines and Positive Ignition Engines for Use in Vehicles. UN/ECE 49/2013. Available online: https://eur-lex.europa.eu/legal-content/EN/TXT/PDF/?uri=CELEX:42013X0624(01)&from=EN (accessed on 12 May 2022).
26. Mendoza Villafuerte, P.; Suarez Bertoa, R.; Giechaskiel, B.; Riccobono, F.; Bulgheroni, C.; Astorga, C.; Perujo, A. NO_X, NH_3, N_2O and PN real driving emissions from a Euro VI heavy-duty vehicle. Impact of regulatory on-road test conditions on emissions. *Sci. Total Environ.* **2017**, *609*, 546–555. [CrossRef]
27. Commission Regulation (EU) No 582/2011 of 25 May 2011, Implementing and Amending Regulation (EC) No 595/2009 of the European Parliament and of the Council with Respect to Emissions from Heavy Duty Vehicles (Euro VI) and Amending Annexes I and III to Directive 2007/46/EC of the European Parliament and of the Council. EU COM/582/2011. Available online: https://eur-lex.europa.eu/legal-content/EN/TXT/HTML/?uri=CELEX:32011R0582&from=EN#ntr4-L_2011167FI.01008101-E0004 (accessed on 6 May 2022).

28. Van den Brink, P.J.; McDonald, C.M. Influence of the fuel hydrocarbon composition on nitric oxide conversion in 3-way catalysts. The NO_x/aromatics effect. *Appl. Catal. B Environ.* **1995**, *6*, 97–103. [CrossRef]
29. Matam, S.K.; Otal, E.H.; Aguirre, M.H.; Winkler, A.; Ulrich, A.; Ulrich, A.; Rentsch, D.; Weidenkaff, A.; Ferri, D. Thermal and chemical aging of model three-way catalyst Pd/Al2O3 and its impact on the conversion of CNG vehicle exhaust. *Catal. Today* **2012**, *184*, 237–244. [CrossRef]
30. Winkler, A.; Dimopoulos, P.; Hauert, R.; Bach, C.; Aguirre, M. Catalytic activity and aging phenomena of three-way catalysts in a compressed natural gas/gasoline powered passenger car. *Appl. Catal. B Environ.* **2008**, *84*, 162–169. [CrossRef]
31. Jääskeläinen, H. Three Way Catalysts for Methane. DieselNet Technology Guide. Revision 2017.12. Available online: https://dieselnet.com/tech/catalyst_methane_three-way.php (accessed on 21 June 2022).
32. Auvinen, P.; Nevalainen, P.; Suvanto, M.; Oliva, F.; Llamas, X.; Barciela, B.; Sippula, O.; Kinnunen, N.M. A detailed study on regeneration of SO_2 poisoned exhaust gas after-treatment catalysts: In pursuance of high durability and low methane, NH_3 and N_2O emissions of heavy-duty vehicles. *Fuel* **2021**, *291*, 120223. [CrossRef]
33. Zheng, Q.; Farrauto, R.; Deeba, M.; Valsamakis, I. Part I: A Comparative Thermal Aging Study on the Regenerability of Rh/Al2O3 and Rh/CexOy-ZrO_2 as Model Catalysts for Automotive Three Way Catalysts. *Catalysts* **2015**, *5*, 1770–1796. [CrossRef]
34. Lou, D.; Ren, Y.; Li, X.; Zhang, Y.; Sun, X. Effect of Operating Conditions and TWC Parameters on Emissions Characteristics of a Stoichiometric Natural Gas Engine. *Energies* **2020**, *13*, 4905. [CrossRef]
35. Commission Regulation (EU) 2016/1718 of 20 September 2016 Amending Regulation (EU) No 582/2011 with Respect to Emissions from Heavy-Duty Vehicles as Regards the Provisions on Testing by Means of Portable Emission Measurement Systems (PEMS) and the Procedure for the Testing of the Durability of Replacement Pollution Control Devices. Available online: https://eur-lex.europa.eu/legal-content/EN/TXT/HTML/?uri=CELEX:32016R1718&from=EN (accessed on 12 May 2022).
36. Systemic. Stormossen Ab/Oy (Vaasa/Korsholm, Finland). *SYSTEMIC Circular Solutions for Biowaste*. Available online: https://systemicproject.eu/wp-content/uploads/Stormossen_fact-sheet-Associated-plants_20191003.pdf (accessed on 2 May 2022).
37. Stormossen. Annual Report 2020. Available online: https://ar2020.stormossen.fi/annual_report/vuosikertomus-2020/tulostus/ (accessed on 2 May 2022).
38. TUV. Biogas to Biomethane Technology Review. Contract Number: IEE/10/130, Deliverable Reference: Task 3.1. TUV—Vienna University of Technology. Institute of Chemical Engineering, Research Division Thermal Process Engineering and Simulation. 2012. Available online: https://www.membran.at/downloads/2012_BioRegions_BiogasUpgradingTechnologyReview_ENGLISH.pdf (accessed on 2 May 2022).
39. Ardolino, F.M.; Cardamone, G.F.; Parrillo, F.; Arena, U. Biogas-to-biomethane upgrading: A comparative review and assessment in a life cycle perspective. *Renew. Sustain. Energy Rev.* **2021**, *139*, 110588. [CrossRef]
40. Uusitalo, V.; Havukainen, J.; Manninen, K.; Höhn, J.; Lehtonen, E.; Rasi, S.; Soukka, R.; Horttanainen, M. Carbon footprint of selected biomass to biogas production chains and GHG reduction potential in transportation use. *Renew. Energy* **2014**, *66*, 90–98. [CrossRef]
41. JRC. *Well-to-Wheels Analysis of Future Automotive Fuels and Powertrains in the European Context, Well-to-Tank Report Version 4.a*; European Commission Joint Research Centre (JRC): Brussels, Belgium, January 2014. [CrossRef]
42. StatFin. Fuel Classification. Statistics Finland. 2021. Available online: https://www.stat.fi/tup/khkinv/khkaasut_polttoaineluokitus.html (accessed on 3 May 2022).
43. Majer, S.; Oehmichen, K.; Kircshmeyr, F.; Scheidl, S. *Calculation of GHG Emission Caused by Biomethane. Biosurf. Fuelling Biomethane. Deliverable 5.3*; European Union: Brussels, Belgium, 2016. [CrossRef]
44. Bauer, F.; Hulteberg, C.; Persson, T.; Tamm, D. Biogas Upgrading—Review of Commercial Technologies. Swedish Gas Technology Centre, SGC Report 2013:270. Available online: https://www.sgc.se/ckfinder/userfiles/files/SGC270.pdf (accessed on 20 May 2022).
45. EEA. Indicators. Greenhouse Gas Emission Intensity of Electricity Generation in Europe. European Environment Agency. 2021. Available online: https://www.eea.europa.eu/ims/greenhouse-gas-emission-intensity-of-1 (accessed on 2 May 2022).
46. Söderena, P.; Nylund, N.-O.; Mäkinen, R. City Bus Performance Evaluation. VTT Technical Research Centre of Finland. VTT Report No. VTT-CR-00544-19. 2019. Available online: https://cris.vtt.fi/ws/portalfiles/portal/26400446/City_bus_performance_evaluation.pdf (accessed on 10 June 2022).

clean technologies

MDPI

Article

Impact of Battery Energy System Integration in Frequency Control of an Electrical Grid with Wind Power

Anukriti Pokhriyal [1,*], José Luis Domínguez-García [2,*] and Pedro Gómez-Romero [1]

1 Catalan Institute of Nanoscience and Nanotechnology, ICN2 (CSIC-BIST), Campus de la UAB, 08193 Barcelona, Spain
2 Catalonia Institute for Energy Research, IREC, Jardins de les Dones de Negre, Sant Adrià de Besòs, 08930 Barcelona, Spain
* Correspondence: anukriti.pokhriyal@icn2.cat (A.P.); jldominguez@irec.cat (J.L.D.-G.)

Abstract: The rapid increase in the integration of renewable energy sources into the electrical grid is posing new challenges for the dynamic responses due to the global inertia reduction. In this regard, the impact on frequency stability of such reduction and the potential support from wind power have been investigated lately. However, it is well-known that the variability of wind power and its reduced inertia may not be enough to handle the power unbalance. Energy storage systems (e.g., batteries) may provide the required additional flexibility to ensure proper response. In this paper, an analysis of how the different control loops for frequency support on wind power and batteries interact and perform is presented. To gain insights from the different impacts, a sensitivity analysis comparison for frequency regulation through (i) inertia and droop control of a variable speed wind turbine and (ii) battery is performed. The analysis was carried out through simulations using the well-known 4 generator 2-area model adapted to include a wind farm. From a battery perspective, its ramping capabilities were varied to provide frequency regulation. The paper shows how, depending on the control parameters and battery size, the frequency response may even become unstable due to interactions of the various pieces of technology. Thus, it shows that coordinated actions, control optimization and grid status among different actors on the grid (as battery and wind) is required for stable operation.

Keywords: wind power plant; frequency regulation; battery energy storage system; droop control; inertia control; control interaction

Citation: Pokhriyal, A.; Domínguez-García, J.L.; Gómez-Romero, P. Impact of Battery Energy System Integration in Frequency Control of an Electrical Grid with Wind Power. *Clean Technol.* **2022**, *4*, 972–986. https://doi.org/10.3390/cleantechnol4040060

Academic Editor: Damien Guilbert

Received: 14 June 2022
Accepted: 16 September 2022
Published: 9 October 2022

1. Introduction

The steady increase in adoption of renewable energy sources (RES) to reduce emissions and make electrical grids sustainable worldwide has led to a large increase in wind power penetration. With offshore technology becoming commercial, the share of wind power is set to expand further. Due to this increasing contribution of RES, especially wind, there is concern over the reducing inertia and balancing the disturbances caused their variable nature. Wind turbines (WTs) are not capable of providing inertia, as they are decoupled from the grid through converters [1]. In addition, many nuclear and thermal power plants are being shut down worldwide, and the rising costs and supply-chain issues surrounding gas imports have increased concerns about the future of grid stability [2]. Inertia is essential in the grid, as it helps avoid changes in frequency by providing power over a short period of time. It provides the system with enough time to not destabilize before additional generation is completed or a load is shed [3].

Inertia control has generally been performed by governor control of synchronous machines and swing or backup generators. Depending on the country, the frequency is either 50 or 60 Hz, and during generation or load imbalance, the frequency starts deviating from these base values. The rate of change of frequency (ROCOF) is given by the swing equation:

$$\frac{df}{dt} = \frac{f_0}{2H_{sys}S_G}(P_m - P_e) \quad (1)$$

where df/dt is the ROCOF; f_0 is the system frequency; H_{sys}, S_G, P_m and P_e are the system inertia constant, generator rated power, mechanical power and electrical power, respectively.

Traditionally, wind power plants (WPPs) do not participate in frequency regulation, as older permanent magnetic synchronous generator (PMSG) turbines operate in maximum power point tracking (MPPT) mode and their speed cannot be altered to vary their output power. Coordinated PMSG wind turbine control strategies have been developed to emulate the combined inertia response of WTs and generator to provide a short-term frequency response, but are still not preferred [4]. However, newer variable-speed wind turbines (VSWTs) can be controlled externally through several mechanisms to provide an inertia response [5–9]. For example, inertia control injects energy into the system based on the ROCOF, and droop control works with the difference between frequency levels [10,11]. On the other hand, deloading control shifts the WT operating point from the optimal to a reduced power level. The deloading is performed through speed and pitch angle control [12–14]. Reference [15] provides a control scheme for an isolated power system using optimum deloading techniques through static converters and pitch control, and rotor speed and active power variations. However, the frequency nadir with just integral control loop provides a marginal improvement over that of a primary frequency response without the control loop.

These WT control strategies can be combined and integrated with other generators in the grid to implement better frequency controls. Reference [16] proposes a coordinated control strategy with time-varying gains of inertia and droop control loops. The control loop gains depend on the desired frequency-response time, and initial gains are based on wind speed and operating conditions of each WT in the farm. Reference [17] formulates a coordination strategy between WT droop control and synchronous machines such that the conventional generators activate power injection from the start of frequency drop. This control leads to a lower frequency deviation value. Nevertheless, there is no improvement in the total time required to bring back the frequency to the threshold. Reference [18] provides a hierarchical control system structure for coordinating WTs and reducing the total number of turbines used for frequency regulation. Three layers of wind farm management are used, along with wind farm control and WT control, to share information among the wind farms, traditional power plants and system dispatching center. The WT control is responsible for tracking the reference active power value and transmitting information, such as output power, available frequency control power, wind speed, rotor speed, pitch angle and other operation status indicators of the WTG to the wind farm management layer. However, for the two active power control methods, rotor speed control and pitch angel control, the pitch angle control is slower and leads to more mechanical wear compared with the rotor speed control. Thus, it is considered that the reserve power quality provided by rotor speed control is better than the pitch control. However, the sensitivity of the WT response due to the variations of the inertia and droop constant was not studied in detail. With numerous WT control models, it is important to not only compare them but also evaluate how they perform with various degrees of sensitivity in different scenarios. In [19], a comparison is made between the performance of an actual WPP and a detailed model containing different gains of the inertial controller. However, gains are not varied by large values, and the approach is not compared with alternative methods of frequency control.

Lately, energy storage systems, especially batteries, have emerged as a popular option to provide primary frequency support. Batteries are not just capable of responding within a few seconds, but can also continuously supply power over longer periods. This could help eliminate the need for secondary or tertiary reserves in many cases [20–22]. Moreover, batteries can be charged with the excess energy generated from the WPPs. In fact, the possibility of providing frequency regulation combined with renewable energy arbitrage has been one of the key areas of interest for wind power plant operators [23].

As mentioned earlier, inertia in a power conventional power grid is provided by synchronous generators in thermal power plants through governor control. That is, in the event of a frequency drop, additional power can be supplied by the synchronous machines to balance the system. However, these are slow, as they take a few seconds to respond. Meanwhile, output from electrochemical energy storage technologies such as batteries and supercapacitors is almost instantaneous, and energy can be provided over a longer time duration as compared to inertia and droop responses of WTs. Numerous studies with battery energy storage system (BESS) control strategies have effectively shown suitability for frequency regulation in largely wind-powered systems [24–33]. Reference [24] depicts a BESS and WT pitch angle control. The pitch control is activated only when battery state of charge (SOC) is below 40%, and the synchronous generator responds at 30% SOC, charging the battery and bringing SOC levels back to 50%. Reference [25] assesses the impacts of various aggregated and distributed energy storage technologies in a wind farm. Reference [26] presents lithium-ion battery (LiB) control to limit damping power associated with inter area low frequency oscillations in a wind power plant. Reference [28] shows an SOC-control-based scheme and response time between generators and BESS to control wind frequency deviations. Reference [31] considers a WPP and a BESS together as a virtual power plant (VPP) to provide both energy and frequency regulation in energy markets. The cooperation strategy deploys a BESS during insufficient generation from the WPP, considering its cycle number and energy change functions to improve battery lifetime. Battery and supercapcitor combinations have also been widely studied for frequency regulation applications to provide faster responses and reduce battery size and overall costs [30,34–40].

Thus, different but complementing strategies can be used to provide better grid support. Simultaneous wind–battery control can be an effective strategy to avoid frequency imbalances, as it could reduce reliance on just one type of support. However, in-depth sensitivity analysis on the impact of different WT control strategies and their comparison with a BESS need to be studied further. In this paper, we explore WT control strategies that incorporate a battery to study their individual impacts on frequency regulation in the event of a grid disturbance. We also analyze the combined effect of the two controls and vary their response capabilities to encourage better control strategies with improved frequency regulation.

2. Materials and Methods

Frequency Support from Wind Turbine Control and BESS

As mentioned, in the event of rapid frequency changes, the WT is capable of regulating itself through several means. In our study, we used inertia and droop control methods, as they are capable of injecting sufficient momentary power into the grid without deloading the WT and causing stress on the turbine blades.

Inertia Response: In the case of VSWTs, the rotating mass is mechanically decoupled from the grid and the kinetic energy (KE) cannot provide a direct inertia response like the synchronous generators for frequency balancing. However, doubly-fed induction generators (DFIGs) can be operated to slow down and release this KE into the grid for a few seconds and provide inertia into the system for 10 s. This provides sufficient time for the synchronous generator to activate its governor control to further balance the frequency within 30 s (primary control). Afterwards, slower secondary and tertiary controls are activated to bring the frequency levels back to threshold values [41].

$$E_k = \frac{1}{2}J\omega^2 \tag{2}$$

where E_k is the kinetic energy, J is the moment of inertia of the wind rotor and ω is the rotational speed.

$$H = \frac{J\omega^2}{2S} \tag{3}$$

The inertia constant H represents the time duration for which the wind generator car provide nominal power (S) by using kinetic energy.

Droop Control: It is a popular method used by wind generators to imitate the governors of conventional generators by regulating the active power output from a WT proportionally to frequency change.

$$\Delta P = \frac{f_{nom} - f_{meas}}{R} \quad (4)$$

where f_{nom}, f_{meas} and R are nominal frequency, measured frequency and the droop constant respectively.

The default inertia constant was selected as 5.04 and the droop constant as 0.05. These were later modified to observe the best response (please see the Figure 1).

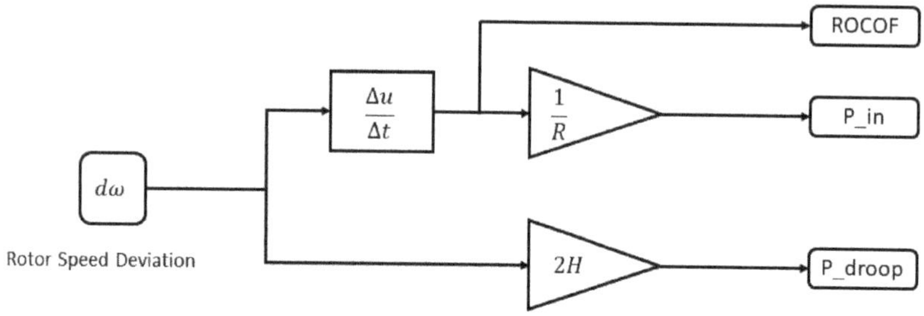

Figure 1. Frequency regulation block with inertia and droop controls.

Battery Control: Unlike synchronous generators, batteries can inject and absorb power from the grid within seconds. The battery control algorithms can be modified to inject appropriate power into the grid as soon as there is a frequency drop. This can be done by responding to (i) frequency deviations in the grid and (ii) the ROCOF in the system.

The battery SOC at a given time (t) is the ratio of available energy to the rated energy of the battery.

$$SOC(t) = \frac{E(t)}{E_{rated}} \quad (5)$$

The operation of the BESS is limited by:

$$SOC^{min} \leqslant SOC(t) \leqslant SOC^{max} \quad (6)$$

where $E(t)$ is the energy stored in the battery at time t and E_{rated} is the rated energy of the battery. The battery parameters were set according to Table 1 [30,42–44]. We assumed an energy rating higher than what is needed in order for the BESS to be able to supply sufficient energy for the whole simulated time period.

Table 1. BESS parameters.

Parameter	Value
Rated Power	300 MW
Rated Capacity	50 MWh
Efficiency	95%
SOC^{max}	80%
SOC^{min}	20%

3. Case Study and Sensitivity Analysis

To compare the response sensitivity of the WT control and a battery, an existing Kundur two-area system was used and modified. The Simscape–MATLAB model simulated a grid system with a synchronous machine as a swing generator (M2), a DFIG wind turbine and a load in Area 1. An additional load of 200 MW was connected via a breaker that switched at 50 and 100 s, creating a frequency drop and rise, respectively. In Area 2, there were two synchronous machines (M3 and M4) and a load. A constant wind speed was assumed over the restricted simulation time (please see the Figure 2 and Table 2).

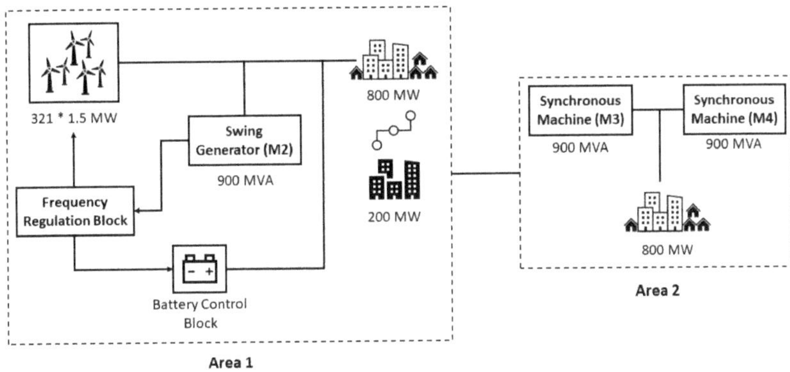

Figure 2. Two-area model.

Table 2. Parameters of the two-area model

Parameter	Value
Wind Turbine Nominal Power	1.5 MW
Number of wind turbines	321
Wind speed	10 m/s
Grid Frequency	60 Hz
Initial Load	1600 MW
Rated Power of Synchronous Turbines	900 MVA

The sensitivity of the response to the frequency disturbance was measured for both under-frequency and over-frequency. Different cases were studied to observe the frequency response by the WT and the BESS.

3.1. Case I: No Frequency Control

The response of the WT was studied without any frequency control to establish the base condition. At 50 s the load switch was connected, which led to the addition of 200 MW to the system. Consequently, there was a large frequency drop, and the frequency tried to settle after a few seconds. The frequency nadir was observed at 59.4 Hz. At 100 s, the additional load was disconnected, and the frequency increased but settled at around 60 Hz after a few seconds. Since there was no external frequency regulation, the synchronous generators (M2, M3, M4) responded to provide the extra power at after a few seconds to compensate for the frequency drop (please see the Figure 3).

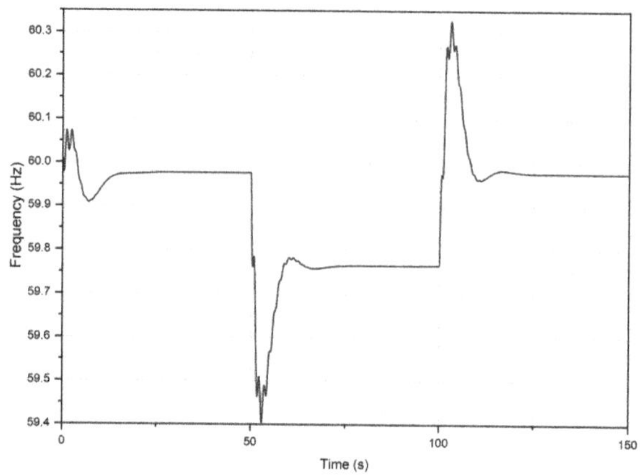

(**a**) Frequency

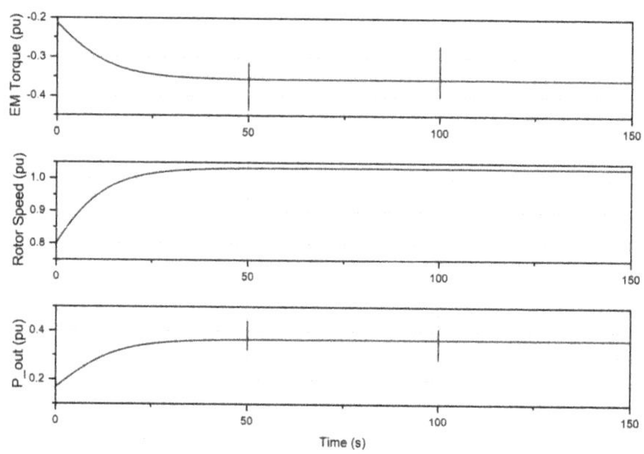

(**b**) Wind Turbine Characteristics

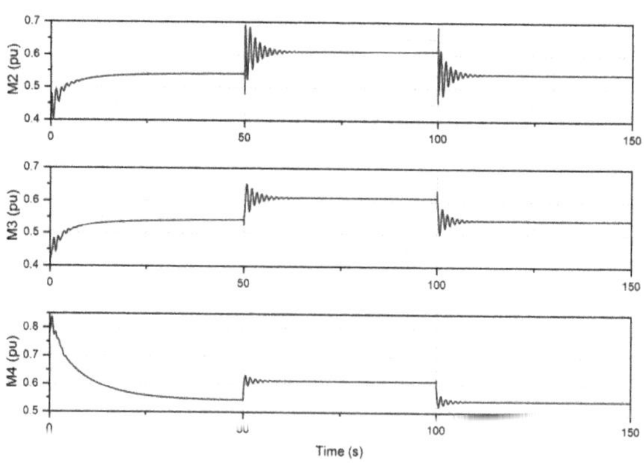

(**c**) Synchronous Machines

Figure 3. Behavior of grid frequency, WTs and synchronous machines during no frequency control conditions.

3.2. *Case II: Inertia and Droop Control*

The inertia and droop constant were varied to study the best response that the WT can provide by itself. We compared the frequency responses of the default values of inertia constant (H = 5.04) and droop constant (R = 0.05) and {varied their values to check system stability. In all scenarios, at 50 s the frequency settled at a higher value compared to Case I, and the frequency nadir was greater with the default inertia and droop constant (A). When the inertia constant was increased and the droop constant reduced, the output power from the WT for frequency regulation also increased. The lowest frequency drop was observed with C and D. However, since additional power from the WT can be supplied only for a few seconds, there was a second drop in the frequency (C and D), and it finally settled at a lower value than that in default constant scenario (A) and could also cause further instability. From Table 3 and Figure 4,we can conclude that in the case of a frequency response though only inertia and droop control, scenario B would be the best option. The wind turbine control clearly prevents a large initial drop, giving time to the synchronous machines to balance the frequency.

Table 3. Inertia and droop constant values for frequency control through WT.

	Inertia Constant (H)	Droop Constant (R)
A	5.04	0.05
B	10.08	0.025
C	13.104	0.019
D	0	0.019

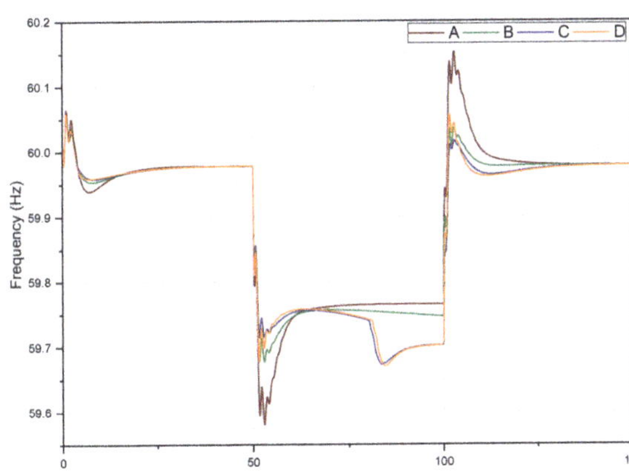

(**a**) Frequency

Figure 4. *Cont.*

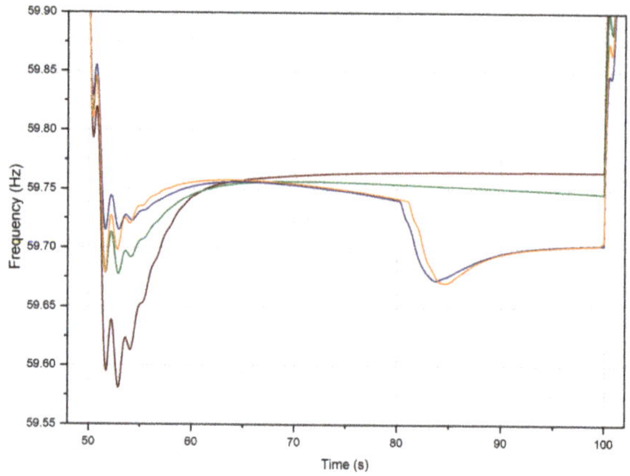

(b) Frequency

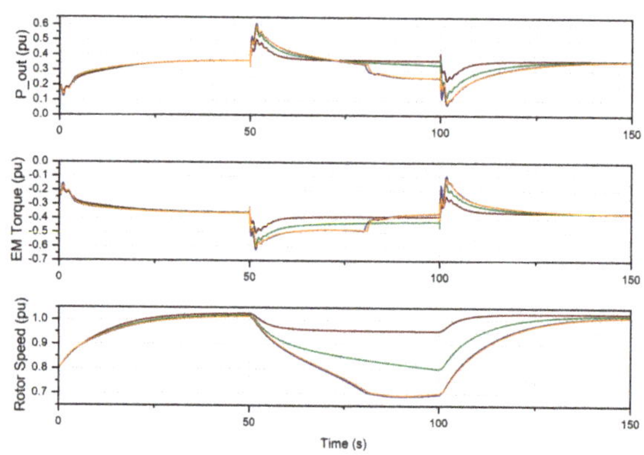

(c) Wind Turbine Characteristics

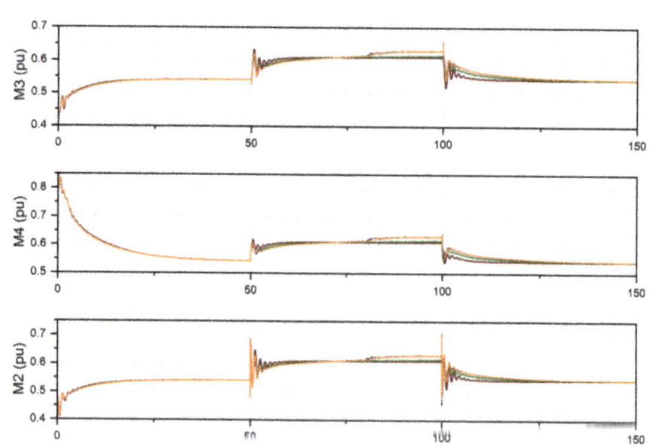

(d) Synchronous Machines

Figure 4. Variations in grid frequency, WTs' and synchronous machines' responses during different inertial and droop control scenarios.

3.3. *Case III: Battery Control*

In this case, inertia and droop controls were turned off, and only the battery provided frequency support. The battery responded according to both frequency deviation and ROCOF, and the battery-response-controlling gains were varied. For lower gains (A), the battery had a slow response to the frequency drop, and the synchronous machines also participated in balancing. In the best case scenario (C), at 50 s, the battery responded very quickly by supplying power instantaneously and prevented a frequency drop. Moreover, the frequency was stabilized faster and at a higher value than in Case II. No significant increase in output from synchronous machines was observed, so we conclude that they did not participate in providing frequency support, and all the power was supplied by the battery. At 100 s, the battery absorbed the excess energy and stabilized the frequency at very close to 60 Hz. Thus, optimal gains can be selected to increase or reduce sensitivity of the battery towards frequency changes (please see the Figure 5 and Table 4).

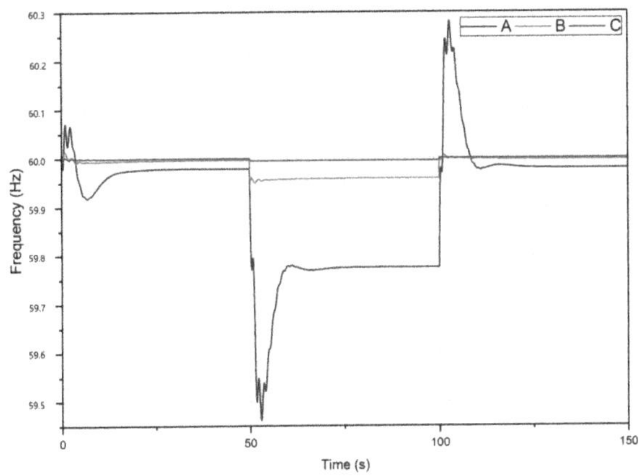

(**a**) Frequency

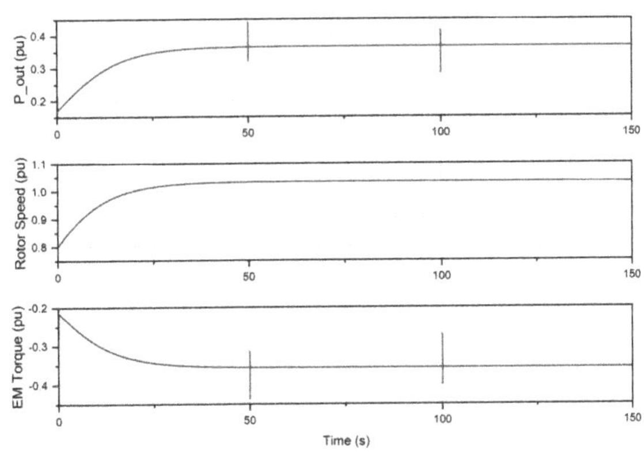

(**b**) Wind Turbine Characteristics

Figure 5. *Cont.*

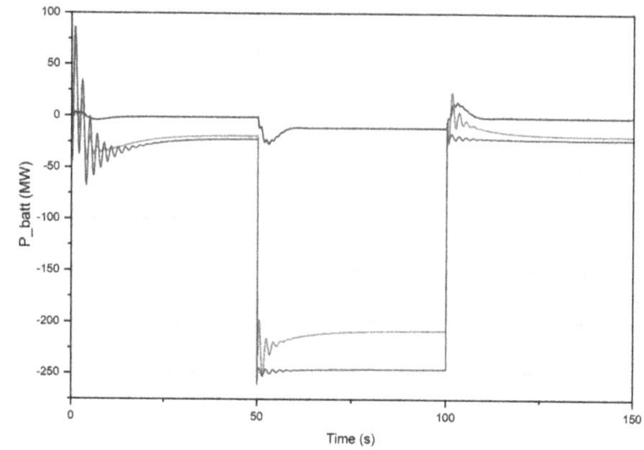

(**c**) Battery Output Power

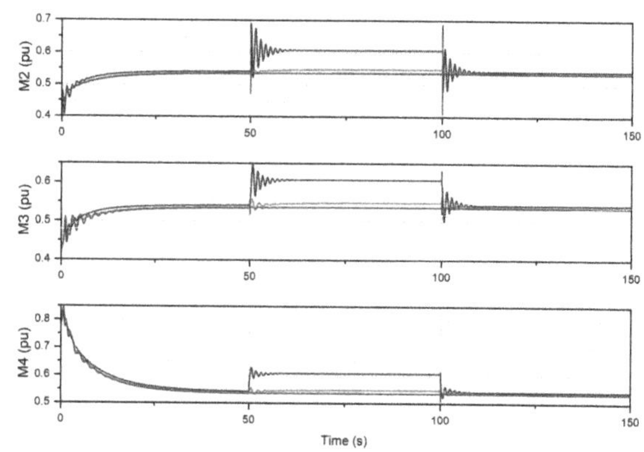

(**d**) Synchronous Machines

Figure 5. Variations in grid frequency, WTs', and synchronous machines' responses during different battery control scenarios.

Table 4. Battery gain multipliers.

	Frequency Deviation Multiplier	ROCOF Multiplier
A	$50e^6$	$10e^6$
B	$50e^8$	$10e^8$
C	$50e^9$	$10e^9$

3.4. Case IV: Inertia, Droop and Battery Control

Both inertia and droop control and battery control were turned on. The best H and R values from Case II were selected to allow maximum possible support from the wind turbine first and the rest through the battery. The battery ramp rates were varied similarly to Case III. In A, the battery control gains were too low for it to respond. The frequency regulations was solely provided by the wind turbine and the synchronous machines, and the output power from the battery was zero. On the other hand, when battery control gains were extremely high (C), neither wind nor synchronous machines contributed at all to frequency regulation.

If we reduce the battery's sensitivity (that is, reduce its response capabilities), we can observe that the inertia and droop control will kick in first. Meanwhile, the battery will also respond by providing less power. The contribution of the synchronous machines is negligible, showing that the wind turbine control and battery can be sufficient for handling frequency imbalances (please see the Table 5 and Figure 6).

Table 5. WT control and BESS gain parameters.

	H	R	Frequency Deviation Multiplier	ROCOF Multiplier
A	13.104	0.019	$50e^6$	$10e^6$
B	13.104	0.019	$50e^8$	$10e^8$
C	13.104	0.019	$50e^9$	$10e^9$
D	0	0.019	$50e^9$	$10e^9$

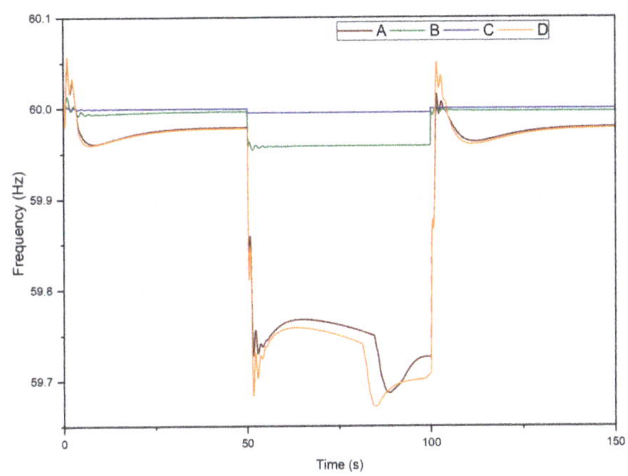

(**a**) Frequency

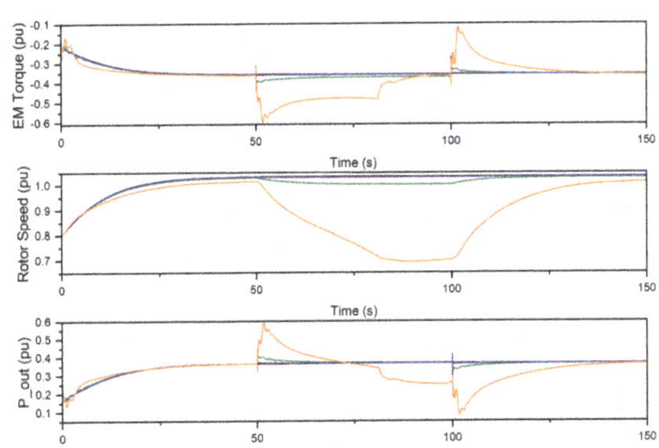

(**b**) Wind Turbine Characteristics

Figure 6. *Cont.*

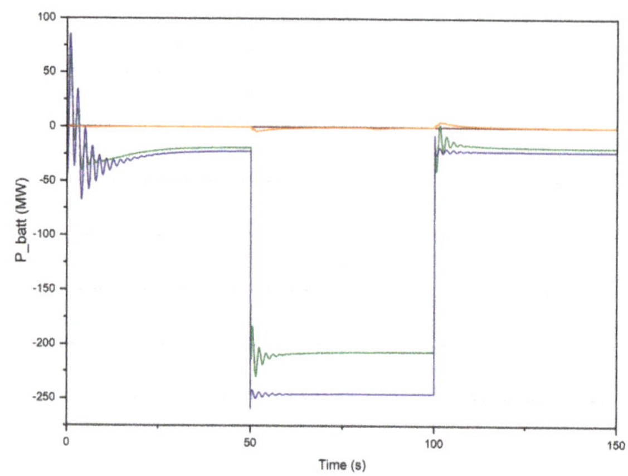

(c) Battery Output Power

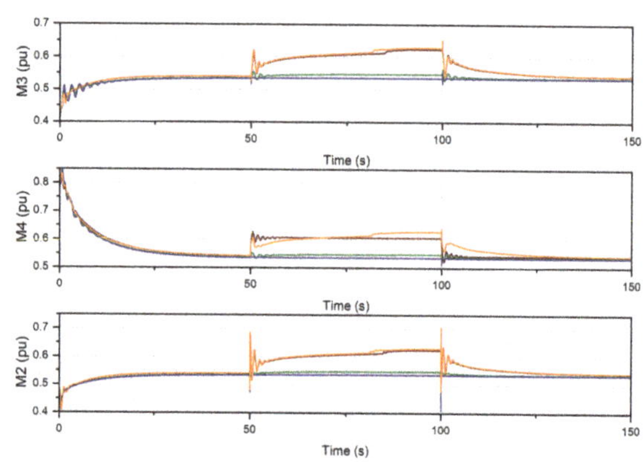

(d) Synchronous Machines

Figure 6. Variations in grid frequency, WTs', and synchronous machines' responses for different wind-battery combined control parameters.

4. Results and Discussion

The individual and hybrid wind-battery frequency control schemes allow better regulation than conventional generators. The response depends on the control parameters and varies depending on the sensitivity of frequency regulation strategies.

In Case II, a highly sensitive WT control strategy leads to a possible further destabilization of the frequency, as seen in Figure 4a. During controls C and D, the first frequency drop is much lower than in Case A at 59.72 or 59.67 Hz, respectively. However, after t = 80.5 s, the WT cannot provide any more extra power, and the frequency sees a second drop of 59.67 Hz for both. In such a scenario, a slight compromise in frequency nadir can provide a more stable response by the wind turbine.

In Case III, when only a BESS is connected, depending on the control, it either maintains the frequency very close to 60 Hz or for very low sensitivity, drops till 59.45 Hz. For the medium-sensitivity control scenario (Case B), the BESS only allows a drop till 59.55 Hz; see Figure 5a. Highly sensitive control would lead to constant battery engagement with grid and could impact its overall health. While the long-term performance of a BESS is outside the scope of this study, it is imperative to note that the battery lifetime is greatly affected by many factors. Various studies have shown that modifying battery cycling, the depth of discharge levels [45] and even charge–discharge protocols can improve battery life [46–50]. Li-ion is the most commonly used battery technology, and recent studies have demonstrated improvements in their cyclability [51,52]. However, they still generally have short cycle lives (10^3 cycles) as compared to supercapacitors (10^6 cycles) [53] and need to be improved. Future energy systems are likely to have more frequency disturbances due to their large shares of variable renewables, which will lead to increased demand from the battery, and its overall lifetime would be impacted. More studies are needed to understand the relationships between different battery applications and its lifetime.

Therefore, the optimum strategy would be combining the best responses of both the WT and the battery to avoid reliance on a single means of control (Case IV). Figure 6a,b show that the battery's response can be limited to allow WTs to simultaneously respond and support each other. The operators can optimize the sensitivity depending on the WT and battery ratings and the level of support required from them [54]. This would help reduce the mechanical stress on the WTs, improving battery life and reduce the need for conventional spinning reserves.

5. Conclusions

A control scheme for the hybrid system was developed up to allow both a WT and a battery to provide frequency regulation support. This prevents the battery from responding to every fluctuation in frequency. The battery clearly is more sensitive to frequency variations than the inertia and droop control. However, very high sensitivity would affect the battery's health over long period of time. In contrast, reduced BESS sensitivity prevents proper frequency support, and the synchronous machines have to provide the inertia. An oversized BESS would lead to higher costs, but an undersized one would not be able to provide adequate response. Therefore, the BESS needs to be properly sized to provide the power needed in cases of frequency drop. For the wind turbine control, far less sensitive inertia and droop control is not sufficient, as the frequency drop is larger. However, highly sensitive inertia and droop control may cause a second frequency drop, thereby increasing the time for frequency stabilization or even further destabilizing the grid. We conclude that a battery may be deployed along with inertia and droop control for maximum support, but sizing and response sensitivity would be key to utilizing the battery efficiently. Depending on the control, additional synchronous machines may not be needed to provide inertia support in the event of a frequency drop.

Author Contributions: Conceptualization, A.P. and J.L.D.-G.; methodology, A.P. and J.L.D.-G.; formal analysis, A.P.; writing—original draft preparation, A.P.; writing—review and editing, J.L.D.-G. and P.G.-R.; funding acquisition, P.G.-R. All authors have read and agreed to the published version of the manuscript.

Funding: This research was funded by H2020-MSCA-COFUND-2016 (DOC-FAM, Grant Agreement No. 754397).

Data Availability Statement: Not applicable.

Conflicts of Interest: The authors declare no conflict of interest.

References

1. Ratnam, K.S.; Palanisamy, K.; Yang, G. Future low-inertia power systems: Requirements, issues, and solutions—A review. *Renew. Sustain. Energy Rev.* **2020**, *124*, 109773. [CrossRef]
2. Europe Faces Gas Supply Disruption after Russia Imposes Sanctions. *Al Jazeera* **2022**. Available online: https://www.aljazeera.com/news/2022/5/12/europes-gas-supply-crisis-grows-after-russia-imposes-sanctions (accessed on 12 May 2022).
3. Denholm, P.; Mai, T.; Kenyon, R.W.; Kroposki, B.; O'Malley, M. *Inertia and the Power Grid: A Guide Without the Spin*; Technical Report NREL/TP-6A20-73856; National Renewable Energy Laboratory: Golden, CO, USA, 2020.
4. Wu, Z.; Gao, W.; Wang, J.; Gu, S. A coordinated primary frequency regulation from Permanent Magnet Synchronous Wind Turbine Generation. In Proceedings of the 2012 IEEE Power Electronics and Machines in Wind Applications, Denver, CO, USA, 16–18 July 2012; pp. 1–6. [CrossRef]

5. Attya, A.; Dominguez-Garcia, J.; Anaya-Lara, O. A review on frequency support provision by wind power plants: Current and future challenges. *Renew. Sustain. Energy Rev.* **2018**, *81*, 2071–2087. [CrossRef]
6. Fernández-Guillamón, A.; Gómez-Lázaro, E.; Muljadi, E.; Ángel Molina-García. Power systems with high renewable energy sources: A review of inertia and frequency control strategies over time. *Renew. Sustain. Energy Rev.* **2019**, *115*, 109369. [CrossRef]
7. Wu, Z.; Gao, W.; Gao, T.; Yan, W.; Zhang, H.; Yan, S.; Wang, X. State-of-the-art review on frequency response of wind power plants in power systems. *J. Mod. Power Syst. Clean Energy* **2018**. [CrossRef]
8. Conroy, J.F.; Watson, R. Frequency Response Capability of Full Converter Wind Turbine Generators in Comparison to Conventional Generation. *IEEE Trans. Power Syst.* **2008**, *23*, 649–656. [CrossRef]
9. Akram, U.; Nadarajah, M.; Shah, R.; Milano, F. A review on rapid responsive energy storage technologies for frequency regulation in modern power systems. *Renew. Sustain. Energy Rev.* **2020**, *120*, 109626. [CrossRef]
10. Díaz-González, F.; Hau, M.; Sumper, A.; Gomis-Bellmunt, O. Participation of wind power plants in system frequency control: Review of grid code requirements and control methods. *Renew. Sustain. Energy Rev.* **2014**, *34*, 551–564. [CrossRef]
11. Gonzalez-Longatt, F.M. Effects of the synthetic inertia from wind power on the total system inertia: Simulation study. In Proceedings of the 2012 2nd International Symposium On Environment Friendly Energies And Applications, Newcastle Upon Tyne, UK, 25–27 June 2012; pp. 389–395. [CrossRef]
12. Yao, W.; Lee, K.Y. A control configuration of wind farm for load-following and frequency support by considering the inertia issue. In Proceedings of the 2011 IEEE Power and Energy Society General Meeting, Detroit, MI, USA, 24–28 July 2011; pp. 1–6. [CrossRef]
13. Zhu, X.; Wang, Y.; Xu, L.; Zhang, X.; Li, H. Virtual inertia control of DFIG-based wind turbines for dynamic grid frequency support. In Proceedings of the IET Conference on Renewable Power Generation (RPG 2011), Edinburgh, UK, 6–8 September 2011; pp. 1–6. [CrossRef]
14. Li, P.; Hu, W.; Hu, R.; Huang, Q.; Yao, J.; Chen, Z. Strategy for wind power plant contribution to frequency control under variable wind speed. *Renew. Energy* **2019**, *130*, 1226–1236. [CrossRef]
15. de Almeida, R.G.; Pecas Lopes, J.A. Participation of Doubly Fed Induction Wind Generators in System Frequency Regulation. *IEEE Trans. Power Syst.* **2007**, *22*, 944–950. [CrossRef]
16. Wu, Y.K.; Yang, W.H.; Hu, Y.L.; Dzung, P.Q. Frequency Regulation at a Wind Farm Using Time-Varying Inertia and Droop Controls. *IEEE Trans. Ind. Appl.* **2019**, *55*, 213–224. [CrossRef]
17. Mauricio, J.M.; Marano, A.; Gomez-Exposito, A.; Martinez Ramos, J.L. Frequency Regulation Contribution Through Variable-Speed Wind Energy Conversion Systems. *IEEE Trans. Power Syst.* **2009**, *24*, 173–180. [CrossRef]
18. Shao, C.; Li, Z.; Hao, R.; Qie, Z.; Xu, G.; Hu, J. A Wind Farm Frequency Control Method Based on the Frequency Regulation Ability of Wind Turbine Generators. In Proceedings of the 2020 5th Asia Conference on Power and Electrical Engineering (ACPEE), Chengdu, China, 4–7 June 2020; pp. 592–596. [CrossRef]
19. Hu, Y.L.; Wu, Y.K. Inertial Response Identification Algorithm for the Development of Dynamic Equivalent Model of DFIG-Based Wind Power Plant. *IEEE Trans. Ind. Appl.* **2021**, *57*, 2104–2113. [CrossRef]
20. Poullikkas, A. A comparative overview of large-scale battery systems for electricity storage. *Renew. Sustain. Energy Rev.* **2013**, *27*, 778–788. [CrossRef]
21. Hesse, H.C.; Schimpe, M.; Kucevic, D.; Jossen, A. Lithium-Ion Battery Storage for the Grid—A Review of Stationary Battery Storage System Design Tailored for Applications in Modern Power Grids. *Energies* **2017**, *10*. [CrossRef]
22. Palizban, O.; Kauhaniemi, K. Energy storage systems in modern grids—Matrix of technologies and applications. *J. Energy Storage* **2016**, *6*, 248–259. [CrossRef]
23. Behnam Zakeri, S.S. Electrical energy storage systems: A comparative life cycle cost analysis. *Renew. Sustain. Energy Rev.* **2015**, *42*, 569–596. [CrossRef]
24. Dang, J.; Seuss, J.; Suneja, L.; Harley, R.G. SOC feedback control for wind and ESS hybrid power system frequency regulation. In Proceedings of the 2012 IEEE Power Electronics and Machines in Wind Applications, Denver, CO, USA, 16–18 July 2012; pp. 1–7. [CrossRef]
25. Li, W.; Joos, G. Comparison of Energy Storage System Technologies and Configurations in a Wind Farm. In Proceedings of the 2007 IEEE Power Electronics Specialists Conference, Orlando, FL, USA, 17–21 June 2007; pp. 1280–1285. [CrossRef]
26. Sun, J.; Sheng, L.; Sun, Y.; Zhou, Z.; Fu, R. Stability Simulation Analysis of a Hybrid Wind-Battery System. In *Theory, Methodology, Tools and Applications for Modeling and Simulation of Complex Systems*; Zhang, L., Song, X., Wu, Y., Eds.; Springer: Singapore, 2016; pp. 154–163.
27. Liang, L.; Zhong, J.; Jiao, Z. Frequency Regulation for a Power System with Wind Power and Battery Energy Storage. In Proceedings of the 2012 IEEE International Conference on Power System Technology (POWERCON), Auckland, New Zealand, 30 October–2 November 2012.
28. Toge, M.; Kurita, Y.; Iwamoto, S. Supplementary load frequency control with storage battery operation considering SOC under large-scale wind power penetration. In Proceedings of the 2013 IEEE Power Energy Society General Meeting, Vancouver, BC, Canada, 21–25 July 2013; pp. 1–5. [CrossRef]
29. Ono, T.; Arai, J. Frequency Control with Dead Band Characteristic of Battery Energy Storage System for Power System Including Large Amount of Wind Power Generation. *Electr. Eng. Jpn.* **2013**, *185*, 1–10. [CrossRef]

30. Li, W.; Joos, G. A power electronic interface for a battery supercapacitor hybrid energy storage system for wind applications. In Proceedings of the 2008 IEEE Power Electronics Specialists Conference, Rhodes, Greece, 15–19 June 2008; pp. 1762–1768. [CrossRef]
31. He, G.; Chen, Q.; Kang, C.; Xia, Q.; Poolla, K. Cooperation of Wind Power and Battery Storage to Provide Frequency Regulation in Power Markets. *IEEE Trans. Power Syst.* **2017**, *32*, 3559–3568. [CrossRef]
32. Singarao, V.Y.; Nimmagadda, S.; Rao, V. Economic Comparison of Deloaded-Wind and Wind-Battery Systems. In Proceedings of the 2014 Sixth Annual IEEE Green Technologies Conference, Corpus Christi, TX, USA, 3–4 April 2014; pp. 126–131. [CrossRef]
33. Tan, J.; Zhang, Y. Coordinated Control Strategy of a Battery Energy Storage System to Support a Wind Power Plant Providing Multi-Timescale Frequency Ancillary Services. *IEEE Trans. Sustain. Energy* **2017**, *8*, 1140–1153. [CrossRef]
34. Li, W.; Joos, G.; Belanger, J. Real-Time Simulation of a Wind Turbine Generator Coupled With a Battery Supercapacitor Energy Storage System. *IEEE Trans. Ind. Electron.* **2010**, *57*, 1137–1145. [CrossRef]
35. Mendis, N.; Muttaqi, K.M.; Perera, S. Management of Battery-Supercapacitor Hybrid Energy Storage and Synchronous Condenser for Isolated Operation of PMSG Based Variable-Speed Wind Turbine Generating Systems. *IEEE Trans. Smart Grid* **2014**, *5*, 944–953. [CrossRef]
36. Hajiaghasi, S.; Salemnia, A.; Hamzeh, M. Hybrid energy storage system for microgrids applications: A review. *J. Energy Storage* **2019**, *21*, 543–570. [CrossRef]
37. Fang, J.; Li, H.; Tang, Y.; Blaabjerg, F. On the Inertia of Future More-Electronics Power Systems. *IEEE J. Emerg. Sel. Top. Power Electron.* **2019**, *7*, 2130–2146. [CrossRef]
38. Tani, A.; Camara, M.B.; Dakyo, B. Energy Management in the Decentralized Generation Systems Based on Renewable Energy—Ultracapacitors and Battery to Compensate the Wind/Load Power Fluctuations. *IEEE Trans. Ind. Appl.* **2015**, *51*, 1817–1827. [CrossRef]
39. Fang, J.; Tang, Y.; Li, H.; Li, X. A Battery/Ultracapacitor Hybrid Energy Storage System for Implementing the Power Management of Virtual Synchronous Generators. *IEEE Trans. Power Electron.* **2018**, *33*, 2820–2824. [CrossRef]
40. Zhang, S.; Mishra, Y.; Shahidehpour, M. Fuzzy-Logic Based Frequency Controller for Wind Farms Augmented With Energy Storage Systems. *IEEE Trans. Power Syst.* **2016**, *31*, 1595–1603. [CrossRef]
41. Mohammad Dreidy, H. Mokhlis, S.M. Inertia response and frequency control techniques for renewable energy sources: A review. *Renew. Sustain. Energy Rev.* **2017**, *69*, 144–155. [CrossRef]
42. Zhan, S.; Hou, P.; Enevoldsen, P.; Yang, G.; Zhu, J.; Eichman, J.; Jacobson, M.Z. Co-optimized trading of hybrid wind power plant with retired EV batteries in energy and reserve markets under uncertainties. *Int. J. Electr. Power Energy Syst.* **2020**, *117*, 105631. [CrossRef]
43. Atieh, A.; Charfi, S.; Chaabene, M. Chapter 8—Hybrid PV/Batteries Bank/Diesel Generator Solar-Renewable Energy System Design, Energy Management, and Economics. In *Advances in Renewable Energies and Power Technologies*; Yahyaoui, I., Ed.; Elsevier: Amsterdam, The Netherlands, 2018; pp. 257–294. [CrossRef]
44. Luo, X.; Wang, J.; Dooner, M.; Clarke, J. Overview of current development in electrical energy storage technologies and the application potential in power system operation. *Appl. Energy* **2015**, *137*, 511–536. [CrossRef]
45. Han, X.; Ouyang, M.; Lu, L.; Li, J.; Zheng, Y.; Li, Z. A comparative study of commercial lithium ion battery cycle life in electrical vehicle: Aging mechanism identification. *J. Power Sources* **2014**, *251*, 38–54. [CrossRef]
46. Zhang, S.S. The effect of the charging protocol on the cycle life of a Li-ion battery. *J. Power Sources* **2006**, *161*, 1385–1391. [CrossRef]
47. Zhang, N.; Yang, N.C.; Liu, J.H. Optimal Sizing of PV/Wind/Battery Hybrid Microgrids Considering Lifetime of Battery Banks. *Energies* **2021**, *14*. [CrossRef]
48. Stroe, D.I.; Świerczyński, M.; Stan, A.I.; Teodorescu, R.; Andreasen, S.J. Accelerated Lifetime Testing Methodology for Lifetime Estimation of Lithium-Ion Batteries Used in Augmented Wind Power Plants. *IEEE Trans. Ind. Appl.* **2014**, *50*, 4006–4017. [CrossRef]
49. Bouchhima, N.; Gossen, M.; Schulte, S.; Birke, K.P. Lifetime of self-reconfigurable batteries compared with conventional batteries. *J. Energy Storage* **2018**, *15*, 400–407. [CrossRef]
50. Alramlawi, M.; Gabash, A.; Mohagheghi, E.; Li, P. Optimal operation of hybrid PV-battery system considering grid scheduled blackouts and battery lifetime. *Sol. Energy* **2018**, *161*, 125–137. [CrossRef]
51. Hou, Z.; Zhang, L.; Chen, J.; Xiong, Y.; Zhang, X.; Qian, Y. An aqueous rechargeable lithium ion battery with long cycle life and overcharge self-protection. *Mater. Chem. Front.* **2021**, *5*, 2749–2757. [CrossRef]
52. Chen, H.; Cong, T.N.; Yang, W.; Tan, C.; Li, Y.; Ding, Y. Progress in electrical energy storage system: A critical review. *Prog. Nat. Sci.* **2009**, *19*, 291–312. [CrossRef]
53. Xiong, R.; Chen, H.; Wang, C.; Sun, F. Towards a smarter hybrid energy storage system based on battery and ultracapacitor—A critical review on topology and energy management. *J. Clean. Prod.* **2018**, *202*, 1228–1240. [CrossRef]
54. *Wind Energy and On-Site Energy Storage. Exploring Market Opportunities*; Technical Report; Wind Europe: Brussels, Belgium, 2017.

energies

Article

Study on the Impact of Clean Power Investment on Regional High-Quality Economic Development in China

Xianrong Zhuang and Lingying Pan *

Business School, University of Shanghai for Science and Technology, Shanghai 200093, China
* Correspondence: panly@usst.edu.cn; Tel.: +86-18221483268

Abstract: In 2017, the 19th CPC National Congress proposed to "establish a sound economic system of green, low-carbon and circular development", which indicates the direction of high-quality economic development in the new era of China. Clean power investment is a powerful way to promote high-quality economic development by adopting non-fossil-energy utilization and low-emission technologies, as well as creating new jobs. Meanwhile, large-scale investment and a long investment return period result in negative effects on local economies. To better understand the effect of clean power investment, this paper selects panel data of thirty provinces in China from 2010 to 2019 to establish a spatial Durbin model to explore the impact of clean power investment on regional high-quality economic development. The results show that inter-regional high-quality economic development shows significant spatial auto-correlation characteristics. Clean power investment has not only a positive direct effect on high-quality economic development but also generates positive spatial spillover effects. Human capital, degree of government intervention, and urbanization rate have positive effects on regional high-quality economic development, while they play a suppressed role on neighboring regions.

Keywords: clean power investment; high-quality development; spatial Durbin model

Citation: Zhuang, X.; Pan, L. Study on the Impact of Clean Power Investment on Regional High-Quality Economic Development in China. *Energies* **2022**, *15*, 8364. https://doi.org/10.3390/en15228364

Academic Editors: Fuqiang Wang, Chao Shen, Dong Li and Zhonghao Rao

Received: 12 October 2022
Accepted: 4 November 2022
Published: 9 November 2022

1. Introduction

In September 2020, President Xi Jinping made a solemn commitment at the United Nations General Assembly that China will strive to peak CO_2 emissions by 2030 and achieve carbon neutrality by 2060, reflecting China's determination and confidence in restructuring the relationship between the environment and the economy [1]. It is necessary to improve energy technology and optimize energy structure to achieve carbon peaking and carbon neutrality goals [2]. While making efficient use of coal for cleaning, we should vigorously develop clean energy, promote the optimal combination of traditional fossil and clean energy, build a diversified energy production and consumption structure, and constantly improve the green development ability of energy. Clean energy development and utilization can significantly reduce fossil energy consumption, increase the proportion of clean energy consumption, and change the energy supply structure in which fossil energy accounts for the vast majority. With fossil energy, clean energy can reduce greenhouse gas emissions, reduce the emissions of various pollutants such as industrial smoke and solid waste, and play a significant and positive role in the environment and society. Thus, the state has issued several policies to encourage governments at all levels, large state-owned enterprises, small and medium-sized enterprises, and private individuals to vigorously develop clean energy industries, and has achieved remarkable results. From the perspective of power generation structure, according to the data from *China Electricity Statistics Yearbook*, clean energy power generation accounts for about 34.5% in 2021 (mainly including hydropower, wind power, photovoltaic, biomass power generation, and nuclear power), among which hydropower accounts for 16.00% of the national power generation, wind power accounts for 7.83%, and solar power accounts for 3.9% (see Figure 1). The "Fourteenth Five-Year Plan for Renewable

Energy Development" released in 2022 pointed out the direction of renewable development: optimizing the development mode and development of renewable energy on a large scale; promoting storage and consumption, and use of renewable energy in a high proportion; adhering to innovation drive and developing renewable energy with high quality.

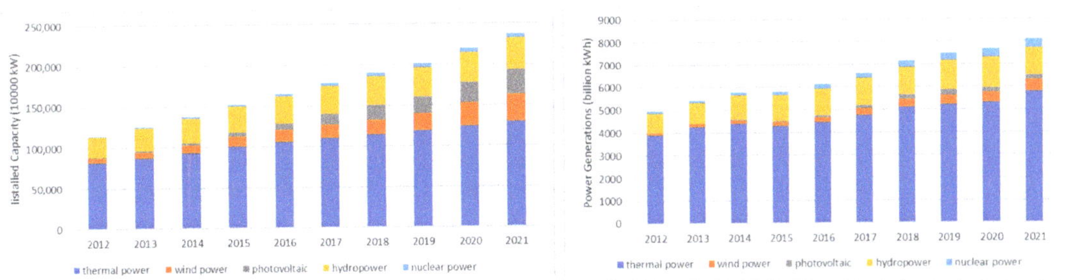

(a) (b)

Figure 1. Installed power capacity and power generation in China (2012–2021). (**a**) The installed capacity of thermal power, wind power, photovoltaic, hydropower and nuclear power by years from 2012 to 2021. (**b**) The power generation from 2012 to 2021.

Green development is an essential part of high-quality economic development in the new era which requires a lot of clean power investment. All kinds of government enterprises' investment in the clean power industry will bring the upgrading of equipment, technical personnel, technology, management, and product. Clean power investment has a considerable impact on the adjustment, transformation, upgrading, and optimization of national industrial structure and guiding green consumption, and it provides an impetus for accelerating and leading high-quality economic development. Therefore, based on sorting out the previous literature, clarifying the profound connotations, interactions, mechanisms, and effects of clean power investment and high-quality economic development, this paper puts forward the countermeasures of clean power investment leading to high-quality economic development, which has vital practical significance for realizing high-quality economic development.

2. Literature Review

Some studies have examined clean power investment and its impact on high-quality economic development. The main research directions are as follows: first, the definition, quantification, and influencing factors of high-quality economic development; second, the economic and social effects of clean power investment; third, some scholars have explored the impact path and empirical research of clean power investment on high-quality economic development.

2.1. Definition, Quantification, and Influencing Factors of High-Quality Economic Development

High-quality economic development was put forward for the first time by President Xi Jinping at the 19th Party Congress, which means that China's economy will pay more attention to balance, effectiveness, sustainability, and sharing, thus triggering an upsurge in studying high-quality economic development [3].

High-quality development is the combination of effectiveness, sufficiency, coordination, innovation, sharing, and stability of economic development with improving total factor productivity and realizing economic endogenous, ecological, and sustainable development [4–6]. Yu et al. (2018) summarized the reality and predicament of China's high-quality economic development from the four dimensions of industry development, innovation, openness, and people's lives [7]. It is the advanced state of economic development quality and the upgraded version of China's economic development. The measurement of high-quality economic development needs to establish a relatively complete index system to measure the level of high-quality economic development Scholars put forward a series of index systems according to their respective understanding of the connotation of high-quality economic development. Wang (2020) calculated the total factor productivity of 108 cities in the Yangtze River Economic Belt by the Malmquist-Luenberger index and Dagum Gini coefficient method, and analyzed the temporal and spatial evolution law and regional gap of economic quality, efficiency change, and technological change. It was concluded that technological progress is the power source of economic quality optimization and promotion [8]. Yu (2022) defines high-quality development in terms of technological innovation, system reform, industrial transformation, upgrading, and development efficiency. The research shows that all the above factors have spatial spillover effects, and technological innovation, system reform, transformation, and upgrading have significant direct and indirect effects on high-quality economic development [9]. Li et al. (2022) constructed 12 secondary and 22 tertiary evaluation indexes from five aspects: economic growth rate, economic growth stability, innovation efficiency, sharing of economic development achievements, and green development, and formed the evaluation index system of high-quality economic development level in China [6]. Wei (2021) subdivided the meaning of high-quality development into economic development, innovation efficiency, environmental impact, ecological services, and people's livelihood. In addition, the direct effect of technological innovation, transformation, and upgrading is stronger than the indirect effect [10]. According to former literature, although the specific indicators chosen by each scholar to measure the high-quality development level of the economy are different, they can be generally summarized into five basic dimensions: "innovation, coordination, green, openness, and sharing". Li (2019) and Wei (2018) both measure the high-quality development level of the local economy from these five basic dimensions [11,12]. There are plenty of factors that affect high-quality economic development, and many scholars have studied it from the perspectives of the digital economy [13,14], environmental regulation [15], foreign direct investment [16], and green innovation [17].

2.2. *Research on the Economic and Social Effects of Clean Power Investment*

According to the definition of the Energy Law of the People's Republic of China (draft) (2020), clean energy is defined as energy with zero or low emissions of environmental pollutants and greenhouse gases such as carbon dioxide in the process of development, utilization, and use, mainly including hydro-energy, nuclear energy, wind energy, solar energy, biomass energy, geothermal energy, hydrogen energy, and ocean energy. As far as "clean power investment" is concerned, there is no consistent definition in existing studies. Man et al. (2013) pointed out that energy investment meets the following conditions: "In order to obtain future benefits, investors invest in fixed assets in the fields of energy production, circulation and consumption" [18]. Considering the central role of enterprises in social production and economic growth, most of the research focuses on the perspective of enterprises. Specifically, Xu and Bai (2009) defined renewable energy investment as "enterprise investment in the field of renewable energy" [19]. Zhang et al. (2015) defined it as "cash paid by renewable energy enterprises for the construction of fixed assets, intangible assets and other long-term assets" [20]. These definitions are different in the subject and object of renewable energy investment. Considering that the role of government and enterprises in clean power investment cannot be ignored, this paper defines clean power investment as the sum of government investment and investment of listed clean power companies.

Scholars focus on the role that clean power investment plays in economic development. Green technology investment can have long-term or short-term positive effects on economic development [21]. It is suggested that the government should increase investment in clean energy [22]. The green investment will promote not only clean energy consumption but also economic growth [23]. Clean power development can boost the local economy by creating employment opportunities, increasing taxes, and reducing energy costs [24]. However, some scholars have come to the opposite conclusion through research. Green investment and renewable energy policies are not conducive to improving the sustainability of enterprises in highly market-oriented areas, especially electronic companies, labour-intensive companies, and companies with low equity concentration. Clean power investment will improve environmental sustainability at the expense of economic growth [25].

Besides the economic effect, clean power investment also has a very positive effect on the social dimension. In terms of improving the environment, clean power investment will inhibit carbon emissions [26] and greenhouse gas emissions, helping to improve environmental quality [27,28]. In the early stage of clean power investment, carbon emissions will increase. In the medium term, clean power investment will begin to play a role in emission reduction. In the later period, clean power investment may lead to an increase in carbon emissions again [29].

The process of replacing fossil fuels with clean energy can increase employment opportunities for many unemployed people. In 2018, about 15% of all jobs created globally came from the clean energy sector, which will create 84 million jobs worldwide by 2050 [30,31]. China's effective clean power investment and green projects will help the Belt and Road Initiative's neighboring countries optimize energy structure, enhance energy security, promote global low-carbon construction, and have a far-reaching impact on the human long-term living environment [32,33]. The results show that, with traditional fossil energy, clean energy power generation will create more jobs, and the global energy transformation will have a comprehensive positive impact on the future stability and growth of the world economy [34,35]. A large number of research results show that clean power investment not only has a positive effect on the environment, but also plays a positive role in creating jobs, increasing taxes, reducing energy costs, optimizing energy structure, and other social effects.

However, there are risks in clean power investment, including economic and commercial risks, market, policy, technological, environmental, and social risks [36]. Organizational efficiency and cost efficiency are the most important factors in clean power investment projects. On the other hand, moderate risks should be considered in these projects. To reduce risks, necessary measures should be taken to improve communication and cost control among departments [37].

2.3. Clean Power Investment on High-Quality Economic Development

At present, there are few articles about the impact of clean power investment on high-quality economic development. The existing literature mainly studies the relationship between green finance, green credit, and green investment and high-quality economic development, and few studies focus on the factor of clean power investment. Minna et al. (2018) pointed out that the combination of green investment and innovation will have a positive impact on high-quality economic development [38]. Ana et al. (2018) takes the incident of insufficient power supply in sub-Saharan Africa as an example, and analyzes that green investment has a positive impact on high-quality economic development [39]. Zhang (2020), based on Durbin's model test, found that green investment can promote the high-quality economic development in the process of marketization, and the effect is remarkable [40]. Shahani et al. (2020) refers to the European Green Investment Agreement of EUR 1 billion published by the European Union, and comes to the conclusion that green investment has a positive effect on high-quality economic development [41]. Chen et al. (2021) studied the relationship between green financial investment, industrial structure upgrading, and high-quality economic development, and found that both green financial investment and industrial structure upgrading are the direct driving forces of high-quality economic development, and that industrial structure upgrading has some intermediary effects between green financial investment and high-quality economic development [42].

Former literature has provided abundant information and help for the research of this paper. Scholars' research focuses on the measurement and influencing factors of high-quality economic development and the economic and social effects of clean power investment, but seldom refine the research object into clean power investment and bring it into a unified analysis structure to analyze the relationship between it and high-quality economic development. Therefore, this paper comprehensively evaluates the high-quality economic development level of each province with entropy TOPSIS (Technique for Order Preference by Similarity to an Ideal Solution) method from the five dimensions of innovation, coordination, greenness, openness, and sharing, and further empirically analyzes the impact of clean power investment on high-quality economic development with the spatial Durbin model.

3. Impacting Mechanism

At the Fifth Plenary Session of the 18th Central Committee of the Communist Party of China, President Xi Jinping systematically discussed the five development concepts of innovation, coordination, greenness, openness, and sharing. This paper intends to subdivide high-quality economic development into five development concepts. This chapter will explain how clean power investment acts on the five development concepts to promote high-quality economic development.

3.1. Clean Power Investment on Innovative Development

Innovative development is the primary driving force for high-quality development, and science and technology are the primary productive forces. The development of the modern world economy is more and more dependent on innovation. The achievements of scientific and technological innovation act on various factors such as production, sales, and management, and exert a "multiplier effect" to promote high-quality economic development. Clean power-related technologies are relatively new, challenging, and promising, with the possibility of overtaking on corners. Currently, China's innovative achievements are mainly concentrated at the level of enterprise and university. Investment in the clean power industry can inject a large number of funds into clean power companies and enhance their vitality. To enhance competitiveness, clean power companies will use the funds for technology research and development to attract high-caliber talents, and produce more energy-saving and low-consumption products. The scientific and technological innovation results generated by clean power investment can provide cleaner and greener production for further production and provide a driving force for high-quality economic development. This paper selects three indicators: R&D investment intensity, R&D personnel equivalent, and the number of domestic invention patents granted per 10,000 people to measure the level of innovative development in each province.

3.2. Clean Power Investment on Coordinated Development

Coordinated development is the endogenous characteristic of high-quality economic development. China is striving to build a situation of industrial coordination, regional coordination, and urban-rural coordination. The demand for clean and low-carbon energy structure with government enterprises will guide the funds invested in clean energy from high-pollution and energy-consuming industries to more environmentally friendly and energy-saving industries. It will increase the scale and business scope of environmental protection industries, drive the rationalization and upgrading of the whole industrial structure, promote the integration and development of advanced manufacturing industries related to clean power technologies with the digital economy and real economy, and promote high-quality economic development. Due to different natural endowments, western regions such as Qinghai, Gansu, and Xinjiang are rich in wind energy and solar energy. Compared with the middle and eastern regions, they are more suitable for developing clean power industries, such as wind energy and solar energy. Most of the investment in clean energy will flow into underdeveloped regions such as the western region to build more power bases, thus increasing local employment, creating tax revenue, and narrowing the economic gap between undeveloped and developed regions. Clean power investment not only plays a role in power supply construction, but also in infrastructure construction such as power grid construction, building a modern clean energy system in rural areas, meeting the needs of agricultural water conservancy machinery and farming equipment, narrowing the gap between urban and rural areas, and improving the lives of rural residents. Therefore, this paper chooses the income gap between urban and rural areas, the rationalization of industrial structure, and the upgrading of industrial structure as indicators to measure coordinated development.

3.3. Clean Power Investment on Green Development

Green development is a common form of high-quality economic development. Decades of rapid economic development and the accumulation of low-carbon technologies can restore the destroyed ecological environment, and people's demand for green life is becoming stronger and stronger. The process of clean power investment will form brand-new green productivity. The production process shows the characteristics of clean production, such as saving energy and raw materials, and producing less waste. The produced products are recyclable and low in pollution, and environmental protection is carried out at the same time of production. The increase of clean power investment can significantly reduce the consumption of high-emission fossil energy such as coal combined with improving the energy structure, reducing the discharge of all kinds of industrial wastes, and improving the living environment and the air quality. Therefore, this paper selects five indicators: per capita industrial waste gas emissions, per capita coal consumption, forest coverage, per capita solid waste production, and the ratio of total environmental pollution control to GDP to comprehensively measure the green development level.

3.4. Clean Power Investment on Open Development

Open development is the necessary way for an economy to develop with high quality. To achieve high-quality economic development, domestic enterprises should not only actively go abroad but also vigorously introduce outstanding enterprises from other countries to supplement the problems of insufficient domestic investment and low investment quality and give full play to their catfish effect. It is also necessary to actively develop foreign investment, give full play to comparative advantages, expand markets, and accumulate experience in developing regions such as Asia, Africa, and Latin America, which can not only increase employment, expand labor export, and transfer some backward production capacity, but also enhance the value of domestic brands, expand national influence, and lay a solid foundation for future product export. Concerning clean power and low-carbon technologies, Western developed countries started earlier and domestic related industries caught up and realized the advantage of backwardness. At present, domestic companies in clean power fields such as photovoltaic, large hydropower station construction, nuclear power station, offshore wind energy, etc.. have a technical level no less than that of foreign countries. More and more Chinese clean power companies set up factories and build large clean power projects abroad. Therefore, this paper selects the ratio of total import and export of goods to GDP and the ratio of foreign direct investment to GDP as two indicators to measure the level of open development.

3.5. Clean Power Investment on Shared Development

Shared development is the fundamental purpose of high-quality development. The achievements of economic development should be shared by all people and the life quality of all people should be improved. China is establishing a public service system integrating urban and rural areas, speeding up the construction of water, electricity, gas, network, and other infrastructure in rural areas, and moderately tilting to rural areas in education, medical care, old-age care, culture, sports, etc. Clean power investment is a part of social public service investment and is concentrated in underdeveloped areas such as the west, which can effectively narrow the difference between urban and rural public facilities. In this paper, the loop length of 35 KV and above transmission lines and the ratio of social service expenditure to GDP are selected as two indicators to measure shared development level.

Through the research on the mechanism of clean power investment in five aspects of high-quality economic development, we know that clean power investment plays a positive role in five aspects: innovation, coordination, greenness, openness, and sharing The present situation of high-quality economic development and clean power investment will be further introduced below.

4. Materials and Methods

As mentioned in the third chapter, this paper plans to build a comprehensive index system based on the five development concepts to measure the high-quality economic development level of 30 provinces (excluding Tibet, Hong Kong, Macao, and Taiwan). This paper chooses the Entropy Weight Method to calculate the weights of specific indicators. Its basic idea is to determine the objective weight according to the variation degree of indexes. Specifically, the smaller the information entropy of an index, the greater the variation of the index, the more information it provides, the greater its role in the comprehensive evaluation system, and the greater its corresponding weight. The specific calculation steps are as follows:

Firstly, we should standardize the original data. The corresponding data dimensions of each index layer are different. In order to make each index comparable, the data should be standardized first, and the calculation formula is as follows:

$$U_{\mathrm{ij}} = \frac{x_{ij} - \mathrm{min}x_{ij}}{\mathrm{max}_{ij} - \mathrm{min}x_{ij}}(i = 1, \ldots, n; j = 1, \ldots, m) \tag{1}$$

Secondly, we can calculate the proportion of x_{ij}:

$$U'_{ij} = \frac{U_{ij}}{\sum_{i=1}^{n} U_{ij}} \tag{2}$$

Thirdly, we calculate information entropy, and the formula is as follows:

$$E_j = -\ln(n)^{-1}\sum_{i=1}^{n} U'_{ij}\ln(U'_{ij}) \tag{3}$$

Finally, we calculate the corresponding weight of each indicator according to the index weight formula:

$$W_j = \frac{1 - E_j}{\sum_{j=1}^{m}(1 - E_j)} \tag{4}$$

The final results and positive and negative attributes (the positive attribute indicate that the higher the indicator value, the better, while the negative attribute is the opposite) are shown in Table 1:

Table 1. Measurements of high-quality economic development.

The First Layer	The Second Layer	Positive or Negative	Weights
Innovative development	R&D investment intensity	+	0.066
	Domestic invention patents granted per 10,000 people	+	0.194
	R&D personnel full-time equivalents	+	0.119
Coordinated development	Urban-rural income gap	−	0.031
	Rationalization of industrial structure	+	0.059
	Premiumization of industrial structure	+	0.068
Green development	Industrial waste gas emissions per capita	−	0.030
	Coal consumption per capita	−	0.021
	Forest coverage	+	0.055
	Solid waste generation per capita	−	0.022
	Total environmental pollution control to GDP ratio	+	0.047
Open development	Ratio of total exports and imports of goods to GDP	+	0.120
	FDI to GDP ratio	+	0.067
Shared development	Loop length of 35 KV and above transmission lines	+	0.047
	Ratio of social service expenditure to GDP	+	0.056

The weight calculated according to the entropy weight method is further passed through the TOPSIS method to get the scores of high-quality economic development.

Firstly, we calculate the weighting matrix of evaluation indicator Z_{ij}:

$$Z_{ij} = (z_{ij})_{n*m} = W_j * x_{ij} \tag{5}$$

Then, we find the best ideal solution A_j^+ and the worst ideal solution A_j^-:

$$A_j^+ = (\max z_1, \ldots, \max z_m) \tag{6}$$

$$A_j^- = (\min z_1, \ldots, \min z_m) \tag{7}$$

Next, we calculate the distance to the best and worst ideal solution to each evaluation object, and the calculation formula is as follows:

$$D_i^+ = \sqrt{\sum_{j=1}^{m} \left(A_j^+ - z_{ij}\right)^2} \tag{8}$$

$$D_i^- = \sqrt{\sum_{j=1}^{m} \left(A_j^- - z_{ij}\right)^2} \tag{9}$$

Lastly, we can calculate the score of each object:

$$C_i = \frac{D_i^-}{D_i^+ + D_i^-} \tag{10}$$

After the above calculations, we get the scores of high-quality economic development level of each province from 2010 to 2019. Due to the space limitation, only odd-numbered year results and ten-year average scores are displayed. The specific results are shown in Table 2.

Table 2. Score of economic high-quality development by province.

Region	2011	2013	2015	2017	2019	Average
Beijing	0.478	0.485	0.517	0.552	0.543	0.506
Tianjin	0.320	0.335	0.349	0.289	0.283	0.313
Hebei	0.248	0.240	0.248	0.266	0.273	0.252
Shanxi	0.207	0.220	0.238	0.209	0.224	0.220
Inner Mongolia	0.236	0.242	0.247	0.251	0.241	0.240
Liaoning	0.311	0.303	0.237	0.257	0.254	0.278
Jilin	0.251	0.245	0.240	0.243	0.248	0.244
Heilongjiang	0.280	0.270	0.266	0.263	0.251	0.264
Shanghai	0.379	0.369	0.378	0.377	0.381	0.376
Jiangsu	0.334	0.346	0.352	0.357	0.372	0.349
Zhejiang	0.328	0.339	0.360	0.362	0.386	0.352
Anhui	0.259	0.285	0.288	0.287	0.282	0.276
Fujian	0.301	0.303	0.300	0.298	0.299	0.298
Jiangxi	0.297	0.290	0.297	0.298	0.299	0.295
Shandong	0.269	0.284	0.291	0.300	0.301	0.285
Henan	0.240	0.249	0.256	0.262	0.268	0.251
Hubei	0.274	0.275	0.282	0.279	0.282	0.276
Hunan	0.268	0.276	0.283	0.288	0.297	0.280
Guangdong	0.385	0.398	0.390	0.391	0.432	0.399
Guangxi	0.273	0.287	0.292	0.289	0.291	0.283
Hainan	0.293	0.289	0.293	0.284	0.285	0.288
Chongqing	0.282	0.289	0.253	0.253	0.259	0.264
Sichuan	0.280	0.287	0.290	0.286	0.290	0.285
Guizhou	0.291	0.281	0.262	0.252	0.251	0.266
Yunnan	0.272	0.273	0.277	0.259	0.244	0.265
Shaanxi	0.245	0.307	0.282	0.280	0.296	0.281
Gansu	0.277	0.291	0.279	0.271	0.252	0.272
Qinghai	0.242	0.214	0.218	0.228	0.199	0.217
Ningxia	0.216	0.216	0.234	0.214	0.178	0.209
Xinjiang	0.240	0.260	0.265	0.260	0.217	0.245

From Table 2, it is obvious that Beijing has a distinctive leading edge in high-quality economic development, ranking first for ten consecutive years in China. The dominant position of Beijing is mainly due to its innovation, environmental protection, and optimization of industrial structure [43–45]. Innovation is the primary power source for Beijing's high-quality economic development. In 2019, the investment intensity of R&D funds in Beijing was 6.3%, which has a good lead over 4.01% of the second place in Shanghai. Thanks to numerous well-known universities and research institutions, the number of invention patents granted per 10,000 people in Beijing increased from 5.71 in 2010 to 24.26 in 2019, 4.24 times that of a decade ago, further expanding Beijing's dominant position in innovation. Beijing is the capital of China, which has high requirements for ecological environment, and some enterprises with high energy consumption and high pollution have moved to other places, so its green development level is high. Beijing's tertiary industry accounted for 83.52% of GDP, and the ratio of the tertiary industry to the secondary industry was 5.17 in 2019, ranking both first in China, with the most optimized industrial structure. Other provinces can increase investment in innovation, strengthen environmental protection, and promote the upgrading of industrial structure from the experience of Beijing.

According to the division of China's Ministry of Finance, this paper divides 30 provincial administrative regions into three regions: East, Middle, and West. The eastern region includes Beijing, Tianjin, Hebei, Liaoning, Shanghai, Jiangsu, Zhejiang, Fujian, Shandong, Guangdong, and Hainan, and the middle region includes Shanxi, Jilin, Heilongjiang, Anhui,

Jiangxi, Henan, Hubei, and Hunan, while the western region includes Inner Mongolia, Guangxi, Chongqing, Sichuan, Guizhou, Yunnan, Shaanxi, Gansu, Qinghai, Ningxia, and Xinjiang. And the scores of high-quality economic development in the three major regions of China are shown in Figure 2:

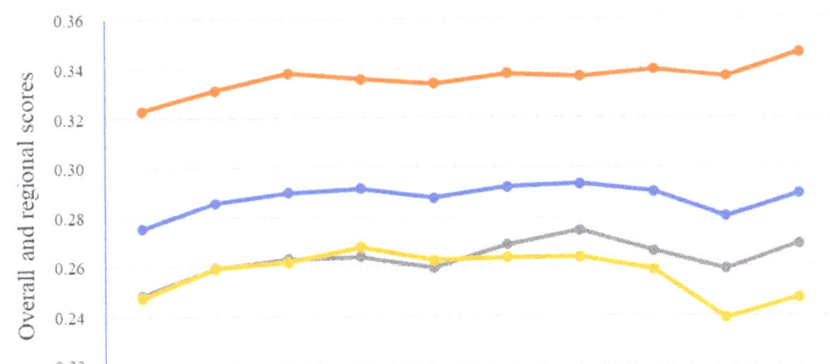

	2010	2011	2012	2013	2014	2015	2016	2017	2018	2019
all	0.275	0.286	0.290	0.292	0.288	0.292	0.293	0.290	0.280	0.289
east	0.323	0.331	0.338	0.336	0.334	0.338	0.337	0.339	0.337	0.346
middle	0.249	0.260	0.263	0.264	0.259	0.269	0.274	0.266	0.259	0.269
west	0.247	0.259	0.262	0.268	0.262	0.264	0.264	0.259	0.239	0.247

Figure 2. Scores of regional high-quality economic development from 2010 to 2019.

By observing the data in Table 2 and Figure 2, it can be seen that the level of high-quality economic development in the eastern region during the decade from 2010 to 2019 was significantly higher than the national average and the middle and western regions. The score of the eastern region increased from 0.323 in 2010 to 0.346 in 2019, with an increase of 7.2%. In the eastern part of China, Beijing scored more than 0.5, ranking first among 30 provinces and cities, and Jiangsu, Zhejiang, Shanghai, and Guangzhou also performed very well. The eastern region was the first to implement the policy of reform and opening-up; the central government's introduction of a series of policies and measures is conducive to the development of the eastern coastal areas and towns. After decades of reform and opening-up, the eastern region has a high degree of marketization. Additionally, private enterprises which have the most innovations in the eastern region are growing vigorously, so it has developed industrial industries, talent reserves, education and medical care, advanced management experience, and technology.

From 2010 to 2014, the level of high-quality economic development in the middle and western regions was almost similar, rising slowly; from 2015 to 2019, the gap between the middle region and the western region began to widen. During this period, the middle region developed rapidly, and the level of high-quality economic development improved, while the level of high-quality economic development in the western region even declined slightly. In 2010, the average score of the eastern region was 0.323, which was higher than the national average of 0.275, the western average of 0.247, and the middle average of 0.249. This shows that the eastern region has always been ahead of other regions, not only at an economic level but also at the national leading level in the five major developments [46–49]. As is clear from Figure 2, after several years of steady improvement the level of high-quality economic development decreased significantly for the middle and western regions in the years from 2016–2018. The western region, since the adoption of the strategy of developing Western China in 1999, was developing rapidly with abundant natural resources and national policies. However, limited by resource endowment, economic foundation, weak private enterprises, and institutional environment, western high-quality economic development has stagnated or even declined. From 31 December 2015 to 4 February 2016, the central environmental protection inspector pilot was launched in Hebei. In the next two years, the action will have achieved full coverage across China. The policy of environmental regulation forced the transformation or closure of high-pollution and high-emission enterprises. Inevitably, this will have a short-term shock on the economy, especially in the west, whose economic structure is dominated by the energy industry [50]. In any case, the low level of industrial structure, lack of innovation motivation, low level of education, and high emission of pollutants have led to the decline of high-quality economic development from 2016 to 2018. As for the middle region, the strength of large population and low labor cost once brought advantages to the development of agriculture and heavy industry. However, after entering the new era, the labor cost is constantly increasing and the development space of labor-intensive industries is gradually shrinking. In addition, the middle region is rich in mineral resources such as coal, oil, iron, nonferrous metals, etc., which have brought advantages to the development of heavy industries, such as metal smelting, machinery, chemical industry, etc. However, high-quality development requires more environmental protection and energy efficiency; the importance of primary energy and raw materials in industrial development is decreasing, and the resource advantages of the middle region are weakening. Weak innovation ability, low level of urbanization, large income gap between urban and rural residents, poor level of openness, and a large number of rural poverty have all led to the fact that the high-quality economic development level of the middle region cannot be steadily improved.

In order to further improve the regional economic high-quality development level, the middle and western regions can implement the innovation-driven strategy vigorously by strengthening the capacity of scientific and technological innovation, intensifying the dominant position of enterprise innovation, and promoting the transformation of scientific and technological achievements. In addition, accelerating industrial transformation and upgrading is an efficient way to promote high-quality development. They can build characteristic industrial clusters with competitive advantages according to local conditions, and promote the common development of surrounding areas through the diffusion effect and correlation effect of leading cities, so as to solve the problem of uneven development in various parts of the region and enhance the coordination of regional development.

Developing clean power is the core of future energy investment. Clean technologies such as wind power and photovoltaic power generation are still the most economical power generation options in many countries. The focus of wind power is shifting to offshore. Last year, more than 20 GW of offshore wind power was put into production, with an investment of about USD 40 billion. China still ranks first, accounting for more than half of the newly added wind energy. It is worth mentioning that from 2019 to 2021, wind power investment soared due to the impact of the affordable access policy. China's clean power investment in recent years is shown in Figure 3.

Liu (2011) divided green investment into fixed assets investment, industrial pollution control investment, and financial environmental protection expenditure [51]. The renewable energy investment data selected by Chen (2020), Pan (2020), and Yang (2020) are the cash paid for the construction of fixed assets, intangible assets, and other long-term assets of renewable energy enterprises [52–54]; Zhu (2019) selected electric power companies listed on Shanghai Stock Exchange or Shenzhen Stock Exchange with clean power investment business to analyze their investment [55]. The author observes that the measurement of clean power investment by other scholars is mostly based on the cash paid by listed clean energy companies to build fixed assets, intangible assets, and other long-term assets, ignoring the investment of the government and large central enterprises in clean energy. Especially considering the huge investment in the early stage of clean power investment and the long recovery period, which requires the intervention of funds from the government and large power enterprises, the government and large power enterprises pay less attention to economic benefits and pay more attention to social responsibility and environmental benefits, so we cannot ignore the role of the government and large power enterprises in clean power investment. Therefore, based on previous studies, this paper measures the amount of clean power investment in each province with the cash paid by listed clean energy companies to build fixed assets, intangible assets, and other long-term assets, plus the amount of investment completed by the government and large power enterprises in clean energy power supply construction. The clean power investment of government and large-scale power enterprises is the sum of the completed investment in the construction of four clean energy sources: nuclear energy, wind energy, solar energy, and hydropower in various regions in *Almanac of China's Water Power*. The total data are shown in Table 3.

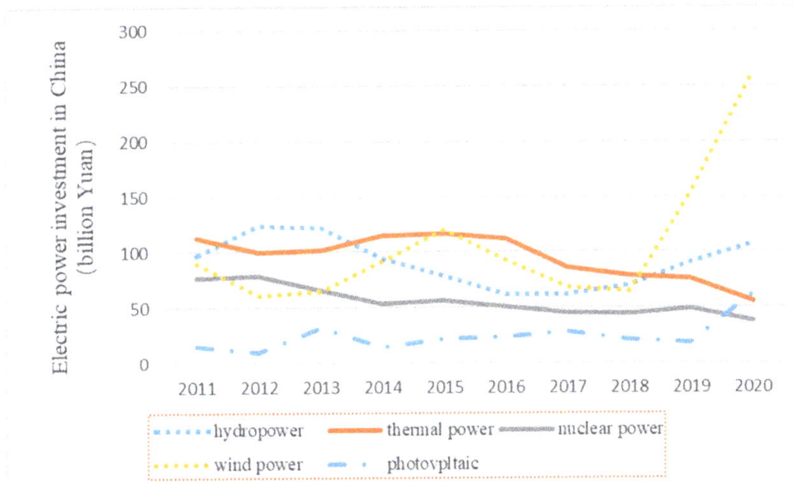

Figure 3. Electric power investment in China from 2011 to 2020.

In this paper, the total investment amount of clean power in each province is the sum of the investment amount of listed companies of clean power and the investment amount of clean power plant construction. Among them, clean power plant construction investment includes investment in hydropower, wind power, nuclear power, and solar power. Due to the vast territory, diverse geological conditions and uneven distribution of natural resources in China, different provinces develop clean power according to their own characteristics. Specifically, Xinjiang, Inner Mongolia, and other western regions, are rich in wind energy and solar energy, and a large amount of water resources in the southwest region are suitable for development and utilization to build large hydropower stations. China's nuclear power stations are all built in the eastern coastal areas because they require a large amount of water for cooling and safety considerations. Similarly, offshore wind energy can only be developed and utilized in provinces near the sea. Due to the uneven distribution of natural resources in different provinces, the investment in various types of clean power is extremely discrepant, resulting in extremely unstable data. Moreover, from the mechanism of clean power investments on high-quality economic development, different types of clean power investment do not significantly affect the development results, but rather, the investment scale is the important factor that affects this effect. Therefore, this paper sums up the investment of various clean power types and then carries out the follow-up measurement analysis.

Table 3. Total clean power investment by region (billion CNY).

Region	2010	2011	2012	2013	2014	2015	2016	2017	2018	2019	Average
Beijing	26.06	29.41	30.79	39.00	50.77	51.27	51.62	45.50	50.05	48.36	42.28
Tianjin	2.25	2.90	2.15	2.41	2.64	2.48	2.77	5.36	7.92	7.15	3.80
Hebei	11.70	10.04	4.45	4.61	7.15	9.85	11.99	14.07	12.40	15.25	10.15
Shanxi	4.83	8.14	5.46	7.53	11.06	12.42	16.00	12.94	6.87	11.29	9.65
Inner Mongolia	30.56	18.34	11.07	14.68	16.87	15.53	6.51	5.84	5.30	12.27	13.70
Liaoning	18.74	16.71	14.44	10.41	4.30	6.40	5.05	7.88	9.17	10.08	10.32
Jilin	10.34	4.29	3.03	3.61	4.59	7.94	6.70	5.56	8.11	7.43	6.16
Heilongjiang	5.31	4.38	3.40	3.06	2.92	1.80	2.30	2.12	1.75	2.18	2.92
Shanghai	11.04	8.87	8.43	8.65	11.43	10.52	10.72	16.51	21.24	20.56	12.80
Jiangsu	6.26	8.62	8.03	12.25	16.88	25.59	33.06	34.12	36.57	39.87	22.13
Zhejiang	15.46	15.52	11.55	9.11	10.45	11.23	8.55	13.84	13.42	10.70	11.98
Anhui	4.01	4.01	3.75	3.77	3.81	5.08	7.58	7.30	6.91	6.90	5.31
Fujian	16.37	17.99	19.81	16.75	11.87	12.80	13.18	14.65	15.40	19.94	15.88
Jiangxi	0.56	0.66	1.99	2.11	2.33	2.84	3.65	3.37	4.64	7.42	2.96
Shandong	27.65	27.87	29.26	8.47	26.12	24.11	30.05	27.69	26.87	25.44	25.35
Henan	1.19	1.42	0.63	0.40	1.90	2.23	3.06	6.29	5.42	10.69	3.32
Hubei	10.24	4.43	3.63	4.40	4.41	7.30	10.97	9.04	8.58	7.72	7.07
Hunan	5.09	6.01	6.19	5.14	4.37	3.68	4.26	5.45	4.40	6.25	5.08
Guangdong	29.51	43.65	38.62	15.03	34.48	32.04	37.05	38.24	30.76	44.27	34.37
Guangxi	3.84	6.50	7.83	22.39	5.84	6.58	7.77	5.68	8.40	11.15	8.60
Hainan	2.18	3.58	4.70	6.09	4.75	3.73	2.72	4.47	4.51	3.58	3.02
Chongqing	2.72	1.88	1.06	1.97	4.56	5.27	5.44	5.16	2.86	3.57	3.45
Sichuan	38.31	50.13	61.36	57.13	45.17	38.73	35.73	40.96	47.53	43.15	45.82
Guizhou	6.10	7.69	6.07	7.75	6.19	5.95	3.52	3.37	2.31	3.58	5.25
Yunnan	13.64	28.27	45.37	46.28	40.83	36.99	26.74	12.02	8.91	25.43	28.45
Shaanxi	1.74	1.58	2.54	3.64	2.66	7.51	9.93	8.93	7.82	10.82	5.72
Gansu	17.07	7.10	9.11	11.14	17.15	10.04	1.76	1.42	1.38	1.32	7.25
Qinghai	2.70	8.65	2.98	6.13	2.15	4.68	3.19	2.55	3.90	5.53	4.25
Ningxia	3.96	12.98	3.88	5.49	7.30	11.16	4.95	3.49	1.42	5.32	5.66
Xinjiang	10.05	13.58	15.94	31.99	15.83	31.60	21.97	9.89	14.61	24.11	18.96

It can be seen from Table 3 that clean power investment has a high degree of dispersion among provinces. For example, in 2019, the total amount of clean power investment in Sichuan Province reached CNY 43.15 billion, which was 32.6 times that of Gansu Province, which had the least amount of investment. This is because there is an objective gap between the innate resource endowment and the technical level of each province. Northwest China is sparsely populated and suitable for developing clean power, and most of the government's investment is concentrated in this area. Jiangsu, Zhejiang, and Shanghai regions have advanced clean energy technologies in China and can invest in large-scale power projects all over the country. These economically developed regions have a large number of clean energy superior companies, so they invest more in clean energy. Within each province, the investment amount in different years can also be nearly ten times different. For example, Gansu's investment reached its peak in 2010 and 2014, with an investment amount of more than CNY 17 billion, while after 2016, the annual investment amount was just over CNY one billion. This phenomenon is due to the huge investment in the early stage of the construction of large-scale power generation bases. Once the power generation bases are built and put into operation, the construction costs are greatly reduced, and only a small amount of operating costs are needed. Therefore, the investment in clean power in some provinces fluctuates greatly in different years.

It should be noted that the "investment" in this paper not only represents the investment of building clean energy power projects in a certain place, but also includes the clean energy investment of a certain place in other areas. It is particularly noteworthy that there is only a thimbleful of clean energy resources in Beijing. It can be seen that the investment in clean energy power construction in Beijing in the past ten years is only CNY 180 million, while the total investment in clean energy is as high as CNY 42.3 billion, most of which are the investments of large energy enterprises located in Beijing. Although Beijing has very few natural resources of clean energy, a number of large energy groups and listed companies are gathered there by virtue of its political, geographical, and educational advantages. These have sufficient capital, technology, talents, and management experience, which enables Beijing to make efficient investments in areas rich in clean energy resources across the country and make up for the disadvantages of resource-rich areas, such as lack of clean energy-related talents, insufficient investment funds in the early stage, outdated project construction management mode, and backward technology of clean energy development and utilization. R&D and updating of clean energy technology require a large amount of capital investment and scientific and technological talents, which is the most dominant aspect of Beijing. After long-term scientific research investment, a large number of cutting-edge technological updates will be generated, which will further reduce the project construction cost and improve the competitiveness of large energy companies. After completing the expensive construction in the early stage, the benefits of clean energy projects will last for decades, and a large part of the profits will belong to Beijing, where the company's headquarters are located. Because innovation and the competitive advantage will bring the leading position in the industry to the enterprise, the investment income of the enterprise will continue to invest in clean energy technology, which will play a positive feedback role.

In other words, clean energy investment projects also have a positive impact on resource-rich areas such as Xinjiang, Inner Mongolia, and Southwest China. The investment period of large hydropower stations, nuclear power plants, wind farms, and photovoltaic bases is about ten years, with an investment of tens of billions. A large number of construction workers and engineers are needed in the early stage of the project construction, and operators and maintenance workers are needed in the later stage. This has a great positive impact on attracting local employment workers, improving the living quality of local workers, and promoting consumption upgrading. The electricity produced by constructed clean energy fields can be used locally or exported to areas with power shortage such as the east, so as to raise local tax revenue and promote high-quality economic development. Governments in resource-rich areas can vigorously promote the development of large-scale local energy enterprises, absorb the successful experience of Beijing and Shanghai, which are relatively mature in finance and investment, formulate preferential policies to encourage local enterprises to raise funds from the whole country and even the whole world, intensify innovation, increase the attraction of relevant talents, and promote the coordinated development of upstream and downstream enterprises in clean energy industry. Resource-rich areas can rely on their own abundant energy to form their own advantageous industries, and leave the development fruits in the local area as much as possible, so as to promote the high-quality economic development with comprehensive development of clean energy industry.

5. Empirical Findings

In the third chapter, this paper systematically analyzes the mechanism of clean power investment on the high-quality development of the regional economy. In addition, in the fourth chapter, the high-quality economic development level and total clean power investment of 30 provinces from 2010 to 2019 are calculated, and the dynamic changes of high-quality economic development in the whole country and regions are analyzed. Based on the above mechanism analysis and calculation results, this chapter will empirically test and analyze the spatial effect of clean power investment on regional high-quality economic development.

5.1. Methodology and Data

Many factors affect the high-quality development of an economy. Published papers have analyzed the influencing factors from many angles. Referring to previous studies, this paper selects human capital, government intervention, and urbanization rate as control variables. Human capital (HC) is expressed by the number of years of education Human capital will stimulate local economic growth, and a high-quality labor force can produce a higher level of production. Therefore, it is expected that the promotion of human capital can effectively promote high-quality economic development. Government intervention (GI) is measured by the ratio of fiscal expenditure to GDP. On the one hand, the government's various intervention measures in the market economy can effectively alleviate market failure and promote high-quality economic development. On the other hand, the government's excessive intervention in the market economy may lead to the mismatch of market resources and hinder high-quality economic development. Therefore, the impact of government intervention on high-quality economic development cannot be directly determined. Urbanization rate (UR) is the ratio of the urban population to the total population. From the perspective of economic theory, cities have comparative advantages in improving regional productivity, employment rate, and space utilization efficiency, and urbanization is conducive to enhancing the regional carrying capacity of economic development advantages. The urbanization rate will promote high-quality economic development. On the other hand, urbanization development may lead to a widening income gap between urban and rural areas and increasing energy consumption, which will hurt high-quality economic development.

Based on the analysis of the mechanism of clean power investment on regional economic growth in the previous article, the following panel model is constructed as:

$$HQD_{it} = \alpha_0 + \alpha_1 \ln CEI_{it} + \alpha_2 HC_{it} + \alpha_3 GI_{it} + \alpha_4 UR_{it} + \epsilon_{it} \tag{11}$$

Among them, the explained variable HQD is the score of high-quality economic development level calculated by the entropy TOPSIS method. The explanatory variable CEI is shown in the above section. According to the observation data, the new energy investment in each province fluctuates greatly. To solve the heteroscedasticity problem, the CEI of clean power investment is logarithmic. The control variables are HC, GI, and UR, which respectively represent human capital, government intervention, and urbanization rate. In this equation, i represents region, t represents year, α_0 denotes intercept term, α_1 denotes regression coefficients of core explanatory variables, α_2, α_3 and α_4 represent regression coefficients of human capital, government intervention, and urbanization rate of control variables, respectively. ϵ_{it} represents random error.

In this paper, the panel data of 30 provincial administrative regions in China from 2010 to 2019 are used for empirical analysis. The main sources of data in this paper are China Statistical Yearbook, China Hydropower Yearbook, China Energy Statistical Yearbook, China Industrial Statistical Yearbook, provincial statistical yearbooks, etc. The relevant data are also compiled from official data. The missing data in the yearbook shall be filled by interpolation or median method. Descriptive statistics of related variables are shown in Table 4:

Table 4. Descriptive statistics of variables.

Variable	Observations	Mean	Standard Deviation	Minimum	Maximum
HQD	300	0.288	0.061	0.177	0.552
CEI	300	127.62	127.80	3.988	613.64
HC	300	9.088	0.929	6.764	12.782
GI	300	24.55	10.23	10.582	62.836
UR	300	57.73	12.607	33.81	89.6

5.2. Relevant Test

In this paper, the data of 30 provincial administrative regions in the past ten years are selected, where n is 30 and t is 10. They are short panel data, so there is no need for a stationarity test. The results of the fixed effect model and random effect model were obtained by the Hausman test. The test results show that the p value is 0, so the fixed effect model is selected for empirical analysis.

Table 5 shows the fixed effect regression results based on the panel model, which shows the impact of China's overall clean power investment on the high-quality development of the regional economy. It can be seen that all variables are significant, but the regression coefficients are small. First, there is a huge difference in clean power investment in different provincial regions and the level of high-quality development is quite different, so the overall correlation degree is not high. Second, China's energy pattern is unbalanced in distribution and development. The east is a large energy consumer, and the west has a large production capacity. This leads to the fact that clean power investment is mostly concentrated in areas with abundant wind energy and solar energy resources in the west, and the produced electric energy is transported to the east through "West-to-East Power Transmission", which further promotes the high-quality development of the more developed areas in the east and widens the high-quality development gap between the east and the west.

Table 5. Regression results of non-spatial fixed effect model.

Explaining Variables	Explained Variable HQD	
	Coefficient	t-Value
lnCEI	0.007 **	2.64
HC	0.014 *	2.03
GI	0.001 **	2.18
UR	−0.0009 *	−1.71
_cons	0.156 ***	2.90
N	300	

Note: ***, ** and * represent passing the significance test at 1%, 5%, and 10% levels, respectively.

5.3. Spatial Model Setting

The spatial weight matrix represents the spatial distance of each region and indirectly reflects the degree of interdependence among regions, which is the premise of spatial econometric analysis. According to the characteristics of the research content, this paper constructs the economic geographical distance matrix and then makes a comparative analysis of the spatial effect between clean power investment and the high-quality development of a regional economy. A spatial weight matrix is a matrix that reflects the dependence of individuals in space. The general form of the matrix is:

$$W = \begin{pmatrix} w_{11}\ w_{12} & \cdots & w_{1n} \\ \vdots & \ddots & \vdots \\ w_{n1}\ w_{n2} & \cdots & w_{nn} \end{pmatrix}$$

The spatial matrix depicts the spatial dependence among N individuals, in which w_{ij} indicates the degree of influence of individual i on j. Considering that a large part of the high-quality development of the regional economy is related to economic development, this paper takes the economic geographical distance matrix, a spatial matrix that considers both geographical factors and economic factors, to analyze the research object of this paper at the spatial level. As the degree of convenience of communication is more and more inconvenient, this paper takes the reciprocal of the square of the economic distance between two places as the weight of geographical distance, and then combines it with economic distance, in which the economic distance parameter is set as the reciprocal of the difference of GDP per capita in ten years from 2010 to 2019; d_{ij} is the distance between the two provinces, calculated by the longitude and latitude of the two provinces. The matrix has been standardized, and the form of the economic and geographical distance matrix is as follows:

$$w_{ij} = \begin{cases} \frac{1}{|\overline{Y_i} - \overline{Y_j}|} \times \frac{1}{d_{ij}^2} & i \neq j \\ 0 & i = j \end{cases} \tag{12}$$

To explore the possible spatial spillover effect of clean power investment on high-quality economic development, this paper uses the spatial Durbin model to test, and the specific formula is as follows:

$$\begin{aligned} HQD_{it} &= \rho WHQD_{it} + \alpha_0 + \alpha_1 lnCEI_{it} + \alpha_2 HC_{it} + \alpha_3 GI_{it} + \alpha_4 UR_{it} \\ &+ \lambda_1 WlnCEI_{it} + \lambda_2 WHC_{it} + \lambda_3 WGI_{it} + \lambda_4 WUR_{it} + \mu_i + \varphi_i + \varepsilon_{it} \end{aligned} \quad (13)$$

Among them, W is the economic geographical distance matrix, ρ is the spatial autoregressive coefficient, α_0 represents the intercept term that does not change with the individual,α_i denotes the estimated coefficient of each explanatory variable, λ_i is the spatial interaction coefficient of the explained variable, μ_i denotes the individual effect, φ_i denotes the time effect, ε_{it} denotes the random error term.

5.4. Empirical Test

5.4.1. Analysis of Spatial Correlation

Use the global Moran index to judge whether there is a spatial correlation between the high-quality economic development level of 30 provinces in China. Moran index is calculated as follows:

$$I = \frac{\sum_{i=1}^{n}\sum_{j=1}^{n} w_{ij}(x_i - \bar{x})(x_j - \bar{x})}{S^2 \sum_{i=1}^{n}\sum_{j=1}^{n} w_{ij}} \quad (14)$$

A negative value of the Moran index indicates negative spatial autocorrelation, a positive value indicates positive spatial autocorrelation, and a value equal to 0 indicates no spatial autocorrelation. The specific calculation results are shown in Table 6:

Table 6. Results of global spatial correlation of high-quality economic development.

Year	Moran's Index	Year	Moran's Index
2010	0.374 ***	2015	0.486 ***
2011	0.388 ***	2016	0.305 ***
2012	0.395 ***	2017	0.232 ***
2013	0.420 ***	2018	0.232 ***
2014	0.469 ***	2019	0.246 ***

Note: *** represents passing the significance test at 1% level.

It can be seen that the Moran's index in the past ten years is greater than 0.2 and is significantly positive at a 1% significance level, indicating that high-quality economic development has a positive spatial correlation and presents the characteristics of spatial concentration distribution. Further observation shows that the Moran index is at a high level from 2010 to 2015, which indicates that the high-quality spatial dependence of regional economy in this period is strong. After 2016, the Moran index shows a downward trend, and the spatial dependence weakened. At the end of 2013, the Third Plenary Session of the 18th CPC Central Committee determined the general policy of comprehensively deepening reform. After two or three years of difficult reform, the results began to appear. The coordinated development of urban and rural areas has taken on a new look. New strategic measures for developing the western region, revitalizing the northeast, raising the middle region, and taking the lead in the eastern region have been continuously launched. The three strategies of "the Belt and Road Initiative" construction, coordinated development of the Beijing-Tianjin-Hebei and Yangtze River Economic Belt have been solidly promoted, and Xiong'an new area has been planned and established. Several national and regional central cities have developed rapidly, and new growth poles and belts are gradually being formed. This reduces the old problem of uncoordinated development among regions, and thus weakens the degree of spatial dependence.

5.4.2. Spatial Model Econometric Analysis

Furthermore, the LM test, Wald test, and LR test are selected to verify whether the spatial Durbin model can degenerate into a spatial lag model or spatial error model, and the Hausman test is used to select the fixed effect or random effect (the results are shown in Table 7). Therefore, it is judged that the spatial Durbin model cannot degenerate into a spatial error model or spatial lag model, and it is necessary to select a fixed effect for subsequent analysis.

Table 7. Test results.

Test	Statistics	p-Value
LM lag	0.100	0.752
Robust LM lag	3.198 *	0.074
LM error	2.013	0.156
Robust LM error	5.111 **	0.024
LR (H0: SAR nested in SDM)	21.59 ***	0.0002
LR (H0: SEM nested in SDM)	20.77 ***	0.0004
Wald	14.91 ***	0.0049
Hausman	20.65 **	0.014

Note: ***, ** and * represent passing the significance test at 1%, 5% and 10% levels respectively.

For the fixed effect, it is necessary to further examine the use of the individual fixed effect, time fixed effect, or double fixed effect. LR test results are shown in Table 8:

Table 8. LR test results.

	Statistics	p-Value
LR (H0:ind nested in both)	47.09 ***	0.0000
LR (H0: time nested in both)	597.84 ***	0.0000

Note: *** represent passing the significance test at 1% level.

In the tests of individual fixed effect, time fixed effect, and double fixed effect, the results showed that the original hypothesis was rejected at a 1% significance level, so the time-space double fixed model was chosen.

Based on the above comprehensive test, this paper selects the fixed effect of the spatial Durbin model (SDM) and the double fixed model of time and space to make a spatial econometric analysis of the clean power investment and the high-quality development of the regional economy and compares it with the mixed OLS regression without considering the spatial effect. The regression results of the model are shown in Table 9.

By observing AIC and BIC, it can be seen that the fitting degree of the model considering spatial effect is higher, so it can be concluded that the effect of spatial effect on high-quality economic development should be considered.

According to the results in Table 9, the spatial autoregressive coefficient of high-quality economic development is −0.1698, which is significant at the level of 5%, indicating that there is a negative spatial spillover effect in high-quality economic development. The possible reason is that the "siphon effect" exceeds the "spillover effect", and the high-quality development of central cities in developed provinces gathers more resources from neighboring areas, such as high-tech and innovative talents, and investment funds, resulting in provinces with higher economic development levels. Clean power investment, human resource level, government intervention, and urbanization rate all have significant positive effects on the high-quality development of the local economy, while they all inhibit the neighboring provinces.

Table 9. Spatial Durbin model regression results.

Variables	OLS	SDM
ln CEI	0.0072 **	0.0069 ***
	(0.0027)	(0.0017)
HC	0.0139 *	0.0100 *
	(0.0068)	(0.0056)
GI	0.0012 **	0.0015 ***
	(0.0005)	(0.0005)
UR	−0.0009 *	0.0016 **
	(0.0006)	(0.0008)
W×ln CEI		0.0190 ***
		(0.0035)
W×HC		−0.0643 ***
		(0.0105)
W×GI		−0.0028 ***
		(0.0008)
W×UR		−0.0041 ***
		(0.0012)
rho		−0.1698 **
AIC	−1679.465	−1736.668
BIC	−1664.649	−1699.631

Note: ***, ** and * represent passing the significance test at 1%, 5%, and 10% levels, respectively. The standard errors are in brackets.

5.4.3. Decomposition of Spatial Effect

As the autoregressive coefficient in the spatial Durbin model cannot fully reflect the influence of independent variables on dependent variables, this paper, referring to LeSage and Pace's method, decomposes the influence of independent variables on dependent variables into direct effects, indirect effects, and total effects by the partial differential method. The effect decomposition results are shown in Table 10.

Table 10. Spatial effect decomposition of spatial Durbin model.

Variables	Direct Effects	Indirect Effects	Total Effects
ln CEI	0.006158 ***	0.016171 ***	0.016171 ***
	(3.55)	(5.09)	(7.14)
HC	0.011907 **	−0.04292 ***	−0.04292 ***
	(2.16)	(−4.41)	(−3.17)
GI	0.001649 ***	−0.00282 ***	−0.00282
	(3.68)	(−3.68)	(−1.56)
UR	0.001779 **	−0.00394 ***	−0.00394 ***
	(2.12)	(−3.28)	(−2.96)

Note: *** and ** represent passing the significance test at 1% and 5% levels, respectively.

The effect directly reflects the impact of clean power investment and other variables on the high-quality development level of the province's economy. It can be seen that all variables are positive at least at a 5% significance level. Clean power investment is significantly positive at the level of 1%; that is, for each additional unit of clean power investment, the high-quality development level of the local economy will increase by 0.6 percentage points. This is because clean power investment has a positive impact on high-quality economic development in five directions: innovation, coordination, green, openness, and sharing. For each additional unit of human resources, the high-quality development level of the local economy will increase by 1.1 percentage points, and the improvement of the education level will significantly increase the quality of workers. When the number of college students changes to a certain extent, it will change qualitatively, resulting in a large number of innovative achievements. Citizens with higher education generally have a greener and more inclusive attitude toward life, which can promote the economy to turn to a higher-quality development direction. Government intervention and urbanization rate are also significant positive effects on the high-quality development of the local economy, but they are not as big as clean power investment and human capital. Adding one unit can promote the high-quality development of the local economy by about 0.1 units.

The indirect effect after decomposition is the spatial spillover effect, which shows the influence of local related variables on the neighboring areas. The indirect effect regression coefficient of clean power investment is significantly positive at the level of 1%; that is, for every unit of local clean power investment, the high-quality economic development level of neighboring provinces will increase by 1.6 percentage points. This is because clean power investment is mostly concentrated in economically developed provinces or areas with abundant wind energy, light energy, and water energy resources. The increase of a province's own clean power investment will drive the coordinated development of related upstream industries and downstream industries in neighboring areas, promote economic exchanges between regions, and exert the agglomeration effect of clean energy industries, thus promoting the high-quality economic development of neighboring provinces. However, human resources, government intervention, and urbanization rates are all significantly negative at the level of 1%. Taking human resources as an example, every unit of education year in this area will inhibit the high-quality economic development level of neighboring areas by 4.3 percentage points. The possible reason is that the improvement of education years reflects not only the local education level but also the local comprehensive level. The improvement of the comprehensive level will attract more high-quality talents to develop their careers in this area, thus promoting the economy, science and technology, and environment.

5.5. Robust Test

The robustness of the spatial econometric model can be tested by changing the explained variables. In this paper, the per capita GDP (PGDP) is used to replace the score of the high-quality economic development index to indicate the high-quality economic development level. Re-perform the bi-directional fixing effect based on the spatial Durbin model and decompose the spatial effect to get the results in Table 11.

Table 11. Decomposition results of spatial effects based on different explained variables.

Variables	HQD			PGDP		
	Direct Effects	Indirect Effects	Total Effects	Direct Effects	Indirect Effects	Total Effects
ln CEI	0.0061 ***	0.0161 ***	0.0161 ***	0.0911 *	0.3350 ***	0.4261 ***
	(3.55)	(5.09)	(7.14)	(1.48)	(2.79)	(3.43)
HC	0.0119 **	−0.0429 ***	−0.0429 ***	0.3072	−0.8616 **	−0.5544
	(2.16)	(−4.41)	(−3.17)	(1.58)	(−2.32)	(−1.42)
GI	0.0016 ***	−0.0028 ***	−0.0028	−0.1976 ***	−0.1017 ***	−0.2993 ***
	(3.68)	(−3.68)	(−1.56)	(−12.6)	(−3.68)	(−10.6)
UR	0.0017 **	−0.0039 ***	−0.0039 ***	−0.1480 ***	−0.2126 ***	−0.3606 ***
	(2.12)	(−3.28)	(−2.96)	(−5.08)	(−4.86)	(−12.3)

Note: ***, ** and * represent passing the significance test at 1%, 5%, and 10% levels, respectively.

From Table 11, it can be seen that most variables are significant and their positive and negative are unchanged, so the previous spatial econometric test results are relatively stable.

6. Conclusions and Policy Suggestions

In this paper, the entropy TOPSIS method is used to measure the high-quality economic development level of 30 provincial administrative regions (excluding Tibet, Hong Kong, Macao, and Taiwan), and the current situation of high-quality economic development and clean power investment in each province is analyzed. Then, the high-quality economic development level is studied by the spatial autocorrelation model. The global spatial autocorrelation model shows that the spatial correlation first rises and then falls, while the local spatial autocorrelation model shows that the Yangtze River Delta, Pearl River Delta, and Beijing-Tianjin regions are in a "high-gathering" state, and the overall high-quality economic development level gradually weakens from east to west. Then, the spatial Durbin model is used to analyze the factors that affect high-quality economic development. The results show that clean power investment not only promotes the high-quality economic development of the local area but also has a significant and more positive impact on the high-quality economic development of neighboring areas. However, the improvement of human resources, government intervention, and urbanization rate has a significant positive effect on the increase of high-quality development level of the local economy but has an opposite inhibitory effect on neighboring provinces.

According to the above conclusions, we can get the following policy suggestions: (1) We should pay more attention to clean power investment and increase support for clean power investment. By increasing investment in clean power, the clean power industry will be organically combined with new infrastructure, 5G, artificial intelligence, and UAV technology, giving full play to digital advantages, and promoting the further development of upstream and downstream related industries of the clean power industry with low energy consumption and high-efficiency development mode, resulting in a more prominent spillover effect to promote the high-quality development of the regional economy. (2) Governments can strengthen the flow of technology, talents, and management experience among regions, take full advantage of the leading role and radiation effect of the Yangtze River Delta, Pearl River Delta, and Beijing-Tianjin advantageous developed regions, and try to get rid of the negative influence of local protectionism and other undesirable forces. Exporting high-quality talents, an experienced labor force, and high tech to neighboring areas will drive the rapid development of underdeveloped areas in the middle and western regions and improve their high-quality economic development level. (3) According to the different resource endowments of different places, invest in clean energy of appropriate scale according to local conditions. Do not over-invest to cause a large number of undesirable phenomena, such as abandoning wind and light. Enhance the synergy of high-quality economic development among regions, jointly research and develop clean energy technologies within regions, improve productivity, enhance the mobility of spatial elements, and promote clean power investment to further exert the spatial spillover effect. (4) Make the most of advantages of the central government to make overall plans. In education, medical care, culture, sports, infrastructure, etc., the central government should increase its support to further strengthen and improve the financial transfer system, continue to support backward areas, reduce regional disparities, and enable the overwhelming majority of people to enjoy the fruits of economic development. (5) Governments at all levels will incorporate rural energy construction into economic and social development planning, and strengthen support for rural energy in poverty-stricken areas. Encourage financial institutions to innovate financing methods and service models, focus on supporting county rural energy industry development and energy infrastructure construction as green financial services, and give differentiated support to high-quality rural energy projects in terms of loan access, term, and interest rate, to narrow the development gap between urban and rural areas.

Author Contributions: Writing—original draft, X.Z.; Writing—review & editing, L.P. All authors have read and agreed to the published version of the manuscript.

Funding: The authors gratefully acknowledge the funding from the Science and Technology Commission of Shanghai Municipality (Grant No. 21692105000) and National Natural Science Foundation of China (Program NO. 71704110).

Data Availability Statement: The data presented in this study are available on request from the corresponding author. The data are not publicly available due to diversity and complexity of data sources.

Conflicts of Interest: The authors declare no conflict of interest.

References

1. Xi Jinping delivers important speech at 75th UN General Assembly general debate. *People's Daily*, 23 September 2020. [CrossRef]
2. Xu, G.; Schwarz, P.; Yang, H. Adjusting energy consumption structure to achieve China's CO_2 emissions peak. *Renew. Sustain. Energy Rev.* **2020**, *122*, 109737. [CrossRef]
3. Xi Jinping Winning to build a well-off society across the board and seizing the great victory of socialism with Chinese characteristics in the new era. *People's Daily*, 28 October 2017.
4. Ren, B.; Wen, F. Judgment Criteria, Determinants and Realization Ways of China's High-quality Development in the New Era. *Reform* **2018**, *4*, 5–16.
5. Ye, Z.; Li, F.; Wang, N. The Stage Achievements and Difficulties of High-quality Development of China's Regional Economy. *Reg. Econ. Rev.* **2022**, 52–58. [CrossRef]
6. Li, Z.; Zhu, Q. The temporal and spatial characteristics of China's high-quality economic development level and its influencing factors. *Stat. Decis.* **2022**, *38*, 95–99. [CrossRef]
7. Yu, Y.; Hu, S. The realistic dilemma and basic path of high-quality economic development in China: Literature review. *Macro Qual. Res.* **2018**, *6*, 117. [CrossRef]
8. Wang, X.; Xu, X. The temporal and spatial evolution of high-quality economic development in the Yangtze River Economic Belt and the regional gap. *Econ. Geogr.* **2020**, *40*, 5–15. [CrossRef]
9. Yu, H.; Wei, L. Spatial econometric analysis of effect of New economic momentum on China's high-quality development. *Res. Int. Bus. Financ.* **2022**, *61*, 101621. [CrossRef]
10. Pan, W.; Wang, J. High-quality development in China: Measurement system, spatial pattern, and improvement paths. *Habitat Int.* **2021**, *118*, 102458. [CrossRef]
11. Li, M.; Ren, B. Comprehensive evaluation and path selection of high-quality development in China in the new era. *Financ. Sci.* **2019**, *5*, 26–40.
12. Wei, M.; Li, S. Measurement of China's high-quality economic development level in the new era. *Res. Quant. Econ. Technol. Econ.* **2018**, *35*, 3–20. [CrossRef]
13. Zhang, W.; Zhao, S.; Wan, X.; Yao, Y. Study on the effect of digital economy on high-quality economic development in China. *PLoS ONE* **2021**, *16*, e0257365. [CrossRef]
14. Ding, C.; Liu, C.; Zheng, C.; Li, F. Digital Economy, Technological Innovation and High-Quality Economic Development: Based on Spatial Effect and Mediation Effect. *Sustainability* **2022**, *14*, 216. [CrossRef]
15. Li, H.; Zhou, M.; Xia, Q.; Hao, X.; Wang, J. Has Central Environmental Protection Inspection Promoted High-Quality Economic Development?—A Case Study from China. *Sustainability* **2022**, *14*, 11318. [CrossRef]
16. Li, X.; Lu, Y.; Huang, R. Whether foreign direct investment can promote high-quality economic development under environmental regulation: Evidence from the Yangtze River Economic Belt, China. *Env. Sci. Pollut. Res.* **2021**, *28*, 21674–21683. [CrossRef]
17. Li, C.; Wan, J.; Xu, Z.; Lin, T. Impacts of Green Innovation, Institutional Constraints and Their Interactions on High-Quality Economic Development across China. *Sustainability* **2021**, *13*, 5277. [CrossRef]
18. Man, X.Y.; Zhu, X.J.; Chen, J. Research on the index system of energy investment statistics. *Stat. Res.* **2013**, *30*, 25–29.
19. Xu, B.; Bai, T. Inflfluencing factor, leading superiority and policy effect of renewable energy industry investment of China's listed companies. *Chin. Soft Sci.* **2009**, 284–289. Available online: https://www.docin.com/p-1592604424.html (accessed on 11 October 2022).
20. Zhang, D.Y.; Cao, H.; Zou, P. Exuberance in China's renewable energy investment: Rationality, capital structure and implications with firm level evidence. *Energy Pol.* **2015**, *95*, 468–478. [CrossRef]
21. He, L.; Zhang, L.; Zhong, Z.; Wang, D.; Wang, F. Green credit, renewable energy investment and green economy development: Empirical analysis based on 150 listed companies of China. *J. Clean. Prod.* **2019**, *208*, 363–372. [CrossRef]
22. Zhang, D.; Kong, Q. Renewable energy policy, green investment, and sustainability of energy firms. *Renew. Energy* **2022**, *192*, 118–133. [CrossRef]
23. Zeng, S.; Jiang, C.; Ma, C.; Su, B. Investment efficiency of the new energy industry in China. *Energy Econ.* **2018**, *70*, 536–544. [CrossRef]
24. Ross, E.; Day, M.; Ivanova, C.; McLeod, A.; Lockshin, J. Intersections of disadvantaged communities and renewable energy potential: Data set and analysis to inform equitable investment prioritization in the United States. *Renew. Energy Focus* **2022**, *41*, 1–14. [CrossRef]
25. Zahoor, Z.; Khan, I.; Hou, F. clean power investment and financial development as determinants of environment and sustainable economic growth: Evidence from China. *Env. Sci. Pollut. Res.* **2022**, *29*, 16006–16016. [CrossRef] [PubMed]
26. Murshed, M.; Ahmed, Z.; Alam, S.; Mahmood, H.; Rehman, A.; Dagar, V. Reinvigorating the role of clean energy transition for achieving a low-carbon economy: Evidence from Bangladesh. *Env. Sci. Pollut. Res.* **2021**, *28*, 67689–67710. [CrossRef]
27. Hassan, S.T.; Batool, B.; Sadiq, M.; Zhu, B. How do green energy investment, economic policy uncertainty, and natural resources affect greenhouse gas emissions? A Markov-switching equilibrium approach. *Environ. Impact Assess. Rev.* **2022**, *97*, 106887. [CrossRef]
28. Chen, H.; Shi, Y.; Zhao, X. Investment in renewable energy resources, sustainable financial inclusion and energy efficiency: A case of US economy. *Resour. Policy* **2022**, *77*, 102680. [CrossRef]

29. Zhang, M.; Yang, Z.; Liu, L.; Zhou, D. Impact of renewable energy investment on carbon emissions in China—An empirical study using a nonparametric additive regression model. *Sci. Total Environ.* **2021**, *785*, 147109. [CrossRef]
30. Gollakota, A.R.; Shu, C.-M. COVID-19 and energy sector: Unique opportunity for switching to clean energy. *Gondwana Res.* **2022**. [CrossRef]
31. Dvořák, P.; Martinát, S.; Van der Horst, D.; Frantál, B.; Turečková, K. Renewable energy investment and job creation; a cross-sectoral assessment for the Czech Republic with reference to EU benchmarks. *Renew. Sustain. Energy Rev.* **2017**, *69*, 360–368. [CrossRef]
32. Gu, A.; Zhou, X. Emission reduction effects of the green energy investment projects of China in belt and road initiative countries. *Ecosyst. Health Sustain.* **2020**, *6*, 1747947. [CrossRef]
33. Li, Z.; Kuo, T.-H.; Siao-Yun, W.; Vinh, L.T. Role of green finance, volatility and risk in promoting the investments in Renewable Energy Resources in the post-covid-19. *Resour. Policy* **2022**, *76*, 102563. [CrossRef]
34. Ram, M.; Aghahosseini, A.; Breyer, C. Job creation during the global energy transition towards 100% renewable power system by 2050. *Technol. Forecast. Soc. Chang.* **2019**, *151*, 119682. [CrossRef]
35. Kul, C.; Zhang, L.; Solangi, Y.A. Assessing the renewable energy investment risk factors for sustainable development in Turkey. *J. Clean. Prod.* **2020**, *276*, 124164. [CrossRef]
36. Zhou, P.; Luo, J.; Cheng, F.; Yüksel, S.; Dinçer, H. Analysis of risk priorities for renewable energy investment projects using a hybrid IT2 hesitant fuzzy decision-making approach with alpha cuts. *Energy* **2021**, *224*, 120184. [CrossRef]
37. Pollin, R.; Heintz, J.; Garrett-Peltier, H. *The Economic Benefits of Investing in Clean Energy: How the Economic Stimulus Program and New Legislation Can Boost U.S. Economic Growth and Employment*; Published Studies; Political Economy Research Institute, University of Massachusetts at Amherst: Amherst, MA, USA, 2009.
38. Saunila, M.; Ukko, J.; Rantala, T. Sustainability as a driver of green innovation investment and exploitation. *J. Clean. Prod.* **2018**, *179*, 537–561. [CrossRef]
39. Pueyo, A. What constrains renewable energy investment in Sub-Saharan Africa? A comparison of Kenya and Ghana. *World Dev.* **2018**, *109*, 85–100. [CrossRef]
40. Zhang, M. Research on the spatial effect of green investment on high-quality economic development in the process of marketization-an empirical analysis based on spatial Durbin model. *J. Guizhou Univ. Financ. Econ.* **2020**, *4*, 89–100. [CrossRef]
41. Shahani Ashwin, J.; Clarke Amy, J. EU announces European Green Deal 1 billion investment. *MRS Bull.* **2020**, *45*, 888.
42. Chen, Y.; Tian, F.; Zhou, T. JResearch on the Effect of Green Finance on High-quality Economic Development. *Univ. Soc. Sci. Ed.* **2022**, 1–13. Available online: http://kns.cnki.net/kcms/detail/50.1023.c.20210513.1530.002.html (accessed on 9 October 2022).
43. Fang, L.; Jia, P.; Jiang, W. Beijing Economic High-quality Development Report. *Sci. Technol. Think Tank* **2022**, 2–9. [CrossRef]
44. Zhao, H.; Lu, G. Analysis of the high-quality development of Beijing-Tianjin-Hebei economy and its coupling coordination. *J. Tianjin Univ. Commer.* **2022**, *42*, 32–39. [CrossRef]
45. Feng, M.; Guo, H. Evaluation of Beijing's high-quality economic development based on the new development concept. *Urban Issues* **2021**, 4–11. [CrossRef]
46. Fang, D.; Ma, W. Measurement and temporal and spatial characteristics of high-quality development among provinces in China. *Reg. Econ. Rev.* **2019**, 61–70. [CrossRef]
47. Xu, X.; Han, P. Digital Economy and High-Quality Development of Manufacturing Industry. In *Business Intelligence and Information Technology*; Hassanien, A.E., Xu, Y., Zhao, Z., Mohammed, S., Fan, Z., Eds.; BIIT 2021. Lecture Notes on Data Engineering and Communications Technologies; Springer: Cham, Switzerland, 2022; Volume 107. [CrossRef]
48. Yang, Y.; Ren, L.; Du, Z.; Tong, G. Measurement and spatiotemporal analysis of high-quality development of China's industry. *PLoS ONE* **2021**, *16*, e0259845. [CrossRef] [PubMed]
49. Huang, X.; Cai, B.; Li, Y. Evaluation Index System and Measurement of High-quality Development in China. *Rev. Cercet. Interv. Sociala* **2020**, *68*, 163–178. [CrossRef]
50. Tu, Z.; Deng, H.; Chen, R.; Gan, T. Environmental and economic benefits of central environmental protection inspectors: Evidence from Hebei province. *Econ. Rev.* **2020**, 3–16. [CrossRef]
51. Liu, Z. An Empirical Study on the Effect of Green Investment on China's Economic Growth. *Bus. Res.* **2011**, 146–150. [CrossRef]
52. Chen, J.; Deng, X. An Empirical Analysis of Renewable Energy Investment and Green Economy Development. *East China Econ. Manag.* **2020**, *34*, 100–106. [CrossRef]
53. Ting, P. *Study on the Influence of New Energy Investment on Green Industrial Development*; China University of Mining and Technology: Beijing, China, 2020. [CrossRef]
54. Yang, X. *Research on the Influence of Green Institutional Environment on Renewable Energy Investment*; China University of Mining and Technology: Beijing, China, 2020. [CrossRef]
55. Zhu, L. *Research on Clean Energy Investment Efficiency and Influencing Factors of China's Electric Power Enterprises*; North China Electric Power University: Beijing, China, 2019. [CrossRef]

energies

MDPI

Article

Computational Tool to Support the Decision in the Selection of Alternative and/or Sustainable Refrigerants

Guilherme Jesus [1], Martim L. Aguiar [1,2] and Pedro D. Gaspar [1,2,*]

[1] Department of Electromechanical Engineering, Univerity of Beira Interior, Rua Marquês de D'Ávila e Bolama, 6201-001 Covilhã, Portugal
[2] C-MAST—Center for Mechanical and Aerospace Science and Technologies, Rua Marquês de D'Ávila e Bolama, 6201-001 Covilhã, Portugal
* Correspondence: dinis@ubi.pt

Abstract: There have been consequences regarding the increment of the greenhouse effect, such as the rise in the planet's global temperature, and climate change. Refrigerants have an important contribution to the aforementioned environmental impact. In particular, hydrofluorocarbons (HFCs) contribute to the destruction of the ozone layer and the increase of the greenhouse effect. Protocols, international agreements, and legislation were developed to slow down the emission of greenhouse gases. Prohibition and definition of deadlines for the gradual elimination of various refrigerants have been proposed to replace them with others that are environmentally sustainable. Soon, the refrigeration sector will have to replace some refrigerants with others that are alternative and/or sustainable with minimal or zero environmental impact. A computational tool to support decision-making regarding the selection of alternative and/or sustainable refrigerant to replace the old one is developed to be used by refrigeration companies, manufacturers, and installers. A suggestion of refrigerants with reduced environmental impact is provided, ensuring similar thermal performance and energy efficiency, considering the safety level and renovation cost of the installation and refrigerant itself. This decision support system (DSS) uses an objective function that includes the technical specifications and properties of alternative and sustainable refrigerants. The computational tool is applied in the agri-food sector in three case studies. The results show not only the consistency of the computational tool, but also its flexibility, objectivity, and simplicity. Its use allows companies to choose refrigerants with reduced environmental impact, reduced or zero ozone depletion potential and global warming impact, thus contributing to environmental sustainability.

Keywords: computational tool; HFCs; objective function; environmental impact; alternative refrigerants; sustainable refrigerants; sustainability

Citation: Jesus, G.; Aguiar, M.L.; Gaspar, P.D. Computational Tool to Support the Decision in the Selection of Alternative and/or Sustainable Refrigerants. *Energies* **2022**, *15*, 8497. https://doi.org/10.3390/en15228497

Academic Editors: Fuqiang Wang, Chao Shen, Dong Li and Zhonghao Rao

Received: 15 October 2022
Accepted: 12 November 2022
Published: 14 November 2022

1. Introduction

Refrigerants are substances or mixtures of substances, which are used in refrigeration systems and air conditioners as mediator fluids, responsible for performing heat transfers, undergoing a reversible phase transition, i.e., from the liquid state to the gaseous state and vice-versa [1]. The heating and cooling equipment involved uses electricity as a source of energy, which is still dependent on fossil fuels, subsequently causing many environmental problems on a global scale. This heavy reliance on conventional energy for the functioning of cooling equipment contributes to the increase in the prices of these energy sources [2].

Refrigeration plays a key role in sustainable development as it has many applications in different areas of daily life. The most commonly used refrigerators and air conditioning systems employ the traditional vapor compression re-chilling system (VCRS). However, the energy consumption of this type of equipment is very high and its working substances (refrigerants) create environmental problems that need to be solved urgently [3].

VCRS systems are responsible for about 30% of the total energy consumption worldwide and this amount can increase when system malfunctions occur, such as refrigerant leaks, however, due to their high coefficient of performance (COP), the use of VCRS will continue to expand worldwide (especially in developing countries) [4].

Chlorofluorocarbon (CFC), hydrochlorofluorocarbon (HCFC) and hydrofluorocarbon (HFC) fluids are chemically synthetic substances, generally used as working fluids due to their excellent thermodynamic and chemical properties. Despite the aforementioned advantages, CFCs and HCFCs contain chlorine which reacts with ozone and gradually destroys the atmospheric ozone layer [4].

In 1974, scientists Rowland and Molina published a paper describing the decline in the thickness of the ozone layer, particularly over the continent of Antarctica. Following the presentation of this study, the 'Vienna Convention for the Protection of the Ozone Layer' was organized to take strong measures for the protection of the ozone layer and served as the basis for the 'Montreal Protocol' [5].

The Portuguese agri-food industry is essentially composed of micro, small and medium enterprises [6]. This sector is highly fragmented due to the great variety of products and production processes, but, given its heterogeneity, it is a sector with huge potential to increase its energy efficiency due to delays in the implementation of energy efficiency measures and renewable energy sources, misuse and waste of energy caused by the great diversity of processes [7]. The environmental issues of global warming are currently forcing supermarket owners to adopt alternative technologies that offer a lower cooling load, with consequent reduction of energy consumption and a lower environmental impact. In order to respond efficiently to the problem, natural refrigerants have been receiving increased attention for their use in supermarket refrigeration systems, especially the use of R744, carbon dioxide, in the low-temperature circuit of cascade systems which have become a commercial alternative [8]. Increasing energy demand as well as rising prices have sounded the alarm for the scientific community as well as policymakers to look for other cheaper and available energy sources so that conventional energy use can be checked. The demand for refrigeration and air conditioning using conventional energy can now be reduced to some extent by using solar energy, biogas, biomass, and geothermal energy [2]. Optimization of any thermodynamic process could be a better option from the point of view of energy conservation. There are many parameters that can affect the performance of cooling equipment and a complete study based on the second law of thermodynamics corresponds to the standard methodology to optimize the design of systems for better performance. However, cooling equipment powered by renewable energy also needs to be evaluated in economic terms in order to understand the feasibility and viability of these systems, since this methodology gives more flexibility in realizing a more efficient system [2].

1.1. The Problem under Study and Its Relevance

The use of refrigerants such as CFCs, HCFCs and HFCs has significant detrimental environmental impacts, such as stratospheric ozone depletion and global warming as they contribute about 70% of the man-made ozone depletion potential (ODP) chemicals in the atmosphere [9]. Due to ODP and global warming potential (GWP), several refrigerants have already been banned or deadlines set for their elimination, namely the CFCs that were banned in the Montreal Protocol (1987) [1]. Frozen ready-to-eat products that can be easily prepared and consumed are the big bet of the moment and, in addition, the increase in global temperature has consequently fired the air conditioning market, as refrigerants are better known [10]. Given this growing adherence to refrigeration systems, conventions have been held and protocols signed with the aim of phasing out refrigerants with negative environmental impact. The authorization and provisional solution (until 2040) to use HFCs, which are alternatives to CFCs, was given under the Kyoto Protocol. As the deadline approaches, the search for environmentally friendly alternative refrigerants has become a challenge for researchers working in this area. Natural refrigerants are one of the most desirable alternatives as they have inherently zero ozone depletion (ODP) as well as negligible contribution to global warming increase (GWP) [11]. In the current scenario, all refrigeration systems should also be analyzed, giving importance to economic and environmental issues in addition to thermodynamic aspects. To protect the environment from further degradation, the refrigerants used in any system should have zero ODP and very low GWP seeking to achieve sustainability in the processes [12].

In the context of agro-related industries, from agriculture to retail, and due to the evolution of computational resources, there has been the development of decision support systems (DSS) based on mathematics, statistics and artificial intelligence, to support energy efficiency, production optimization, environmental impact and sustainable management [13]. Some DSS have been developed for irrigation decision-making and water management [14–21], crop yield estimation [22], fruit diseases [23], energy consumption and performance of agri-food facilities [24–27], food logistics and distribution [28,29], commercialization time of perishable food products [30] and their pricing [31,32].

The main objective of this study lies in the development of a computational tool to support decision-making in the selection of refrigerants with reduced environmental impact, ensuring similar thermal performance and energy efficiency, considering the cost of the refrigerant and upgrading the refrigeration system. This tool must have capabilities such as ease of access, as well as the ability to understand, analyze, and interpret the results. After surveying the most diverse characteristics associated with alternative sustainable refrigerants, the model was applied in companies, giving them the best possible solution for their specific case in terms of the replacement of their refrigerant. With the increased demand for efficient, economical, and safe cooling systems for the rapidly growing chilled/frozen food industry, optimizing low temperature refrigeration systems using environmentally friendly refrigerants has now become one of the most important goals to achieve [33].

1.2. Evolution of the Refrigerants

Halogenated refrigerants used in cooling systems currently pose a threat to the environment when vented into the atmosphere because of their ODP and GWP [3]. The first large scale environmental impact caused by the activity of the refrigeration-based industries was ozone depletion. Chlorine-based refrigerants are stable enough to reach the stratosphere, where chlorine atoms act as a catalyst to destroy the stratospheric ozone layer, which protects the Earth's surface from ultraviolet (UV) radiation, altering the dynamic balance of ozone formation and consumption and causing its depletion [34]. Table 1 shows, in chronological order, the measures taken in light of the events held to date and demonstrates the set of restrictions imposed on the use of various refrigerants [35].

Table 1. Historical evolution of refrigerants taking into account their environmental impact.

Year	Event	Measure
1985	Vienna Convention	Recognition of the various consequences of CFC use and demonstration of great concern by major companies
1987	Montreal Protocol	Regulation of the production and consumption of "ozone-depleting substances", focusing particularly on CFC gases, which have a high ozone-depleting potential and, in addition, a high global warming potential
1990	London Amendment	Phase-out definition of all CFC, halon and carbon tetrachloride based refrigerant gases in developed and developing countries.
1992	UNFCCC	Inclusion of HCFCs in the list of "ozone-depleting gases" in a phase-out process, in this case only for developed countries, using commonly used refrigerants such as R22 and R123

Table 1. *Cont.*

Year	Event	Measure
1997	Kyoto Protocol	The HCFC phase-out is extended to all countries and the methyl bromide phase-out is scheduled for 2005 and 2015, in developed and developing countries respectively
1999	Beijing Amendment	Tighter controls on HCFC production and marketing
2015	Paris Agreement	Proposed an early freeze date to reduce the damage caused by refrigerants but this was not accepted by all countries Formulation of a long-term low greenhouse gas emission development strategy ("Long-term Strategy")
2016	Kigali Amendment to the Montreal Protocol	Phase-down definition of hydrofluorocarbons (HFCs) due to their high GWP value

Halogenated refrigerants, such as CFCs, HCFCs, and HFCs, are chemical compounds obtained from methane and ethane hydrocarbons by replacing hydrogen atoms with chlorine and fluorine atoms. If the hydrogen atoms are replaced by a halocarbon, it is fully halogenated. When halogenated refrigerants are leaked from equipment during operation or by accident, they contribute to the depletion of the ozone layer and to global warming [4].

Although leakage is usually in small quantities, it is a major source of greenhouse gas (GHG) emissions due to the high GWP of these refrigerants. Moreover, refrigerant released from equipment leads to insufficient system charge and negatively affects the performance of the equipment, resulting in high energy consumption. On the other hand, HFCs contain no chlorine or bromine, but are greenhouse gases that affect the overall temperature of the earth's surface. In short, all these refrigerants contribute significantly to environmental impact and climate change. To effectively meet global environmental issues, all these refrigerants need to be replaced by others with reduced ODP and GWP [1]. In addition, the performance of heat exchangers must be improved to increase efficiency and therefore reduce the indirect emissions of GHG, associated with the energy consumption of the refrigeration systems. One of the alternatives is to replace these halogenated refrigerants with natural refrigerants, such as hydrocarbons (HCs) [1].

1.3. Substitution Strategies

As a result of the Montreal Protocol, CFC production was completely eliminated in developed countries in 1996 and in 2010 in developing countries, consequently, CFCs have been replaced by less harmful HCFC refrigerants. HCFCs are expected to be completely eliminated by the end of 2030 in developed countries and by 2040 in developing countries [3]. Developed countries then started using HFCs, which have no impact on the ozone layer but still have high GWP. Proposals to decrease HFCs are also being discussed under the Montreal Protocol [1]. Countries have generally been aggressive and effective in implementing protocols and their subsequent amendments to slow and reverse the consequences of the presence of refrigerants with high ODP, specifically those containing chlorine and bromine, in the stratosphere [36].

To achieve the desired effect, protocols have been issued over the years, namely in 2006, 2009, and 2014, in which, although the articles have changed slightly, the scope is the same. They all have, without exception, the objective of phasing out refrigerants with a harmful effect on the ozone layer. Regulation 842/2006 of the European Parliament and of the Council of 17 May 2006 was issued with the aim of "containing, preventing and thereby reducing emissions of the fluorinated greenhouse gases covered by the Kyoto Protocol. It applies to the constant fluorinated greenhouse gases ... " [37].

Regulation No. 1005/2009 of the European Parliament and of the Council of 16 September 2009 was issued mainly for substances that deplete the ozone layer, the scope of which is related to establishing "rules concerning production, import, export, market placement, recovery, recycling, reclamation and destruction of substances that deplete the ozone layer, the communication of information about these substances and the import, export, placing on the market and use of products and equipment containing or relying on these substances" [38].

Regulation No. 517/2014 of the current European Parliament and of the Council of 16 April and 2014 was issued in order to further protect the environment by reducing emissions of fluorinated greenhouse gases [39]. To this end:

- It establishes rules on the containment, use, recovery, and destruction of fluorinated greenhouse gases and on related ancillary measures;
- It imposes conditions on the placing on the market of specific products and equipment containing, or whose functioning relies upon, fluorinated greenhouse gases;
- It imposes conditions on the specific uses of fluorinated greenhouse gases;
- It establishes quantitative limits on the placing on the market of hydrofluorocarbons.

Regarding the evolution of greenhouse gas emissions in Portugal, and according to the most recent update of the National Inventory of Emissions 2021 (for the year 2019), GHG emissions, without accounting for emissions from land use change and forests (LULUCF), are estimated at about 63.6 Mt CO_2e, representing an increase of about 8.1% compared to 1990 and a decrease of 5.4% compared to 2018. Considering the LULUCF sector, total emissions in 2019 are estimated at 55.8 MtCO_2e, corresponding to a decrease of 7.2% compared to 1990 and a decrease of 7.9% to 2018 [40]. After the rapid growth experienced during the 1990s, national emissions slowed down in the early 2000s, with a subsequent decrease in national emissions, particularly after 2005. These trends largely reflect the evolution of the Portuguese economy, which was characterized by strong growth associated with increased energy demand and mobility in the 1990s, followed by a situation of stagnation and recession, especially in the period 2009–2013. In the following years, there was a reversal of that trend. In 2019, GDP recorded a positive variation of 2.2%, slowing down compared to the previous year in which it grew by 2.8%, but maintaining a growth that has been verified since 2014. Total emissions in 2019 decreased by 5.4% compared to 2018, with this reduction being mostly associated with the "energy industries" sector, which registered a sharp drop of 27.2% compared to 2018, as a result of the higher proportion of renewable energy in the national electricity production, associated with the replacement of coal by natural gas in thermoelectric production, and greater use of electricity imports [41].

Both in Portugal and throughout the European continent, a roadmap is being followed for the transition into a competitive low-carbon economy in 2050. This roadmap has been outlined as an economically advantageous way to achieve the necessary emission reductions in the European Union by 2050. Non-CO_2 emissions, including fluorinated greenhouse gases, but excluding emissions from agriculture, are to be reduced by 72–73% by 2030 and 70–78% by 2050 compared to 1990 levels. Based on the full application of existing Union legislation, emissions in 2030 are projected to be 104 Mt CO_2e, which requires a further decrease of approximately 70 Mt CO_2e. The European Parliament then decreed in the same resolution the need to opt for alternative refrigerants by phasing out hydrofluorocarbons on an international scale also following the Montreal Protocol [41].

1.4. Properties of Alternative Refrigerants

Refrigerants play a key role in the refrigeration cycle, influencing its operation and all equipment in the installation, as they allow the heat formed in the cold zone to be absorbed and transferred to a hot zone (outside environment) through evaporation and condensation processes [42]. The thermodynamic properties of a refrigerant are essential to predict the behavior of a refrigeration system and its performance. Excellent thermodynamic proprieties involve a boiling point just below ambient temperature, a critical temperature above ambient temperature, a high normal boiling point, and a high heat of vaporization [4]. There is no refrigerant that gathers all of these desirable properties, this means that when a certain fluid is applied in a certain type of refrigeration installation, it is not always recommended for use in another, even if it is equivalent. The perfect refrigerant is the one that brings together the greatest possible number of qualities, relative to a given purpose and objective [42].

Table 2 specifies a set of proper properties of refrigerants [35].

Table 2. Desirable requisites and properties of refrigerants.

Requirement	Properties
Chemistry	Stable and inert
Health, Safety and Environment	Non-toxic
	Non-flammable
	Low GWP
Thermal	High latent heat
	Critical point and boiling point appropriate for the application
	Low specific heat in vapor state
	Low viscosity
	High thermal conductivity
Others	Reasonable solubility/miscibility with lubricants
	Low melting point
	Easy leak detection
	Low cost

It is a very long list of qualities and none of the refrigerants can be considered ideal and adaptable to all applications. However, special attention needs to be paid to the selection of the most sustainable refrigerant for a given application based on an overall assessment [43]. It is crucial to raise awareness and consciousness for the use of environmentally friendly refrigerants, such as hydrocarbons (HCs), hydro-fluorophenolines (HFO), R744 (carbon dioxide) and R717 (ammonia), as alternatives so that it may be possible to reduce ozone depletion and global warming, and so making refrigeration systems using these refrigerants futureproof. Within these refrigerants, natural refrigerants stand out for the lower environmental impacts that they have and for being more appropriate within the perspective of sustainable technological development [1].

According to the laws established in the European Parliament regarding bans on certain refrigerants in the coming years, as of 2020 R404A will no longer be used in new units and facilities in the European Union and, from that date, facilities may not be loaded with more than 10.2 kg of R404A. In recent years, a range of alternative HFC blends to R404A has emerged, such as R407A, R407F, R442A, etc., all with corresponding trade names. From 2022 onwards, a new F-gas regulation prohibits the use of HFCs in refrigerated cabinets and centralized installations of more than 40 kW [44].

2. Materials and Methods

In order to carry out the development of the computational tool, there is a prior need to characterize in detail the refrigerants in question, analyzing those which are abolished, those which will have the same fate in the near future and finally those alternative refrigerants with reduced GWP and high sustainability. Always keeping in mind the legislation imposed by the European Commission which, over the years, has adopted measures in order to progressively eliminate the production and emission of greenhouse gases and to protect our planet.

2.1. Materials

Many refrigerants have been phased out and those with significant GWP not yet phased out will eventually exit the market in accordance with EU regulations. However, companies still using such refrigerants need to change the refrigerants in their plants and industrial processes soon to meet regulations. To do so, they will have to use refrigerants with low or zero GWP. According to Annex IV of Regulation N°517/2014 issued in 2014, and which is still in force, alternative refrigerants that appeal to sustainability and could be used in industrial and commercial refrigeration processes are as follows:

- R-1234ze (hydrofluorophelines);
- R-170 (Ethane);
- R-290 (Propane);
- R-600a (Isobutane);
- R-717 (Ammonia);
- R-744 (Carbon Dioxide);
- R-1150 (Ethylene);
- R-1270 (Propylene).

Although there is no exact formula when choosing a refrigerant, there are some details that should be considered when making the decision, so that it is possible to acquire the ideal option for a specific case. Therefore, the following factors should be considered in the decision-making process [45]:

- Availability;
- Cost-benefit;
- Quality;
- Safety.

Since the objective of this study culminates with the development of a decision support tool for the consumer for a refrigerant that appeals to sustainability, the technical and thermodynamic data must be known, described, and presented. Next, Table 3 presents the various characteristics of each of these alternative refrigerants that will serve as a basis for the decision system in order to optimize the process of selecting the "ideal" refrigerant for companies in the sector.

Table 3. Operating conditions for alternative and/or sustainable refrigerants.

	Refrigeration			
Refrigerant	**Very Low Temperature**	**Low Temperature**	**Medium Temperature**	**High Temperature**
R-1234ze			✓	✓
R-170	✓	✓		
R-290		✓	✓	✓
R-600a			✓	✓
R-717		✓	✓	✓
R-744		✓	✓	✓
R-1150	✓	✓		
R-1270		✓	✓	✓

Once we have gathered the properties of the coolants that are about to be accepted by companies (consumers) in terms of safety, quality, cost-effectiveness, availability, thermodynamic characteristics, and their contribution to sustainability, we can build a decision support tool that will report the best possible choice for the user, always considering guidelines and needs to maximize satisfaction.

In summary, when comparatively analyzing the refrigerants based on the characteristics and thermodynamic properties that are described in Tables 3 and 4, characteristics must be taken into account in order for the decision support tool to realize the desired goal. This goal should culminate in the selection of the ideal refrigerant for the user in question, based on their economic and financial guidelines and on issues of energy efficiency, cooling capacity, availability of the refrigerant and on the refrigerant's contribution to sustainability.

2.2. Methods

The tool was developed in Excel with the aim of making it accessible to the user insofar as all the operations for the user will be deductive and simple and will allow them to reach a conclusion in a few steps, all of them justified based on user choices, preferences, and guidelines. Figure 1 shows the user interface, which, after the fields are filled in, offers a recommendation of the ideal refrigerant, giving three options in descending order of priority. The decision support system then consists of three parts. The first part requires the identification of the user, the company, and role. The second part consists of a questionnaire with the following questions:

- Do you use any of these old refrigerants?
- Regarding the operating temperature, what do you want?
- Which lubricating oil do you use?
- How concerned are you about the cost/benefit of replacing the refrigerant?
- What is your concern regarding the quality of operation when replacing the refrigerant?
- What is your concern regarding the safety of refrigerant replacement?
- What is your concern regarding availability of refrigerant replacement?
- What is your concern regarding the environmental impact of refrigerant replacement?

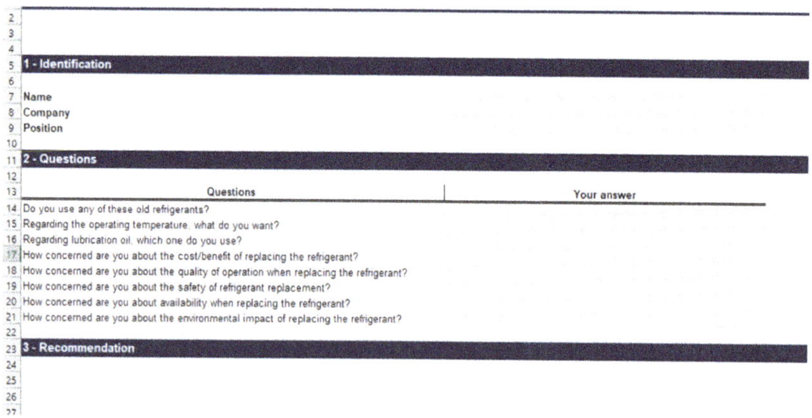

Figure 1. Decision support tool layout.

To make it simple to answer the questions imposed by the program, there are dropdown boxes in the cells where the answers are, in which all the possibilities are presented to facilitate the filling-in process, as shown in Figure 2.

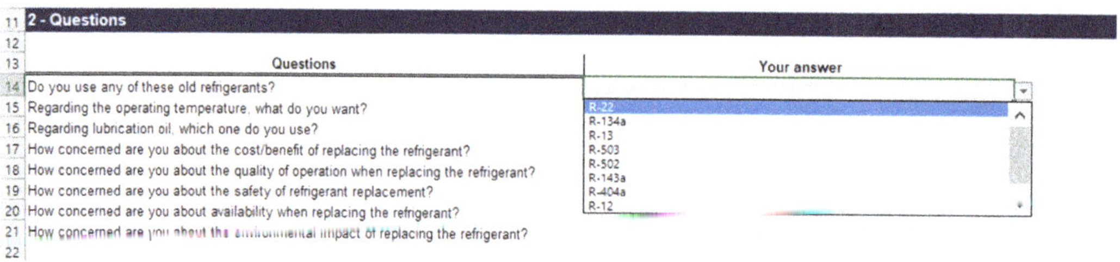

Figure 2. Example of dropdown boxes in the tool.

Table 4. Thermodynamic properties of alternative and/or sustainable refrigerants.

	ODP	GWP	GWP (100)	Boiling Point 1atm (°C)	Critical Temperature (°C)	Critical Pressure (MPa)	Compatible Lubricants	Toxicity	Flammability	Security	Biodegradable	Cost of the Refrigerant	Cost of Installations	Cooling Capacity	Efficiency	Availability
R-1234ze	0	6	6	−19	109.4	3.60	POE	No	Yes	A2	Yes	High	Medium	Medium	Medium	Reduced
R-170	0	6	6	−89	32.3	4.87	Mineral/AB/POE	No	Yes	A3	Yes	Reduced	High	Medium	High	High
R-290	0	3	3	−42	97.0	4.30	Mineral/AB/POE	No	Yes	A3	Yes	Reduced	High	Medium	High	High
R-600a	0	3	3	−12	135.0	3.60	Mineral/AB/POE	No	Yes	A3	Yes	Reduced	High	Medium	High	High
R-717	0	0	0	−33	132.0	11.30	Mineral	Yes	Yes	B2	Yes	Reduced	Very High	Medium	High	Very High
R-744	0	1	1	−57	31.0	7.40	POE	No	No	A1	Yes	Reduced	Very High	Very High	Medium	Very High
R-1150	0	4	4	−104	9.2	50.00	Mineral/AB/POE	No	Yes	A3	Yes	Reduced	High	Medium	High	High
R-1270	0	2	2	−48	91.0	46.00	Mineral/AB/POE	No	Yes	A3	Yes	Reduced	High	Medium	High	High

The third and last stage is associated with the recommendations provided, based on all the data, the restrictions, and conditions regarding alternative refrigerants. The tool provides the user with recommendations, sorting three refrigerant options by priority of use. The answer is given based on an objective function that analyses the answers given by the user to the eight questions, transforming the qualitative answers into data and numerical coefficients. The result is also dependent on restrictions associated with the old refrigerants (making the answer almost straightforward), the lubricating oils used and the temperature conditions in the cooling process.

The decision support tool boils down to the application of a function that depends on five coefficients (varying according to the user's answers to the last five questions), constraints, and data that result from the constant properties of each of the refrigerants, as given by Equation (1). This equation is calculated for each of the eight alternative refrigerants, with higher results meaning more adequate refrigerants for the application in question, considering the user's preferences. Finally, the top three coolants are presented to the user in order of recommendation.

$$R = (C_1 \times x_1 + C_2 \times x_2 + C_3 \times x_3 + C_4 \times x_4 + C_5 \times x_5) \times y_1 \times y_2 \times y_3 \quad (1)$$

The expression can be divided into three parts in which:

- *C*'s—coefficients;
- *X*'s—refrigerant properties;
- *Y*'s—constraints of the refrigerants.

2.2.1. Coefficients

When the user answers the last five questions that determine the value of the coefficients C_1, C_2, C_3, C_4, and C_5, he introduces qualitative values that correspond to the expressions: "None"; "Reduced"; "Medium"; "High", and "Very High", that is, the coefficients do not correspond to constant values, as they are altered according to the preferences and concerns of those who answer the questionnaire. Since these values are qualitative, there is a need to transform them into numerical data to enable the resolution of the objective function.

This transformation is based on the study conducted by [46], in which the author aims to transform qualitative dependent variables into quantitative ones to culminate with an equivalent conclusion. To achieve such an effect, a sensory analysis is performed through the responses given by consumers. The researcher imposes a certain set of questions on consumers by means of a computational tool from which qualitative data essential for solving the problem is extracted.

After the resolution of the questionnaire by the user, Fonseca [46] processes the data into quantitative values, attributing to each of the answers a value in the scale of 1 to 5 in which 1 corresponds to the worst and 5 to the best result. This way it is possible to express the responses graphically, where they were found to grow linearly. Although this is a study with a different context, the procedure to convert qualitative answers into quantitative values can be adapted to solve an objective function to transmit the best recommendation for users.

Similar to the study presented, the developed decision support tool imposes on the user common questions and a set of possible answers (qualitative data), which will be transformed into a scale in which the values will be separated by the same interval (quantitative data). Following this context, to make Equation (1) possible, a scale from 0 to 1 is attributed as shown in Table 5, to transform qualitative data into quantitative data to quantify the coefficients related to cost/benefit (C_1), operation performance (C_2), safety (C_3), availability (C_4) and environmental impact (C_5). The higher the quantitative value associated with the coefficients, the greater the user's concern in this regard.

Table 5. Scale defined for the objective function coefficients.

Qualitative Value (User Response)	Quantitative Value (User Response)
None	0.00
Reduced	0.25
Medium	0.50
High	0.75
Very High	1.00

2.2.2. Refrigerant Properties

The data referring to the alternative refrigerants that were used in the tool are described and qualitatively characterized in Table 5. These properties are presented as constant values in the objective function which will be multiplied by the previously mentioned coefficients. The characteristics of the refrigerants that serve as analysis and that influence the function are the following:

- Cost of both the refrigerant and the installations;
- Refrigeration quality, analyzing its efficiency and capacity;
- Safety;
- Availability of human and material resources;
- Environmental impact addressing the GWP (100 years).

Similar to the coefficients, the properties to be used in the objective function were transformed into numerical values through a scale set from 0 to 1. However, for the different properties, different scales were considered. Since the objective of the tool is to optimize the objective function to find out which are the best refrigerants, it is necessary to pay special attention to aspects such as cost and environmental impact. In these cases, the higher their qualitative value (described in Table 5) the lower their respective value in numerical terms must be. Thus, the scales were defined as shown in Table 6.

Table 6. Scale defined for the objective function properties.

Quantitative Value	Value (Cost)	Value (Quality of Operation)	Value (Safety)	Value (Availability)
Null	1.00–0.00 = 1.00	0.00	0.00	0.00
Reduced	1.00–0.25 = 0.25	0.25	0.25	0.25
Medium	1.00–0.50 = 0.50	0.50	0.50	0.50
High	1.00–0.75 = 0.75	0.75	0.75	0.75
Very High	1.00–1.00 = 0.00	1.00	1.00	1.00

Finally, the property that addresses the environmental impact, despite being described in Table 4 in numerical terms, was entitled to a specific scale, because, as previously indicated, the higher this factor, the lower must be its value to be used in the objective function. Thus, in this way the scale for the GWP data will be based in Equation (2), wherein the coefficient x_5 refers to the value associated with the GWP of each of the alternative refrigerants.

$$1 - \left(\frac{x_5}{10}\right) \quad (2)$$

2.2.3. Restriction of the Refrigerants

In last place are the refrigerant restrictions that concern the following factors:

- Possibility of directly replacing the refrigerant;
- Temperature conditions used in the refrigeration process;
- Type of lubricant.

These three answers are limiting factors for the result, as they narrow down the range of possibilities for the suggested replacement refrigerant.

As far as the restriction of old refrigerants is concerned, it must be considered that the eight alternative refrigerants being studied and analyzed in this tool can be direct replacements. For the first question, the user has nine possible answers, eight refrigerants that will be abolished in the future, and none of the above.

Depending on the answers given, the tool will direct the objective function in a certain direction. The relationships between the nine possible answers to the first question and the replacement refrigerants are as follows:

- R-22 => R-290, R-600a, R-717, R-744;
- R-134a => R-1234ze, R-600a, R-717, R-744;
- R-13 => R-170, R-1150;
- R-503 => R-170, R-1150;
- R-502 => R-1270;
- R-143a => R-1270;
- R-404a => R-744;
- R-12 => R-600a;
- None => R-1234ze, R-170, R-290, R-600a, R-717, R-744, R-1150, R-1270.

Once again, the answers are presented in qualitative format, and once again there is the need to transform them into numerical data. In this case it is not a scale, but if and else statements that are applied. The if statement is known as the decision statement because it allows for a certain condition or expression. The code inside the if function is executed if the condition is true. However, the code inside the if function is disregarded if the condition is false [47]. In this way, results can assume two logical values, 1 if true and 0 if false. As regards the tool itself, and more specifically the objective function, whenever the user selects an answer, it will assume the logical value of 1. Therefore, in this case, the value of y1 of equation 1 will be quantified. For instance, when the user mentions using R-22 in their facilities, the objective function will only return 1 for refrigerants R-290, R-600a, R-717, and R-744 (y_1 = 1), since the others will be 0 as the result of false conditions, thus shortening the range of possibilities for the recommended refrigerant.

Regarding the temperature range in the refrigeration process, this was established based on Table 3, whereby the user, when answering the second question, defines the operating temperatures and consequently restricts the refrigerants that do not operate on those conditions. Following this context, and once again using the same type of programming established in the previous constraint, should the user enter the answer in the respective dropdown box, it will assume the logical value of 1 if the temperatures are within the range or 0 otherwise; this will be the parameter y_2 of Equation (1). Depending on the answers provided to the tool, the objective function will be directed in a particular direction. Therefore, the relationships between the four possible answers to the second question and the substitution refrigerants are as follows:

- Very Low => R-170, R-1150;
- Low => R-290, R-717, R-744, R-1270;
- Medium => R-1234ze, R-290, R-600a, R-717, R-744, R-1270;
- High => R-1234ze, R-290, R-600a, R-717, R-744, R-1270.

Basically, if the user is given the example of assuming he operates in average cooling temperature conditions, the objective function will be equated only for refrigerants R-1234ze, R-290, R-600a, R-717, R-744, and R-1270, since y_2 = 1 and the others are null, again reducing the set of possible recommendations.

As per the previous constraints, the type of lubricant used is converted to a quantitative value by the same methodology, further reducing the set of possible answers. Once an option in the respective dropdown box is selected, the answer will be associated with the logical value of 1, automatically quantifying the parameter y_3. Depending on the answers provided to the tool, the objective function will be directed in a particular direction.

Therefore, the relationships between the three possible answers to the third question and the replacement refrigerants are as follows:

- Mineral => R-170, R-290, R-600a, R-717, R-1150, R-1270;
- AB => R-170, R-290, R-600a, R-1150, R-1270;
- POE => R-1234ze, R-170, R-290, R-600a, R-744, R-1150, R-1270.

In other words, if by chance the user shows the need to use mineral oil as a lubricant in their refrigeration plants, there will only be compatibility with the R-170, R-290, R-600a, R-717, R-1150, and R-1270 refrigerants, insofar as the objective function will only be considered for these, because their $y_3 = 1$, reducing the set of possible recommendations for replacement refrigerant by the third and final time. Once descriptions of all the components of the tool have been made and all the calculations that culminate in the final recommendation of the replacement refrigerant for the company have been shown, the tool was applied in three case studies for a better understanding of how it works and how it helps users. As the users access the tool and fill out the questionnaire step by step, the final answers associated with solving the objective function are determined and presented, depending on the user's preferences, values, and needs.

3. Case Studies

Over the years, compliance with legislation by companies in the industrial refrigeration sector has led them (once banned from using certain refrigerants) to replace the outlawed refrigerants with others that have similar performance and energy efficiency. However, after contact with the sector, it is possible to verify that thy are not yet aware of the importance of sustainability, so that adoption of the refrigerants referred to as alternative and/or sustainable mentioned in Table 4 will soon be the solution for several companies in their refrigeration processes. One of the sectors that proves this fact is the Portuguese agri-food industry, which in 2016 was limited to a set of refrigerants with high GWP. The refrigerants R-32, R-404a, R-22, R-422, and R-449 represent a large share in the percentage distribution of refrigerant used in Portugal. These refrigerants will be discontinued, giving rise to concern for the replacement of refrigerants by other environmentally friendly ones [48]. Awareness of and sensitivity to the use of more ecological refrigerants will become crucial to make refrigeration systems that use these refrigerants futureproof, since they present a minimum environmental impact and because they are governed by the perspective of sustainable and efficient technological development and growth. To consolidate what is presented above, companies in the sector were questioned and three practical cases were obtained to be studied and applied in the decision support tool so that it might give the best possible recommendation to the companies and transmit the best solutions considering the situation they are facing. Therefore, the cases to be studied, analyzed, and applied in the tool concern the search for alternative refrigerants that will replace R-449, R-422 and R-32, based on the preferences and guidelines of the companies themselves, which, once they provide information for the tool in Excel, will obtained feasible answers and recommendations.

3.1. Case Study One

On contacting the refrigeration company, the information was obtained that the refrigerant in use is R-449, which is characterized by its low environmental impact and low GWP compared to HFCs. Its thermodynamic properties are balanced, it is used in low and medium temperature conditions in commercial and industrial applications and the lubricating oil used in the installations is miscible with POE oils [49]. These coolant data are key to obtaining and transmitting a recommendation from the tool to the user. Firstly, to gather all the possible conditions for solving the problem, the company was asked in the DSS about the refrigerant used to date (in this case R-449), the temperature conditions (averages) and the lubricating oil (POE). Following these first three questions, the parameters associated with the constraints were quantified as shown in Figure 3.

	Questions	Your answer
10		
11	**2 - Questions**	
12		
13	**Questions**	**Your answer**
14	Do you use any of these old refrigerants?	R-449
15	Regarding the operating temperature, what do you want?	Medium
16	Regarding lubrication oil, which one do you use?	POE

Figure 3. Filling in the parameters of the restrictions in the case of R-449.

The next and last step before the recommendation is resolved boils down to five questions imposed on the user of the tool about their degree of concern for:

- Cost/Benefit;
- Quality of Operation;
- Safety;
- Availability;
- Environmental Impact.

Following these questions, which are again based solely on the user's opinion in the context of the refrigeration process used in their company, the following answers were obtained in the respective Excel cells as shown in Figure 4. The objective function coefficients were thus quantified, giving rise to the single solution for this case, replacement by R-744 (carbon dioxide).

17	How concerned are you about the cost/benefit of replacing the refrigerant?		High
18	How concerned are you about the quality of operation when replacing the refrigerant?		Very High
19	How concerned are you about the safety of refrigerant replacement?		Medium
20	How concerned are you about availability when replacing the refrigerant?		Medium
21	How concerned are you about the environmental impact of replacing the refrigerant?		Medium
22			
23	**3 - Recommendation**		
24			
25		1st	R-744
26			

Figure 4. Solution of practical case one using average temperatures.

According to the data collected from this company, there is extreme concern about the replacement of its coolant by another with excellent properties in terms of capacity and efficiency in refrigeration. Furthermore, there is concern that this does not lead the company to spend large sums of money on the acquisition of the coolant or on changing facilities. Regarding the other factors, no major concerns were presented, giving rise to the answer presented. In case the company, under its conditions, wanted low temperatures in the cooling process, with the same responses to the other parameters, the recommendation would end up being the same, as shown in Figure 5.

11	**2 - Questions**		
12			
13	**Questions**		**Your answer**
14	Do you use any of these old refrigerants?		R-449
15	Regarding the operating temperature, what do you want?		Medium
16	Regarding lubrication oil, which one do you use?		POE
17	How concerned are you about the cost/benefit of replacing the refrigerant?		High
18	How concerned are you about the quality of operation when replacing the refrigerant?		Very High
19	How concerned are you about the safety of refrigerant replacement?		Medium
20	How concerned are you about availability when replacing the refrigerant?		Medium
21	How concerned are you about the environmental impact of replacing the refrigerant?		Medium
22			
23	**3 - Recommendation**		
24			
25		1st	R-744
26			
27			

Figure 5. Solution for practical case one using low temperatures.

In short, the R-744 (carbon dioxide) refrigerant was selected as the recommendation for this company, as it ends up being its direct substitute, given that out of the eight possibilities it is the only one with similar thermodynamic properties, temperature conditions and lubricating oil as the R-449, despite the potential need to change the refrigeration installation (compressors, piping, valves, etc.).

3.2. Case Study Two

The procedure is the same as in the first practical case, differing only in the refrigerant used and to be changed by the company, designated R-422. This is characterized by its application in low, medium, and high temperature conditions in refrigeration (industrial, commercial, and domestic) and conditioning (residential and commercial) applications. Furthermore, it is highly compatible with both traditional and new generation lubricants and is therefore miscible with AB, mineral and POE oils [50]. In contrast to the first case study where the recommended refrigerant was the same for the two existing possibilities, in this second case study the company in its replacement process has nine different possibilities as it can choose between three temperature conditions and three types of lubricating oil. The company was also asked about their level of concern regarding cost, quality of operation, safety, availability, and environmental impact when replacing the coolant whose answers are shown in Figure 6.

17	How concerned are you about the cost/benefit of replacing the refrigerant?	Very High
18	How concerned are you about the quality of operation when replacing the refrigerant?	Very High
19	How concerned are you about the safety of refrigerant replacement?	Medium
20	How concerned are you about availability when replacing the refrigerant?	Medium
21	How concerned are you about the environmental impact of replacing the refrigerant?	Medium
22		

Figure 6. Filling in the parameters of the restrictions in the case of R-422.

Subsequently, the dropdown boxes for temperature conditions and lubrication oils were also filled in, and, depending on the data provided by the user, different recommendations were obtained when solving the objective function by the support system. This way, the company has at its disposal a range of possible solutions described in a decision tree as shown in Figure 7.

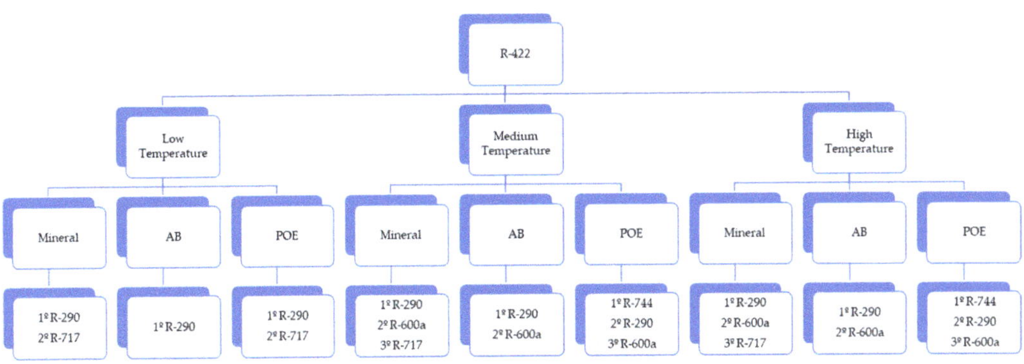

Figure 7. Decision tree for R-422 refrigerant replacement possibilities.

Depending on the company's intentions regarding the temperature conditions and the oil used for lubricating the installations, it can be seen from Figure 7 that the recommendations vary and, thanks to the versatility of the tool itself, the three best alternative and/or most sustainable refrigerants for the specific case in question are provided, whenever possible, considering the various constraints presented. When analyzing the results shown in Figure 7, one can see a trend towards the recommendation of R-290 (propane), regardless of operating temperature and lubricating oils.

3.3. Case Study Three

Finally, another company in the industrial refrigeration sector was contacted and questioned about the replacement of refrigerants, once again registering the sector's lack of awareness on the subject. The company in question still uses the R-32 refrigerant, which will also be discontinued in the future. It is characterized by being miscible with POE lubricant, is used in low-temperature conditions, and has a relatively high GWP of 550, contributing to ozone layer depletion [51]. Again, the computational tool was presented, and the methodology was the same as in the previous cases, with the answers to the questions imposed by DSS represented in Figure 8.

2 - Questions

Questions	Your answer
Do you use any of these old refrigerants?	R-32
Regarding the operating temperature, what do you want?	Low
Regarding lubrication oil, which one do you use?	POE
How concerned are you about the cost/benefit of replacing the refrigerant?	Very High
How concerned are you about the quality of operation when replacing the refrigerant?	Very High
How concerned are you about the safety of refrigerant replacement?	Medium
How concerned are you about availability when replacing the refrigerant?	Medium
How concerned are you about the environmental impact of replacing the refrigerant?	Very High

3 - Recommendation

1st	R-290

Figure 8. Answers from the company for the replacement of R-32.

Similar to practical cases one and two, this company is essentially concerned with the costs involved in purchasing the refrigerant and changing the refrigeration equipment, as well as with maximizing cooling capacity and efficiency. However, since the company is still using a refrigerant that will be discontinued in the future, there is a need to replace the refrigerant with one with excellent thermodynamic properties, temperature conditions, similar or compatible lubricant, and low environmental impact. For this purpose, none is better than R-290 (propane), because, according to DSS itself, this refrigerant is the direct substitute for R-32, given the characteristics of both.

In conclusion, the tool has been successfully applied in the three cases studied, thanks to its flexibility and accessibility for the user. In addition to helping solve some of the problems faced by companies during the phase-out of GWP and ODP refrigerants, the tool has the ability to advise companies on the possibility of imposing other conditions on temperatures and types of oil, always seeking the maximum satisfaction of the user of the DSS, providing them clearly and objectively with all possible recommendations for the future replacement refrigerant.

4. Discussion

Although the tool provides the best recommendation for the user, answering the initial problem with solutions for the specific cases presented by the companies in the sector, it can be improved and optimized. The developed DSS is consistent and flexible throughout its process, transmitting clear and objective answers, being in turn easy to use and understand for the user. However, the decision support system can eventually enhance its parameters considering the following factors: form; quantity; and quality.

After the transformation of qualitative data into quantitative data through specific scales, these are now represented through discrete numerical values. One of the improvement perspectives for the future of DSS is the implementation of continuous data with the purpose of providing even more data for each of the refrigerants, consequently substantiating the answers and recommendations presented for the sustainable refrigerants to be an option for the companies in the sector.

In relation to the amount of data available to the decision support system, for each of the coolants, DSS uses three constraints (condition of the coolants previously used, operation temperature, and type of oil used for lubrication), five coefficients (cost, operation quality, safety, availability, environmental impact) and seven constant data associated to these coefficients to be determined by the user. To increase the flexibility of the tool, to make it more solid and secure in the answers that it provides, this tool can present a possibility of expanding the number of parameters to be used by the objective function, namely the restrictions, coefficients, and the properties inherent to the alternative refrigerants.

Regarding quality, the demonstration of the recommendations to the user can also be improved. Whenever the result is presented, only the names of the sustainable refrigerants to be recommended are displayed (in order). However, this same presentation can be accompanied by a detailed, concise description of the refrigerant in question. For this purpose, a dropdown box could be implemented, similarly to those already presented to the user.

In short, the optimization of these three factors will lead to an increase in the consistency, efficiency, and accuracy of the results of the decision support tool.

5. Conclusions

5.1. General Conclusions

The industrial refrigeration sector has become increasingly focused on the importance of and need to apply safe designs, with appropriate safety devices, in accordance with international safety legislation. Currently, the use of systems containing HFCs implies compliance with emission targets and rules that, in case of non-compliance, are subject to sanctions. These targets and rules aim to contribute to the reduction of global warming.

Over the years and with the implementation of numerous protocols, namely those of Kyoto and Montreal, and in accordance with the agreements imposed, halogenated refrigerants have been progressively eliminated in a move towards sustainability. This move aims to protect the planet from the harmful emissions that culminate in the degradation of the ozone layer and contribute to global warming.

The replacement of refrigerant fluids that are some of the main manmade polluting agents present in the atmosphere—through their contribution to the destruction of the ozone layer and the increase of the greenhouse effect—by others that appear to be less harmful to the environment requires an extensive in-depth study. The study carried out for this research was focused not only on the direct consequences of these refrigerants on the environment, but also on the analysis of the energy and exergetic performance of the refrigerants in general. Following this context, the general study of several alternatives for replacement, in the short and long term, for the most used refrigerants was defined.

For that purpose, during the work, the general aspect of refrigeration was approached, with the presentation of the main techniques used to produce cold, giving special attention to the cycle under study: the vapor compression refrigeration cycle. Subsequently, the general characteristics and properties of refrigerants are discussed, allowing their rapid classification according to the type and family of refrigerant fluid. The main events that have marked the history of the evolution of refrigerants are described, with emphasis on the legislation published in this area.

After the issuing of regulations N° 842/2006, N° 1005/2009, N° 517/2014 by the European Parliament, not only all refrigerants to be eliminated considering their destructive content for the ozone layer, but also alternative refrigerants with zero ODP and reduced GWP were mentioned, thus appealing to the much sought-after sustainability. In turn, the main objective of this work lies essentially on the implementation of these same refrigerants in companies based on a decision support tool, to the extent that users according to their perspectives, values and wishes will have the answer to their future replacement of halogenated refrigerants.

In short, the computational tool involves an Excel database that evaluates both qualitative and quantitative thermodynamic properties associated with an objective function that depends on the constraints of each of the alternative and/or sustainable refrigerants under analysis and the user's answers to the questions imposed on it. This decision support system gives extreme importance to the user's opinion and specific case for the recommendation of a well-suited refrigerant.

5.2. Specific Conclusions

The execution of the DSS depends fundamentally on constant parameters and specific conditions relative to each of the alternative refrigerants that constitute the possible answer for the substitution process. The objective function programmed in the computational tool is equated based on the restrictions and properties associated with the eight fluids under analysis. Depending on the coefficients to be introduced by the user, the decision system will dictate three recommendations by order of relevance. These restrictions and properties, unique to each of the refrigerants, are obtained and determined based on an evaluation of qualitative and quantitative data previously acquired, namely through the comparison of these alternative refrigerants with the halogenated ones already abolished or that will be discontinued in the future. This same scientific research is supported by the literature review carried out. To enable the resolution of the equation, the qualitative data was transformed into quantitative data through regular scales established for all necessary cases.

Although the sustainable refrigerants present excellent thermodynamic properties, reduced cost when compared to HCFCs and present themselves as excellent solutions for industrial cooling process, they show, however, certain ambiguities that require further research with qualitative and quantitative analysis, in order to develop appropriate recommendations. The development of the study and of the tool leads to a specific and not generalized conclusion due to the simple fact that each of the replacement refrigerants is not the ideal solution for all cases, but rather for a specific situation, because although they all have excellent characteristics and properties, they are associated with different operating conditions in the refrigeration process, lubrication oils and in many cases can be the direct replacement refrigerants for the old ones, which were used until then, without the need for major changes in the pre-existing installations.

Author Contributions: Conceptualization, P.D.G.; methodology, P.D.G. and M.L.A.; validation, P.D.G. and M.L.A.; formal analysis, P.D.G., M.L.A. and G.J.; investigation, M.L.A. and G.J.; resources, P.D.G., M.L.A. and G.J.; data curation, M.L.A. and G.J.; writing—original draft preparation, G.J.; writing—review and editing, P.D.G. and M.L.A.; supervision, P.D.G.; project administration, P.D.G.; funding acquisition, P.D.G. All authors have read and agreed to the published version of the manuscript.

Funding: This research was funded as part of the project "PrunusPós—Optimization of processes for the storage, cold conservation, active and/or intelligent packaging and food quality traceability in post-harvested fruit products", project n.° PDR2020-101-031695, Partnership n.°87, initiative n.°175, promoted by PDR 2020 and co-funded by FEADER within Portugal 2020.

Data Availability Statement: More information on the project PrunusPós can be found at https://www.prunuspos.pt/ (accessed on 10 November 2022).

Acknowledgments: This work was supported in part by the Fundação para a Ciência e Tecnologia (FCT) and C-MAST (Centre for Mechanical and Aerospace Science and Technologies), under project UIDB/00151/2020.

Conflicts of Interest: The authors declare no conflict of interest.

References

1. Gaspar, P.; Aguiar, L. Refrigerantes Naturais: Tendências do Mercado, políticas e tecnologias na indústria agroalimentar portuguesa. In Proceedings of the X Congresso Ibérico-VIII Congresso Iberoamericano de las Ciencias y Técnicas del Frío (CYTEF), Pamplona, Spain, 1–3 July 2020.
2. Gupta, A.; Anand, Y.; Tyagi, S.; Anand, S. Economic and thermodynamic study of different cooling options: A review. *Renew. Sustain. Energy Rev.* **2016**, *62*, 164–194. [CrossRef]
3. Verde, M.; Harby, K.; de Boer, R.; Corberán, J.M. Performance evaluation of a waste-heat driven adsorption system for automotive air-conditioning: Part Imodeling and experimental validation. *Energy* **2016**, *116*, 526–538. [CrossRef]

4. Harby, K. Hydrocarbons and their mixtures as alternatives to environmental unfriendly halogenated refrigerants: An updated overview. *Renew. Sustain. Energy Rev.* **2017**, *73*, 1247–1264. [CrossRef]
5. Norman, J. Rowland & Molina Suggest that CFCs Deplete the Ozone Layer. 2022. Available online: https://www.historyofinformation.com/detail.php?id=2721 (accessed on 5 March 2022).
6. Ascenção, P.O. COMPETE 2020 Alavancou 113 Milhões de Euros de Investimento no Setor Agro-Alimentar, o 2.º Maior Empregador em Portugal. 2016. Available online: https://www.compete2020.gov.pt/destaques/detalhe/Setor_agroalimentar_COMPETE2020 (accessed on 23 May 2022).
7. Cardoso, B.J.; Lamas, F.B.; Gaspar, A.R.; Ribeiro, J.B. Refrigerants used in the Portuguese food industry: Current status. *Int. J. Refrig.* **2017**, *83*, 60–74. [CrossRef]
8. Rocha, R. Análise de Desempenho Termodinâmico de Ciclos de Refrigeração Cascata para Temperaturas Ultrabaixas. Master's Thesis, Universidade Federal do Rio de Janeiro, Rio de Janeiro, Brazil, 2022.
9. UNEP. Handbook for the International Treaties for the Protection of the Ozone Layer. Ozone Secretariat United Nations Environment Programme. Sixth Edition. 2003. Available online: http://www.unep.org/ozonoe (accessed on 1 September 2022).
10. Lopes, P. *Válvulas de Expansão: Comparação da Aplicação de Válvulas de Expansão Eletrônica e Termostática em Sistemas Frigoríficos*; Engenharia Mecânica; Centro Universitário do Sul de Minas: Minas Gerais, Brazil, 2019.
11. Singh, K.K.; Kumar, R.; Singh, A. Thermo-economic optimization of environment-friendly refrigerants for cascade refrigeration. *Arab. J. Sci. Eng.* **2021**, *46*, 12235–12252. [CrossRef]
12. Roy, R.; Mandal, B. Thermo-economic analysis and multi-objective optimization of vapour cascade refrigeration system using different refrigerant combinations: A comparative study. *J. Therm. Anal. Calorim.* **2020**, *139*, 3247–3261. [CrossRef]
13. Alibabaei, K.; Gaspar, P.D.; Lima, T.; Campos, R.M.; Girão, I.; Monteiro, J.; Lopes, C.M. A review of the challenges of using deep learning algorithms to support decision-making in agricultural activities. *Remote Sens.* **2022**, *14*, 638. [CrossRef]
14. Zhang, J.; Zhu, Y.; Zhang, X.; Ye, M.; Yang, J. Developing a long short-term memory (LSTM) based model for predicting water table depth in agricultural areas. *J. Hydrol.* **2018**, *561*, 918–929. [CrossRef]
15. Zinkernagel, J.; Maestre-Valero, J.F.; Seresti, S.Y.; Intrigliolo, D.S. New technologies and practical approaches to improve irrigation management of open field vegetable crops. *Agric. Water Manag.* **2020**, *242*, 106404. [CrossRef]
16. Alibabaei, K.; Lima, T.M.; Gaspar, P.D. Modeling evapotranspiration using Encoder-Decoder Model. In Proceedings of the 2020 International Conference on Decision Aid Sciences and Applications (DASA'20), Sakheer, Bahrain, 8–9 November 2020; Volume 2020, pp. 132–136. [CrossRef]
17. Ara, I.; Turner, L.; Harrison, M.T.; Monjardino, M.; deVoil, P.; Rodriguez, D. Application, adoption and opportunities for improving decision support systems in irrigated agriculture: A review. *Agric. Water Manag.* **2021**, *257*, 107161. [CrossRef]
18. Alibabaei, K.; Gaspar, P.D.; Lima, T. Modeling soil water content and reference evapotranspiration from climate data using Deep Learning methods. *Appl. Sci.* **2021**, *11*, 5029. [CrossRef]
19. Bwambale, E.; Abagale, F.K.; Anornu, G.K. Smart irrigation monitoring and control strategies for improving water use efficiency in precision agriculture: A review. *Agric. Water Manag.* **2022**, *260*, 107324. [CrossRef]
20. Alibabaei, K.; Gaspar, P.D.; Assunção, E.; Alirezazadeh, S.; Lima, T. Irrigation with a deep reinforcement learning model—Case study on a site in Portugal. *Agric. Water Manag.* **2022**, *263*, 107480. [CrossRef]
21. Alibabaei, K.; Gaspar, P.D.; Assunção, E.; Alirezazadeh, S.; Lima, T.M.; Soares, V.N.G.J.; Caldeira, J.M.L.P. Comparison of on-policy deep reinforcement learning A2C with off-policy DQN in irrigation optimization: A case study at a site in Portugal. *Computers* **2022**, *11*, 104. [CrossRef]
22. Alibabaei, K.; Gaspar, P.D.; Lima, T.M. Crop yield estimation using deep learning based on climate big data and irrigation scheduling. *Energies* **2021**, *14*, 3004. [CrossRef]
23. Assunção, E.; Diniz, C.; Gaspar, P.D.; Mesquita, R.; Proença, H. Decision-making support system for fruit diseases classification using Deep Learning. In Proceedings of the 2020 International Conference on Decision Aid Sciences and Applications (DASA'20), Sakheer, Bahrain, 8–9 November 2020; Volume 2020, pp. 652–656. [CrossRef]
24. Nunes, J.; Neves, D.; Gaspar, P.D.; Silva, P.D.; Andrade, L.P. Predictive tool of energy performance of cold storage in agrifood industries: The Portuguese case study. *Energy Convers. Manag.* **2014**, *88*, 758–767. [CrossRef]
25. Silva, P.D.; Gaspar, P.D.; Nunes, J.; Andrade, L.P. Specific electrical energy consumption and CO_2 emissions assessment of agrifood industries in the central region of Portugal. *Appl. Mech. Mater.* **2014**, *675–677*, 1880–1886. Available online: http://hdl.handle.net/10400.6/7270 (accessed on 5 September 2022). [CrossRef]
26. Zocca, R.; Gaspar, P.D.; Silva, P.D.; Santos, F.C.; Andrade, L.P.; Nunes, J. Decision-making computationally aided in the management of energy sources used in agrifood industries. *Energy Procedia* **2019**, *161*, 100–107. [CrossRef]
27. Zocca, R.; Lima, T.M.; Gaspar, P.D.; Charrua-Santos, F. Computational tool to foster systematic thinking and sustainable environmental conscience in the selection of energy sources systems in agrifood companies. In Proceedings of the 48th International Conference on Computers and Industrial Engineering (CIE 48), Auckland, New Zealand, 2–5 December 2018.
28. Mendes, A.; Cruz, J.; Saraiva, T.; Lima, T.M.; Gaspar, P.D. Logistics strategy (FIFO, FEFO or LSFO) decision support system for perishable food products. In Proceedings of the 2020 International Conference on Decision Aid Sciences and Applications (DASA'20), Sakheer, Bahrain, 8–9 November 2020; pp. 173–178. [CrossRef]
29. Gomes, D.E.; Iglésias, M.I.D.; Proença, A.P.; Lima, T.M.; Gaspar, P.D. Applying a genetic algorithm to an m-TSP: Case study of a decision support system for optimizing a beverage logistics vehicles routing problem. *Electronics* **2021**, *10*, 2298. [CrossRef]

30. Ananias, E.; Gaspar, P.D.; Soares, V.N.G.J.; Caldeira, J.M.L.P. Artificial intelligence decision support system based on artificial neural networks to predict the commercialization time by the evolution of peach quality. *Electronic* **2021**, *10*, 2394. [CrossRef]
31. Maciel, V.; Matos, C.; Lima, T.M.; Gaspar, P.D. Decision support system to assign price rebates of fresh horticultural products on the basis of quality decay, Chapter 23. In *Computational Management: Applications of Computational Intelligence in Business Management*; Patnaik, S., Tajeddini, K., Jain, V., Eds.; Springer International Publishing: Cham, Switzerland, 2021; pp. 487–497, ISBN 978-3-030-72928-8. [CrossRef]
32. Pina, M.; Gaspar, P.D.; Lima, T.M. Decision support system for dynamic pricing of horticultural products based on the quality decline due to microbial growth. *Appl. Syst. Innov.* **2021**, *4*, 80. [CrossRef]
33. Ramos, A. A Investigação e a Inovação na Fabricação de Equipamentos de frio Industrial em Portugal. Master's Thesis, Instituto Superior de Engenharia de Lisboa (ISEL), Lisbon, Portugal, 2016.
34. ATM Revolution. Earth The Power of the Planet. Camadas da Atmosfera Terrestre. 2009. Available online: https://atm-revolution.blogs.sapo.pt/3418.html (accessed on 15 May 2021).
35. Saldanha, P. Dimensionamento de um Circuito Frigorífico a CO_2 com Evaporadores Inundados. Master's Thesis, Faculdade de Engenharia, Universidade do Porto, Porto, Portugal, 2019.
36. Paul, S.; Sarkar, A.; Mandal, B.K. Environmental Impacts of Halogenated Refrigerants and Their Alternatives: Recent Developments. *Int. J. Emerg. Technol. Adv. Eng.* **2013**, *3*, 400–409.
37. UE. *Regulamento (CE) N.º 842/2006 do Parlamento Europeu e do Conselho Relativo a Determinados Gases Fluorados com Efeito de Estufa*; A. M. Guerra, "Apontamentos de Produção de Frio e Refrigeração"; Instituto Superior de Engenharia de Lisboa (ISEL): Lisbon, Portugal, 2014.
38. UE. *Regulamento (CE) N.º 1005/2009 do Parlamento Europeu e do Conselho Relativo às Substâncias que Empobrecem a Camada de Ozono*; Jornal Oficial da União Europeia: Porto, Portugal, 2009.
39. UE. *Regulamento (EU) N.º 517/2014 do Parlamento Europeu e do Conselho Relativo aos Gases Fluorados com Efeito de Estufa e que Revoga o Regulamento (CE) n.o 842/2006*; Jornal Oficial da União Europeia: Porto, Portugal, 2014.
40. Agência Portuguesa do Ambiente. Políticas de Mitigação. 2021. Available online: http://www.apambiente.pt/index.php?ref=16&subref=81&sub2ref=117 (accessed on 13 November 2021).
41. Agência Portuguesa do Ambiente. Protocolo de Quioto. 2021. Available online: http://www.apambiente.pt/index.php?ref=16&subref=81&sub2ref=119&sub3ref=500 (accessed on 14 May 2021).
42. Boa, J. Análise Energética de Equipamentos de uma Fábrica de Lacticínios. Master's Thesis, Universidade da Beira Interior, Covilhã, Portugal, 2012.
43. Bandarra, D.E.P. *Uso de Fluidos Alternativos em Sistemas de Refrigeração e Ar Condicionado: Artigos Técnicos. Tendências do Uso de Fluidos Refrigerantes Alternativos em Sistema de Ar Condicionado Automotivo*; Instituto do Meio Ambiente e dos Recursos Naturais Renováveis, Ministério do Meio Ambiente: Brasília, Brazil, 2011.
44. INTARCON. F-Gas or the Prohibition of Fluorinated Gases. 2021. Available online: https://www.intarcon.com/en/f-gas/ (accessed on 16 March 2021).
45. Dufrio, R. Fluído Gás Refrigerante: Saiba Como Escolher a Melhor Opção. Dufrio Refrigeração, 29/07/2021. 2021. Available online: https://www.dufrio.com.br/blog/ar-condicionado/fluido-gas-refrigerante-saiba-como-escolher-a-melhor-opcao/ (accessed on 30 June 2022).
46. Fonseca, M. *Transformação de Variáveis Qualitativas em Quantitativas; Um Estudo de Caso*; Dissertação para a obtenção do grau de Químico com atribuições Tecnológicas; Instituto de Química da Universidade Federal do Rio de Janeiro: Rio de Janeiro, Brazil, 2017.
47. Rosa, D. Instruções If... Else em C Explicadas. 2022. Available online: https://www.freecodecamp.org/portuguese/news/instrucoes-if-else-em-c-explicadas/ (accessed on 12 August 2022).
48. Pavkovic, B. Refrigerants—Part 2: Past, present and future perspectives of refrigerants in air-conditioning applications. *REHVA Eur. HVAC J.* **2013**, *50*, 28–33.
49. Aldifrio Gás Refrigerante R-449 (Opteon™ XP40) Substituto R-404A/R-507. 2022. Available online: https://aldifrio.com/index.php/2018/05/01/novo-refrigerante-concebido-substituir-r-404a-r-507/ (accessed on 5 September 2022).
50. Tazzetti Fluidos Refrigerantes. R422A. 2022. Available online: https://www.tazzetti.com/pt-pt/products-and-services/fluidos-refrigerantes/R422A (accessed on 5 September 2022).
51. ACR. Que Problema con el Refrigerante de los Coches. 2008. Available online: https://www.acrlatinoamerica.com/20080325456/noticias/empresas/ique-problema-con-el-refrigerante-de-los-coches.html (accessed on 5 September 2022).

energies

MDPI

Article

PV-CrackNet Architecture for Filter Induced Augmentation and Micro-Cracks Detection within a Photovoltaic Manufacturing Facility

Muhammad Hussain [1], Hussain Al-Aqrabi [1,2,*] and Richard Hill [1]

[1] Department of Computer Science, Centre for Industrial Analytics, School of Computing and Engineering, University of Huddersfield, Queensgate, Huddersfield HD1 3DH, UK
[2] Faculty of Computer Information System (CIS), Higher Colleges of Technology, University City, Sharjah P.O. Box 7947, United Arab Emirates
* Correspondence: h.al-aqrabi@hud.ac.uk

Abstract: Photovoltaic cell manufacturing is a rigorous process involving many stages where the cell surface is exposed to external pressure and temperature differentials. This provides fertile ground for micro-cracks to develop on the cell surface. At present, domain experts carry out a manual inspection of the cell surface to judge if any micro-cracks are present. This research looks to overcome the issue of cell data scarcity through the proposed filter-induced augmentations, thus providing developers with an effective, cost-free mechanism for generating representative data samples. Due to the abstract nature of the cell surfaces, the proposed augmentation strategy is effective in generating representative samples for better generalization. Furthermore, a custom architecture is developed that is computationally lightweight compared to state-of-the-art architectures, containing only 7.01 million learnable parameters while achieving an F1-score of 97%.

Keywords: defect detection; micro-cracks; photovoltaics; smart manufacturing; quality inspection

Citation: Hussain, M.; Al-Aqrabi, H.; Hill, R. PV-CrackNet Architecture for Filter Induced Augmentation and Micro-Cracks Detection within a Photovoltaic Manufacturing Facility. *Energies* **2022**, *15*, 8667. https://doi.org/10.3390/en15228667

Academic Editor: Alon Kuperman

Received: 18 October 2022
Accepted: 13 November 2022
Published: 18 November 2022

1. Introduction

Global emissions and their mitigation are a shared concern across the globe, giving rise to the field of renewable energy, from which solar power is accepted as one of the practically feasible options, deployable at all levels of society. Manifesting its signification and mitigation of C02 emissions, an example is presented of a solar deployment project based in California, where 113,533 domestic solar deployments have resulted in the reduction of 696,544 metric tons of CO_2 emissions [1].

Similar to other manufacturing procedures, photovoltaic (PV) cell production is a rigorous and delicate process, vulnerable to the emergence of defects such as micro-cracks. Micro-cracks are a common defect produced by unwarranted mechanical or thermal stress during fabrication [2]. Micro-cracks can also be difficult to detect with the human eye. Hence, electroluminescence (EL) imaging is utilized for the detection of micro-cracks [3] in multi-crystalline PV cells [4].

Presently, quality inspection within PV manufacturing facilities is a manual process involving domain experts, analyzing EL images of PV cell surfaces at various manufacturing stages. This not only leads to increased costs for recruiting experts or third-party inspection organizations but also increases inspection time and can increase the error rate due to human bias. This provides a segue into the case for implementing computer vision as a non-invasive mechanism integrated into the existing quality inspection process for assisting with the detection of defective PV cells in a timely and cost-effective manner [5]. This research addresses the issue of data scarcity for EL-based PV samples representative of manufacturing facilities and presents a highly generalized CNN for micro-crack detection with EL-based PV samples.

1.1. Literature Review

Studying the present literature concerning the quality inspection of PV cells via automation, it can be stated that active research is taking place, with researchers exploring the use of deep learning, in particular computer vision, for PV fault detection.

Starting with machine learning, Hussain et al. [6] present a framework for deciding if the use of machine/deep learning is required for detecting PV faults compared to conventional statistical methods. The authors utilize 'kMeans' for obtaining labels for the unclassified dataset before training on a wide range of machine learning algorithms as well as a multilayer perceptron neural network. The authors report respectable performance for most algorithms, ranging from 94 to 100%. Although the performance is impressive, the research lacks in differentiating on surface faults rather than focusing on post-deployment disconnection faults.

Akram et al. [7] present a deep learning architecture based on convolutional neural networks (CNN) for the detection of defects on the PV cell surface. After developing and training an 'isolated model' based on EL PV cell samples, the authors implement transfer learning with the aim of tuning the trained architecture for infrared PV cell surface samples, reporting an impressive accuracy of 99.23%. With regards to the size of the dataset, it can be classified as small, i.e., less than 800 images. However, further inspection into the PV cell surface faults shows the faults had to be artificially generated onto the PV cell surface, which could raise questions on the representativeness and true generalization of the trained architecture.

Ahmad et al. [8] develop a CNN-inspired architecture for fault detection with EL extract samples of the PV cell surface. The authors report a respectable accuracy of 91.58%. The authors lay emphasis on the importance of data inspection and representative augmentations before delving deeper into their proposed architecture. The presented architecture consists of eight convolutional blocks followed by a single fully connected layer. Looking deeper into the development logic for the internal convolutional layers, many filters were implemented within each convolutional block, i.e., 32 filters each for the first 4 layers, followed by 64 for the next two layers, with 128 in the final two layers. The increased number of convolutional filters would have a significant impact on the computational load of the architecture if explored as part of the research.

Dunderdale et al. [9] propose a feature-oriented, deep-learning strategy for the detection of defective PV cell surfaces. For the purpose of benchmarking, the dataset is trained on the VGG-16 [10] and MobileNet [11] architectures. Furthermore, the authors provide a comparison between the de facto ADAM optimizer against the stochastic gradient descent (SGD). The latter provided an accuracy of 85.8% for the VGG-16 architecture, while a poor result was recorded when utilizing the ADAM as the optimizer (27.4%). Conversely, MobileNet achieved the highest accuracy of 89.5% based on the application of generic augmentations, namely, sample rotations, when utilizing ADAM as the optimizer. Although an explanation for these results was not provided, MobileNet is understood to be a more computationally lightweight architecture used in object detection applications [12] due to the utilization of depth-wise convolutions, which lessen computations by as much as ninefold [13].

Pierdicca et al. [14] select the VGG-16 network for the detection of defective PV cell surfaces. The rationale for selecting VGG-16 as opposed to computationally more friendly architectures is given as simplicity of implementation and development. Though ongoing development in the field of deep learning by various companies has led to frameworks such as PyTorch (Facebook) and TensorFlow (Google). These frameworks make the development and testing of various state-of-the-art architectures simpler, enabling users to develop architectures that are both higher performant and computationally lightweight at the same time. Computational data with regards to the trained architecture would have provided a broader spectrum for evaluating the VGG-16 architecture, unearthing factors such as high convergence time compared to other architectures implementing various strategies for speeding the convergence process, such as batch normalization [15].

Deitsch et al. [16] demonstrate PV fault detection of multiple defects on PV cell surfaces via machine learning (SVM) and deep learning (CNN). Focusing on the CNN, the proposed architecture achieved an accuracy of 88.42% based on the implementation of transfer learning. The tuning of the architecture is performed in two stages. Firstly, the weights of the fully connected layer are randomly initialized with ADAM selected as the optimizer. This is followed by the random initiating of weights for the previous convolutional layers with respect to the fully connected layer, this time selecting SGD-M as the optimizer.

Hussain et al. [17] propose the detection of micro-cracks with PV cell surfaces through the development of a custom CNN architecture, achieving an overall F1-score of 98.8%. The authors provide an in-depth comparison of the proposed architecture across a wide range of metrics consisting of architectural, computational, and post-deployment performance. Furthermore, the comparison is extended to SOTA architectures, with the developed architecture achieving the highest performance in 4 out of 5 metrics.

Tang et al. [18] present a CNN-inspired architecture for the detection of faults within EL-based PV cell surfaces. Interestingly, the authors implement a generative adversarial network (GAN) as their data augmentation strategy for introducing representative variance. The fact that a GAN is essentially an architecture itself, which is computationally demanding, makes its rationale is not clear especially when generic augmentations can be applied at a fraction of the computational load compared to the GAN. The authors report an accuracy of 83% post-implementation of GAN for generating new data samples, again questioning the selection of GAN for augmentation purposes.

Summing up the literature, it can be observed that although research in the automation of PV fault detection is actively being pursued, there is a lack of practical considerations. In particular, there is a dearth of literature addressing faults originating within PV manufacturing facilities focusing on not only the lack of representative data scaling but also the development of computationally lightweight architectures that can be integrated into existing quality inspection protocols to assist and enhance EL-based PV fault detection at an early stage before PV Modules are shipped to client sites.

1.2. Paper Contribution

This research is focused on two primary objectives and hence makes two contributions. Procuring EL-based PV cell surface images from within a PV manufacturing site is significantly more difficult than post-deployment data due to logistical and access restrictions within manufacturing facilities. However, as PV cells are more at risk of developing defects as a result of manual handling and development processes, it is paramount for researchers to access quality data that can be utilized to train architectures for deployment within manufacturing facilities. To address the issue of data scarcity, our first contribution comes in the form of filter-induced augmentations (FIA). FIA makes use of convoluted filter outputs without optimized backpropagation for generating representative PV cell surface samples. The FIA is based on the development of a custom CNN architecture consisting of two convolutional blocks and two fully connected layers.

Secondly, the custom CNN architecture is utilized for training by enabling the backpropagation of weights via the SGD-M optimizer. The developed architecture contains a lightweight footprint with only 7.01 million learnable parameters. In order to manifest the efficiency of the developed architecture and FIA mechanism, the results are presented in an iterative manner, comparing the original dataset, generic applied augmentations, and the proposed FIA-generated dataset.

2. Methodology

2.1. Dataset

The dataset utilized for undertaking this research comprised of PV-cell images procured at the PV manufacturing facility, manually inspected and labeled by domain experts.

Table 1 presents the status of the dataset consisting of the following two classes: normal and defective. Based on the number of samples within each class, it can be concluded that the dataset is significantly small in sample quantity for developing a robust, highly generalized architecture, that can differentiate between the two classes.

Table 1. Original dataset.

Class	Samples
Normal	140
Defective	200

Figure 1 presents a sample set of (A) normal and (B) defective PV cell surfaces. Before applying any augmentations with the objective to scale the original dataset, it was essential to study and comprehend the visual differentiation features of the two classes as well as the degree of variance at an internal class level. Simply by inspecting Figure 1, we can deduce various considerations regarding the level of variance, global and internal differential features as well as external factors impacting resultant surface images. Starting with global-level variance, it can be observed that there is an element of surface heterogeneity with regard to the texture. For example, the far left and right images for the normal class Figure 1A present a clearer PV cell surface image as opposed to the center image. This is a significant observation, as the textural differential within the normal class, may lead to the developed architecture falsely generalizing on the assumption that only clear surface images belong to the normal class.

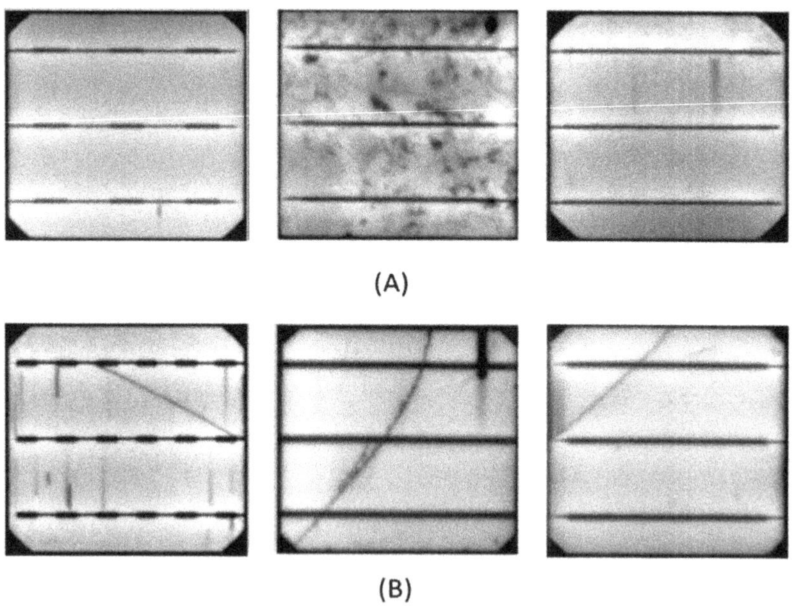

Figure 1. Data inspection (**A**) normal (**B**) Defective.

When comparing the normal and defective classes, the significance of internal and external factors on the resultant cell image can be appreciated. Taking Figure 1A far right as an example compared to Figure 1B far right, it can be observed that an element of shading or poor filter quality has induced pixel shading on a certain locality of the normal cell. When comparing this to the defective cell, the similarity between the micro-crack and the shading is obvious and hence an increased chance of misclassification.

As evident from all sample images within Figure 1, the fundamental component for a PV cell surface is the busbar, facilitating energy flow from the cell surface. However, even this mandatory component is manifested in various configurations. For example, Figure 1A,B far left images present intermittent busbar configuration whilst Figure 1B center presents the complete 'cross-surface' configuration, and Figure 1A center presents a 'cut-off' busbar configuration. Furthermore, the starkness of the busbar also varies, leading to the potential misclassification of busbar configurations that consist of lighter pixel intensities as micro-cracks and hence a defective cell.

2.2. Generic Domain Augmentations

Post data inspection, a hypothesis was formed stating that; While there is variance within the present dataset at both global and internal class levels, it could be addressed via representative data modeling as opposed to random application of augmentations for simply increasing the size of the dataset. Hence, the size of the dataset post augmentations was capped at 777 samples from the initial size of 340 images. The capping of the dataset was also representative of the practical limitations of PV data procurement from within PV manufacturing facilities due to restricted access and lack of open-source data.

The augmentations for data transformation belonged to one of the following two categories: translational invariance and translational equivariance. Translational invariance was represented as follows:

$$f(t(a)) = f(a) \tag{1}$$

where f equates to the function for the image (a) and t equates to the applied transformation.

Translational invariance was reserved for utilization during the designing of the architectures internal layers as it would not have any physical implications on the dataset i.e., increase dataset size as evident via the equation. Explaining further, translational invariance attempts to preserve regional transformations through aggregation. Thus, it was decided that this type of transformation resulting in aggregation of local features would be more useful during the design stage of the architecture as information propagates deeper through the architectural layers. For instance, the application of max-pooling between the convolution layers would create regional invariance by accepting only the max value from each feature map. Therefore, translational equivariance was selected as the framework for data scaling.

$$f(t(a)) = t(f(a)) \tag{2}$$

where f equates to the function for the image (a) and t equates to the applied transformation.

Comparing the above equation with that for translational invariance it can be seen that translational equivariance would transform the input image with respect to the kind of transformation t, applied. Each augmentation presented in the subsequent sections was selected based on its probability of occurrence within a PV manufacturing complex due to internal and external factors such as varying production line configurations, EL camera specifications, etc.

Orientation-Based Scaling (Generic)

PV cells go through various stages from silicon ingots to wafer slicing and hence are processed on various production line configurations. Additionally, the EL quality inspection can happen at various stages of PV manufacturing. The aforementioned production-based processes can lead to varying orientations of the resultant PV cell surface images procured such as varying orientation, hence the vertical flipping was selected amongst other techniques to generate representative samples, accordingly, as shown in Figure 2.

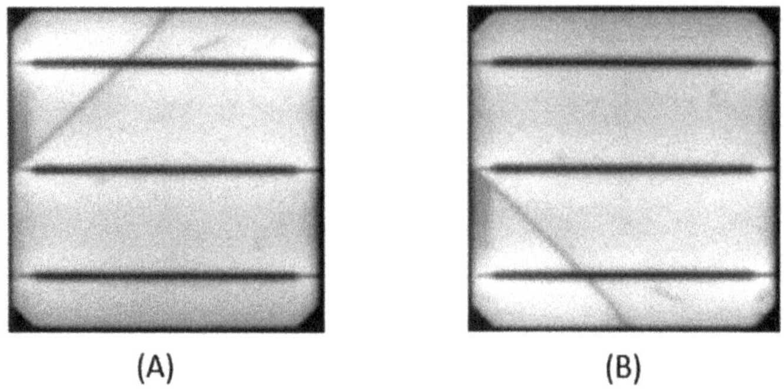

Figure 2. Vertical flipping (A) input (B) output.

Figure 3 presents the application of the horizontal orientation. The rationale was similar to that of vertical orientation i.e., a practical case due to production line or EL camera configuration. It is important to note that the augmentations were not applied indiscriminately as this would result in cases of duplication. For example, implementing 'horizontal' technique to a normal cell image with an approximately uniform surface would result in a replication of the input image due to the symmetrical configuration of the busbars.

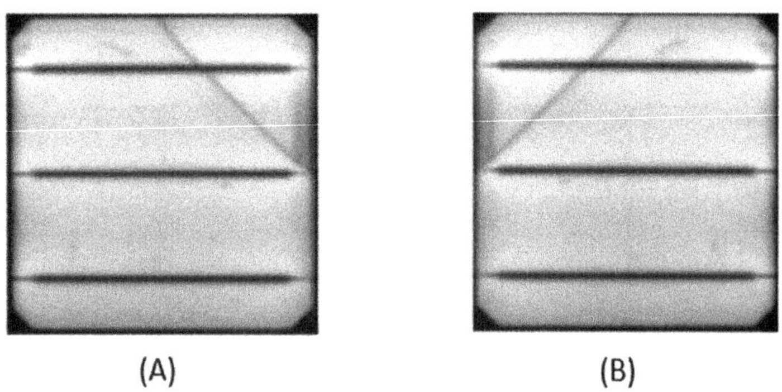

Figure 3. Horizontal flipping (**A**) input (**B**) output.

The tertiary augmentation technique selected was pixel-shifting with respect to the image width. The inspiration could again be traced back to the device-induced variance and production lines configuration variations from factory to factory. The procedure of EL image acquisition requires PV cell to be shielded from external light. Hence EL shields are utilized but may not always be in perfect alignment with respect to the PV cell, resulting in border cut-off discrepancies in the acquired image.

Additionally, based on considerations such as location, inspection procedures, and EL camera specifications, a certain level of shift may be noted in the obtained images. The same underlying principle provides justification for implementing 'pixel-shifting' with respect to the height. Figure 4 presents the (A) input image, (B) width-shifting, and (C) height-shifting implementations. It can be observed that the use of both techniques was not excessive, but rather representative of ground realities i.e., input images were not pixel-shifted by a significant percentage such as 50%.

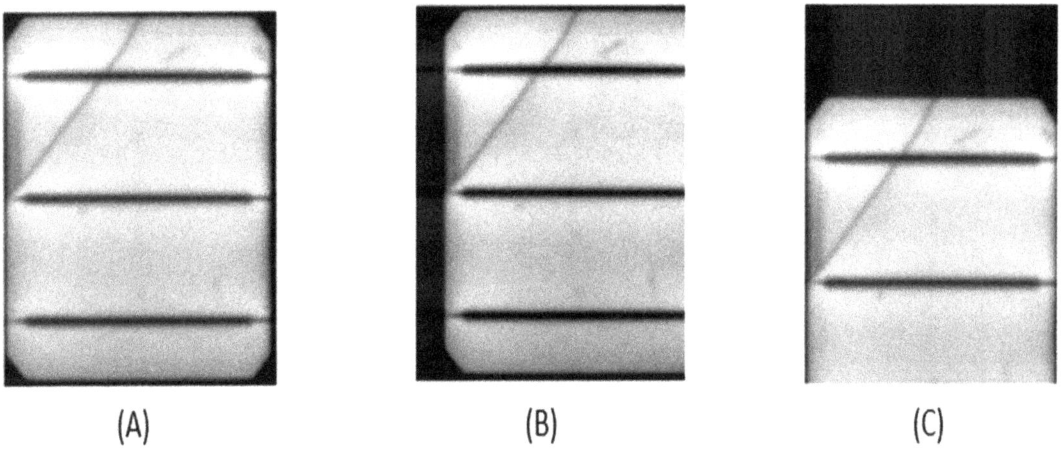

Figure 4. Pixel-based shifting (**A**) input (**B**) width shift (**C**) height shift.

The status of the transformed dataset with respect to the number of samples within each class is presented in Table 2. The fact that post augmentations the dataset contained a total of 787 samples, shows that indiscriminate use of augmentations for significantly increasing the dataset was not the motive.

Table 2. Transformed dataset.

Class	Samples
Normal	282
Defective	787

2.3. *Proposed PV-CrackNet Architecture*

The research was based on two distinct research objectives. Firstly, the development of a lightweight CNN architecture that is compatible for deployment onto computationally constrained edge devices and secondly to address the issue of EL-based PV data procurement from within PV manufacturing facilities. Both objectives directed our research towards the development of a custom CNN architecture rather than implementing transfer learning, as evidenced in the results section. The development of a custom architecture would enable the suppression of the architectural and computational complexities associated with CNNs by reducing the internal convolutional blocks and the number of filters implemented within each block.

Figure 5 presents the internal architectural block diagram of the proposed architecture. It can be observed that only two convolutional blocks were implemented. This was due to the fact that during the data inspection section, discussed earlier, it was concluded that, though variance existed at global and internal class levels, the level of diversity in the variance was not high, mostly limited to surface textural differentials, light intensities, hardware induced and busbar configurations. Hence, by implementing a smaller number of convolutional blocks with carefully tuned filters, there was a high probability of capturing the underlying differentiating features between the two classes.

Layer	Output Shape	Total learnable Parameters
Input	3,224x224	---
Conv-01	12,222x222	336
Batch norm	12,222x222	24
ReLu	12,222x222	---
Max-pool	12,111x111	---
Conv2	24,109x109	2,616
Batch norm	24,109,109	48
ReLu	24,109x109	---
Max-pool	24,54x54	---
Fc1	100 neurons	9,998,500
ReLu	---	---
Fc2	50 neurons	5,050
ReLu	---	---
Output	2 neurons	102
Total Learnable Parameters		**7.01 Million**

Figure 5. Internal block diagram for PV-CrackNet.

The first convolutional block consisted of 12 filters with each filter's dimensions defined as 3 × 3 pixels. It may be argued as to why an odd number was selected for the filter dimensions. The justifications for this were that as opposed to an even-dimensional filter (e.g., 2 × 2 pixels) an odd filter provides a center pixel for filter output encoding. In the case of an even filter due to the lack of a center pixel, aliasing errors would occur. The dimensions of the resultant feature maps from the initial convolutional block were ascertained via.

$$n_{out} = [\frac{n_{in} - 2_p - k}{s}] + 1 \quad (3)$$

where n_{out} = resultant features, n_{in}= Nu. of input features, p = padding dimensions, k = kernel dimensions, s = stride

Equation (3) provided the number of output feature maps that were utilized as input to the second convolutional block consisting of 24 filters. Moreover, from the block diagram in Figure 5, it can be noticed that the number of filters was doubled during the transition from convolutional block one to convolutional block two. The rationale behind this was that the initial layer would carry out a highly abstract feature extraction process looking for features such as lines and edges. Whilst the second convolutional block would go deeper into the extraction of useful features to provide an input to the two fully connected layers and hence require more filters.

Max-pooling was also utilized as a translational invariant component, enabling the aggregation of local features, and reducing positional dependencies. ReLu was selected as the activation due to it having an edge over its predecessors (Sigmoid and TanH) in addressing the issue of vanishing gradients. From the resultant dimensions post the application of Max-pooling, it can be observed that there was a 50% reduction in output dimensions. This was due to the max-pooling being defined with a stride of 2, per the equation above. The number of learnable parameters was calculated via,

$$(n \times m \times I + 1) \times k \tag{4}$$

where n, m = filter sizes, I = input features, k = resultant feature maps.

The second i.e., last convolutional block was flattened before being utilized as input to the first fully connected layer, hence the filter dimensions presented in the above equations were no longer required for calculating the resultant number of learnable parameters, resulting in,

$$(In + 1) \times On \tag{5}$$

where In = input neurons, On = output neurons.

As per the internal architectural block diagram, the proposed architecture with respect to the convolutional blocks and fully connected layers consisted of 7.01 million parameters. Figure 6 presents a more abstract view of the proposed PV-CrackNet architecture consisting of two convolutional blocks followed by two fully connected layers.

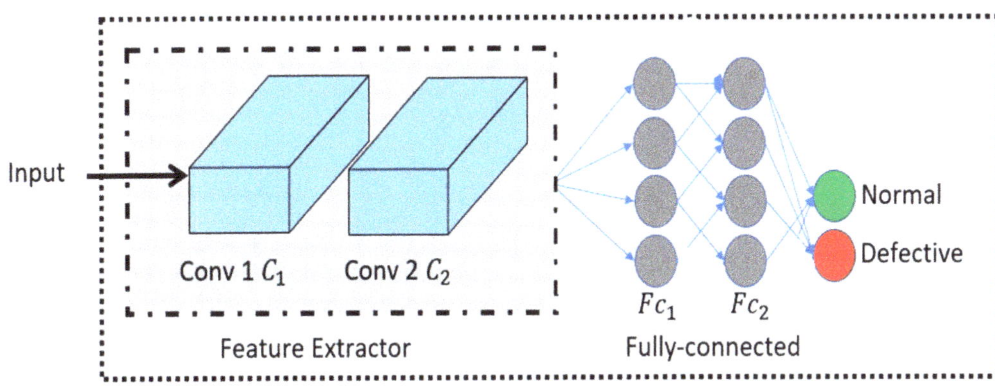

Figure 6. Proposed PV-CrackNet abstract view.

2.4. Filter-Induced Augmentations (FIA)

As mentioned earlier, the second part of the research objective was to address the issue of data scarcity. Quality data procurement especially in the form of EL-based PV cell images at present can be seen as a difficult task as evident from the lack of open-source datasets, hence hindering the development of automated quality inspection solutions within PV quality inspection. Although generic augmentations have been proposed in an earlier section, as evident from the results section, these augmentations although were justified in terms of their representativeness to production-based variance, were not sufficient in providing the architecture with a truly representative dataset for efficient generalization.

Based on the above premise, FIA was proposed to address the issue of generating representative data samples. The inspiration was derived from the fact that earlier convolutional layers learn more abstract representations. These may not be suitable for highly complex feature-intensive image domains such as facial recognition, however as the EL process essentially provides an abstract resultant image with limited variance (discussed earlier), coupled with the fact that the PV-CrackNet contained only two convolutional blocks, there was a possibility that the resultant feature maps could be utilized as representative samples. In order to achieve the extraction of gradients with respect to the input image, the gradients at the prediction layer were back propagated with being passed onto the loss function, essentially disabling the backpropagation post-optimization. The raw gradient via backpropagation containing the same number of pixels as the input image provided a resultant image that could be visually inspected and added to the generated dataset if it was deemed to be representative of production floor conditions. The process flow for FIA is presented in Figure 7.

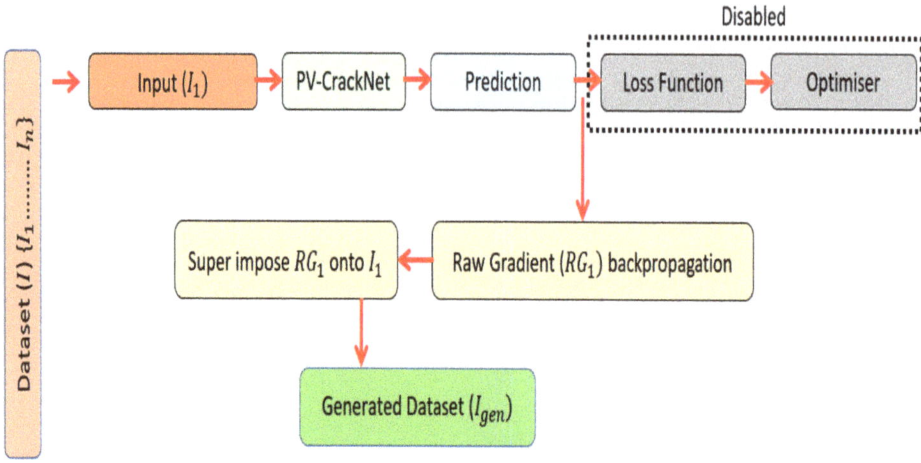

Figure 7. Filter-induced augmentation flow.

As shown in Figure 7, the conventional backpropagation route was disabled and instead the raw gradients were projected back onto the input image for comparison.

Figure 8 presents a comparison of an input image with that of the generated output via FIA (A) input image, (B) generated image. The generated image can be labeled as representative of real variance caused by varying EL filter specifications. Furthermore, notice that the key underlying feature distribution of the input image is intact.

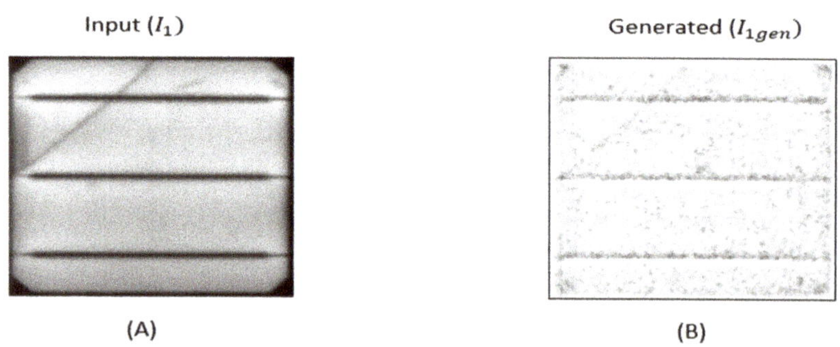

Figure 8. FIA output (**A**) input (**B**) resultant PV cell.

2.5. Reactivating Optimizer

After completing the process of data generation via FIA, the optimizer was reactivated, whilst the FIA procedure was disabled, as shown in Figure 9. Stochastic gradient descent with momentum (SGD-M) was selected as the optimizer for updating the weights with respect to the prediction loss via backpropagation. The SGD-M can be termed as an extension of the plain gradient descent as it replaces the gradient with the moving average of the gradient. Gradient descent-based updating is expressed as,

$$w_t = w_{t-1} - a g_{t-1} \tag{6}$$

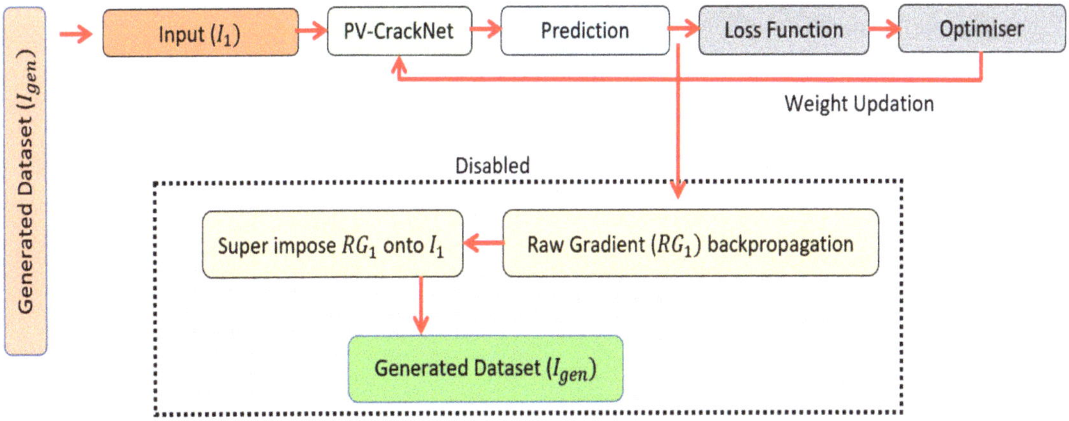

Figure 9. Optimizer reactivation.

Replacing the gradient with moving average of gradient over time,

$$w_t = w_{t-1} - av_{t-1} \tag{7}$$

Moving average is calculated,

$$v_t = \beta v_{t-1} + (1-\beta)g_{t-1} \tag{8}$$

Moving average of the gradient was computed via the equation above, 'g_t' contained the gradient from the previous step. Furthermore, the equation manifests the fact that the entire gradient was not utilized, rather gradient 'mixing' with the moving average was implemented until reaching βv_{t-1}. The moving average v_0 begins as a vector of zeros, gradually accumulating the moving average of the gradients whilst combining $v_t = \beta v_{t-1} + (1-\beta)g_{t-1}$ at each step. Consequently, the present gradient was not actioned upon, rather historical gradients were registered and deposited within v_t . Momentum (β) was a hyper-parameter set by default to 0.9,

$$v_0 = 0 \tag{9}$$

$$v_1 = \beta v_0 + (1-\beta)g_0 = (1-\beta)g_0 \tag{10}$$

$$v_2 = \beta v_1 + (1-\beta)g_1 = (1-\beta)g_0 + (1-\beta)g_1 \tag{11}$$

The activation of the optimizer enabled the training of the architecture, whilst the data generation process was disabled, shown in Figure 9.

3. Results

3.1. Hyper-Parameters

This section of the research presents a comparison of the proposed PV-CrackNet architecture's performance on the original, generic, and FIA-augmented datasets. Google Colaboratory was selected as the virtual training environment due to its free GPU access for speeding up the training process. The access to GPU was however time constrained due to the subscription to the free tier.

The hyperparameters defined at a global level for providing a fair performance comparison are presented in Table 3. As mentioned earlier, due to the limited GPU access, the number of epochs was capped at 40. Additionally, to assist with faster training, the learning rate was set to 0.02.

Table 3. Hyperparameters.

Batch Size	32
Epochs	40
Optimizer	SGD-M
Learning Rate	0.02

3.2. *Original Dataset Performance*

Although the data in its original form was significantly small in size and, more importantly, representative variance, to provide a fair comparison, the architecture was trained on the raw dataset, and the results are presented in Figure 10. Figure 10A presents the loss, and through its interpretation, it can be said that the architecture had stopped any useful learning after a couple of epochs.

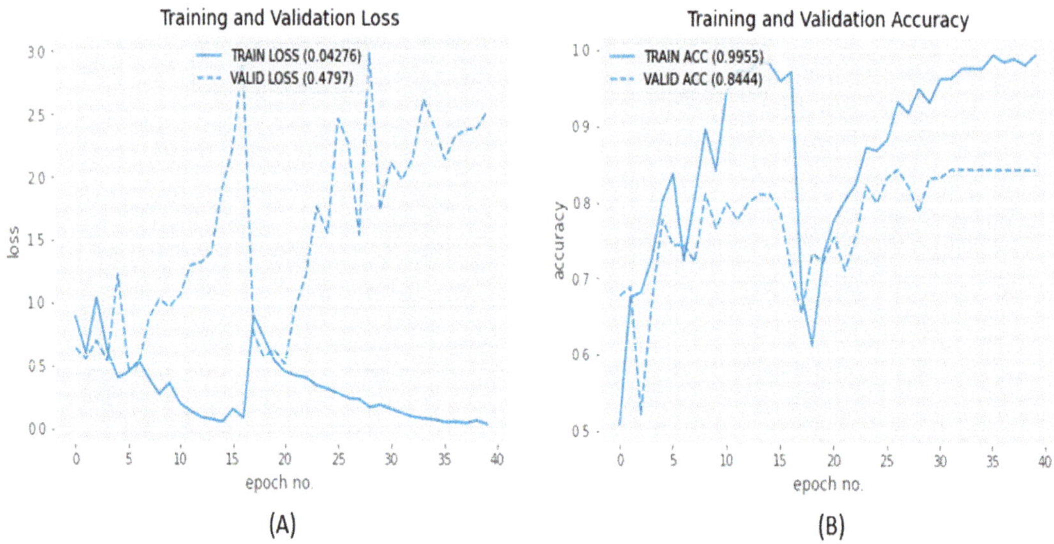

Figure 10. Original data performance (**A**) loss (**B**) accuracy comparison.

Furthermore, the significant increase between the training and validation loss is evident from the fact that the dataset in its present form was not sufficient. When observing the training and validation accuracy, the increasing differential again hints at insufficient data supply.

It may be argued that the architecture had not been given adequate training time, hence increasing the training time (epochs) could have yielded better results. The argument is countered by pointing out from Figure 10B that the validation accuracy stagnates after around 30 epochs, this stagnation remains until the training is complete, demonstrating that training time was not an issue but rather a lack of representative data.

3.3. *Generic-Augmented Dataset Performance*

The next iteration was the application of generic yet representative augmentations. Figure 11 presents the performance of the trained architecture with respect to the translational equivariance-based augmentations presented in the methodology section. These consisted of readily available augmentation techniques within Pytorch, namely, horizontal and vertical rotations, and image shifting.

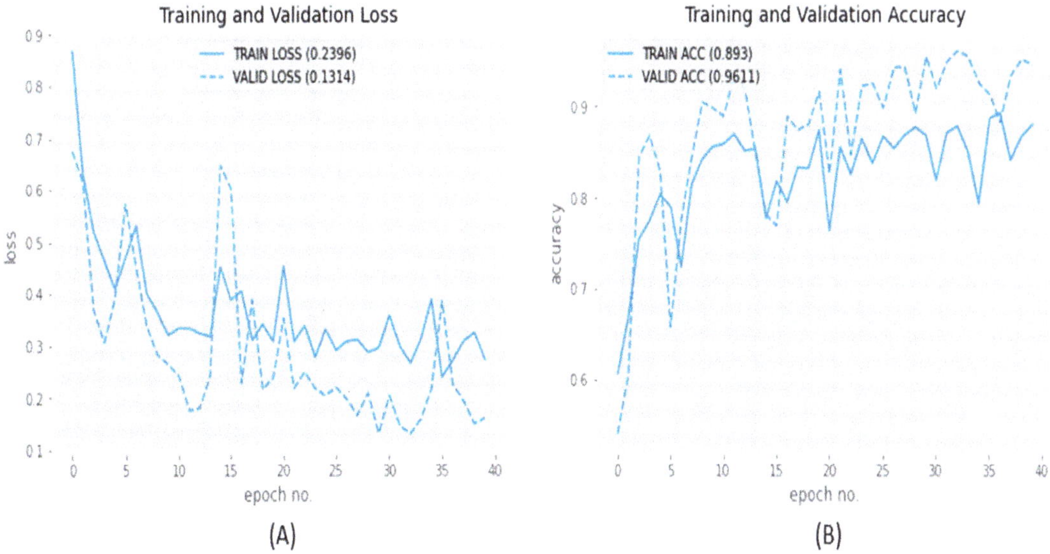

Figure 11. Generic-augmented data performance (**A**) loss (**B**) accuracy comparison.

A clear improvement in performance can be observed, with the validation loss (Figure 11A) reducing to 0.1314 as compared to 0.4797 for the original data. Similarly, the validation accuracy (Figure 11B) improved to 96.1% as compared to 84.4% for the original data. Furthermore, the differential between the two respective accuracies, known as overfitting, also decreased from 15.1% (original data) to 6.8%.

A critical analysis shows that although the degree of overfitting had decreased by 8.4%, the difference between the training and validation accuracies stood at 6.8%, indicating that further improvements to the model could be made via a more representative dataset.

3.4. Filter-Induced Augmentations Dataset Performance

The final iteration, aimed at improving the performance whilst further suppressing the degree of overfitting, was based on the proposed FIA mechanism. As evident from the generic augmentation testing, the validation accuracies had surpassed 90%, hence the evaluation was broadened to include not only the accuracy but also the precision, recall, and F1-score. The performance of the architecture via FIA is presented in Table 4.

Table 4. FIA performance.

Training Acc	99.11%
Validation Acc	97.42%
Precision	98%
Recall	96%
F1-score	97%

It is evident from Table 4 that the FIA produced the highest performance. Starting with the performance with respect to the degree of overfitting, the difference between the training and validation stood at its lowest point (1.69%), compared to 15.1% (original data) and 6.8% (generic augmented data), as shown in Figure 12. The fact that the validation accuracy was the highest at 97.42% and the degree of overfitting was at its lowest (1.69%) demonstrates that the proposed FIA mechanism was effective in providing a representative dataset, which was well-generalized upon by the proposed PV-CrackNet architecture.

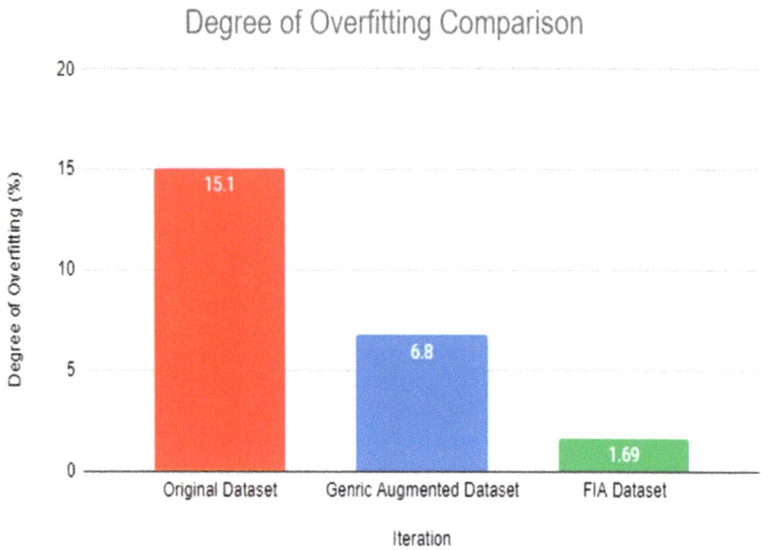

Figure 12. Degree of overfitting comparison.

To provide a more robust analysis, the precision, recall, and F1-score were also inspected. Precision was the highest at 98%, whilst the recall stood at a respective 96%. Overall, the proposed PV-CrackNet performance could be gauged via the F1-score at an impressive 97%.

4. Discussion

The results section demonstrated the effectiveness of the proposed FIA mechanism for generating representative samples and resulting in a highly generalized architecture. However, the generalization of the architecture was not the only objective of this research. The second objective was based on the development of a lightweight architecture in order to be computationally efficient.

To evaluate the effectiveness of the developed PV-CrackNet architecture, the computational complexities of various state-of-the-art architectures were evaluated. The evaluation was based on the number of learnable parameters as shown in Table 5.

Table 5. Computational comparison.

Architecture	Parameters (M)
PV-CrackNet	7.01
VGG-19 [19]	143.67
ResNet-18 [20]	11.69
AlexNet [21]	61.0
GoogleNet [22]	13.0

It can be clearly observed via Table 5, that the proposed architecture contained the least number of parameters at 7.01 million, whilst the VGG-19 was the most computationally demanding with 143.67 million learnable parameters. The number of learnable parameters would have a direct implication on the training process, as a higher number of learnable parameters would require more training time/resources for the architecture to converge.

AlexNet was the second-most computationally demanding with 61 million parameters. This was interesting, as its internal block composition with respect to the convolutional layers was the most similar to the proposed PV-CrackNet. Although AlexNet contained only five convolutional blocks followed by three fully connected layers, its internal filter configurations had an adverse impact on the computational complexity of the network. For example, the initial convolutional block contained 11 × 11-pixel filters as opposed to the PV-CrackNet initiating with 3 × 3-pixel filters. As a result, the convolutional process resulted in an increased number of computations.

5. Conclusions

In conclusion, it can be stated with confidence that the presented research was highly successful in achieving the two research objectives. The first objective i.e., addressing the issue of EL-based data scarcity, was addressed via the FIA mechanism. The effectiveness of the FIA in generating highly representative samples was demonstrated by the fact that it reached the highest validation accuracy of 97.42% with the lowest degree of overfitting at 1.69%.

The second objective was achieved by demonstrating, via a comparison of various state-of-the-art architectures with the proposed PV-CrackNet with respect to the number of learnable parameters. The developed architecture again provided the highest performance, consisting of only 7.01 million parameters, evidencing the claim of developing a computationally lightweight architecture.

Commenting on the marketing potential, the proposed PV-CrackNet can be integrated into existing quality inspection processes within PV manufacturing facilities for enhancing their production output in terms of efficiency. Additionally, third-party PV quality inspection firms can utilize PV-CrackNet for automating defect detection and providing their clients with a higher degree of confidence in the shipped PV panels.

Furthermore, due to the abstract nature of images within various other domains, the proposed FIA mechanism can be implemented for introducing representative variance whilst addressing the issue of data scarcity, i.e., retina-based exudate detection data enhancements and defect accentuation [23] or industrial infrastructure defect detection [24].

Author Contributions: Conceptualization, M.H.; Formal analysis, M.H.; Investigation, M.H. and H.A.-A.; Methodology, M.H.; Project administration, H.A.-A. and R.H.; Visualization, H.A.-A. and R.H.; Writing—original draft, M.H.; Writing—review & editing, H.A.-A. and R.H. All authors have read and agreed to the published version of the manuscript.

Funding: This research received no external funding.

Conflicts of Interest: The authors declare no conflict of interest.

References

1. Kabir, E.; Kumar, P.; Kumar, S.; Adelodun, A.A.; Kim, K.H. Solar Energy: Potential And Future Prospects. *Renew. Sustain. Energy Rev.* **2018**, *82*, 894–900. [CrossRef]
2. Mahmoud, D.; Peter, M. Development of Novel Solar Cell Micro Crack Detection Technique. *IEEE Trans. Semicond. Manuf.* **2019**, *32*, 277–285.
3. Mahmoud, D.; Violeta, H.; Mark, D.; Bruce, M. Effect of micro cracks on photovoltaic output power: Case study based on real time long term data measurements. *Micro Nano Lett.* **2017**, *12*, 803–807.
4. Frazão, M.; Silva, J.A.; Lobato, K.; Serra, J.M. Electroluminescence of silicon solar cells using a consumer grade digital camera. *Measurement* **2017**, *99*, 7–12. [CrossRef]
5. Fu, Y.; Ma, X.; Zhou, H. Automatic detection of multi-crossing crack defects in multi-crystalline solar cells based on machine vision. *Mach. Vis. Appl.* **2021**, *32*, 60. [CrossRef]
6. Hussain, M.; Al-Aqrabi, H.; Hill, R. Statistical Analysis and Development of an Ensemble-Based Machine Learning Model for Photovoltaic Fault Detection. *Energies* **2022**, *15*, 5492. [CrossRef]
7. Akram, M.W.; Li, G.; Jin, Y.; Chen, X.; Zhu, C.; Ahmad, A. Automatic detection of photovoltaic module defects in infrared images with isolated and develop-model transfer deep learning. *Sol. Energy* **2020**, *198*, 175–186. [CrossRef]
8. Ahmad, A.; Jin, Y.; Zhu, C.; Javed, I.; Maqsood, A.; Akram, M.W. Photovoltaic cell defect classification using convolutional neural network and support vector machine. *IET Renew. Power Gener.* **2020**, *14*, 2693–2702. [CrossRef]
9. Dunderdale, C.; Brettenny, W.; Clohessy, C.; van Dyk, E.E. Photovoltaic defect classification through thermal infrared imaging using a machine learning approach. *Prog. Photovolt. Res. Appl.* **2020**, *28*, 177–188. [CrossRef]
10. Tun, N.L.; Gavrilov, A.; Tun, N.M.; Aung, H. Remote Sensing Data Classification Using A Hybrid Pre-Trained VGG16 CNN-SVM Classifier. In Proceedings of the IEEE Conference of Russian Young Researchers in Electrical and Electronic Engineering (ElConRus), Moscow, Russia, 26–29 January 2021.

11. Zhang, J.; Feng, Y. Advanced Chinese Character Detection for Natural Scene Based on EAST. *J. Phys. Conf. Ser.* **2020**, *1550*, 032050. [CrossRef]
12. Hussain, M.; Chen, T.; Hill, R. Moving toward Smart Manufacturing with an Autonomous Pallet Racking Inspection System Based on MobileNetV2. *J. Manuf. Mater. Process.* **2022**, *6*, 75. [CrossRef]
13. Khan, Z.Y.; Niu, Z. CNN with depthwise separable convolutions and combined kernels for rating prediction. *Expert Syst. Appl.* **2021**, *170*, 114528. [CrossRef]
14. Pierdicca, R.; Malinverni, E.S.; Piccinini, F.; Paolanti, M.; Felicetti, A.; Zingaretti, P. Deep Convolutional Neural Network for Automatic Detection of Damaged Photovoltaic Cells. *Int. Arch. Photogramm. Remote Sens. Spat. Inf. Sci.* **2018**, *XLII-2*, 893–900. [CrossRef]
15. Napoletano, P.; Piccoli, F.; Schettini, R. Anomaly Detection in Nanofibrous Materials by CNN-Based Self-Similarity. *Sensors* **2018**, *18*, 209. [CrossRef]
16. Deitsch, S.; Christlein, V.; Berger, S.; Buerhop-Lutz, C.; Maier, A.; Gallwitz, F.; Riess, C. Automatic classification of defective photovoltaic module cells in electroluminescence images. *arXiv* **2019**, arXiv:1807.02894v3. [CrossRef]
17. Hussain, M.; Chen, T.; Titrenko, S.; Su, P.; Mahmud, M. A Gradient Guided Architecture Coupled with Filter Fused Representations for Micro-Crack Detection in Photovoltaic Cell Surfaces. *IEEE Access* **2022**, *10*, 58950–58964. [CrossRef]
18. Tang, W.; Yang, Q.; Xiong, K.; Yan, W. Deep learning based automatic defect identification of photovoltaic module using electroluminescence images. *Sol. Energy* **2020**, *201*, 453–460. [CrossRef]
19. Yap, X.Y.; Chia, K.S.; Tee, K.S. A Portable Gas **Pressure** Control and Data Acquisition System using Regression Models. *Int. J. Electr. Eng. Inform.* **2021**, *13*, 242–251. [CrossRef]
20. Gao, M.; Song, P.; Wang, F.; Liu, J.; Mandelis, A.; Qi, D. A Novel Deep Convolutional Neural Network Based on ResNet-18 and Transfer Learning for Detection of Wood Knot Defects. *J. Sens.* **2021**, *2021*, 4428964. [CrossRef]
21. Krizhevsky, A.; Sutskever, I.; Hinton, G.E. ImageNet classification with deep convolutional neural networks. *Commun. ACM* **2017**, *60*, 84–90. [CrossRef]
22. Szegedy, C.; Liu, W.; Jia, Y.; Sermanet, P.; Reed, S.; Anguelov, D.; Erhan, D.; Vanhoucke, V.; Rabinovich, A. Going deeper with convolutions. In Proceedings of the 2015 IEEE Conference on Computer Vision and Pattern Recognition (CVPR), Boston, MA, USA, 7–12 June 2015. [CrossRef]
23. Hussain, M.; Al-Aqrabi, H.; Munawar, M.; Hill, R.; Parkinson, S. Exudate Regeneration for Automated Exudate Detection in Retinal Fundus Images. *IEEE Access* **2022**. [CrossRef]
24. Hussain, M.; Al-Aqrabi, H.; Munawar, M.; Hill, R.; Alsboui, T. Domain Feature Mapping with YOLOv7 for Automated Edge-Based Pallet Racking Inspections. *Sensors* **2022**, *22*, 6927. [CrossRef] [PubMed]

sustainability

Article

Study on Properties of Regenerated Fluorinated Polyurethane Rigid Foam Prepared by Degrading Waste Polyurethane

Da-Sheng Zhang [1], Xiao-Hua Gu [1,2,3,*], Si-Wen Liu [4], Yan Liu [2], Qing-Yun Zhou [1], Shang-Wen Zhu [2] and Yan-Wei Zhu [3]

1 School of Material Science and Engineering of Qiqihar University, Qiqihar 161006, China
2 School of Energy and Building Environment of Guilin University of Aerospace Technology, Guilin 541004, China
3 State Key Laboratory for Modification of Chemical Fibers and Polymer Materials, College of Materials Science and Engineering, Donghua University, Shanghai 201620, China
4 College of Innovative Material & Energy, Hubei University, Wuhan 430062, China
* Correspondence: gxh218@163.com; Tel.: +86-181-21138868

Citation: Zhang, D.-S.; Gu, X.-H.; Liu, S.-W.; Liu, Y.; Zhou, Q.-Y.; Zhu, S.-W.; Zhu, Y.-W. Study on Properties of Regenerated Fluorinated Polyurethane Rigid Foam Prepared by Degrading Waste Polyurethane. *Sustainability* **2022**, *14*, 15685. https://doi.org/10.3390/su142315685

Academic Editors: Fuqiang Wang, Chao Shen, Dong Li and Zhonghao Rao

Received: 14 October 2022
Accepted: 21 November 2022
Published: 25 November 2022

Abstract: Polyurethane (PU) has become one of the most widely used materials in the industrial field due to its excellent performance and wide range of applications. The increasing consumption of polyurethane materials has resulted in significant polyurethane waste. We can recycle waste polyurethane to obtain recycled polyurethane, but to make the recycled polyurethane meet different performance requirements in terms of utility and cost and enable the recycled polyether polyol to be further applied to high-end applications, it is necessary to improve the use value of recycled polyether polyol. In this paper, self-made Fluorodiol was added to the degradation process of waste polyurethane to obtain fluorinated regenerated polyether polyol. Recycled fluorinated polyurethane with excellent performance was prepared using fluorine-containing recycled polyether polyol. The regenerated fluorinated polyether polyols were characterized by infrared spectroscopy, viscosity, hydroxyl value, and GPC molecular weight distribution. The density, apparent morphology, water absorption, mechanical strength, thermal conductivity, and thermal stability of RFPU rigid foams with different fluorine contents were studied by scanning electron microscopy. The results show that when the addition of Fluorodiol was 8% of the mass of waste polyurethane, the density was 41.2 kg/m^3, the immersion loss rate was 2.125%, the compressive strength was 0.315 Mpa, and the thermal conductivity was 0.0227 W/m·K. The RFPU prepared by Fluorodiol has low surface energy, good compressive strength, hydrolysis resistance, and is expected to be widely used in special insulation materials. Thus, the sustainable recycling of polyurethane is achieved.

Keywords: fluoroalcohol; waste polyurethane; recycled polyurethane; recycling; sustainable material

1. Introduction

Polyurethane (PU) is a polymer with a carbamate segment repeating structural unit prepared by an isocyanate and polyol reaction. Polyurethane is now one of the most widely used materials. Polyurethane foam, which accounts for up to 80% of the polyurethane industry, has been widely used in various industries, including building and refrigerator insulation, car seats, and cushions [1–4]. Due to the continuous increase in the consumption of polyurethane materials, a large amount of waste polyurethane has accumulated. However, due to the excellent stability of polyurethane, the material is not easily degraded by microorganisms or air under natural conditions; it is insoluble, and it is difficult to treat [5–9]. Moreover, more and more states explicitly prohibit its burial or incineration, and it is difficult to recycle, resulting in a waste of living space and chemical resources [1,4,10–14]. Therefore, how to deal with the degradation and reuse of waste PU in a green and efficient manner is crucial to determine [15–18]. The recycling prospect of waste polyurethane is very broad and has great potential research value and significance.

Fluorinated polyurethane is a polymer material with special functions. Since Lovelace invented the first fluorine-containing polyurethane patent in 1958, the synthesis of fluorine-containing polyurethane has attracted widespread attention and become a popular topic of polyurethane research [11,19–24]. In this paper, by including fluorine-containing groups in the degradation process of waste polyurethane, fluorine-containing recycled polyols were prepared and then foamed to prepare fluorine-containing polyurethane rigid foam, which combines the excellent mechanical properties and surface properties of polyurethane and fluorine-containing polymers, giving the material high strength, high thermal stability, high thermal insulation performance, and excellent hydrophobicity and oil resistance [13,20,25–28]. In this paper, a series of fluorine-containing polyurethane rigid foams were prepared using fluorine-containing recycled polyols obtained from the degradation of waste polyurethane foams by ethylene glycol (EG) and diethylene glycol (DEG) composite Fluorodiols as raw materials, and their properties and composition were characterized. This modification method has no relevant literature in the field of waste polyurethane high-value recycling. This method can help realize low energy consumption, low cost, rapid recovery, and high-value utilization of waste polyurethane. This method can explore the sustainability of polyurethane to contribute to the protection of the environment, and is of great significance to the industrial production of waste polyurethane degradation and recovery.

2. Materials and Methods

2.1. Reagents and Apparatus

Waste PU rigid foam, waste directly buried pipe polyurethane foam, Shandong Haier Group, Qingdao, China; ethylene glycol (EG): purity 98%, China Nanjing Gaohua Chemical Group Co., Ltd., Nanjing, China; diethylene glycol (DEG): purity 98%, China Shandong Huasheng Chemical Co., Ltd., Zhanhua, China; isophorone diisocyanate (IPDI): industrial grade, China Shandong Lianchuang Chemical Co., Ltd. Qingdao, China, KOH, purity 98%, China Beijing Kono Chemical Co., Ltd., Beijing, China; double metal catalyst (DMC), industrial grade, China Changzhou Hongyu Chemical Co., Ltd., Changzhou, China; polyether polyol 4110, industrial grade, China Shandong Yinuowei Chemical Co., Ltd., Zibo, China; 4,4′-diphenylmethane diisocyanate (MDI), industrial grade, China Shandong Vanke New Materials Co., Ltd., Dongying, China; organic tin, industrial grade, China Pharmaceutical Shanghai Chemical Reagent Company, Shanghai, China; dimethyl silicone oil, analytically pure, China Shandong Yinuowei Chemical Co., Ltd., Zibo, China; foaming agent Cyclopentane, industrial grade, China Pharmaceutical Shanghai Chemical Reagent Co., Ltd., Shanghai, China; Fluoroalcohol (C8, 35.16%; C10, 53.31%), molecular weight 486, China Yangzhou Modier Electronic Materials Co., Ltd., Yangzhou, China; diethanolamine (DEA), analytically pure; ethyl acetate, analytically pure, Aladdin reagent; toluene, analytically pure, Aladdin reagent.

2.2. Sample Preparation

2.2.1. Fluorodiol Preparation

The first reactor with a thermometer and stirring device was cleaned with nitrogen, and nitrogen was continuously introduced during the reaction. Ethyl acetate and IPDI (2) were added to the first reactor, stirred, and heated to 60 °C until IPDI was completely dissolved. Fluoroalcohol (1) was slowly added to the reactor. After the addition, the reactor was heated to 80 °C for 3 h to obtain perfluoroalkyl isocyanate (3).

The second reactor with a thermometer and stirring device was cleaned with nitrogen, and nitrogen was continuously introduced during the reaction. Ethyl acetate and diethanolamine (4) were added to the second reaction kettle and stirred at 10 °C. After the diethanolamine was completely dissolved, the temperature was increased to 20 °C. The product (3) was dropped into the diethanolamine solution, and the system's temperature was kept below 20 °C during the dropping process. After the addition, the reactor temperature was raised to 25 °C for 2 h. After the reaction, toluene was added to the final reaction system for washing, filtering, and drying, and finally, the Fluorodiol product (5) was obtained. The reaction process is shown in Figure 1.

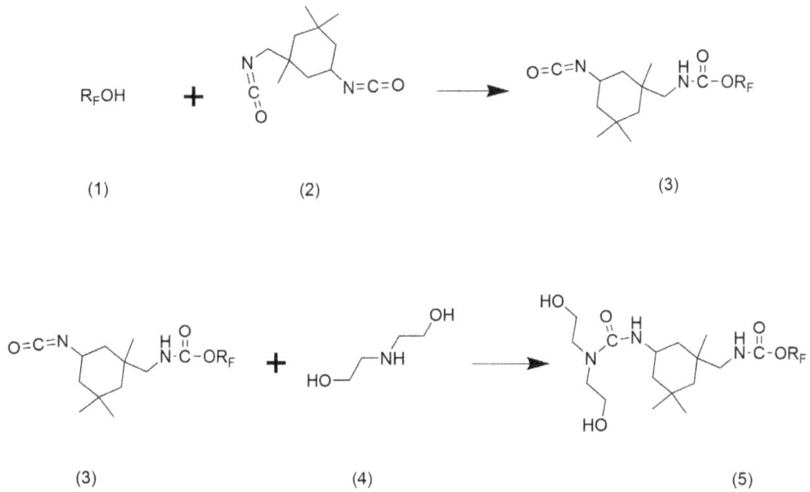

Figure 1. Synthesis schematic diagram of Fluorodiol (R_F are several CF groups on fluoroalcohol).

2.2.2. RFPU Sample Preparation

A total of 50 g waste polyurethane foam was cleaned, dried, and then broken into 1~2 mm powder, which was waste polyurethane powder; fluoroglycol and two-component decomposition crosslinking agents (EG as alcoholysis agent, DEG as alcoholysis agent), catalyst KOH, and DMC were added to the reactor, stirred, and dissolved at 130 °C. After dissolution, 100 g of waste polyurethane powder was added. The temperature was increased to 200 °C for alcoholysis reaction for 2 h, then cooled to room temperature to obtain fluorinated regenerated polyether polyol. The ratio of the mass of waste polyurethane powder to the total mass of Fluorodiol, two-component decomposition crosslinking agent and catalyst is 1:1.2. Five parallel samples with different fluorine contents were prepared, and the addition of Fluorodiol was 2%, 4%, 6%, 8%, and 10% of the mass of waste polyurethane. Figure 2 shows the mechanism of the degradation reaction [4,29–32].

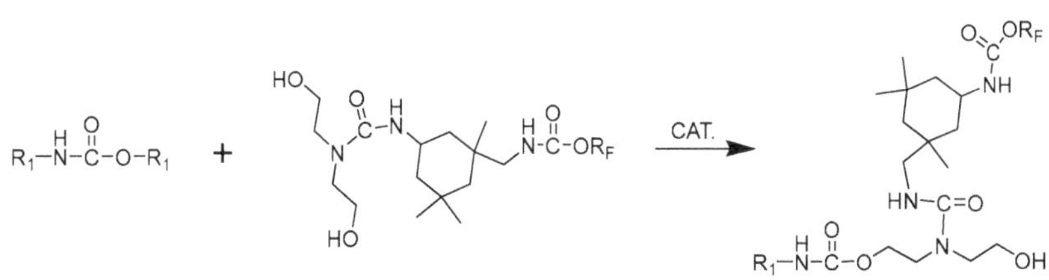

Figure 2. Schematic formula of fluorinated polyether polyol (R_1 is other groups on the long chain of polyurethane; R_F are several CF groups on fluoroalcohol).

Take 20 g fluorine-containing recycled polyether polyol and 20 g polyether polyol 4110 obtained by the above steps: add silicone oil L-600, organotin catalyst, and foaming agent Cyclopentane in turn; stir well; add 45 g MDI; stir quickly; and let stand to allow it to foam naturally to create the fluorine-containing recycled polyurethane material. Figure 3 shows the schematic formula for preparing the RFPU.

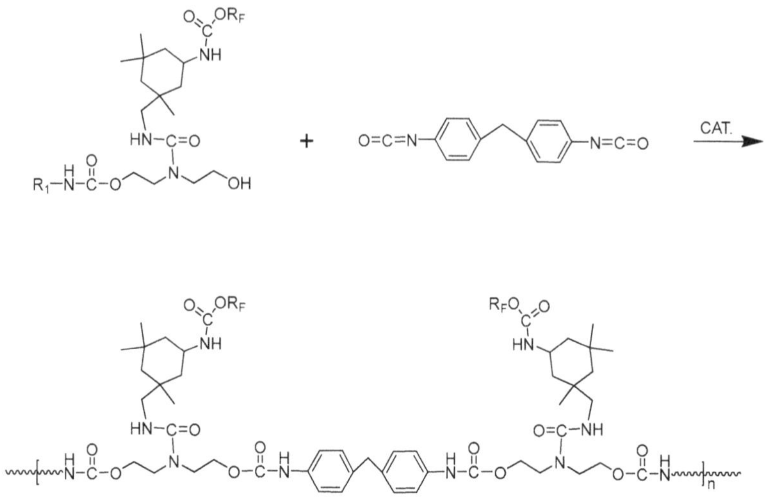

Figure 3. Main schematic formula of fluorinated recycled polyurethane (R_1 is other groups on the long chain of polyurethane; R_F are several CF groups on fluoroalcohol).

2.2.3. Reused Rigid Polyurethane Foam (RRPU) Sample Preparation

The steps of treating waste polyurethane foam are the same as those of RFPU. The two-component decomposition crosslinking agent (EG as alcoholysis agent, DEG as alcoholysis agent), KOH, and DMC were added to the reactor and stirred at 130 °C. After dissolution, 100 g of waste polyurethane powder was added, and the temperature was raised to 200 °C for an alcoholysis reaction for 2 h and cooled to room temperature to obtain recycled polyether polyol. The mass ratio of waste polyurethane powder to the total mass of the two-component decomposition crosslinking agent and catalyst is 1:1.2; the foaming step is the same as that of RFPU, resulting in RRPU.

2.2.4. Fluoroalcohol Comparison Samples

The steps of treating waste polyurethane foam are the same as those of RFPU. A total of 8 g of fluoroalcohol and two-component decomposition crosslinking agent (EG as alcoholysis agent, DEG as alcoholysis agent), KOH catalyst, and DMC were added to the reactor, stirred, and dissolved at 130 °C. After dissolution, 100 g of waste polyurethane powder was added, the temperature was raised to 200 °C for an alcoholysis reaction for 2 h, and the sample was then cooled to room temperature to obtain recycled polyether polyol. The mass ratio of waste polyurethane powder to the total mass of the two-component decomposition crosslinking agent and catalyst is 1:1.2. The foaming step is the same as the RFPU foaming step, resulting in fluoroalcohol comparison samples.

2.2.5. Prue 4100 Sample Preparation

The pure 4100 sample is obtained by following this process: take 40 g of the above steps to obtain the polyether polyol 4110; add silicone oil L-600, organotin catalyst, and foaming agent Cyclopentane; stir evenly; add 45 g MDI; quickly stir; and stand to allow natural foaming.

2.3. Characterization

The infrared spectra of the prepared Fluorodiol, regenerated fluorinated polyether polyol, and RFPU were analyzed using a Fourier transform infrared spectrometer (IR-960, China Tianjin Rui'an Technology Co., Ltd., Tianjin, China).

Use the 'GB/T12008.7-2010' test viscosity method under the experimental conditions of 25 degrees Celsius with an NDJ-1 rotary viscometer to measure the viscosity of fluorinated polyether polyol products.

The GPC molecular weight distribution test was carried out by taking 5 mg of fluorine-containing polyether polyol in a sample bottle and adding 2 mL of spectral-grade N-N dimethylamide.

The hydroxyl value of fluorinated polyether polyol was determined by acid–base titration.

The prepared foam was cut into 50 mm × 50 mm × 50 mm samples, and the compressive strength of the samples at 10% deformation was tested using a material universal testing machine (WSM20KN, China Changchun Intelligent Equipment Co., Ltd., Changchun, China).

The regenerated RFPU rigid foam was cut into thin samples, and the cell of the RFPU structure was observed by scanning electron microscopy (KYKY-EM3900, China Guangzhou Hongsheng Technology Co., Ltd., Guangzhou, China). The sample was prepared into a square of 1 cm × 1 cm × 1 cm. The sample was completely immersed in water for 24 h, and the water-absorption rate was measured. After the sample was completely dried, the immersion loss rate was measured.

Thermogravimetric analysis of recycled PU rigid foam was carried out using a thermogravimetric analyzer (Q5000IR, TA, New Castle County, DE, USA).

The thermal conductivity of recycled PU rigid foam at 25 °C was measured using a thermal conductivity meter (DRPL-III, China Shanghai Jiezhun Instrument Equipment Co., Ltd., Shanghai, China).

3. Results and Discussion

3.1. FTIR Spectra Analysis

3.1.1. FTIR Spectra Analysis of Fluorodiol

Figure 4 shows the FTIR spectra of Fluoroalcohol and Fluorodiol. For Fluoroalcohol, the absorption peak at 3600–2000 cm^{-1} belongs to the stretching vibration of −OH, and the absorption peak at 1242 cm^{-1},1320 cm^{-1} should be attributed to the stretching vibration of C−F [25,33]. Compared with the Fluorodiol, the new absorption peaks at 1726 cm^{-1} and 1464 cm^{-1} correspond to the stretching vibration of C=O and C−N in −NCO [34]. Therefore, it can be concluded that the dihydroxy group has been successfully applied to Fluoroalcohol.

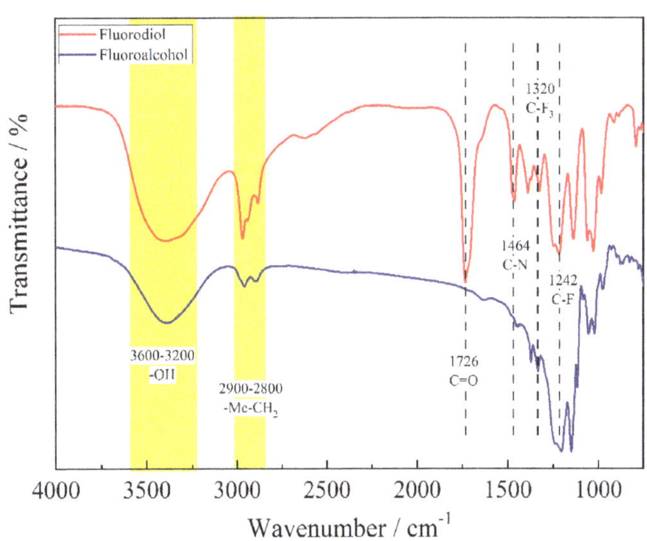

Figure 4. Infrared absorption spectra of Fluoroalcohol and Fluorodiol.

3.1.2. FTIR Spectra Analysis of Fluorinated Regenerated Polyols

It can be seen in Figure 5 that compared with the other three curves (b, c, d), the polyether polyol 4110 (a) has a strong absorption band in the range of 3500–3300 cm^{-1}, which is the stretching vibration peak of the alcohol hydroxyl group. A clear, strong absorption band near 1082 cm^{-1} is attributed to the stretching vibration peak of the polyether polyurethane ether group. It can be concluded that the degradation product is a mixture of polyether polyols and aromatic polyols [5,12,35–37]. The vibration absorption peaks of −C−F near 1180 cm^{-1} and 1128 cm^{-1} can be found on curves c and d, and the saturated vibration absorption peak of 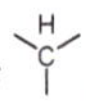 is found at 1352 cm^{-1}.

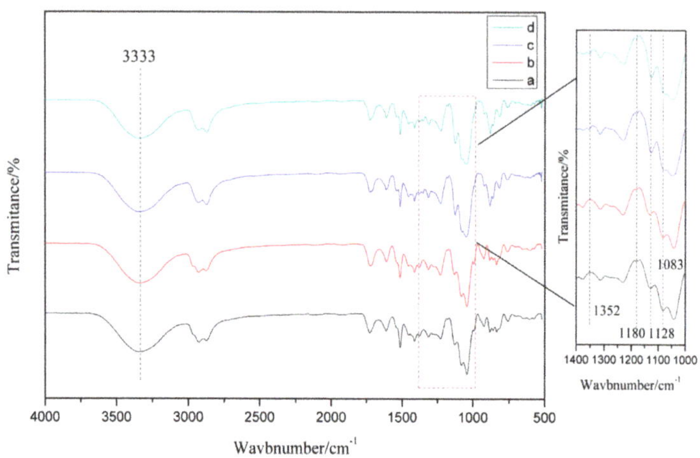

Figure 5. FTIR spectra of regenerated polyols (a: polyether polyol 4110; b: regenerated polyether polyol; c: comparison of fluorinated recycled polyether polyols; d: fluorinated regenerated polyether polyol).

It can be seen in Figure 5 that the Fluorodiol synthesized in this paper is consistent with the product structure in the synthesis path. The curves of four kinds of degradation strips showed strong absorption bands in the range of 3500–3300 cm^{-1}, which were the stretching vibration peaks of the alcohol hydroxyl groups. The strong absorption band near 1740 cm^{-1} is the benzene overtone peak. A clear, strong absorption band near 1082 cm^{-1} is attributed to the stretching vibration peak of the polyether polyurethane ether group [3,29,38–40]. It can be concluded that the degradation product is a mixture of fluorinated polyether polyols and fluorinated aromatic polyols.

3.2. Viscosity Analysis of Regenerated Polyol

It is known that the lower the viscosity of the degraded material obtained after the alcoholysis of waste polyurethane, the more thorough the degradation and the higher the hydroxyl value. The data in Figure 6 show that with the addition of Fluorodiols, the viscosity of the degraded material will increase, and the hydroxyl value will decrease, which is not expected. When the addition of Fluorodiol reaches 11%, the viscosity is greatly increased to 3653.2 mPa· s, at which cannot be foamed at room temperature; the hydroxyl value also decreased significantly to 352.5 mg KOH/g, indicating that the waste polyurethane was not fully degraded to form small molecular alcohols. This situation is due to the fact that the addition of Fluorodiols hindered the degradation of waste polyurethane so that it was not fully degraded; meanwhile, the degradation of a large number of unreacted carbamate bonds led to an increase in viscosity and hydroxyl value decline.

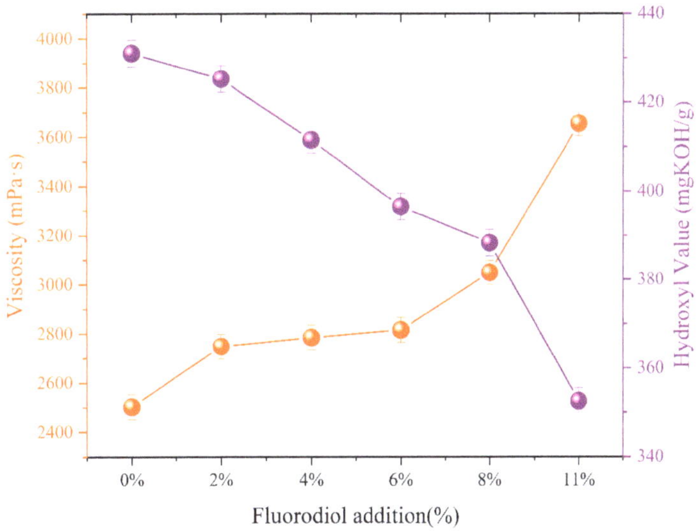

Figure 6. Effect of the addition of different amounts of Fluorodiol on viscosity and hydroxyl value of regenerated fluoropolyether polyols.

3.3. GPC Analysis of Regenerated Polyol

The molecular weight test of regenerated polyols with different amounts of Fluorodiol gradients yielded the following data, as Table 1:

Table 1. Molecular weight distribution of regenerated polyols degraded by different mass concentration gradients of Fluorodiol.

Fluorodiol Addition %	Mn	PDI
Polyether 4110	1104	1.179
0	2465	1.217
2	2682	1.242
4	2734	1.234
6	2744	1.254
8	2812	1.255
11	3455	1.322

The polyurethane chain is decomposed into low-molecular-weight segments due to the alcoholysis agent and catalyst, and its elution time is similar to that of polyether polyol 4110, which is an oligomer. The average molecular weight (Mn) of the recovered polyols was between 4804 and 2465. With the increase in the amount of Fluorodiol, the molecular weight gradually increased, and the elution time gradually decreased. When the addition of Fluorodiol exceeded 8%, the changes in Mn and PDI of recovered polyols increased significantly. This shows that the maximum Fluorodiol amount that can be used when the molecular weight distribution of the prepared fluorine-containing recycled polyol is the most uniform is 8%, which ensures that the waste polyurethane is degraded as fully as possible.

3.4. The Effect of Fluorodiol on the Density of RFPU Rigid Foams

The densities of RFPUs prepared with different amounts of Fluorodiol were measured using the pycnometer method. The results are shown in Figure 7. The results showed that the foam density did not increase significantly with the increase in the amount of Fluorodiol from 0 to 8%, and the foam density was 41.2 kg/m^3 when the amount was

8%. When the amount of Fluorodiol reached 10%, the excessive fluorine content in the polyurethane foaming process produced many defoaming phenomena, making the final density soar to 72.4 kg/m^3.

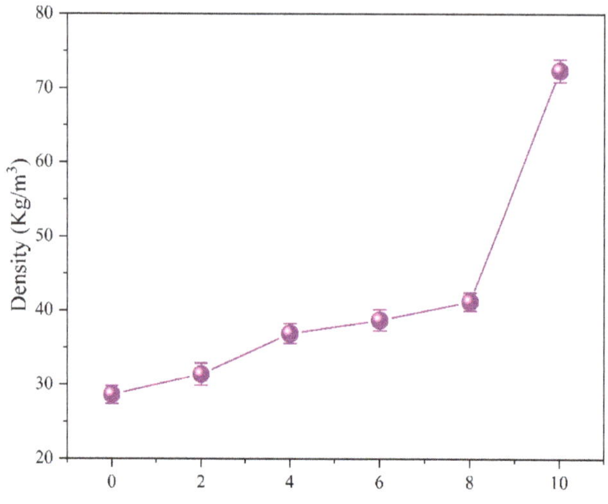

Figure 7. Density curves of RFPU prepared with different Fluorodiol ratios.

3.5. *Effect of Fluorodiols on Water Absorption and Loss Rate of RFPU Foam*

The effect of different amounts of Fluorodiol on the water absorption of RFPU was studied by measuring the water absorption. First, RFPU was made into a sample with a volume of 1 cm^3 and then put into deionized water. The water-absorption rate of RFPU with different Fluorodiol amounts was calculated by measuring the mass of the sample before and after immersion. The results are shown in Figure 8.

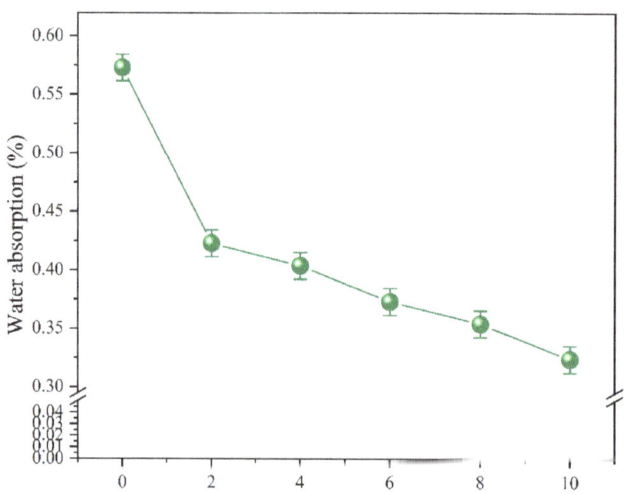

Figure 8. Water-absorption curves of RFPU prepared with different amounts of Fluorodiol.

It can be seen in Figure 8 that the water-absorption decreases with the increase in the amount of Fluorodiol. This is because the fluorine atoms in the Fluorodiol are distributed in a spiral shape on the carbon chain, which can protect the polyurethane molecular chain. At the same time, the polarizability of the C−F bond is small, so the chain segment of the fluorine−containing C−F group easily migrates from the inside of the material to the surface, which causes the surface of the RFPU to accumulate a large number of polyurethane segments with low surface energy, making it difficult for water to spread and wet its surface. At the same time, because Fluorodiol has a six-ring structure group, this group can enhance the strength of the polyurethane chain segment, ensure the strength of the pore wall of each polyurethane cell, and protect the relative independence of the cells. For these two reasons, the water absorption of RFPU decreases with the increase in the amount of Fluorodiol.

It can be seen from the experimental results that the water absorption of the RFPU prepared by adding Fluorodiol decreased significantly, and the loss rate of the RFPU soaked in deionized water was determined, as shown in Figure 9.

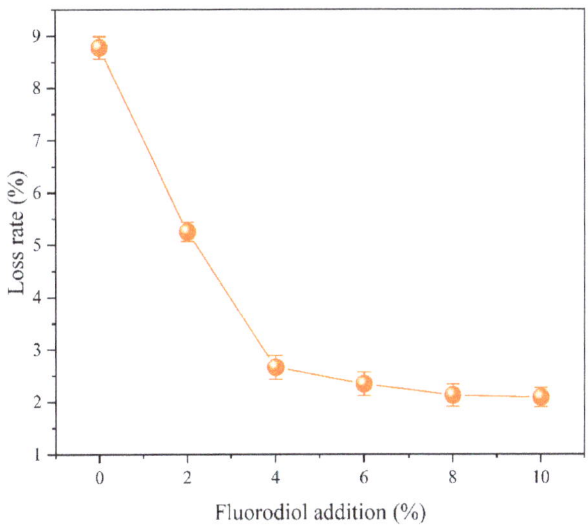

Figure 9. Water immersion loss rate curve of RFPU prepared with different Fluorodiol ratios.

It can be seen from Figure 9 that the loss rate of the PU rigid foam without Fluorodiol was 8.779%, while the loss of the RFPU was significantly reduced after the addition of Fluorodiol, reaching 2.125% when the amount of Fluorodiol was 8%. Similarly, due to the addition of fluorine-containing groups, the fluorine-containing polyurethane chain segment has a strong shielding effect and stability. The surface energy of the RFPU is significantly reduced and stable.

3.6. Effect of Fluorodiol on the Compressive Strength of RFPU Foam

The effect of different fluoroalcohols on the compressive strength of fluorinated recycled PU rigid foam is shown in Figure 10. In Figure 10, it can be seen that when fluonol is added to the system, the compressive strength of the prepared fluorine-containing recycled polyurethane rigid foam decreases significantly. This is because the monohydric alcohol seals the regenerated polyol and reduces the activity of the regenerated polyol. The reaction of the fluorine-containing recycled polyol during foaming is insufficient, and the generated polyurethane segment is short, so the strength of the fluorine-containing recycled polyurethane rigid foam is significantly reduced.

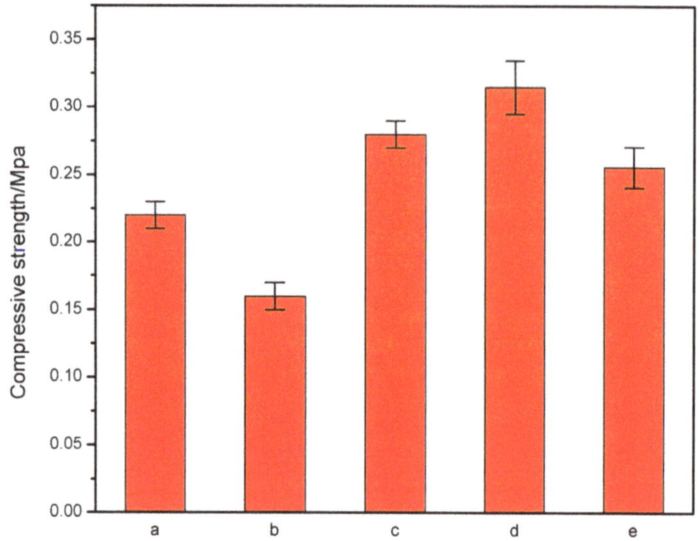

Figure 10. Effect of different fluoroalcohols on compressive strength of RFPU rigid foam (a: RRPU; b: fluoroalcohol comparison samples; c: RFPU (6%); d: RFPU (8%); e: RFPU (10%)).

When Fluorodiol was added to the system, the compressive strength of the prepared fluorine-containing recycled polyurethane rigid foam was significantly improved. When the addition of Fluorodiol was 8%, the strength was the highest, up to 0.315 Mpa. However, when the addition of Fluorodiol reached 10%, the compressive strength of the RFPU decreased significantly because of the incomplete degradation of Fluorodiol. This is because the bond energy of C−F is as high as 485 kJ/mol, which is much larger than that of the C−C bond, which leads to a spiral distribution of adjacent fluorine atoms along the carbon chain, while straight-chain alkanes generally form a zigzag chain configuration. In addition, a rigid six-ring structure was introduced in the synthesis of Fluorodiols, which can greatly improve the rigidity of polyurethane segments. These two structures provide a shielding effect for molecular chain coordination, which reduces the intermolecular force and improves the strength of the chain segment. At the same time, the binding force between the long chains of fluorinated polyurethane is improved, thus improving the compressive strength of the RFPU.

3.7. Analysis of Thermal Conductivity of Fluorine-Containing Recycled Polyurethane

Rigid polyurethane foam is usually used for thermal insulation applications, and thermal conductivity (λ) is a crucial characteristic. The results of the thermal conductivity of regenerated polyurethane prepared from regenerated polyols with different amounts of fluorine and polyurethane prepared using 4110 are shown in Figure 11. The thermal conductivity is related to the apparent density of the foam and the thermal conductivity of the gas is used as the foaming agent. Although the whole foam contains only a small part of the polyurethane matrix, because its λ value is much higher than that of the foaming agent, the foam with higher density often has higher thermal conductivity. Therefore, in a certain range, with the increase in recycled polyols, the thermal conductivity of polyurethane foam increases gradually, and the thermal insulation decreases [41–44]. For example, the thermal conductivity of polyurethane prepared from recycled polyols from 0.0285 W/m·K to 0.0336 W/m·K showed an increase of 17.9%. When the addition of perfluorinated monohydric alcohol made the polyurethane chain segment shorter, the small cell increased the foam density and decreased the thermal conductivity of the foam, but it still did not reach the level prepared by pure 4110. However, the thermal conductivity decreased significantly after the addition of Fluorodiols. When the content reached 8%, the thermal conductivity reached 0.0227 W/m·K, which was 20.3% lower than pure 4110. This is because the special helical structure of Fluorodiols can make different polyurethane segments more firmly connected and increase the distance between each segment, which makes the polyurethane pore wall thicker and stronger. Such pore walls allow polyurethane foam to have larger cells while maintaining strength.

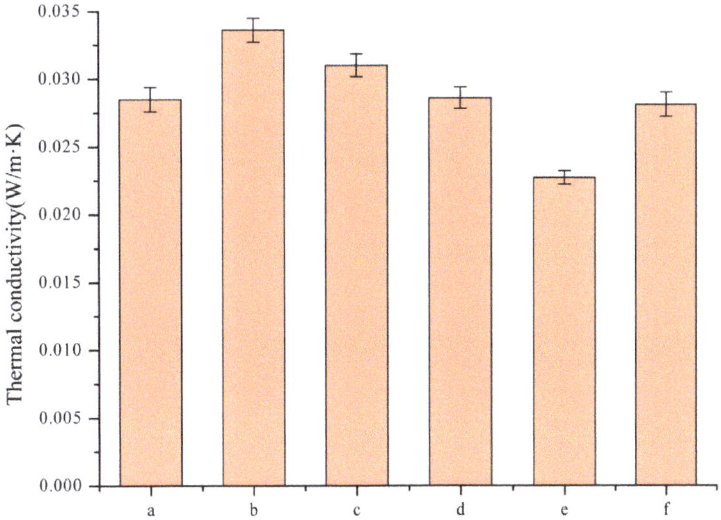

Figure 11. Thermal conductivity of polyurethane with different components: (a) pure 4100 sample; (b) RRPU; (c) fluoroalcohol comparison samples; (d) RFPU (6%); (e) RFPU (8%) and (f) RFPU (10%).

3.8. SEM Analysis of RFPU

In Figure 12a-1,a-2, it can be seen that the RFPU prepared by Fluoroalcohol has an obvious defoaming phenomenon and skeleton rupture phenomenon. This is due to the end-capping phenomenon of the regenerated polyol produced by Fluoroalcohol during the degradation of waste polyurethane, which leads to an insufficient polymerization reaction in the foaming process. The long chain of polyurethane becomes shorter, and the polyurethane is not crosslinked between the molecular links, resulting in the phenomenon of cracking and skeleton rupture.

Figure 12b-1,b-2 show RFPU with 4% Fluorodiol addition, Figure 12c-1,c-2 show RFPU with 6% Fluorodiol addition, and Figure 12d-1,d-2 show RFPU with 8% Fluorodiol addition. Compared with the RFPU prepared by fluorine-containing monohydric alcohol, it can be seen that the RFPU prepared by Fluorodiol has a more complete cell structure; the cell structure is a regular hexagon, and the skeleton is thick and the crosslinking structure is excellent. Compared with the RFPU prepared by Fluoroalcohol, the cell distribution of RFPU prepared by Fluorodiol is more uniform and denser. This good geometric structure and uniform cell distribution can enable the RFPU to have better compressive strength and lower thermal conductivity.

The fluorinated regenerated PU rigid foam prepared with 8% Fluorodiol has a larger cell structure. The good crosslinking degree of the fluorine-containing recycled PU rigid foam provides higher compressive strength for the foam body and can seal the gas well so that the foam has very good heat-insulation and heat-preservation performance.

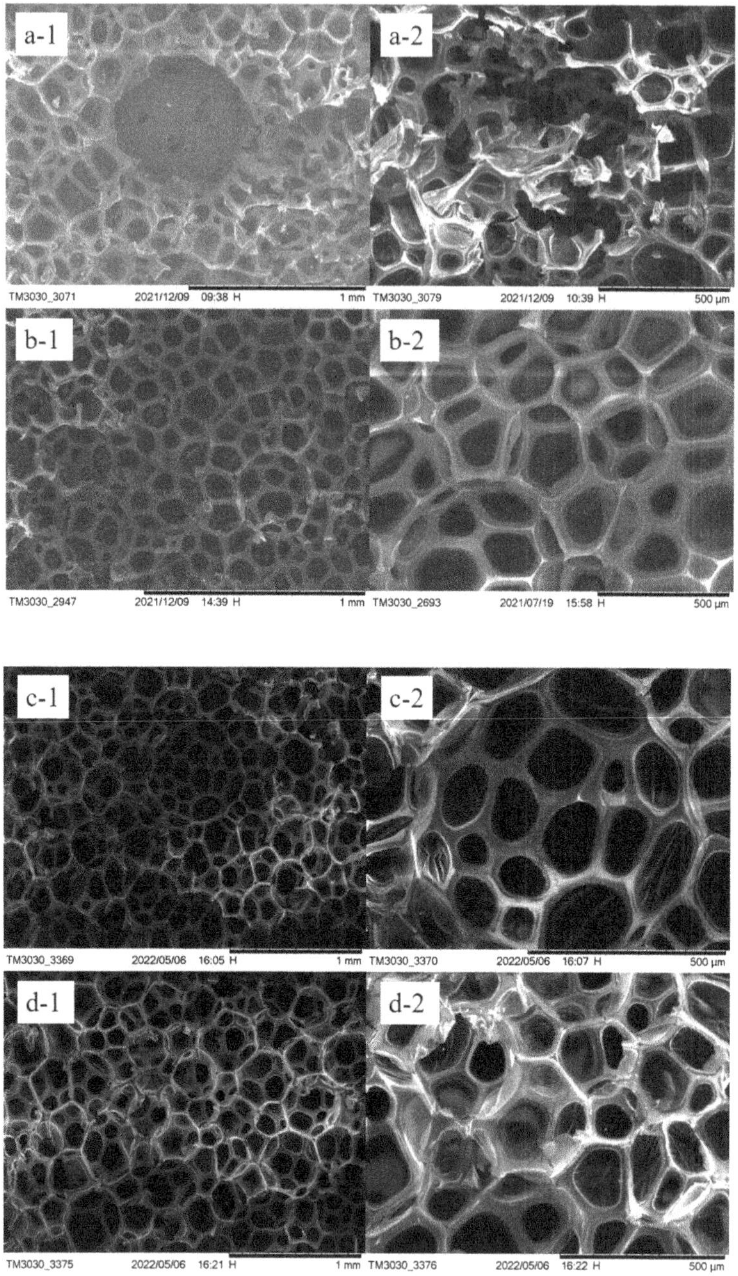

Figure 12. Scanning electron microscopy of RFPU.

3.9. TG Analysis of RFPU

Figure 9 shows the TG curve of the RFPU with different amounts of Fluorodiol. The thermal degradation of the RFPU has two stages: the first stage is caused by the pyrolysis of the hard segment at about 350 °C; the pyrolysis of the soft segment is the cause of the second stage. It is clear that two different degradation steps are involved in Figure 13.

The second halves of the two curves are almost identical because RFPUs with different amounts of Fluorodiol have the same soft segment composition. However, the first halves of the two curves are different. With the increase in the amounts of Fluorodiols, the half-decomposition temperature (T1/2) of the RFPU is higher than that of RPU [45,46]. This shows that introducing Fluorodiol as a modifier improves the thermal stability of the hard segment. The thermal stability of the hard segment determines the lower limit of the performance stability of polyurethane foam at high temperatures. Therefore, when improving the thermal stability of polyurethane foam, the modified hard segment is more useful than the modified soft segment [47].

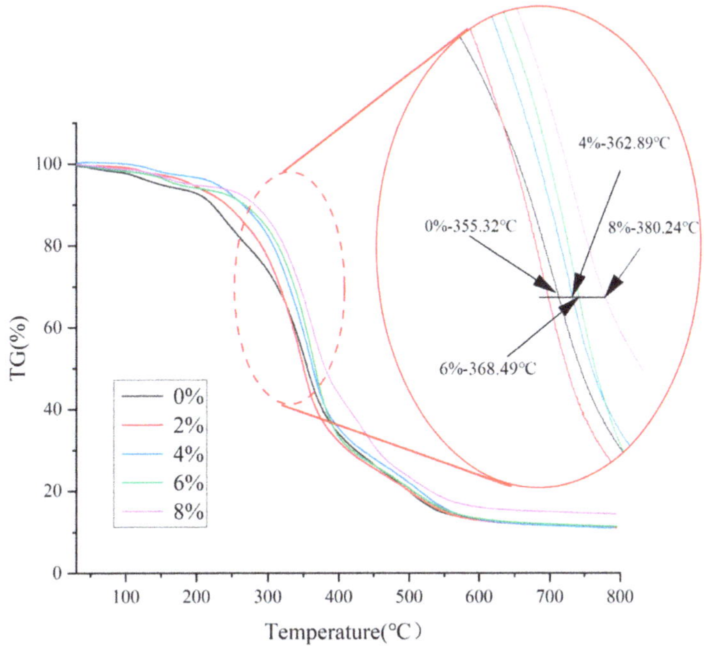

Figure 13. TG curves of RFPU with different amounts of Fluorodiol.

3.10. XPS Analysis of RFPU

The regenerated PU hard foams prepared by different degradation systems were tested by XPS, as shown in Figure 14. Comparing the a and b curves in Figure 14 and detail data in Table 2 shows that the C, N, and O elements in the polyurethane are decreased when the Fluoroalcohol is modified, indicating that Fluoroalcohol has an inhibitory effect on the polyurethane foaming process. This is because, in the foaming process, Fluoroalcohol has only one hydroxyl group, and the polyurethane produced in the foaming process has a long chain blocked by Fluoroalcohol, so the polyurethane foaming process blocked by the final preparation of fluorine-containing recycled polyurethane groups is also diminished [21,48–52]. By comparing b and c curves, it can be seen that the F content of the RFPU prepared by Fluorodiols is significantly higher than that of Fluoroalcohol. This is because Fluorodiols with two hydroxyl groups can be better combined with polyurethane segments to obtain longer polyurethane long chains; by comparing the a and c curves, it can be seen that the N element increases significantly, which indicates that the Fluorodiol contributes to the synthesis of hard polyurethane segments, which improves the thermal stability and mechanical properties of RFPU [53–55].

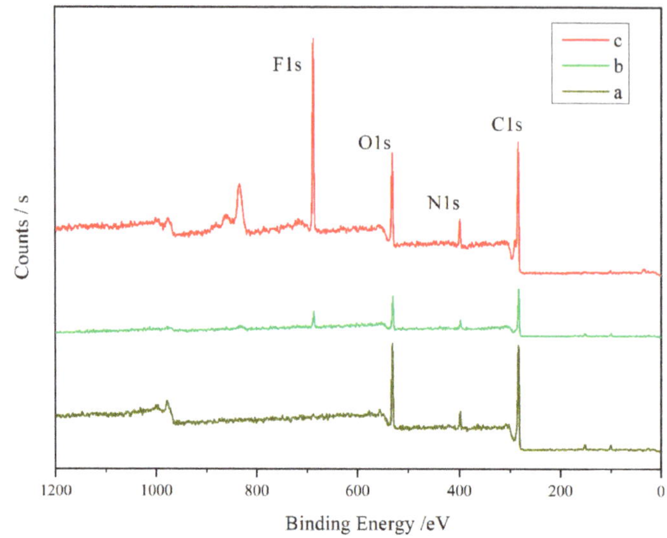

Figure 14. XPS full spectrum of regenerated PU rigid foam prepared by different degradation systems (a: RRPU; b: fluoroalcohol comparison samples; c: RFPU (8% Fluorodiol addition)).

Table 2. Element composition of different polyurethanes.

	C	F	N	O
a	67.99	0.39	7.78	21.05
b	65.30	8.36	5.23	18.25
c	58.93	21.09	5.35	14.63

Since fluorine-containing polyurethane has a low surface energy during the foaming process, the hard segment is pulled to the foam surface to form a film, and this increase in the length of the Fluoroalcohol side chain facilitates the migration of the fluorinated side chain. In addition, since the fluorine side chain is attached to the hard segment of the polyurethane molecular chain, the influence of the hard segment is affected.

4. Conclusions

Waste PU was successfully alcoholized into oligomer polyols using self-made Fluorodiol and EG, DEG, and a two-component alcoholizing agent. RFPU rigid foam composites with excellent properties can be prepared using fluorinated recycled polyols as raw materials. The specific performance of the material is as follows.

The addition of fluorine-containing groups can greatly improve the performance of RFPU. When the addition of Fluorodiol was 8% of the mass of waste polyurethane, the compressive strength was 0.315 MPa, which was 43.2% higher than that of RFPU prepared by the RRPU. The waterproof performance of the RFPU was significantly improved, and the immersion loss rate was 2.125%, which was 75.59% lower than that of the RFPU prepared by RRPU. The thermal conductivity reaches 0.0227 W/m·K, which is 20.3% lower than the commercially available sample, indicating its excellent thermal insulation performance. The pore walls of the prepared RFPU are thick and uniform, and the skeleton geometry is also positive.

The raw materials used in the RFPU are waste polyurethane degradation products, and the properties of the prepared samples are comparable to or even better than commercially available PU materials. Thus, preparing the RFPU successfully demonstrated a high-value utilization of waste polyurethane. This will make polyurethane material a green and sustainable environmental protection material. This material also offers a new choice to help protect the Earth's resources.

Author Contributions: Conceptualization, X.-H.G. and D.-S.Z.; methodology, X.-H.G. and D.-S.Z.; software, D.-S.Z.; validation, D.-S.Z., X.-H.G. and Q.-Y.Z.; formal analysis, D.-S.Z.; investigation, S.-W.L. and Q.-Y.Z.; resources, S.-W.L. and Y.-W.Z.; data curation, D.-S.Z.; writing—original draft preparation, D.-S.Z.; writing—review and editing, D.-S.Z. and X.-H.G.; visualization, D.-S.Z.; supervision, X.-H.G. and S.-W.Z.; project administration, X.-H.G., Y.L. and S.-W.Z.; funding acquisition, X.-H.G. and Y.L. All authors have read and agreed to the published version of the manuscript.

Funding: This research was funded by the Heilongjiang Provincial Department of Education Project, grant number CLKFKT2021Z3, 145109301.

Institutional Review Board Statement: Not applicable for studies not involving humans.

Informed Consent Statement: Not applicable for studies not involving humans or animals.

Data Availability Statement: Not applicable.

Conflicts of Interest: The authors declare no conflict of interest.

References

1. Calvo-Correas, T.; Ugarte, L.; Trzebiatowska, P.J.; Sanzberro, R.; Datta, J.; Corcuera, M.A.; Eceiza, A. Thermoplastic polyurethanes with glycolysate intermediates from polyurethane waste recycling. *Polym. Degrad. Stabil.* **2017**, *144*, 411–419. [CrossRef]
2. Zia, K.M.; Bhatti, H.N.; Bhatti, I.A. Methods for polyurethane and polyurethane composites, recycling and recovery: A review. *React. Funct. Polym.* **2007**, *67*, 675–692. [CrossRef]
3. Wang, X.T.; Shi, Y.; Liu, Y.; Wang, Q. Recycling of waste melamine formaldehyde foam as flame-retardant filler for polyurethane foam. *J. Polym. Res.* **2019**, *26*, 12. [CrossRef]
4. Gu, X.H.; Luo, H.X.; Lv, S.W.; Chen, P. Glycolysis Recycling of Waste Polyurethane Rigid Foam Using Different Catalysts. *J. Renew. Mater.* **2021**, *9*, 1253–1266. [CrossRef]
5. Zhu, P.; Cao, Z.B.; Chen, Y.; Zhang, X.J.; Qian, G.R.; Chu, Y.L.; Zhou, M. Glycolysis recycling of rigid waste polyurethane foam from refrigerators. *Environ. Technol.* **2014**, *35*, 2676–2684. [CrossRef]
6. Shin, S.R.; Kim, H.N.; Liang, J.Y.; Lee, S.H.; Lee, D.S. Sustainable rigid polyurethane foams based on recycled polyols from chemical recycling of waste polyurethane foams. *J. Appl. Polym. Sci.* **2019**, *136*, 9. [CrossRef]
7. Nikje, M.M.A.; Mohammadi, F.H.A. Polyurethane Foam Wastes Recycling under Microwave Irradiation. *Polym. Plast. Technol. Eng.* **2010**, *49*, 818–821. [CrossRef]
8. Nikje, M.M.A.; Garmarudi, A.B.; Idris, A.B. Polyurethane Waste Reduction and Recycling: From Bench to Pilot Scales. *Des. Monomers Polym.* **2011**, *14*, 395–421. [CrossRef]
9. Gomes, M.; Carvalho, E.A.S.; Barreto, G.N.S.; Rodriguez, R.J.S.; Monteiro, S.N.; Vieira, C.M.F. Development of Sustainable Artificial Stone Using Granite Waste and Biodegradable Polyurethane from Castor Oil. *Sustainability* **2022**, *14*, 6380. [CrossRef]
10. Gong, C.H.; Zhang, K.H.; Yang, C.; Chen, J.; Zhang, S.; Yi, C.W. Simple process for separation and recycling of nylon 6 and polyurethane components from waste nylon 6/polyurethane debris. *Text. Res. J.* **2021**, *91*, 18–27. [CrossRef]
11. Godinho, B.; Gama, N.; Barros-Timmons, A.; Ferreira, A. Recycling of different types of polyurethane foam wastes via acidolysis to produce polyurethane coatings. *Sustain. Mater. Technol.* **2021**, *29*, 7. [CrossRef]
12. Wang, C.; Li, X.R.; Du, B.; Li, P.Z.; Lai, X.J.; Niu, Y.H. Preparation and properties of a novel waterborne fluorinated polyurethane-acrylate hybrid emulsion. *Colloid Polym. Sci.* **2014**, *292*, 579–587. [CrossRef]
13. Zhang, R.Z.; Lu, W.; Yan, D.K.; Liu, X.D. Cavitation Erosion Resistant Hydrophobic Fluorinated Polyurethane. *ACTA Polym. Sin.* **2015**, *1*, 808–818.
14. Ge, Z.; Zhang, X.Y.; Dai, J.B.; Li, W.H.; Luo, Y.J. Synthesis, characterization and properties of a novel fluorinated polyurethane. *Eur. Polym. J.* **2009**, *45*, 530–536. [CrossRef]
15. Agrizzi, C.P.; Carvalho, E.A.S.; Gadioli, M.C.B.; Barreto, G.N.S.; de Azevedo, A.R.G.; Monteiro, S.N.; Vieira, C.M.F. Comparison between Synthetic and Biodegradable Polymer Matrices on the Development of Quartzite Waste-Based Artificial Stone. *Sustainability* **2022**, *14*, 6388. [CrossRef]
16. Shen, M.Y.; Kuan, C.F.; Kuan, H.C.; Ke, C.Y.; Chiang, C.L. Flame Retardance and Char Analysis of an Eco-Friendly Polyurethane Hyperbranched Hybrid Using the Sol-Gel Method. *Sustainability* **2021**, *13*, 486. [CrossRef]
17. Hsu, Y.T.; Wang, W.H.; Hung, W.H. Architectural Sustainability and Efficiency of Enhanced Waterproof Coating from Utilization of Waterborne Poly (Siloxane-Imide-Urethane) Copolymers on Roof Surfaces. *Sustainability* **2020**, *12*, 4411. [CrossRef]
18. Hsu, Y.T.; Wang, W.H.; Hung, W.H. Evaluating the Properties of a Coating Material with Polycaprolactone-Degradable Fluorinated Silicon-Containing Waterborne Polyurethane. *Sustainability* **2020**, *12*, 3745. [CrossRef]
19. Jiang, M.; Zhao, X.L.; Ding, X.B.; Zheng, Z.H.; Peng, Y.X. A novel approach to fluorinated polyurethane by macromonomer copolymerization. *Eur. Polym. J.* **2005**, *41*, 1798–1803. [CrossRef]

20. Li, N.; Zeng, F.L.; Wang, Y.; Qu, D.Z.; Zhang, C.; Li, J.; Huo, J.Z.; Bai, Y.P. Synthesis and characterization of fluorinated polyurethane containing carborane in the main chain: Thermal, mechanical and chemical resistance properties. *Chin. J. Polym. Sci.* **2018**, *36*, 85–97. [CrossRef]
21. Shi, X.; Shi, H.X.; Wu, H.K.; Shen, H.M.; Cao, P. Synthesis and properties of novel fluorinated polyurethane based on fluorinated gemini diol. *Polym. Adv. Technol.* **2018**, *29*, 1939–1952. [CrossRef]
22. Ge, Z.; Zhang, X.Y.; Dai, J.B.; Li, W.H.; Luo, Y.J. Synthesis and characterization of fluorinated polyurethane with fluorine-containing pendent groups. *Chin. Chem. Lett.* **2008**, *19*, 1293–1296. [CrossRef]
23. Chen, L.; Hayashi, M.; Takasu, A. Hydrophobicity enhancement of polyurethanes by attaching fluorinated end blocks via ATRP and correlation between surface properties and self-assembly nature. *Polymer* **2019**, *172*, 312–321. [CrossRef]
24. Xu, W.Z.; Lu, B.; Hu, Y.; Yin, J.G.; Zhang, Y.Y. Synthesis and Characterization of Novel Fluorinated Polyurethane Elastomers. *Asian J. Chem.* **2011**, *23*, 2284–2288.
25. Zeng, S.H.; Wang, Q.M.; Chen, P.P.; Xu, Y.; Nie, W.Y.; Zhou, Y.F. Controllable hydrolytic stability of novel fluorinated polyurethane films by incorporating fluorinated side chains. *Prog. Org. Coat.* **2022**, *165*, 106729. [CrossRef]
26. Tonelli, C.; Ajroldi, G.; Marigo, A.; Marega, C.; Turturro, A. Synthesis methods of fluorinated polyurethanes. 2. Effects on morphology and microstructure. *Polymer* **2001**, *42*, 9705–9711. [CrossRef]
27. Zhao, B.; Jia, R.; Zhang, Y.; Liu, D.; Zheng, X. Design and synthesis of antibacterial waterborne fluorinated polyurethane. *J. Appl. Polym. Sci.* **2019**, *136*, 46923. [CrossRef]
28. Zhang, R.Z.; Wang, W.B.; Wang, C.Y.; Tian, W.J.; Hang, J.L.; Hussain, M.I. Effect of Nano Alumina on the Properties of Fluorinated Polyurethane. *Materials* **2019**, *12*, 4120. [CrossRef]
29. Gu, X.H.; Lyu, S.W.; Liu, S.W. Alcoholysis of Waste Polyurethane Rigid Foam and Its Modification with Lignin for Recovery. *J. Renew. Mater.* **2021**, *9*, 1913–1926. [CrossRef]
30. Colomines, G.; Rivas, F.; Lacoste, M.L.; Robin, J.J. Study of polyurethane formulations containing Diols obtained via glycolysis of poly(ethylene terephthalate) (PET) by oligoesters Diols through a reactive extrusion process. *Macromol. Mater. Eng.* **2005**, *290*, 710–720. [CrossRef]
31. Galimzyanova, A.R.; Bakirova, I.N.; Valuev, V.I.; Zenitova, L.A. Mechanism and relationships of chemical degradation of rigid polyurethane foam. *Russ. J. Appl. Chem.* **2005**, *78*, 824–829. [CrossRef]
32. Galeeva, E.I.; Bakirova, I.N. Chemical degradation of elastic foamed polyurethanes under the action of thiodiglycol. *Russ. J. Appl. Chem.* **2007**, *80*, 1741–1744. [CrossRef]
33. Luo, J.B.; Ma, C.; Liao, R.; Wan, J.; Zhang, P.; Zhang, J. Surface and Antibacterial Properties of Polyurethane with Fluorinated Bis-ammonium Salts Attached to Hard Segments. *Chem. J. Chin. Univ. Chin.* **2010**, *31*, 1268–1273.
34. Wang, P.C.; Lu, D.; Wang, H.; Bai, R.K. A New Strategy for the Synthesis of Fluorinated Polyurethane. *Polymers* **2019**, *11*, 1440. [CrossRef]
35. Zhu, M.; Chen, K.; Zhang, Y.F.; Wang, X.R.; Zhou, X.D. Gradient Structure and Surface Property of Fluorinated Polyacrylate and Polyurethane Latex Blend Films. *Polymer* **2014**, *38*, 265–271. [CrossRef]
36. Zhang, X.Q.; Jiang, X.; Li, J.H.; Tan, H.; Zhong, Y.P.; Fu, Q. Surface and bulk properties of poly(ether urethane)s/fluorinated phosphatidylcholine polyurethanes blends. *J. Appl. Polym. Sci.* **2008**, *108*, 548–553. [CrossRef]
37. Tan, H.; Xie, X.Y.; Li, J.H.; Zhong, Y.P.; Fu, Q. Synthesis and surface mobility of segmented polyurethanes with fluorinated side chains attached to hard blocks. *Polymer* **2004**, *45*, 1495–1502. [CrossRef]
38. Smirnova, O.; Glazkov, A.; Yarosh, A.; Sakharov, A. Fluorinated Polyurethanes, Synthesis and Properties. *Molecules* **2016**, *21*, 904. [CrossRef]
39. Ghaderian, A.; Haghighi, A.H.; Taromi, F.A.; Abdeen, Z.; Boroomand, A.; Taheri, S.M.R. Characterization of Rigid Polyurethane Foam Prepared from Recycling of PET Waste. *Period. Polytech. Chem. Eng.* **2015**, *59*, 296–305. [CrossRef]
40. Jiang, M.; Zheng, Z.H.; Ding, X.B.; Cheng, X.; Peng, Y.X. Convenient synthesis of novel fluorinated polyurethane hybrid latexes and core-shell structures via emulsion polymerization process with self-emulsification of polyurethane. *Colloid Polym. Sci.* **2007**, *285*, 1049–1054. [CrossRef]
41. Estravis, S.; Tirado-Mediavilla, J.; Santiago-Calvo, M.; Ruiz-Herrero, J.L.; Villafane, F.; Rodriguez-Perez, M.A. Rigid polyurethane foams with infused nanoclays: Relationship between cellular structure and thermal conductivity. *Eur. Polym. J.* **2016**, *80*, 1–15. [CrossRef]
42. Kirpluks, M.; Kalnbunde, D.; Benes, H.; Cabulis, U. Natural oil based highly functional polyols as feedstock for rigid polyurethane foam thermal insulation. *Ind. Crops Prod.* **2018**, *122*, 627–636. [CrossRef]
43. Hatakeyama, H.; Matsumura, H.; Hatakeyama, T. Glass transition and thermal degradation of rigid polyurethane foams derived from castor oil-molasses polyols. *J. Therm. Anal. Calorim.* **2013**, *111*, 1545–1552. [CrossRef]
44. Kirpluks, M.; Cabulis, U.; Zeltins, V.; Stiebra, L.; Avots, A. Rigid polyurethane foam thermal insulation protected with mineral intumescent mat. *Autex Res. J.* **2014**, *14*, 259–269. [CrossRef]
45. Jia, D.K.; Hu, J.; He, J.Y.; Yang, R.J. Properties of a novel inherently flame-retardant rigid polyurethane foam composite bearing imide and oxazolidinone. *J. Appl. Polym. Sci.* **2019**, *136*, 47943. [CrossRef]
46. Xu, D.F.; Yu, K.J.; Qian, K. Thermal degradation study of rigid polyurethane foams containing tris(1-chloro-2-propyl)phosphate and modified aramid fiber. *Polym. Test.* **2018**, *67*, 159–168. [CrossRef]

47. Reinerte, S.; Avotina, L.; Zarins, A.; Cabulis, U.; Viksna, A. TG/DTA-FTIR as a method for analysis of tall oil based rigid polyurethane foam decomposition gaseous products in a low oxygen environment. *Polym. Degrad. Stabil.* **2020**, *180*, 109313. [CrossRef]
48. Uyama, M.; Kanda, M.; Nishi, Y. Creation of Adhesive Force between Laminated Sheets of Polytetrafluoroethylene (PTFE) and Polyethylene (PE) by Homogeneous Low Energy Electron Beam Irradiation Prior to Hot-Press for Bio-Adaptable Application. *Mater. Trans.* **2014**, *55*, 566–571. [CrossRef]
49. Miyazawa, Y.; Uyama, M.; Ishii, S.; Kanda, M.; Nishi, Y. Creation of Adhesive Force between Laminated Sheets of Polyurethane (PU) and Polytetrafluoruethylene (PTFE) by Homogeneous Low Energy Electron Beam Irradiation Prior to Hot-Press for Bio-Adaptable Application. *Mater. Trans.* **2013**, *54*, 1166–1170. [CrossRef]
50. Xu, W.; Zhao, W.J.; Hao, L.F.; Wang, S.; Pei, M.M.; Wang, X.C. Synthesis and characterization of novel fluoroalkyl-terminated hyperbranched polyurethane latex. *Appl. Surf. Sci.* **2018**, *436*, 1104–1112. [CrossRef]
51. Xu, W.; Zhao, W.J.; Hao, L.F.; Wang, S.; Pei, M.M.; Wang, X.C. Synthesis of novel cationic fluoroalkyl-terminated hyperbranched polyurethane latex and morphology, physical properties of its latex film. *Prog. Org. Coat.* **2018**, *121*, 209–217. [CrossRef]
52. West, J.O.F.; Critchlow, G.W.; Lake, D.R.; Banks, R. Development of a superhydrophobic polyurethane-based coating from a two-step plasma-fluoroalkyl silane treatment. *Int. J. Adhes. Adhes.* **2016**, *68*, 195–204. [CrossRef]
53. Jiang, G.C.; Tuo, X.L.; Wang, D.R.; Li, Q. Synthesis and properties novel polyurethane-hexafluorobutyl methacrylate copolymers. *J. Mater. Sci. Mater. Med.* **2012**, *23*, 1867–1877. [CrossRef] [PubMed]
54. Jeong, H.Y.; Lee, M.H.; Kim, B.K. Surface modification of waterborne polyurethane. *Colloids Surf. A Physicochem. Eng. Asp.* **2006**, *290*, 178–185. [CrossRef]
55. Hinder, S.J.; Watts, J.F.; Lowe, C. Surface and interface analysis of complex polymeric paint formulations. *Surf. Interface Anal.* **2006**, *38*, 557–560. [CrossRef]

sustainability MDPI

Review

Diesel Spray: Development of Spray in Diesel Engine

Djati Wibowo Djamari [1], Muhammad Idris [2], Permana Andi Paristiawan [3], Muhammad Mujtaba Abbas [4], Olusegun David Samuel [5,6,*], Manzoore Elahi M. Soudagar [7,8], Safarudin Gazali Herawan [9], Davannendran Chandran [10], Abdulfatah Abdu Yusuf [11], Hitesh Panchal [12] and Ibham Veza [10,*]

1 Mechanical Engineering Study Program, Sampoerna University, Jakarta 12780, Indonesia
2 PT Perusahaan Listrik Negara (Persero), Engineering and Technology Division, Jakarta 11420, Indonesia
3 Research Center for Metallurgy, National Research and Innovation Agency, South Tangerang 15314, Indonesia
4 Department of Mechanical Engineering, University of Engineering and Technology (New Campus), Lahore 54890, Pakistan
5 Department of Mechanical Engineering, Federal University of Petroleum Resources, P.M.B 1221, Effurun 330102, Nigeria
6 Department of Mechanical Engineering, University of South Africa, Science Campus, Private Bag X6, Florida 1709, South Africa
7 Department of Mechanical Engineering and University Centre for Research & Development, Chandigarh University, Mohali 140413, India
8 Department of Mechanical Engineering, School of Technology, Glocal University, Delhi-Yamunotri Marg, SH-57, Mirzapur Pole, Saharanpur 247121, India
9 Industrial Engineering Department, Faculty of Engineering, Bina Nusantara University, Jakarta 11480, Indonesia
10 Mechanical Engineering, Universiti Teknologi PETRONAS, Seri Iskandar 32610, Malaysia
11 Department of Mechanical Engineering, University of Liberia, Monrovia 1000, Liberia
12 Department of Mechanical Engineering, Government Engineering College, Patan 384265, India
* Correspondence: samuel.david@fupre.edu.ng (O.D.S.); ibham.veza@utp.edu.my (I.V.)

Citation: Djamari, D.W.; Idris, M.; Paristiawan, P.A.; Abbas, M.M.; Samuel, O.D.; Soudagar, M.E.M.; Herawan, S.G.; Chandran, D.; Yusuf, A.A.; Panchal, H.; et al. Diesel Spray: Development of Spray in Diesel Engine. *Sustainability* **2022**, *14*, 15902. https://doi.org/10.3390/su142315902

Academic Editors: Fuqiang Wang, Chao Shen, Dong Li and Zhonghao Rao

Received: 23 October 2022
Accepted: 28 November 2022
Published: 29 November 2022

Abstract: Research and development in the internal combustion engine (ICE) has been growing progressively. Issues such as air pollution, fuel cost, and market competitiveness have driven the automotive industry to develop and manufacture automobiles that meet new regulation and customers' needs. The diesel engine has some advantages over the gasoline or spark ignition engine, including higher engine efficiency, greater power output, as well as reliability. Since the early stage of the diesel engine's development phase, the quest to obtain better atomization, proper fuel supply, and accurate timing control, have triggered numerous innovations. In the last two decades, owing to the development of optical technology, the visualization of spray atomization has been made possible using visual diagnostics techniques. This advancement has greatly improved research in spray evolution. Yet, a more comprehensive understanding related to these aspects has not yet been agreed upon. Diesel spray, in particular, is considered a complicated phenomenon to observe because of its high-speed, high pressure, as well as its high temperature working condition. Nevertheless, several mechanisms have been successfully explained using fundamental studies, providing several suggestions in the area, such as liquid atomization and two-phase spray flow. There are still many aspects that have not yet been agreed upon. This paper comprehensively reviews the current status of theoretical diesel spray and modelling, including some important numerical and experimental aspects.

Keywords: diesel spray; compression ignition engine; combustion; emission; spray process; atomization

1. Introduction

Energy transition and environmental issues on the restrictions of emissions have been paid much more attention in most countries across the world [1–4]. CO_2 emissions have been addressed as the dominant pollutant that contribute to climate change [5]. Hence, engine manufacturers should improve the performance of internal combustion engines by developing efficient engines with lower emissions [6,7]. Regardless of the stringent emissions regulation imposed by several countries, diesel fuel remains the major source of energy for heavy duty engines. The basic explanation for this is because diesel engines have a relatively higher thermal efficiency spark ignition (SI) than gasoline engines, which is especially important for heavy duty purposes.

In the diesel engine, mixture formation between air and fuel plays a vital role in the process of combustion, thus affecting emission qualities [8]. To enhance the combustion and lower the emission, it is essential to have comprehensive knowledge of the mixture formation. Prior to acquiring this knowledge, a thorough understanding of spray atomization must be first understood. Studies on diesel spray remains a crucial part of research in the internal combustion engine. The quality of the ignition is substantially influenced by local and temporal processes, as well as the completeness of combustion. The pollutants in the cylinder interact and break down, generating the components in the exhaust gas. This is specifically true for carbon monoxide (CO) and unburned hydrocarbon (UHC) emissions. The process of fuel injection controls how well the engine performs in terms of combustion and emissions. Note that the injection system from the high-pressure common-rail diesel engine is currently viewed as one of the most viable ways to satisfy the increasingly strict emission standards for diesel engines. This illustrates the importance of diesel spray in the development of the diesel engine.

Spray is a process in which liquid is forced out of a particular container so that it becomes a mass of small-liquid drops. In the direct-injection internal combustion engine, fuel, which is still in its liquid phase, is injected by the injector into a gaseous condition in the cylinder. In a compression ignition engine, such a phenomenon is commonly known as diesel spray. The injected liquid fuel that is now inside the combustion chamber then begins to interact with the in-cylinder air instantaneously, leading to the dispersion of the liquid phase. As this liquid phase vaporizes, a flammable mixture is then generated to produce great power inside the chamber. The whole process is known as the spray evolution process, consisting of liquid-fuel breakup, atomization, and evaporation. A detailed knowledge in spray evolution is important as the mixture preparation, which later determines the quality of the combustion and emission formation, depends greatly on this process.

2. Diesel Combustion Process

Unlike gasoline engines, that use a spark plug to initiate the combustion, diesel engines work based on the principle of autoignition, as shown in Figure 1. Following fuel in the liquid form being injected into the cylinder, this fuel then mixes with the in-cylinder air at a high temperature and pressure, achieved by the high volumetric compression of the air. The typical compression ratios in diesel engines ranges from 15 to 22. Figure 1 shows one complete working cycle of a typical diesel engine via four strokes. The crankshaft must complete two revolutions (720°) per cycle of operation.

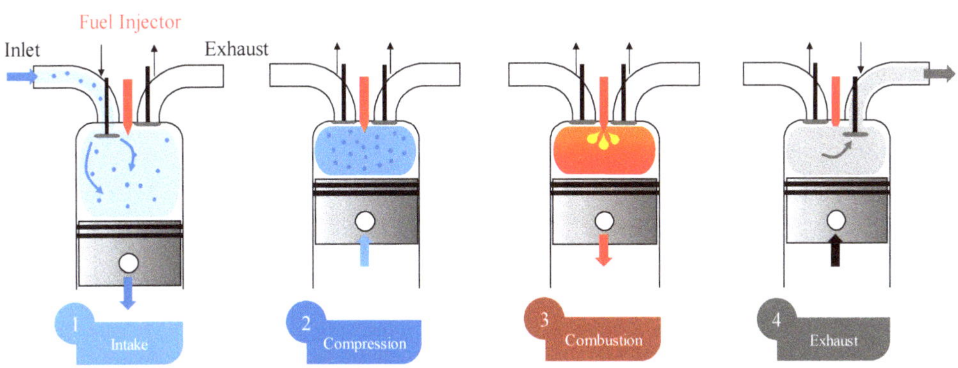

Figure 1. Diesel engine four strokes process.

Diesel engines can operate in high compression ratios, while their counterparts, gasoline engines, operate only in low to medium ratios to avoid engine knock. Due to these relatively higher compression ratios, diesel engines are more fuel efficient, making them a natural choice for heavy-duty purposes. In modern diesel engines that use common-rail technology, a high-pressure injector is used to insert high-velocity liquid fuel into the hot compressed air. The fuel pressure is raised to the desired level using a pump, which moves the liquid fuel from a tank to multiple injectors. As the engine load is subject to variation in the engine speed while the vehicles are operated, the load thus needs to be adjusted by regulating the fuel amount to be injected. Each injector then spreads the liquid fuel in small droplets. As the liquid fuel penetrates through the combustion chamber, this fuel then breaks down to smaller droplets, evaporates, as well as mixes with the in-cylinder gas. This is when the autoignition starts to happen, before the main combustion finally takes place in the chamber.

As mentioned above, the increase in injection pressure is considered as a promising approach to enhance combustion behaviors [9–13]. High-pressure injected diesel fuel could improve the engine performance and reduce particulate matter [14–18]. In addition to that, the increased injection pressure tends to accelerate the vaporization of droplets and the preparation of the combustible mixture [19–23]. As expected, this would reduce the ignition delay. Note that the process of combustion begins with the autoignition of the fuel and air mixture. The characteristics of autoignition are affected by the chemical as well as the physical properties of the fuel. As it is more difficult to observe those two properties macroscopically, the ignition delay is hence used as a key measure of the autoignition. However, in typical combustion, autoignition occurs in two stages that are categorized by these properties: physical and chemical.

The first stage of autoignition is determined by the physical properties of the liquid spray. The delay between the injection start, and the combustible mixture formation, is known as the physical delay. Such a delay is controlled by its physical properties (density, surface tension, and specific heat). Furthermore, the structures of the spray flow in the combustion chamber also have substantial effects on the physical delay. Take small- and large-scale hydrodynamic structures, for example, these types of structures tend to drive the entrainment of the air into the fuel jet in a turbulent environment. The liquid jet starts to break down and the spray begins to evaporate, caused by aerodynamic interactions and ambient air interactions, respectively.

The second stage of the autoignition is determined by the fuel chemical properties. The delay between the mixture formation and its ignition is known as the chemical delay. Such a delay is controlled by its chemical properties i.e., the molecular structure, adiabatic flame temperature, C/H/O ratio, and the sooting tendency. Therefore, the total ignition delay is the combination between physical and chemical delays. As far as the chemical kinetics are concerned, the process of chemical kinetics in the diesel engine is a complicated mechanism as it involves many simultaneous reactions. Although the rate of heat release that causes a rapid rise in pressure and temperature leading to autoignition is easy to observe, the chemical mechanisms involved microscopically are complex. These mechanisms include initiation, propagation, and the termination processes of the chain.

Firstly, the chain begins its initiation reaction by building a group of radical species from the reactants. The radicals then react with each other, as well as with other stable species, to form more radicals, resulting in the propagation of the chain reactions. Of these chain propagation reactions, chain branching reactions are of great importance as they determine the self-propagating flame that leads to combustion. These chain branching reactions can have an explosive effect on a combustion system by producing radicals that control the overall reaction rate. Lastly, the stable products are formed afterwards, which indicate the termination process of the reactions. The overall chain reactions thus consist of a complicated process of simultaneous chemical reactions, which are also subject to the mixture physical state.

In diesel or compression ignition engines, the combustion occurs in two phases following the autoignition of the mixture inside the cylinder [24,25]. They are premixed-controlled and mixing-controlled combustion. The premixed-controlled combustion occurs in a fuel-rich region and produces partially oxidized fuel fragments that diffuse outwardly, leading to the mixing-controlled diffusion flame that surround the premixed core region. The first stage (heat release along with temperature rise) takes place in a fuel-rich premixed region. The overall HRR is primarily controlled by the fuel mixing rate and oxidizer. Typical direct injection in the diesel engine is schematically illustrated in Figure 2.

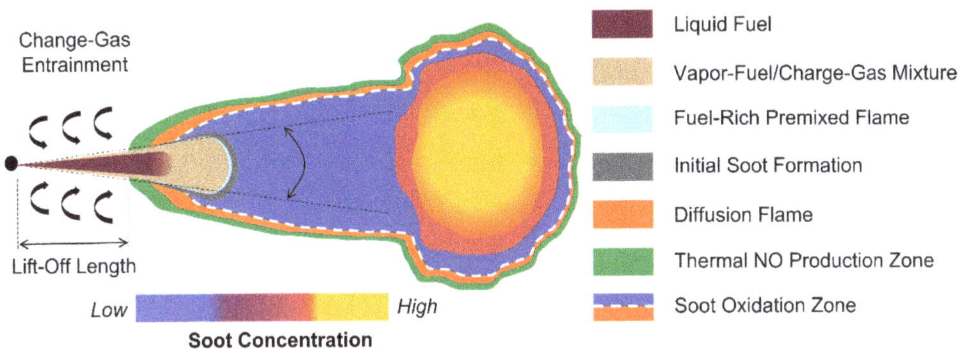

Figure 2. Combustion model of direct injection diesel, reproduced from [26].

As the liquid fuel jet penetrates the cylinder, it entrains the surrounding gas. The liquid spray is then heated up and evaporation occurs at the spray edge. The more air enters the spray, the higher the evaporation rate of the spray. The combustion then starts, while the mixture reaches a particular equivalence ratio. In the region of the diffusion flame, the orange color represents the highest temperature, which is believed to be the primary location of NO production [27]. This region also oxidizes a great amount of soot that is generated in the fuel-rich region. This diffusion layer flame can be identified using the OH* chemiluminescence technique for laser-based optical combustion diagnostics [28–30].

There are a number of strategies to improve the diesel combustion process. The introduction of pilot fuel is believed to be able to reduce the combustion noise as it shortens the premixing process [31]. The pilot fuel could reduce the pressure-rise rate by allowing only a small amount of liquid fuel to be introduced for the premixed-combustion mixture, leading to lower NOx emission [32,33]. Just after this pilot fuel has been introduced, the appropriate amount of fuel should be injected immediately to avoid the formation of soot. A post injection is then needed, avoiding the formation of soot [34–36]. In addition to that strategy, the multiple injections approach was also proposed to obtain the desired MFR shape. The multiple injections strategy allows the injector to have a shorter injection duration for each split injection, allowing fuel to blend sufficiently with the air so that an improved mixture can be achieved. Such a strategy also increases the IMEP, which was found to reduce the emissions. The multiple injections strategy can also prevent the impingement of the spray onto the wall chamber, resulting in low hydrocarbon (HC) pollutant. The multiple injections concept in the diesel engine has nowadays been used widely, due to its several advantages. Its superiority compared to the single injection mode is achieved through three different stages: (i) pre-injection allows a small quantity of fuel to be injected, leading to a reduction in combustion noise and NOx emissions; (ii) the main injection provides optimized torque release without producing excessive particles from the combustion in the chamber; and (iii) post-injection gives a better soot oxidation rate by letting other small fuel mass to be injected after the main injection period.

3. Spray Atomization

Diesel spray is a crucial element of compression ignition engine operation, particularly during the injection of fuel. Many studies have been conducted focusing on spray atomization [37–41]. Several factors are known to be responsible for spray atomization. They were: (i) the aerodynamic forces; (ii) internal turbulent; (iii) boundary mutation at the exit of the nozzle; and (iv) internal flow cavitation. With the help of the optically-accessible-measurement technique, none of the factors mentioned above were found to influence the spray atomization independently. In fact, it is believed that a combination of some dominant factors is actually responsible collectively.

An aerodynamic force is defined as the force that a body experiences from the air (or other gas), where it is submerged. This force results from the relative velocity of the body and the gas. Aerodynamic force is produced by two factors: the normal force, which results from pressure acting on the body's surface; and the shear force, which is a result of the viscosity of the gas, and is also referred to as skin friction. Shear force moves parallel to the surface, while pressure moves perpendicular to the surface. Both factors are regional. The pressure, as well as shear forces summed over the entire exposed region of the body, constitute the body's net aerodynamic force.

When diesel spray leaves the orifice of the nozzle, aerodynamic-triggered break-up gradually contributes to the production of the mixture. It is important to remember that coherent liquid does not instantly break up into small droplets, and that a zone with ligaments and concentrated big droplets is present close to the nozzle exit. The internal turbulence, along with cavitation, have a more significant impact on the disintegration of the fuel during this process than compared to aerodynamic forces, hence it should be incorporated in the primary break-up process. Aerodynamic force causes the secondary break-up, which is the creation of smaller droplets, when the spray increases the break-up length. The balance between surface tension and aerodynamics is crucial to this process. As previously stated, the aerodynamic force, that primarily serves as ambient gas entrainment, dominates the break-up process, particularly for secondary break up. As a result, researchers pay close attention to the air entrainment.

Cavitation occurs as a result of sudden alterations in the shape of the internal nozzle and pockets from low-static pressure. Vapor phase disruptions, internal flow circumstances, wall roughness, as well as any surface-level microscale manufacturing flaws, all contribute to turbulence. Cavitation would raise the liquid velocity at the exit of the nozzle, which will affect how the spray forms and how it is made. Enhanced spray development is generally believed to result in a more complete combustion, less fuel consumption, and fewer emissions. However, because of its impact on the outgoing jet, cavitation can reduce flow effectiveness (discharge coefficient). Additionally, expanding cavitation bubbles on the inside of the orifice can lead to material degradation, reducing the injector's lifespan and effectiveness.

Numerous studies have investigated internal flow cavitation [42–44]. It is believed that the disintegrations of the needle lift in the cavitation bubbles speed up the spray breakup. The cavitation of the nozzle hole could increase the spray angle and decrease the penetration of the spray tip. Interestingly, the nozzle hole inlet geometry could also influence the intensity of cavitation. It is important to remember that liquid fuel disintegrates into ligaments and large droplets at the stage of internal flow, before the spray comes out of the nozzle. This process, known as primary break-up, is caused by the internal flow cavitation and turbulence instability [45–47]. Figure 3 is break-up regimes of droplet.

The break-up mechanisms of these two factors usually occur simultaneously. Turbulent eddies formed the droplets in small aerodynamic condition, such as in nozzle-hole internal flow. It took place while the surface energy was smaller, compared to the radial fluctuation kinetic energy of the eddy flow velocity, which greatly affects the turbulent effect. It is also known that when the flow velocity is high, the cavitation starts to form. The cavitation takes place since the liquid phase vaporizes instantly when the pressure of local static is lower than vapor pressure (the static pressure decreases when the flow velocity increases). Thus, a two-phase flow emerges in the interior of the nozzle hole once the cavitation is formed. This cavitation phenomenon is characterized by the vaporization of low temperature volatile elements.

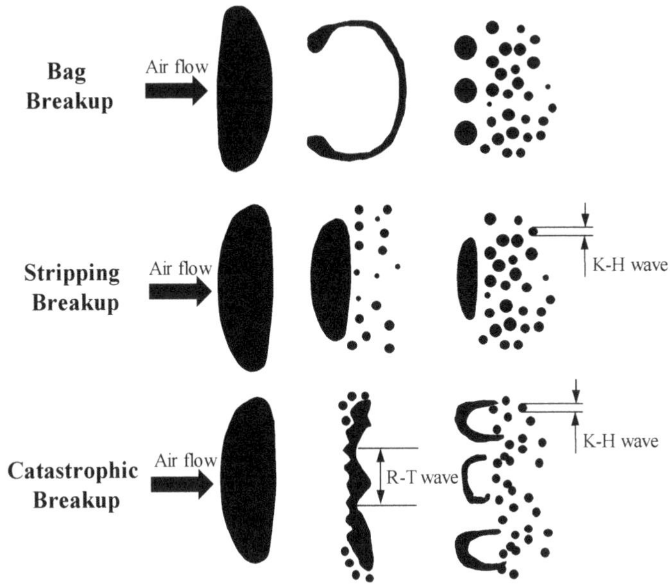

Figure 3. Mechanism of droplet break-up, reproduced from [48].

The spray penetration was time-dependence, being proportional to the evolution timing. Droplets were formed in the spray during the second phase. In this phase, despite the drop in its penetrating velocity, the spray still continued to penetrate further inside the chamber because of energy from the late injection of high momentum fuel. As discussed previously, the break-up (secondary) was primarily initiated by the forces of aerodynamic. It was widely known that aerodynamic forces were greatly dependent on air entrainment. Hence, it is important to investigate the air entrainment in diesel spray. The surrounding air mostly entrained into the spray via the upstream.

The spray and wall interaction in the diesel engine has gained a great deal of attention recently. The impingement of spray on the wall primarily happens in a small-bore engine (high speed) and is believed to have a major effect on mixture formation, combustion, as well as emission. At this impinging condition, the spray atomization undergoes different processes compared to those of free spray. Two main primary differences are the improvement of the secondary break-up process due to smaller droplets formation at a large Weber number, and the reduction of unburned hydrocarbon and soot emission, due to liquid film formation at a low Weber number [49–53]. The interaction of wall and spray, mainly taking place in the small-bore high-speed diesel engine, not only has significant effects on mixture formation, but also on the combustion, as well as the emission process in the cylinder. In order to clarify the effect of spray impingement, an analysis of the two-dimensional piston cavity, along with flat wall impinging spray flame, is usually performed in a high pressure and temperature constant-volume engine-like combustion chamber [54–58]. As a basis for evaluation, free spray flame is normally also conducted.

To recognize the impact of diesel spray on the engine, the most influential factors within the spray structure must be first understood. Those determinant factors are illustrated macroscopically in Figure 4. Near the nozzle tip is the region in which the fuel in the liquid form is dense and uniform [59–61]. Below this dense area, fuel then begins to disperse, and waves begin to emerge at the edge of the spray. Lines of liquid also start to exist and the pitches in the middle of these stripes increase additionally downstream. Afterwards, spray and air inside the chamber interact immediately. As a result of the penetrating spray, the air surrounding it then compresses the spray clusters to be compact. Simultaneously, the spray is transferring its momentum to the air leading to further dispersion of the spray. This is called the air entrainment effect [62–66].

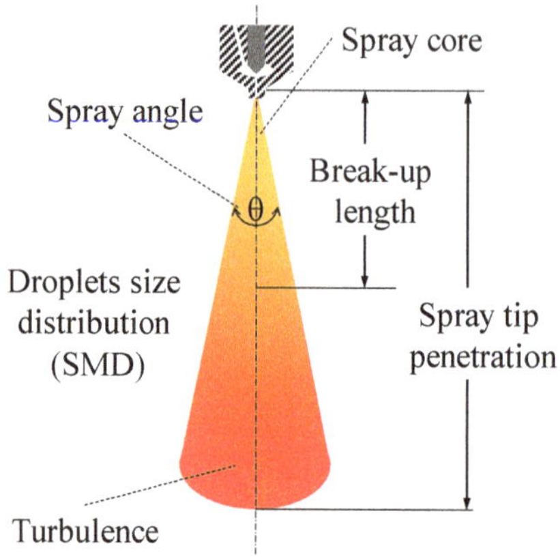

Figure 4. Illustration of diesel spray macroscopically, reproduced from [67].

- Spray core: high dense spray region, the closer to the injector is, the denser it is.
- Spray angle: a bigger spray angle gives wider spatial distribution.
- Break-up length: a part of the liquid in the spray that does not disintegrate.
- Spray tip penetration: the macroscopic development of diesel spray.

Large angle spray is favored and is formed by short spray core and break-up length. Long break-up length, on the other hand, is avoided as it causes a narrow unsteady spray [68–71]. Moreover, due to its longer spray characteristics, the fuel might stick to the inner side of the chamber when it is impinged to the in-cylinder wall, leading to high hydrocarbon and particulate matter emissions. As the formation of the mixture continues, the spray volume is increasing parallel to the increase of an entrained air. As a consequence, the spray penetrates further due to the influences from such air entrainment effect and spray velocity. This phenomenon can be observed and analyzed by optical diagnostic techniques, which will be explained later.

4. Factors Affecting Spray Evolution

Injector technology plays a crucial role in determining the quality of atomization, particularly the injector type and its geometry [40,72–75]. For diesel engines, two types of injectors are normally used, namely, the piezo and the solenoid injector. The piezo injector has a relatively quicker response and opening, which results in shorter injection delay, thus allowing the injected fuel to blend with the air quicker. As a result, mixture formation can be improved leading to better engine performance and reduced emission pollutants. Furthermore, the quicker response of the piezo injector tends to ease the fuel mass delivery control and to enhance the profile of MFR.

In terms of injector geometry, nozzle configuration is classified into three main categories: the mini and micro sac, as well as the valve covered orifice (VCO) [40,76–79]. The difference in the injector nozzle geometry greatly affects the cavitation and evolution of the spray. The illustration of those configurations is displayed in the following Figure 5.

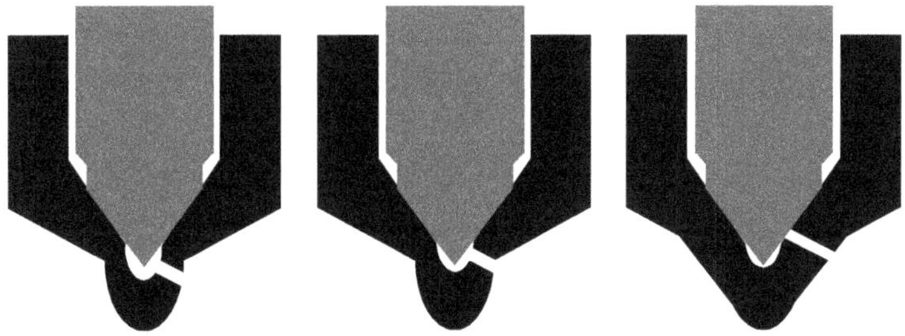

Figure 5. Three types of injector nozzle in diesel engines: (**a**) mini-sac volume; (**b**) micro-sac volume; and (**c**) valve covered orifice (VCO).

The sac refers to the part of an injector that looks like a small bag and contains liquid fuel [51,80–82]. As a result, nozzles with this geometry can lead to unburned hydrocarbon left in the sac. Despite its drawback, the nozzles with sac configuration tend to have superior spray quality compared to VCO nozzle geometry, which have no sac configuration [40,76,77,83,84]. This is because the holes in the sac configuration are positioned just below the needle seat, thus minimizing the loss from throttling. Additionally, VCO is also subject to working conditions and can consequently lead to the oscillation of the needle. In more than one injector, for example, the pressure tends not to be uniformly distributed around the needle in the course of the opening. As a result, the holes-cycles mass flow varies. The VCO injector, however, is able to provide a specific fuel amount and govern the time of injection more accurately, since no sac volume has to be filled. Clearly, the sac and VCO configuration has different effects in the spray evolution. In terms of the cavitation, for example, vortex cavitation occurs in the sac configuration, while conventional cavitation takes place in VOC nozzle geometry.

In order to examine the impact of both the injector hole size and the pressure of the injection, Lacoste [85] evaluated the droplet velocity and SMD with the hole diameters set to be 0.1 and 0.2 mm, whereas the injection pressures were varied at two opposite conditions: low pressure (60 MPa), and high pressure (160 MPa). The results revealed that the smaller hole managed to reduce the droplet size and at the same time increase its velocity. The cavitation was also improved, thus leading to the enhanced turbulence effect. Therefore, the better atomization quality of the spray can be achieved by simultaneously using higher injection pressure and reduced hole size. The nozzle hole convergence can significantly affect the spray behavior. Note that the hole geometry could affect the spray behavior. Different nozzle structure could lead to different spray behavior [86–90].

The injection pressure is critical in determining the spray evolution [91–95]. A high-inertia effect from high-injection pressure results in high-velocity spray, thus creating better dispersion by forming relatively smaller droplets [53,96–99]. Higher-injection pressure could deliver the same amount of fuel relatively faster [100–104]. Moreover, the rise in injection pressure also causes back pressures of the spray that interact with the air [105–109]. This will lead to reduced breakup time and length. As a result, faster propagation and better atomization can be achieved, despite higher fuel concentration at the beginning of the injection. Note that the increased injection pressure can allow the entrainment of the air from the surrounding gas to completely develop, enabling droplets to interact with ambient gas [54,107,110–112].

Ambient pressure, along with density, have a great influence on macroscopic attributes of the spray [113–117]. Additionally, the rise in ambient density could reduce the length of spray penetration as well as widen the angle of the spray cone [39,65,114,118,119]. The ambient temperature primarily affects the spray behavior in terms of its evaporation rate, both microscopically and macroscopically. Hot ambient air inside the cylinder could improve the evaporation of the spray outer periphery and consequently decreasing the spray cone angle [39,107,120–122]. Increasing ambient temperature could lead to enhanced evaporation rate and thus considerably shorten the spray penetration length [123–127].

During the injection start, where the fuel in the liquid form is relatively dense, the spray penetration rate is not enhanced by higher ambient temperature; instead, it is influenced by the gas density [114,128–131]. As the spray penetrates downstream, the liquid fuel becomes less dense, thus increasing both heat as well as momentum transfers by the side of the front periphery. Although lower gas density could raise the penetration rate, the droplet clusters at the front periphery are separated more quickly, resulting in relatively far shorter spray length penetration. The raise in evaporation rate under hot ambient temperature could reduce the SMD. Note that higher ambient air temperature could reduce droplets size at the periphery. The center of the droplet, however, is not influenced by raised temperature owing to the high density.

Higher fuel temperature tends to produce vapor bubble and this phenomenon is known as cavitation [125,132,133]. In the liquid fuel injection behavior, cavitation combined with turbulence can improve spray performance in terms of the liquid atomization, droplets breakup and air entrainment. Under a particular case, such as an extreme condition, cavitation could result in a hydraulic flip that can worsen atomization.

One fuel property that significantly affects the combustion inside the cylinder is viscosity [134–138]. It influences the mixture formation, MFR, and particle size distribution. High viscosity fuel tends to stabilize the spray and delays the fuel dispersion as extra energy is needed to offset the effect of viscous force [139–143]. In contrast, low viscosity fuel has better dispersion resulting in shorter penetration length but wider cone angle [114,130,144–146]. Biodiesel, for instance, could hinder the fuel atomization owing to its high viscosity as well as surface tension content [147–151].

Surface tension is another determinant factor that influences the atomization as it signifies the energy amount needed to increase bubbles [152–155]. While surface tension affects the fuel surface area, volatility influences the spray evolution and mixture formation [114,156–159]. High volatile fuel tends to increase the evaporation rate and lowers SMD thus reducing the spray penetration length [160–165]. Low volatile fuel, on the other hand, has relatively lower evaporation rate, larger droplets size, thus increasing spray penetration length, which in some cases could lead to the wall impingement [166–170].

5. Modeling of Turbulent Diesel Spray and Combustion

The turbulence behavior of diesel spray and the combustion process play a considerable part in influencing the engine performance, combustion, as well as emission formation [171–174]. The physical understanding of these two concepts (diesel spray and combustion) brings a substantial contribution to the modeling in diesel engine combustion. The success of computational models is greatly dependent on how accurate the models can represent the actual combustion process inside the cylinder. Computational fluid dynamic (CFD) models have been proposed, improved, and implemented to capture the physical processes that determine turbulent combustion in diesel engines.

For the atmospheric pressure turbulent flames modeling, until today, most of the turbulent flame models have been approached using simple geometric configurations towards laboratory-scale, in which the flames pressure is assumed to be near atmospheric. This is because in such a condition the thermochemical conditions are well known, and sufficient measurement data are available. To investigate piloted flames, Sandia has generously provided turbulent flame data to facilitate several modeling studies in engine combustion. These data include Sandia C, D, E, and F flames, that are non-luminous, turbulent, and non-premixed. The fuel jet, along with pilot velocities, rise from flame C to F, and so do their local extinction effects. Therefore, while flame D shows mild extinction effects, flame F shows near global extinction effects. The flames provided by Sandia are suitable for modeling studies of turbulence chemistry interactions (TCI). The Damköhler number is often used to characterize the importance of TCI. The number represents the ratio of hydrodynamic to chemical time scales. Low-to-intermediate Damköhler numbers are of particular concern here as they indicate the important phenomena occurs in turbulent and chemistry reactions.

Dispersion, collisions of dispersed phase inter-particle, the modification of continuous phase turbulence, evaporation, mixing, as well as combustion, all take place concurrently in a true turbulent spray flame. It is a huge modelling challenge to deal with all of these intricacies and their relationships. Therefore, it is sensible to aim for advancement in individual sub-areas like breakdown, dispersion, mixing, and combustion, which cannot be completely viewed in isolation. Additionally, the benefits and drawbacks of the general modelling approach must be taken into consideration, including probability density function (PDF), direct numerical simulation (DNS), and large eddy simulation based on Reynolds averaged equations [175].

Several modeling studies have been proposed to improve the PDF-based models by exploring variants of transported modeling, such as composition and velocity-composition-joint PDF, advanced mixing models, improved and faster numerical algorithms, parallelization strategies, radiation modeling, and detailed chemistry. The turbulence modeling and flow structures configurations have been approached with both Reynolds average simulation (RAS) and large eddy simulation (LES). While RAS uses the PDF model, LES uses FDF, which stands for filtered density function. The multi-environment Eulerian-field PDF method, or often shortened MEPDF, has been developed as an alternative to particle-based Monte Carlo methods. The MEPDF method has relatively lower computational cost, while at the same maintain the benefits of simple transported PDF methods.

For high-pressure turbulent spray flames, modeling high-pressure turbulent spray, which can mimic the real flame in internal combustion engines, is a challenging task. Unlike the modeling for stationary turbulent flame, high-pressure turbulent spray flames require accurate high-temporal and spatial resolution. The task is even more difficult due to the transient ignition characteristics involving multiple combustion modes: premixed, mixing-controlled, and kinetically-controlled combustion. In direct-injection engines, the two-phase process of the spray increases the modeling complexity. Furthermore, limited reliable detailed data from experiments for real engine configurations limits the understanding of turbulent spray inside the cylinder.

It is important to remember that most of the diesel combustions are modeled based on measurements at atmospheric pressure [176–179]. In reality, the combustion process occurs at high-pressure involving spray breakup, droplet evaporation, two-phase heat and mass transfer, chemical kinetics, flame propagation, and other transport properties. This obviously requires a different approach as opposed to the near-atmospheric modeling method, considering the turbulent-chemistry interaction to realistically model the turbulent spray.

Therefore, in order to produce realistic predictions for combustion and emission characteristics under such conditions, the model should be supported by a number of experimental results. The experimentation should be conducted to represent detailed thermochemical process to fulfill the prerequisite of advanced combustion of diesel technology. The Engine Combustion Network (ECN) from Sandia National Laboratories, provides the reliable experimental database performed in practical diesel engines. This is carried out to bridge the gap in modeling between the laboratory-scale flame and real modern diesel engines.

Note that the engine performance is considerably responsive to slight variations in the physical state of combustible mixture. Therefore, accurate modeling of the injector and spray behavior are needed to achieve realistic CFD predictions. The spray near the injector, in particular, is of great concern when the injected liquid fuel undergoes a series of instable processes resulted from complex liquid-gas phase interactions. This region often referred to as primary atomization region, is the dense-spray area and has been studied extensively.

Still, comprehensive physical descriptions that affect the formation and growth of the spray have not yet been fully known. Furthermore, the lack of reliable experimental techniques that can sufficiently describe its physical characteristics has restricted the theoretical understanding of the spray formation. Several modeling findings, however, were performed to describe the spray behavior statistically. Yet, no experimental results exist to validate these models. To solve this problem, most modeling strategies have to tune the model coefficients to match measurements further downstream of the injector. In this region, often referred to dilute-spray region, the physical behavior of the spray is relatively easy to understand due to its regularity.

A number of Eulerian- as well as Lagrangian-based methods were successfully employed to model the spray behavior in dilute sprays [180–184]. Such methods are also able to model the two-phase interaction involving droplets (liquid) and surrounding in-cylinder gas. Of the two methods, Lagrangian-based dispersed-phase is more popular for diesel spray modeling. Most of the Lagrangian models analyze the droplet evaporation by assuming the droplet properties to be uniform and the droplet internal circulation to be considerably fast. The interactions between the dispersed-phase droplets and strong turbulent flow within the combustion chamber are important factors to control the evaporation as well as heat transfer rates to and from the droplets. Therefore, assuming the droplets to have uniform properties that change with time is a reasonable approach in the dilute-spray region.

Multi-component droplet evaporation has also been paid attention due to growing interest in the engine development that operate on multi fuels. Initially, multi-component vaporization studies estimated the droplet evaporation as a single spray component of vaporization sequence controlled by volatility differences. Later, it was found that for multi-fuels consisting of different mass diffusivities, the evaporation rates were influenced by each spray component and limited by the low-diffusivity component. In general, the droplet evaporates due to combined effects of both volatility and mass diffusion of each spray components. However, when the mass diffusivity drops to zero, the vaporization rates are no longer subject to volatility of the components.

Most diesel spray modeling neglects the diffusive transport effect inside the droplets. While this assumption is widely accepted in conventional diesel engine condition, in-depth consideration should be taken into account if this assumption to be applied in a low-pressure low-temperature diesel engine. In such condition, the evaporation rate is relatively low so that wall film formation could be formed. As a consequence, the transient heating of the wall film and differential mass diffusion of the liquid components can significantly change the evaporation rates, thus altering the combustion as well as emission behaviors.

All in all, although the advancement of the rapid computational method has significantly improved the numerical predictions of spray behavior, a thorough understanding of the fundamental physical mechanism remains the key factor to the research in engine combustion. The physical phenomena inside the cylinder should be able to be modeled with simple, yet realistic flame configurations without using complex geometries. After being compared with the experimental results, the CFD model could hopefully accurately represent real engine combustion.

6. Optical Measurement Systems

Diesel spray phenomena in the combustion process is difficult to measure. It is considerably challenging to capture spray behavior [185]. In order to gain a deeper comprehension of diesel spray and fuel-air mixing phenomena, several optical techniques have been introduced. The optical diagnostic aims to investigate the spray behavior using non-intrusive method [186]. This method is generally categorized into two types: photography and non-photography. The photography technique intends to observe the macroscopic characteristics such as fuel distribution, spray length, and angle, whereas the non-photographic technique seeks to study the microscopic characteristics.

Specifically, based on the working principle, optical diagnostic techniques are divided into two: traditional diagnostic techniques based on conventional optics and laser-based diagnostic techniques [187]. Figure 6 shows the laser absorption-scattering (LAS) technique basic principle. When the two laser beams go all the way through the spray, the attenuation effect decreases the intensities. In the ultraviolet image, the lower intensity is caused by scattering and absorption of the liquid as well as absorption of the vapor where the impact of the liquid absorption is insignificant [60,188,189].

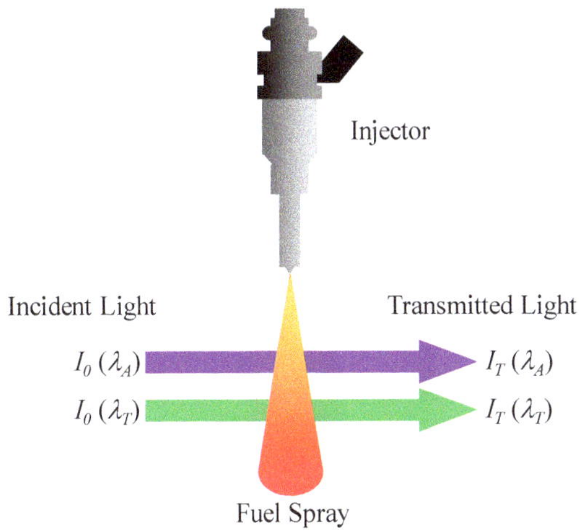

Figure 6. LAS technique basic principle, reproduced from [190].

The diagnostic techniques provide substantial information about the nature of the fuel flow or combustion under high temperature and pressure conditions, thus providing insights into the chemical and physical mechanisms of spray both on macro- and micro-scales. Various optical diagnostic methods have been adopted in previous studies, such as Mie scattering [191,192], Rayleigh scattering [193,194], Raman scattering [195,196], shadow graphy (SG) [191,197], phase doppler interferometry (PDI) [198], Schlieren photography [191,199], laser-induced (exciplex) fluorescence (LIF/LIEF) [123,200], and the combined method [201,202]. Table 1 shows the optical diagnostic techniques used for several phenomena capturing.

Table 1. Optical diagnostics techniques for fuel spray phenomena.

Technique	Application	Phenomena Capturing	References
Mie Scattering	Diesel, gasoline engine	Liquid distribution of spray	Kim et al. [191]
Mie Scattering	Gasoline engine	Droplet size	Mounaïm-Rousselle and Pajot [192]
Raman Scattering	Diesel engine	Spray-wall interaction	Egermann et al. [195]
Raman Scattering	Diesel engine	Spray-wall interaction (near wall condition)	Egermann et al. [196]
Raylaigh Scattering	Diesel engine	Vaporizing diesel spray	Idicheria and Pickett [193]
Raylaigh Scattering	Diesel engine	Liquid penetration	Pickett et al. [194]
Schlieren	Diesel, gasoline engine	Vapor distribution	Kim et al. [191]
Schlieren	Diesel engine	Vaporizing diesel spray	Payri et al. [199]
Shadowgraphy	Diesel, gasoline engine	Vapor distribution (a mare shadow)	Kim et al. [191]
Shadowgraphy	Diesel engine	Diesel—water emulsion fuel injection sprays	Emberson et al. [197]
Phase doppler interferometry (PDI)	Diesel engine	Macroscopic and microscopic spray	Gupta and Agrawal [198]
Laser-induced (exciplex) fluorescence (LIF/LIEF)	Gasoline engine	Mixing process	Bruneaux [200]
Combined (Mie-Rayleigh)	Diesel engine	Evaporation of diesel spray	Adam et al. [201]

The distribution of liquid and vapor phases in diesel spray has gained great attention [126,189,203–205]. In the last two decades, a new technique called particle image velocimetry (PIV) has successfully advanced the research in mixture formation [206–208]. PIV has revealed that the nearby airflow can be classified into three different regions according to the flow property. In the first region (head vortex zone), the gas is initially pushed apart by the head front of the spray. The gas is subsequently recirculated in the recirculation zone (region 2). Finally, in the last region, region 3 (near quasi static zone), the gas is entrained into the following spray zone.

7. Conclusions

Diesel engines have relatively higher efficiency compared to the gasoline engine, achieved by a higher compression ratio by eliminating the throttling losses as commonly found in gasoline engines. This superior benefit of high efficiency in diesel engines comes at the expense of the high level of emissions. The two most significant pollutants of diesel engines are NO_x and PM. The latter stands for particulate matter and is commonly known as soot. Subject to engine load conditions, diesel engines produce a substantial amount of NO_x. Several factors that determine the NO_x formation include thermal NO, fuel-bound nitrogen, prompt NO, and N_2O-intermediate. It occurs in a lean-fuel region, while soot emission, on the other hand, occurs in a rich-fuel region.

Diesel engine performance and pollutant emissions are the outcome of the chemical and physical developments taking place in the cylinder when fuel jets are ignited. Important in-cylinder processes include: the spatial and temporal development of liquid-fuel sprays; fuel vaporization, the mixing of vaporized fuel with in-cylinder gases; the low-temperature autoignition chemistry of fuel components; the high-temperature chemical kinetics of combustion and pollutant formation; as well as late-cycle mixing and burnout processes.

Diesel spray is a crucial event affecting the diesel engine performance, combustion, and emission characteristics. It has received considerable attention not only from automobile manufacturers, but also from academicians across the world. There are several major aspects of diesel spray, which have been considered the dominant factors, including injector cavitation, break-up process, and spray development. Spray cavitation suggest the importance of turbulence phenomenon caused by vapor bubble disruption in an injector, whereas the break-up process indicates the existence of a problem caused by liquid surface instability during the spray development. To observe these events, a number of measurement methods have been proposed, encouraging a new scientific achievement known as diesel spray laser diagnostics.

Many studies have been conducted focusing on spray atomization. Several factors are known to be responsible for the spray atomization. They were: (i) the aerodynamic forces; (ii) internal turbulence; (iii) boundary mutation at the exit of the nozzle; and (iv) internal flow cavitation. With the help of the optically-accessible-measurement technique, none of the factors mentioned above are found to influence the spray atomization independently. In fact, it is believed that the combination of some dominant factors is actually responsible collectively.

It is difficult to determine whether the cavitation has a constructive or adverse effect in engine performance. Although, the cavitation enhances the spray atomization, it also reduces the effective flow area (cross sectional area) which, in fact, deteriorates the combustion under large amount condition. Most researchers are, therefore, trying to reach an agreement on the effects of cavitation on engine performance. Additionally, aerodynamic forces play an important role as the spray exits the nozzle hole [31,40,41]. After the spray passes break-up length, aerodynamic forces cause the smaller droplets formation. This phenomenon is known as the secondary break-up. It relies greatly upon the ratio of aerodynamic and surface tension (as gas phase Weber number). The process of break up with low Weber number occurs in the downstream region, while that with high number takes place in upstream region. It is also important to remember that in diesel engine, spray penetration and angle are of significance on the utility of ambient gas and mixture formation rate. Overlong spray penetration causes several negative consequences such as higher hydrocarbon and carbon monoxide emissions, lower fuel economy, as well as more lubricant consumption due to wall wetting. On the other hand, short spray penetration is known to deteriorate the utilization of chamber gas.

Observing spray behavior in the diesel engine is a challenging task, requiring techniques that involve measurement in high-speed high-pressure and high-temperature condition. There are, however, three parameters to measure the structure of the spray in diesel engine: spray shape, microscopic, and macroscopic characteristics. Besides spray shape and macroscopic parameters, microscopic is of great importance in diesel spray/combustion. Even though numerous studies have investigated these microscopic factors, no definitive theoretical concept has been made regarding the droplet size of diesel spray. These microscopic factors include: (i) mean diameter, (ii) droplets size distribution, and (iii) spatial distribution. For mean diameter, the Sauter mean diameter (SMD) is the most extensively utilized definition. The Sauter mean diameter could indicate the spray average evaporation qualities. For droplets size distribution, smaller droplet size leads to faster evaporation. Yet, smaller size droplets cannot maintain its momentum. As a result, after losing momentum, it fails to accelerate the mixing between fuel and air. For spatial distribution, this important factor affects the quality of diesel spray combustion. Although a number of optical measurements as well as numerical simulations have been conducted in the last two decades, the established evaluation method for spatial distribution of diesel spray has not yet been settled. A combination between the spray's angle, penetration, volume and droplets' spatial number concentration remain to be used to characterize the spatial spray distribution.

Author Contributions: Conceptualization, D.W.D., P.A.P. and O.D.S.; methodology, P.A.P., M.E.M.S. and S.G.H.; software, M.M.A. and H.P.; validation, D.W.D., P.A.P. and S.G.H.; formal analysis, D.W.D., M.I. and O.D.S.; investigation, M.E.M.S. and D.C.; resources, M.M.A., S.G.H. and A.A.Y.; data curation, M.I. and M.M.A.; writing—original draft preparation, D.W.D., M.I., M.E.M.S. and S.G.H.; writing—review and editing, D.C., A.A.Y., H.P. and I.V.; visualization, D.W.D. and H.P.; supervision, M.E.M.S., H.P. and I.V.; project administration, O.D.S., D.C. and A.A.Y.; funding acquisition, D.W.D. and I.V. All authors have read and agreed to the published version of the manuscript.

Funding: The authors would like to thank Sampoerna University for the financial support to this work.

Institutional Review Board Statement: Not applicable.

Informed Consent Statement: Not applicable.

Data Availability Statement: Not applicable.

Conflicts of Interest: The authors declare no conflict of interest.

References

1. Tian, J.; Yu, L.; Xue, R.; Zhuang, S.; Shan, Y. Global low-carbon energy transition in the post-COVID-19 era. *Appl. Energy* **2022**, *307*, 118205. [CrossRef]
2. Zhu, Y.; Zhou, W.; Xia, C.; Hou, Q. Application and Development of Selective Catalytic Reduction Technology for Marine Low-Speed Diesel Engine: Trade-Off among High Sulfur Fuel, High Thermal Efficiency, and Low Pollution Emission. *Atmosphere* **2022**, *13*, 731. [CrossRef]
3. Mohan, S.; Dinesha, P.; Kumar, S. NOx reduction behaviour in copper zeolite catalysts for ammonia SCR systems: A review. *Chem. Eng. J.* **2020**, *384*, 123253. [CrossRef]
4. Damma, D.; Ettireddy, P.R.; Reddy, B.M.; Smirniotis, P.G. A Review of Low Temperature NH3-SCR for Removal of NOx. *Catalysts* **2019**, *9*, 349. [CrossRef]
5. Bytnerowicz, A.; Omasa, K.; Paoletti, E. Integrated effects of air pollution and climate change on forests: A northern hemisphere perspective. *Environ. Pollut.* **2007**, *147*, 438–445. [CrossRef]
6. Veza, I.; Said, M.F.M.; Latiff, Z.A. Improved Performance, Combustion and Emissions of SI Engine Fuelled with Butanol: A Review. *Int. J. Automot. Mech. Eng.* **2020**, *17*, 7648–7666. [CrossRef]
7. Rusli, M.Q.; Said, M.F.M.; Sulaiman, A.M.; Roslan, M.F.; Veza, I.; Perang, M.R.M.; Lau, H.L.N.; Wafti, N.S.A. Performance and Emission Measurement of a Single Cylinder Diesel Engine Fueled with Palm Oil Biodiesel Fuel Blends. *IOP Conf. Series Mater. Sci. Eng.* **2021**, *1068*, 012020. [CrossRef]
8. Fan, B.; Pan, J.; Yang, W.; Chen, W.; Bani, S. The influence of injection strategy on mixture formation and combustion process in a direct injection natural gas rotary engine. *Appl. Energy* **2017**, *187*, 663–674. [CrossRef]
9. Chen, W.; Pan, J.; Yang, W.; Liu, Y.; Fan, B.; Lu, Y.; Otchere, P. Stratified combustion characteristics analysis and assisted-ignition strategy optimization in a natural gas blended diesel Wankel engine. *Fuel* **2021**, *292*, 120192. [CrossRef]
10. GWang, G.; Yu, W.; Li, X.; Yang, R. Influence of fuel injection and intake port on combustion characteristics of controllable intake swirl diesel engine. *Fuel* **2020**, *262*, 116548.

11. Cao, D.N.; Hoang, A.T.; Luu, H.Q.; Bui, V.G.; Tran, T.T.H. Effects of injection pressure on the NOx and PM emission control of diesel engine: A review under the aspect of PCCI combustion condition. *Energy Sources Part A Recover. Util. Environ. Eff.* **2020**, *23*, 2908–2920. [CrossRef]
12. Sener, R.; Yangaz, M.U.; Gul, M.Z. Effects of injection strategy and combustion chamber modification on a single-cylinder diesel engine. *Fuel* **2020**, *266*, 117122. [CrossRef]
13. Chen, H.; Su, X.; He, J.; Zhang, P.; Xu, H.; Zhou, C. Investigation on combustion characteristics of cyclopentanol/diesel fuel blends in an optical engine. *Renew. Energy* **2020**, *167*, 811–829. [CrossRef]
14. Zhang, M.; Hong, W.; Xie, F.; Su, Y.; Liu, H.; Zhou, S. Combustion, performance and particulate matter emissions analysis of operating parameters on a GDI engine by traditional experimental investigation and Taguchi method. *Energy Convers. Manag.* **2018**, *164*, 344–352. [CrossRef]
15. La Rocca, A.; Ferrante, A.; Haffner-Staton, E.; Cairns, A.; Weilhard, A.; Sans, V.; Carlucci, A.P.; Laforgia, D. Investigating the impact of copper leaching on combustion characteristics and particulate emissions in HPCR diesel engines. *Fuel* **2020**, *263*, 116719. [CrossRef]
16. Dhahad, H.A.; Fayad, M.A.; Chaichan, M.T.; Jaber, A.A.; Megaritis, T. Influence of fuel injection timing strategies on performance, combustion, emissions and particulate matter characteristics fueled with rapeseed methyl ester in modern diesel engine. *Fuel* **2021**, *306*, 121589. [CrossRef]
17. Attia, A.M.; Kulchitskiy, A.; Nour, M.; El-Seesy, A.I.; Nada, S.A. The influence of castor biodiesel blending ratio on engine performance including the determined diesel particulate matters composition. *Energy* **2022**, *239*, 121951. [CrossRef]
18. Song, J.; Lee, Z.; Song, J.; Park, S. Effects of injection strategy and coolant temperature on hydrocarbon and particulate emissions from a gasoline direct injection engine with high pressure injection up to 50 MPa. *Energy* **2018**, *164*, 512–522. [CrossRef]
19. Wang, Z.; Dai, X.; Li, F.; Li, Y.; Lee, C.-F.; Wu, H.; Li, Z. Nozzle internal flow and spray primary breakup with the application of closely coupled split injection strategy. *Fuel* **2018**, *228*, 187–196. [CrossRef]
20. Chen, Q.; Wang, C.; Shao, K.; Liu, Y.; Chen, X.; Qian, Y. Analyzing the combustion and emissions of a DI diesel engine powered by primary alcohol (methanol, ethanol, n-butanol)/diesel blend with aluminum nano-additives. *Fuel* **2022**, *328*, 125222. [CrossRef]
21. Ooi, J.B.; Chow, M.R.; Chee, K.M.; Pun, C.H.; Tran, M.-V.; Leong, J.C.K.; Lim, S. Effects of ethanol on the evaporation and burning characteristics of palm-oil based biodiesel droplet. *J. Energy Inst.* **2021**, *98*, 35–43.
22. Wu, F.; Wang, H.; Yu, H.; Zang, X.; Pan, X.; Hua, M.; Jiang, J. Experimental study on the lower explosion limit and mechanism of methanol pre-mixed spray under negative pressure. *Fuel* **2022**, *321*, 124049. [CrossRef]
23. Wang, Y.; Qi, C.; Ning, Y.; Lv, X.; Yu, X.; Yan, X.; Yu, J. Experimental determination of the lower flammability limit and limiting oxygen concentration of propanal/air mixtures under elevated temperatures and pressures. *Fuel* **2022**, *326*, 124882. [CrossRef]
24. Wang, H.; Luo, K.; Hawkes, E.R.; Chen, J.H.; Fan, J. Turbulence, evaporation and combustion interactions in n-heptane droplets under high pressure conditions using DNS. *Combust. Flame* **2021**, *225*, 417–427. [CrossRef]
25. Rajasegar, R.; Niki, Y.; Li, Z.; García-Oliver, J.M.; Musculus, M.P. Influence of pilot-fuel mixing on the spatio-temporal progression of two-stage autoignition of diesel-sprays in low-reactivity ambient fuel-air mixture. *Proc. Combust. Inst.* **2021**, *38*, 5741–5750. [CrossRef]
26. Dec, J.E. A conceptual model of DL diesel combustion based on laser-sheet imaging. *SAE Trans.* **1997**, *106*, 1319–1348.
27. Demarco, R.; Jerez, A.; Liu, F.; Chen, L.; Fuentes, A. Modeling soot formation in laminar coflow ethylene inverse diffusion flames. *Combust. Flame* **2021**, *232*, 111513. [CrossRef]
28. Wang, L.-Y.; Chatterjee, S.; An, Q.; Steinberg, A.M.; Gülder, L. Soot formation and flame structure in swirl-stabilized turbulent non-premixed methane combustion. *Combust. Flame* **2019**, *209*, 303–312. [CrossRef]
29. Li, B.; Zhang, D.; Liu, J.; Tian, Y.; Gao, Q.; Li, Z. A Review of Femtosecond Laser-Induced Emission Techniques for Combustion and Flow Field Diagnostics. *Appl. Sci.* **2019**, *9*, 1906. [CrossRef]
30. Ruan, C.; Chen, F.; Cai, W.; Qian, Y.; Yu, L.; Lu, X. Principles of non-intrusive diagnostic techniques and their applications for fundamental studies of combustion instabilities in gas turbine combustors: A brief review. *Aerosp. Sci. Technol.* **2019**, *84*, 585–603. [CrossRef]
31. Cheng, Q.; Ahmad, Z.; Kaario, O.; Vuorinen, V.; Larmi, M. Experimental study on tri-fuel combustion using premixed methane-hydrogen mixtures ignited by a diesel pilot. *Int. J. Hydrog. Energy* **2021**, *46*, 21182–21197. [CrossRef]
32. Beatrice, C.; Denbratt, I.; Di Blasio, G.; Di Luca, G.; Ianniello, R.; Saccullo, M. Experimental Assessment on Exploiting Low Carbon Ethanol Fuel in a Light-Duty Dual-Fuel Compression Ignition Engine. *Appl. Sci.* **2020**, *10*, 7182. [CrossRef]
33. Abdelaal, M.; Hegab, A. Combustion and emission characteristics of a natural gas-fueled diesel engine with EGR. *Energy Convers. Manag.* **2012**, *64*, 301–312. [CrossRef]
34. Gehmlich, R.; Mueller, C.; Ruth, D.; Nilsen, C.; Skeen, S.; Manin, J. Using ducted fuel injection to attenuate or prevent soot formation in mixing-controlled combustion strategies for engine applications. *Appl. Energy* **2018**, *226*, 1169–1186. [CrossRef]
35. Rao, L.; Zhang, Y.; Kim, D.; Su, H.C.; Kook, S.; Kim, K.S.; Kweon, C.-B. Effect of after injections on late cycle soot oxidation in a small-bore diesel engine. *Combust. Flame* **2018**, *191*, 513–526. [CrossRef]
36. Fayad, M.A.; Al-Salihi, H.A.; Dhahad, H.A.; Mohammed, F.M.; Al-Ogidi, B.R. Effect of post-injection and alternative fuels on combustion, emissions and soot nanoparticles characteristics in a common-rail direct injection diesel engine. *Energy Sources Part A Recovery Util. Environ. Eff.* **2021**, 1–15. [CrossRef]

37. Gad, H.; Ibrahim, I.; Abdel-Baky, M.; El-Samed, A.A.; Farag, T. Experimental study of diesel fuel atomization performance of air blast atomizer. *Exp. Therm. Fluid Sci.* **2018**, *99*, 211–218. [CrossRef]
38. Zhang, P.; Su, X.; Yi, C.; Chen, H.; Xu, H.; Geng, L. Spray, atomization and combustion characteristics of oxygenated fuels in a constant volume bomb: A review. *J. Traffic Transp. Eng. Engl. Ed.* **2020**, *7*, 282–297. [CrossRef]
39. Xia, J.; Huang, Z.; Xu, L.; Ju, D.; Lu, X. Experimental study on spray and atomization characteristics under subcritical, transcritical and supercritical conditions of marine diesel engine. *Energy Convers. Manag.* **2019**, *195*, 958–971. [CrossRef]
40. YSun, Y.; Guan, Z.; Hooman, K. Cavitation in Diesel Fuel Injector Nozzles and its Influence on Atomization and Spray. *Chem. Eng. Technol.* **2019**, *42*, 6–29.
41. Anez, J.; Ahmed, A.; Hecht, N.; Duret, B.; Reveillon, J.; Demoulin, F. Eulerian–Lagrangian spray atomization model coupled with interface capturing method for diesel injectors. *Int. J. Multiph. Flow* **2019**, *113*, 325–342. [CrossRef]
42. Kim, Y.-I.; Kim, S.; Yang, H.-M.; Lee, K.-Y.; Choi, Y.-S. Analysis of internal flow and cavitation characteristics for a mixed-flow pump with various blade thickness effects. *J. Mech. Sci. Technol.* **2019**, *33*, 3333–3344. [CrossRef]
43. Cheng, H.Y.; Bai, X.R.; Long, X.P.; Ji, B.; Peng, X.X.; Farhat, M. Large eddy simulation of the tip-leakage cavitating flow with an insight on how cavitation influences vorticity and turbulence. *Appl. Math. Model.* **2020**, *77*, 788–809. [CrossRef]
44. Mamaikin, D.; Knorsch, T.; Rogler, P.; Wensing, M. Experimental investigation of flow field and string cavitation inside a transparent real-size GDI nozzle. *Exp. Fluids* **2020**, *61*, 154. [CrossRef]
45. Zhou, J.; Andersson, M. An analysis of surface breakup induced by laser-generated cavitation bubbles in a turbulent liquid jet. *Exp. Fluids* **2020**, *61*, 242. [CrossRef]
46. Liu, Z.; Li, Z.; Liu, J.; Wu, J.; Yu, Y.; Ding, J. Numerical Study on Primary Breakup of Disturbed Liquid Jet Sprays Using a VOF Model and LES Method. *Processes* **2022**, *10*, 1148. [CrossRef]
47. Trummler, T.; Schmidt, S.J.; Adams, N.A. Investigation of condensation shocks and re-entrant jet dynamics in a cavitating nozzle flow by Large-Eddy Simulation. *Int. J. Multiph. Flow* **2020**, *125*, 103215. [CrossRef]
48. Chryssakis, C.A.; Assanis, D.N.; Tanner, F.X. Atomization Models. In *Handbook of Atomization and Sprays: Theory and Applications*; Ashgriz, N., Ed.; Springer: Boston, MA, USA, 2011; pp. 215–231.
49. Alozie, N.S.; Ganippa, L.C. *Diesel Exhaust Emissions and Mitigations*; IntechOpen: London, UK, 2019.
50. Zhou, L.; Zhao, W.; Luo, K.H.; Wei, H.; Xie, M. Spray–turbulence–chemistry interactions under engine-like conditions. *Prog. Energy Combust. Sci.* **2021**, *86*, 100939. [CrossRef]
51. Chintagunti, S.J.; Kalwar, A.; Kumar, D.; Agarwal, A.K. Spray Chamber Designs and Optical Techniques for Fundamental Spray Investigations. In *Novel Internal Combustion Engine Technologies for Performance Improvement and Emission Reduction*; Springer: Singapore, 2021; pp. 105–144.
52. Hamdi, F.; Agrebi, S.; Idrissi, M.S.; Mondo, K.; Labiadh, Z.; Sadiki, A.; Chrigui, M. Impact of Spray Cone Angle on the Performances of Methane/Diesel RCCI Engine Combustion under Low Load Operating Conditions. *Entropy* **2022**, *24*, 650. [CrossRef]
53. Zhan, C.; Luo, H.; Chang, F.; Nishida, K.; Ogata, Y.; Tang, C.; Feng, Z.; Huang, Z. Experimental study on the droplet characteristics in the spray tip region: Comparison between the free and impinging spray. *Exp. Therm. Fluid Sci.* **2020**, *121*, 110288. [CrossRef]
54. Peraza, J.E.; Salvador, F.J.; Gimeno, J.; Ruiz, S. ECN Spray D visualization of the spray interaction with a transparent wall under engine-like conditions. Part I: Non-reactive impinging spray. *Fuel* **2022**, *307*, 121699. [CrossRef]
55. Peraza, J.E.; Payri, R.; Gimeno, J.; Martí-Aldaraví, P. ECN Spray D visualization of the spray interaction with a transparent wall under engine-like conditions, Part II: Impinging spray combustion. *Fuel* **2022**, *308*, 121964. [CrossRef]
56. Aizawa, T.; Kinoshita, T.; Akiyama, S.; Shinohara, K.; Miyagawa, Y. Infrared high-speed thermography of combustion chamber wall impinged by diesel spray flame. *Int. J. Engine Res.* **2021**, *23*, 1116–1130. [CrossRef]
57. Yang, K.; Nishida, K.; Yamakawa, H. Effect of split injection ratio on combustion process of diesel spray into two-dimensional piston cavity. *Fuel* **2020**, *260*, 116316. [CrossRef]
58. Mahmud, R.; Kurisu, T.; Nishida, K.; Ogata, Y.; Kanzaki, J.; Akgol, O. Effects of injection pressure and impingement distance on flat-wall impinging spray flame and its heat flux under diesel engine-like condition. *Adv. Mech. Eng.* **2019**, *11*, 1687814019862910. [CrossRef]
59. Bothell, J.K.; Machicoane, N.; Li, D.; Morgan, T.B.; Aliseda, A.; Kastengren, A.L.; Heindel, T.J. Comparison of X-ray and optical measurements in the near-field of an optically dense coaxial air-assisted atomizer. *Int. J. Multiph. Flow* **2020**, *125*, 103219. [CrossRef]
60. Chen, R.; Nishida, K.; Shi, B. Quantitative investigation on the spray mixture formation for ethanol-gasoline blends via UV–Vis dual-wavelength laser absorption scattering (LAS) technique. *Fuel* **2019**, *242*, 425–437. [CrossRef]
61. Li, Y.; Wang, Z.; Kong, Q.; Li, B.; Wang, H. Sulfur dioxide absorption by charged droplets in electrohydrodynamic atomization. *Int. Commun. Heat Mass Transf.* **2022**, *137*, 106275. [CrossRef]
62. Jia, H.; Wei, Z.; Yin, B.; Liu, Z. Analysis of elliptical diesel nozzle spray dynamics using a one-way coupled spray model. *Int. J. Engine Res.* **2021**, 14680874211063352. [CrossRef]
63. Xie, K.; Cui, Y.; Qiu, X.; Wang, J. Experimental study on flame characteristics and air entrainment of diesel horizontal spray burners at two different atmospheric pressures. *Energy* **2020**, *211*, 118906. [CrossRef]
64. Wei, Y.; Li, T.; Zhou, X.; Zhang, Z. Time-resolved measurement of the near-nozzle air entrainment of high-pressure diesel spray by high-speed micro-PTV technique. *Fuel* **2020**, *268*, 117343. [CrossRef]

65. Wei, Y.; Li, T.; Chen, R.; Zhou, X.; Zhang, Z.; Wang, X. Measurement and modeling of the near-nozzle ambient gas entrainment of high-pressure diesel sprays. *Fuel* **2022**, *310*, 122373. [CrossRef]
66. Santos, E.G.; Shi, J.; Gavaises, M.; Soteriou, C.; Winterbourn, M.; Bauer, W. Investigation of cavitation and air entrainment during pilot injection in real-size multi-hole diesel nozzles. *Fuel* **2020**, *263*, 116746. [CrossRef]
67. Hiroyasu, H.; Arai, M. Structures of fuel sprays in diesel engines. *SAE Trans.* **1990**, *99*, 1050–1061.
68. Machicoane, N.; Bothell, J.K.; Li, D.; Morgan, T.B.; Heindel, T.J.; Kastengren, A.L.; Aliseda, A. Synchrotron radiography characterization of the liquid core dynamics in a canonical two-fluid coaxial atomizer. *Int. J. Multiph. Flow* **2019**, *115*, 1–8. [CrossRef]
69. Kong, Q.; Yang, S.; Wang, Q.; Wang, Z.; Dong, Q.; Wang, J. Dynamics of electrified jets in electrohydrodynamic atomization. *Case Stud. Therm. Eng.* **2022**, *29*, 101725. [CrossRef]
70. Berni, F.; Sparacino, S.; Riccardi, M.; Cavicchi, A.; Postrioti, L.; Borghi, M.; Fontanesi, S. A zonal secondary break-up model for 3D-CFD simulations of GDI sprays. *Fuel* **2022**, *309*, 122064. [CrossRef]
71. Koukouvinis, P.; Vidal-Roncero, A.; Rodriguez, C.; Gavaises, M.; Pickett, L. High pressure/high temperature multiphase simulations of dodecane injection to nitrogen: Application on ECN Spray-A. *Fuel* **2020**, *275*, 117871. [CrossRef]
72. Pielecha, I. The influence of petrol injection parameters on the structure of geometry of fuel spray injected from outward-opening injectors. *Fuel* **2018**, *222*, 64–73. [CrossRef]
73. Payri, R.; Gimeno, J.; Martí-Aldaraví, P.; Martínez, M. Transient nozzle flow analysis and near field characterization of gasoline direct fuel injector using Large Eddy Simulation. *Int. J. Multiph. Flow* **2022**, *148*, 103920. [CrossRef]
74. Cui, J.; Lai, H.; Feng, K.; Ma, Y. Quantitative analysis of the minor deviations in nozzle internal geometry effect on the cavitating flow. *Exp. Therm. Fluid Sci.* **2018**, *94*, 89–98. [CrossRef]
75. Monieta, J.; Kasyk, L. Optimization of Design and Technology of Injector Nozzles in Terms of Minimizing Energy Losses on Friction in Compression Ignition Engines. *Appl. Sci.* **2021**, *11*, 7341. [CrossRef]
76. Piscaglia, F.; Giussani, F.; Hèlie, J.; Lamarque, N.; Aithal, S. Vortex Flow and Cavitation in Liquid Injection: A Comparison between High-Fidelity CFD Simulations and Experimental Visualizations on Transparent Nozzle Replicas. *Int. J. Multiph. Flow* **2021**, *138*, 103605. [CrossRef]
77. Yang, S.; Ma, Z.; Li, X.; Hung, D.L.; Xu, M. A review on the experimental non-intrusive investigation of fuel injector phase changing flow. *Fuel* **2020**, *259*, 116188. [CrossRef]
78. Wang, C.; Adams, M.; Jin, T.; Sun, Y.; Röll, A.; Luo, F.; Gavaises, M. An analytical model of diesel injector's needle valve eccentric motion. *Int. J. Engine Res.* **2022**, *23*, 469–481. [CrossRef]
79. Gavaises, M.; Murali-Girija, M.; Rodriguez, C.; Koukouvinis, P.; Gold, M.; Pearson, R. Numerical simulation of fuel dribbling and nozzle wall wetting. *Int. J. Engine Res.* **2022**, *23*, 132–149. [CrossRef]
80. Torres-Garcia, M.; García-Martín, J.F.; Aguilar, F.J.J.-E.; Barbin, D.F.; Alvarez-Mateos, P. Vegetable oils as renewable fuels for power plants based on low and medium speed diesel engines. *J. Energy Inst.* **2020**, *93*, 953–961. [CrossRef]
81. Lee, Z.; Kim, T.; Park, S.; Park, S. Review on spray, combustion, and emission characteristics of recent developed direct-injection spark ignition (DISI) engine system with multi-hole type injector. *Fuel* **2020**, *259*, 116209. [CrossRef]
82. Li, H.; Rutland, C.J.; Pérez, F.E.H.; Im, H.G. Large-eddy spray simulation under direct-injection spark-ignition engine-like conditions with an integrated atomization/breakup model. *Int. J. Engine Res.* **2021**, *22*, 731–754. [CrossRef]
83. Guo, G.; He, Z.; Wang, Q.; Lai, M.-C.; Zhong, W.; Guan, W.; Wang, J. Numerical investigation of transient hole-to-hole variation in cavitation regimes inside a multi-hole diesel nozzle. *Fuel* **2021**, *287*, 119457. [CrossRef]
84. Chouak, M.; Dufresne, L.; Seers, P. Large eddy simulation of a double-injection cycle and the impact of the needle motion on the sac-volume flow characteristics of a single-orifice diesel injector. *Int. J. Engine Res.* **2021**, *22*, 2464–2476. [CrossRef]
85. Julien, L. Characteristics of Diesel Sprays at High Temperatures and Pressures. Ph.D. Thesis, The University of Brighton, Brighton, UK, 2006.
86. Kale, R.; Banerjee, R. Experimental investigation on GDI spray behavior of isooctane and alcohols at elevated pressure and temperature conditions. *Fuel* **2019**, *236*, 1–12. [CrossRef]
87. Zigan, L.; Schmitz, I.; Flügel, A.; Wensing, M.; Leipertz, A. Structure of evaporating single- and multicomponent fuel sprays for 2nd generation gasoline direct injection. *Fuel* **2011**, *90*, 348–363. [CrossRef]
88. Han, J.-S.; Lu, P.-H.; Xie, X.-B.; Lai, M.-C.; Henein, N.A. Investigation of diesel spray primary break-up and development for different nozzle geometries. *SAE Trans.* **2002**, *111*, 2528–2548.
89. Yu, S.; Yin, B.; Bi, Q.; Jia, H.; Chen, C. Effects of gasoline and ethanol on inner flows and swallowtail-like spray behaviors of elliptical GDI injector. *Fuel* **2021**, *294*, 120543. [CrossRef]
90. Som, S.; Longman, D.; Ramírez, A.; Aggarwal, S. A comparison of injector flow and spray characteristics of biodiesel with petrodiesel. *Fuel* **2010**, *89*, 4014–4024. [CrossRef]
91. Battistoni, M.; Grimaldi, C.N. Numerical analysis of injector flow and spray characteristics from diesel injectors using fossil and biodiesel fuels. *Appl. Energy* **2012**, *97*, 656–666. [CrossRef]
92. Kostas, J.; Honnery, D.; Soria, J. A correlation image velocimetry-based study of high-pressure fuel spray tip evolution. *Exp. Fluids* **2011**, *51*, 667–678. [CrossRef]
93. Wu, G.; Zhou, X.; Li, T. Temporal Evolution of Split-Injected Fuel Spray at Elevated Chamber Pressures. *Energies* **2019**, *12*, 4284. [CrossRef]

94. Klein-Douwel, R.J.H.; Frijters, P.J.M.; Somers, L.M.T.; de Boer, W.A.; Baert, R.S.G. Macroscopic diesel fuel spray shadowgraphy using high speed digital imaging in a high pressure cell. *Fuel* **2007**, *86*, 1994–2007. [CrossRef]
95. Chen, L.; Li, G.; Huang, D.; Zhang, Z.; Lu, Y.; Yu, X.; Roskilly, A.P. Experimental and numerical study on the initial tip structure evolution of diesel fuel spray under various injection and ambient pressures. *Energy* **2019**, *186*, 115867. [CrossRef]
96. Ghasemi, A.; Li, X.; Hong, Z.; Yun, S. Breakup mechanisms in air-assisted atomization of highly viscous pyrolysis oils. *Energy Convers. Manag.* **2020**, *220*, 113122. [CrossRef]
97. Zhang, W.; Liu, H.; Liu, C.; Jia, M.; Xi, X. Numerical investigation into primary breakup of diesel spray with residual bubbles in the nozzle. *Fuel* **2019**, *250*, 265–276. [CrossRef]
98. Sykes, D.; Turner, J.; Stetsyuk, V.; de Sercey, G.; Gold, M.; Pearson, R.; Crua, C. Quantitative characterisations of spray deposited liquid films and post-injection discharge on diesel injectors. *Fuel* **2021**, *289*, 119833. [CrossRef]
99. Fauchais, P.L.; Heberlein, J.V.; Boulos, M.I. Overview of thermal spray. In *Thermal Spray Fundamentals*; Springer: Berlin/Heidelberg, Germany, 2014; pp. 17–72.
100. Zhang, X.; Ranjith, P.G. Experimental investigation of effects of CO_2 injection on enhanced methane recovery in coal seam reservoirs. *J. CO_2 Util.* **2019**, *33*, 394–404. [CrossRef]
101. Yip, H.L.; Srna, A.; Yuen, A.C.Y.; Kook, S.; Taylor, R.A.; Yeoh, G.H.; Medwell, P.R.; Chan, Q.N. A Review of Hydrogen Direct Injection for Internal Combustion Engines: Towards Carbon-Free Combustion. *Appl. Sci.* **2019**, *9*, 4842. [CrossRef]
102. Boretti, A. Advances in Diesel-LNG Internal Combustion Engines. *Appl. Sci.* **2020**, *10*, 1296. [CrossRef]
103. McTaggart-Cowan, G.; Mann, K.; Huang, J.; Singh, A.; Patychuk, B.; Zheng, Z.X.; Munshi, S. Direct Injection of Natural Gas at up to 600 Bar in a Pilot-Ignited Heavy-Duty Engine. *SAE Int. J. Engines* **2015**, *8*, 981–996. [CrossRef]
104. Hamzehloo, A.; Aleiferis, P. Gas dynamics and flow characteristics of highly turbulent under-expanded hydrogen and methane jets under various nozzle pressure ratios and ambient pressures. *Int. J. Hydrog. Energy* **2016**, *41*, 6544–6566. [CrossRef]
105. Yu, S.; Yin, B.; Deng, W.; Jia, H.; Ye, Z.; Xu, B.; Xu, H. Internal flow and spray characteristics for elliptical orifice with large aspect ratio under typical diesel engine operation conditions. *Fuel* **2018**, *228*, 62–73. [CrossRef]
106. Zhao, J.; Grekhov, L.; Yue, P. Limit of Fuel Injection Rate in the Common Rail System under Ultra-High Pressures. *Int. J. Automot. Technol.* **2020**, *21*, 649–656. [CrossRef]
107. Wu, H.; Zhang, F.; Zhang, Z.; Gao, H. Experimental investigation on the spray characteristics of a self-pressurized hollow cone injector. *Fuel* **2020**, *272*, 117710. [CrossRef]
108. Yu, S.; Yin, B.; Deng, W.; Jia, H.; Ye, Z.; Xu, B.; Xu, H. Experimental study on the diesel and biodiesel spray characteristics emerging from equilateral triangular orifice under real diesel engine operation conditions. *Fuel* **2018**, *224*, 357–365. [CrossRef]
109. Yu, S.; Yin, B.; Deng, W.; Jia, H.; Ye, Z.; Xu, B.; Xu, H. Experimental study on the spray characteristics discharging from elliptical diesel nozzle at typical diesel engine conditions. *Fuel* **2018**, *221*, 28–34. [CrossRef]
110. Dhanji, M.; Zhao, H. Investigations of split injection properties on the spray characteristics using a solenoid high-pressure injector. *Int. J. Engine Res.* **2022**, *23*, 262–284. [CrossRef]
111. Aleiferis, P.; Papadopoulos, N. Heat and mass transfer effects in the nozzle of a fuel injector from the start of needle lift to after the end of injection in the presence of fuel dribble and air entrainment. *Int. J. Heat Mass Transf.* **2021**, *165*, 120576. [CrossRef]
112. Brulatout, J.; Garnier, F.; Seers, P. Interaction between a diesel-fuel spray and entrained air with single- and double-injection strategies using large eddy simulations. *Propuls. Power Res.* **2020**, *9*, 37–50. [CrossRef]
113. Zhan, C.; Feng, Z.; Zhang, M.; Tang, C.; Huang, Z. Experimental investigation on effect of ethanol and di-ethyl ether addition on the spray characteristics of diesel/biodiesel blends under high injection pressure. *Fuel* **2018**, *218*, 1–11. [CrossRef]
114. Algayyim, S.J.M.; Wandel, A.P. Macroscopic and microscopic characteristics of biofuel spray (biodiesel and alcohols) in CI engines: A review. *Fuel* **2021**, *292*, 120303. [CrossRef]
115. Luo, H.; Nishida, K.; Uchitomi, S.; Ogata, Y.; Zhang, W.; Fujikawa, T. Microscopic behavior of spray droplets under flat-wall impinging condition. *Fuel* **2018**, *219*, 467–476. [CrossRef]
116. Hawi, M.; Kosaka, H.; Sato, S.; Nagasawa, T.; Elwardany, A.; Ahmed, M. Effect of injection pressure and ambient density on spray characteristics of diesel and biodiesel surrogate fuels. *Fuel* **2019**, *254*, 115674. [CrossRef]
117. Zhang, P.; Su, X.; Chen, H.; Geng, L.; Zhao, X. Assessing fuel properties effects of 2,5-dimethylfuran on microscopic and macroscopic characteristics of oxygenated fuel/diesel blends spray. *Sci. Rep.* **2020**, *10*, 1427. [CrossRef] [PubMed]
118. Liu, J.; Feng, L.; Wang, H.; Zheng, Z.; Chen, B.; Zhang, D.; Yao, M. Spray characteristics of gasoline/PODE and diesel/PODE blends in a constant volume chamber. *Appl. Therm. Eng.* **2019**, *159*, 113850. [CrossRef]
119. Ulu, A.; Yildiz, G.; Özkol, Ü.; Rodriguez, A.D. Experimental investigation of spray characteristics of ethyl esters in a constant volume chamber. *Biomass Convers. Biorefin.* **2022**, *23*, 1–18. [CrossRef] [PubMed]
120. Shi, C.; Bu, S.; Zhang, L.; Yuan, H.; Xu, W.; Liu, L.; Zhang, Z. Experimental and numerical investigation on the evaporation performance of a cyclone-type spray desalination chamber. *Desalination* **2019**, *467*, 125–135. [CrossRef]
121. Sun, Y.; Alkhedhair, A.M.; Guan, Z.; Hooman, K. Numerical and experimental study on the spray characteristics of full-cone pressure swirl atomizers. *Energy* **2018**, *160*, 678–692. [CrossRef]
122. Badawy, T.; Xu, H.; Li, Y. Macroscopic spray characteristics of iso-octane, ethanol, gasoline and methanol from a multi-hole injector under flash boiling conditions. *Fuel* **2022**, *307*, 121820. [CrossRef]

123. Feng, L.; Sun, X.; Pan, X.; Yi, W.; Cui, Y.; Wang, Y.; Wen, M.; Ming, Z.; Liu, H.; Yao, M. Gasoline spray characteristics using a high pressure common rail diesel injection system by the method of laser induced exciplex fluorescence. *Fuel* **2021**, *302*, 121174. [CrossRef]
124. Shi, Z.; Lee, C.-F.; Wu, H.; Wu, Y.; Zhang, L.; Liu, F. Optical diagnostics of low-temperature ignition and combustion characteristics of diesel/kerosene blends under cold-start conditions. *Appl. Energy* **2019**, *251*, 113307. [CrossRef]
125. Cao, T.; He, Z.; Si, Z.; El-Seesy, A.I.; Guan, W.; Zhou, H.; Wang, Q. Optical experimental study on cavitation development with different patterns in diesel injector nozzles at different fuel temperatures. *Exp. Fluids* **2020**, *61*, 185. [CrossRef]
126. Shi, Z.; Lee, C.-F.; Wu, H.; Li, H.; Wu, Y.; Zhang, L.; Bo, Y.; Liu, F. Effect of injection pressure on the impinging spray and ignition characteristics of the heavy-duty diesel engine under low-temperature conditions. *Appl. Energy* **2020**, *262*, 114552. [CrossRef]
127. Yan, J.; Gao, S.; Liu, W.; Chen, T.; Lee, T.H.; Lee, C.-F. Experimental study of flash boiling spray with isooctane, hexane, ethanol and their binary mixtures. *Fuel* **2021**, *292*, 120415. [CrossRef]
128. Han, M.; Gordon, R.L.; Talei, M.; Lacey, J.S. Ignition of dense, inhomogeneous fuel sprays at elevated pressures and temperatures. *Fuel* **2022**, *321*, 123853. [CrossRef]
129. Wu, S.; Yang, S.; Wooldridge, M.; Xu, M. Experimental study of the spray collapse process of multi-hole gasoline fuel injection at flash boiling conditions. *Fuel* **2019**, *242*, 109–123. [CrossRef]
130. Zhang, Z.; Lu, Y.; Qian, Z.; Roskilly, A.P. Spray and engine performance of cerium oxide nanopowder and carbon nanotubes modified alternative fuel. *Fuel* **2022**, *320*, 123952. [CrossRef]
131. Reuss, D.L.; Kim, N.; Sjöberg, M. The influence of intake flow and coolant temperature on gasoline spray morphology during early-injection DISI engine operation. *Int. J. Engine Res.* **2022**, 14680874221104301. [CrossRef]
132. Duy, T.-N.; Nguyen, V.-T.; Phan, T.-H.; Hwang, H.-S.; Park, W.-G. Numerical analysis of ventilated cavitating flow around an axisymmetric object with different discharged temperature conditions. *Int. J. Heat Mass Transf.* **2022**, *197*, 123338. [CrossRef]
133. Podbevšek, D.; Lokar, Ž.; Podobnikar, J.; Petkovšek, R.; Dular, M. Experimental evaluation of methodologies for single transient cavitation bubble generation in liquids. *Exp. Fluids* **2021**, *62*, 167. [CrossRef]
134. Sanli, H.; Alptekin, E.; Canakci, M. Using low viscosity micro-emulsification fuels composed of waste frying oil-diesel fuel-higher bio-alcohols in a turbocharged-CRDI diesel engine. *Fuel* **2022**, *308*, 121966. [CrossRef]
135. Bari, S.; Hossain, S.; Saad, I. A review on improving airflow characteristics inside the combustion chamber of CI engines to improve the performance with higher viscous biofuels. *Fuel* **2019**, *264*, 116769. [CrossRef]
136. Hamid, M.F.; Idroas, M.Y.; Sa'ad, S.; Yew Heng, T.; Che Mat, S.; Zainal Alauddin, Z.A.; Shamsuddin, K.A.; Shuib, R.K.; Abdullah, M.K. Numerical investigation of fluid flow and in-cylinder air flow characteristics for higher viscosity fuel applications. *Processes* **2020**, *8*, 439. [CrossRef]
137. Cui, Y.; Liu, H.; Geng, C.; Tang, Q.; Feng, L.; Wang, Y.; Yi, W.; Zheng, Z.; Yao, M. Optical diagnostics on the effects of fuel properties and coolant temperatures on combustion characteristic and flame development progress from HCCI to CDC via PPC. *Fuel* **2020**, *269*, 117441. [CrossRef]
138. Venu, H.; Raju, V.D.; Lingesan, S.; Soudagar, M.E.M. Influence of Al_2O_3nano additives in ternary fuel (diesel-biodiesel-ethanol) blends operated in a single cylinder diesel engine: Performance, Combustion and Emission Characteristics. *Energy* **2021**, *215*, 119091. [CrossRef]
139. Jhalani, A.; Sharma, D.; Soni, S.L.; Sharma, P.K.; Sharma, S. A comprehensive review on water-emulsified diesel fuel: Chemistry, engine performance and exhaust emissions. *Environ. Sci. Pollut. Res.* **2019**, *26*, 4570–4587. [CrossRef]
140. Khandavalli, S.; Sharma-Nene, N.; Kabir, S.; Sur, S.; Rothstein, J.P.; Neyerlin, K.C.; Mauger, S.A.; Ulsh, M. Toward Optimizing Electrospun Nanofiber Fuel Cell Catalyst Layers: Polymer–Particle Interactions and Spinnability. *ACS Appl. Polym. Mater.* **2021**, *3*, 2374–2384. [CrossRef]
141. Geo, V.E.; Prabhu, C.; Thiyagarajan, S.; Maiyalagan, T.; Aloui, F. Comparative analysis of various techniques to improve the performance of novel wheat germ oil—An experimental study. *Int. J. Hydrog. Energy* **2020**, *45*, 5745–5756.
142. Broumand, M.; Albert-Green, S.; Yun, S.; Hong, Z.; Thomson, M.J. Spray combustion of fast pyrolysis bio-oils: Applications, challenges, and potential solutions. *Prog. Energy Combust. Sci.* **2020**, *79*, 100834. [CrossRef]
143. Dafsari, R.A.; Lee, H.J.; Han, J.; Park, D.-C.; Lee, J. Viscosity effect on the pressure swirl atomization of an alternative aviation fuel. *Fuel* **2019**, *240*, 179–191. [CrossRef]
144. Thongchai, S.; Lim, O. Influence of Biodiesel Blended in Gasoline-Based Fuels on Macroscopic Spray Structure from a Diesel Injector. *Int. J. Automot. Technol.* **2019**, *20*, 701–711. [CrossRef]
145. Liu, F.; Li, Z.; Wang, Z.; Dai, X.; He, X.; Lee, C.-F. Microscopic study on diesel spray under cavitating conditions by injecting fuel into water. *Appl. Energy* **2018**, *230*, 1172–1181. [CrossRef]
146. Das, S.K.; Kim, K.; Lim, O. Experimental study on non-vaporizing spray characteristics of biodiesel-blended gasoline fuel in a constant volume chamber. *Fuel Process. Technol.* **2018**, *178*, 322–335. [CrossRef]
147. Ashikhmin, A.E.; Khomutov, N.A.; Piskunov, M.V.; Yanovsky, V.A. Secondary Atomization of a Biodiesel Micro-Emulsion Fuel Droplet Colliding with a Heated Wall. *Appl. Sci.* **2020**, *10*, 685. [CrossRef]
148. Patiño-Camino, R.; Cova-Bonillo, A.; Lapuerta, M.; Rodríguez-Fernández, J.; Segade, L. Surface tension of diesel-alcohol blends: Selection among fundamental and empirical models. *Fluid Phase Equilibria* **2022**, *555*, 113363. [CrossRef]
149. Singh, G.; Pham, P.; Kourmatzis, A.; Masri, A. Effect of electric charge and temperature on the near-field atomization of diesel and biodiesel. *Fuel* **2019**, *241*, 941–953. [CrossRef]

150. Panchasara, H.; Ashwath, N. Effects of Pyrolysis Bio-Oils on Fuel Atomisation—A Review. *Energies* **2021**, *14*, 794. [CrossRef]
151. Pham, P.X.; Nguyen, K.T.; Pham, T.V.; Nguyen, V.H. Biodiesels Manufactured from Different Feedstock: From Fuel Properties to Fuel Atomization and Evaporation. *ACS Omega* **2020**, *5*, 20842–20853. [CrossRef]
152. Antonov, D.V.; Kuznetsov, G.V.; Strizhak, P.A.; Fedorenko, R.M. Micro-explosion of droplets containing liquids with different viscosity, interfacial and surface tension. *Chem. Eng. Res. Des.* **2020**, *158*, 129–147. [CrossRef]
153. Park, S.; Park, K. Principles and droplet size distributions of various spraying methods: A review. *J. Mech. Sci. Technol.* **2022**, *36*, 4033–4041. [CrossRef]
154. Emerson, P.; Crockett, J.; Maynes, D. Thermal atomization during droplet impingement on superhydrophobic surfaces: Influence of Weber number and micropost array configuration. *Int. J. Heat Mass Transf.* **2021**, *164*, 120559. [CrossRef]
155. Shlegel, N.; Tkachenko, P.; Strizhak, P. Influence of viscosity, surface and interfacial tensions on the liquid droplet collisions. *Chem. Eng. Sci.* **2020**, *220*, 115639. [CrossRef]
156. Hoang, A.T.; Le, A.T.; Pham, V.V. A core correlation of spray characteristics, deposit formation, and combustion of a high-speed diesel engine fueled with Jatropha oil and diesel fuel. *Fuel* **2019**, *244*, 159–175. [CrossRef]
157. Zhang, G.; Shi, P.; Luo, H.; Ogata, Y.; Nishida, K. Investigation on fuel adhesion characteristics of wall-impingement spray under cross-flow conditions. *Fuel* **2022**, *320*, 123925. [CrossRef]
158. Zhuang, Y.; Chi, H.; Huang, Y.; Teng, Q.; He, B.; Chen, W.; Qian, Y. Investigation of water spray evolution process of port water injection and its effect on engine performance. *Fuel* **2020**, *282*, 118839. [CrossRef]
159. Sathiyamoorthi, R.; Sankaranarayanan, G.; Munuswamy, D.B.; Devarajan, Y. Experimental study of spray analysis for Palmarosa biodiesel-diesel blends in a constant volume chamber. *Environ. Prog. Sustain. Energy* **2021**, *40*, e13696. [CrossRef]
160. Biswal, A.; Kale, R.; Balusamy, S.; Banerjee, R.; Kolhe, P. Lemon peel oil as an alternative fuel for GDI engines: A spray characterization perspective. *Renew. Energy* **2019**, *142*, 249–263. [CrossRef]
161. Chen, Y.; Liu, S.; Guo, X.; Jia, C.; Huang, X.; Wang, Y.; Huang, H. Experimental Research on the Macroscopic and Microscopic Spray Characteristics of Diesel-PODE3-4 Blends. *Energies* **2021**, *14*, 5559. [CrossRef]
162. Wu, H.; Zhang, F.; Zhang, Z. Fundamental spray characteristics of air-assisted injection system using aviation kerosene. *Fuel* **2021**, *286*, 119420. [CrossRef]
163. Suraj, C.; Sudarshan, G.; Anand, K.; Sundararajan, T. Effects of autooxidation on the fuel spray characteristics of Karanja biodiesel. *Biomass Bioenergy* **2021**, *149*, 106084. [CrossRef]
164. Yan, J.; Gao, S.; Zhao, W.; Lee, T.H.; Lee, C.-F. Experimental study of sprays with isooctane, hexane, ethanol and their binary mixtures under different flash boiling intensities. *Int. J. Heat Mass Transf.* **2021**, *179*, 121715. [CrossRef]
165. Mei, S.S.; Rahman, A.A.A.; Abidin, M.S.Z.; Mazlan, N.M. d2 Law and Penetration Length of Jatropha and Camelina Bio-Synthetic Paraffinic Kerosene Spray Characteristics at Take-Off, Top of Climb and Cruise. *Aerospace* **2021**, *8*, 249.
166. Wang, Y.; Zhuang, Y.; Yao, M.; Qin, Y.; Zheng, Z. An experimental investigation into the soot particle emissions at early injection timings in a single-cylinder research diesel engine. *Fuel* **2022**, *316*, 123288. [CrossRef]
167. Hwang, J.; Weiss, L.; Karathanassis, I.K.; Koukouvinis, P.; Pickett, L.M.; Skeen, S.A. Spatio-temporal identification of plume dynamics by 3D computed tomography using engine combustion network spray G injector and various fuels. *Fuel* **2020**, *280*, 118359. [CrossRef]
168. Yi, P.; Li, T.; Wei, Y.; Zhou, X. Experimental and numerical investigation of low sulfur heavy fuel oil spray characteristics under high temperature and pressure conditions. *Fuel* **2021**, *286*, 119327. [CrossRef]
169. Zhou, Z.-F.; Liang, L.; Murad, S.H.M.; Camm, J.; Davy, M. Investigation of fuel volatility on the heat transfer dynamics on piston surface due to the pulsed spray impingement. *Int. J. Heat Mass Transf.* **2021**, *170*, 121008. [CrossRef]
170. He, X.; Li, Y.; Liu, C.; Sjöberg, M.; Vuilleumier, D.; Liu, F.; Yang, Q. Characteristics of spray and wall wetting under flash-boiling and non-flashing conditions at varying ambient pressures. *Fuel* **2020**, *264*, 116683. [CrossRef]
171. Bao, J.; Qu, P.; Wang, H.; Zhou, C.; Zhang, L.; Shi, C. Implementation of various bowl designs in an HPDI natural gas engine focused on performance and pollutant emissions. *Chemosphere* **2022**, *303*, 135275. [CrossRef] [PubMed]
172. Hoang, A.T. Combustion behavior, performance and emission characteristics of diesel engine fuelled with biodiesel containing cerium oxide nanoparticles: A review. *Fuel Process. Technol.* **2021**, *218*, 106840. [CrossRef]
173. Rajak, U.; Nashine, P.; Verma, T.N.; Pugazhendhi, A. Performance, combustion and emission analysis of microalgae Spirulina in a common rail direct injection diesel engine. *Fuel* **2019**, *255*, 115855. [CrossRef]
174. Hoang, A.T.; Le, A.T. A review on deposit formation in the injector of diesel engines running on biodiesel. *Energy Sources Part A Recover. Util. Environ. Eff.* **2019**, *41*, 584–599. [CrossRef]
175. Jenny, P.; Roekaerts, D.; Beishuizen, N. Modeling of turbulent dilute spray combustion. *Prog. Energy Combust. Sci.* **2012**, *38*, 846–887. [CrossRef]
176. Huang, H.; Lv, D.; Zhu, J.; Zhu, Z.; Chen, Y.; Pan, Y.; Pan, M. Development of a new reduced diesel/natural gas mechanism for dual-fuel engine combustion and emission prediction. *Fuel* **2019**, *236*, 30–42. [CrossRef]
177. Pang, K.M.; Jangi, M.; Bai, X.-S.; Schramm, J.; Walther, J.H.; Glarborg, P. Effects of ambient pressure on ignition and flame characteristics in diesel spray combustion. *Fuel* **2019**, *237*, 676–685. [CrossRef]
178. Finesso, R.; Hardy, G.; Mancarella, A.; Marello, O.; Mittica, A.; Spessa, E. Real-time simulation of torque and nitrogen oxide emissions in an 11.0 L heavy-duty diesel engine for model-based combustion control. *Energies* **2019**, *12*, 460. [CrossRef]

179. Rubio, J.A.P.; Vera-García, F.; Grau, J.H.; Cámara, J.M.; Hernandez, D.A. Marine diesel engine failure simulator based on thermodynamic model. *Appl. Therm. Eng.* **2018**, *144*, 982–995. [CrossRef]
180. Yu, F. Numerical Studies of Nuclear Containment Spray Process by Stochastic Field Method and CGCFD Approach. Ph.D. Thesis, Institut für Thermische Energietechnik und Sicherheit, Karlsruhe, Germany, 2020.
181. Boel, E.; Koekoekx, R.; Dedroog, S.; Babkin, I.; Vetrano, M.R.; Clasen, C.; Van den Mooter, G. Unraveling Particle Formation: From Single Droplet Drying to Spray Drying and Electrospraying. *Pharmaceutics* **2020**, *12*, 625. [CrossRef]
182. Sharma, M.; Goyal, D.K.; Kaushal, G.; Grover, N.K.; Bansal, A.; Goyal, K. CFD and experimental study of slurry erosion wear in Hydro-machinery. *Mater. Today Proc.* **2022**, *62*, 7581–7594. [CrossRef]
183. Han, S.; Zhang, R.; Song, Y.; Xing, J.; Zhou, L.; Li, L.; Zhang, H.; Du, X. Numerical study of swirl cooling enhancement by adding mist to air: Effects of droplet diameter and mist concentration. *Appl. Therm. Eng.* **2022**, *211*, 118475. [CrossRef]
184. Venkatachalam, P.; Sahu, S.; Anupindi, K. Investigation of cross-stream spray injection and wall impingement in a circular channel for SCR application. *Therm. Sci. Eng. Prog.* **2022**, *32*, 101229. [CrossRef]
185. Fansler, T.D.; Parrish, S. Spray measurement technology: A review. *Meas. Sci. Technol.* **2014**, *26*, 012002. [CrossRef]
186. Leipertz, A.; Wensing, M. Modern optical diagnostics in engine research. *J. Phys. Conf. Ser.* **2007**, *85*, 012001. [CrossRef]
187. Xu, H.E.; Yue, W.U.; Xiao, M.A.; Yanfei, L.I.; Yunliang, Q.I.; Zechang, L.I.U.; Yifan, X.U.; Yang, Z.H.O.U.; Xiongwei, L.I.; Cong, L.I.U.; et al. A review of optical diagnostic platforms and techniques applied in internal combustion engines. *Shiyan Liuti Lixue (J. Exp. Fluid Mech.)* **2020**, *34*, 1–52.
188. Qi, W.; Zhang, Y. Quantitative measurement of binary-component fuel vapor distributions via laser absorption and scattering imaging. *Appl. Phys. B* **2019**, *125*, 127. [CrossRef]
189. Zhou, Y.; Wei, Z.; Zhu, Q.; Cao, Y.; Zhang, Y. Quantitative characterization on cyclic variation of mixture formation for flash boiling sprays. *Energy* **2022**, *257*, 124808. [CrossRef]
190. Jin, Y.; Wu, Q.; Zhai, C.; Kim, J.; Luo, H.-L.; Ogata, Y.; Nishida, K. Evaporating characteristics of diesel sprays under split-injection condition with a negative dwell time. *Energetic Mater. Front.* **2021**, *2*, 265–271. [CrossRef]
191. Kim, D.; Park, S.S.; Bae, C. Schlieren, Shadowgraph, Mie-scattering visualization of diesel and gasoline sprays in high pressure/high temperature chamber under GDCI engine low load condition. *Int. J. Automot. Technol.* **2018**, *19*, 1–8. [CrossRef]
192. Mounaïm-Rousselle, C.; Pajot, O. Droplet sizing by Mie scattering interferometry in a spark ignition engine. *Part. Part. Syst. Charact. Meas. Descr. Part. Prop. Behav. Powders Other Disperse Syst.* **1999**, *16*, 160–168. [CrossRef]
193. Idicheria, C.A.; Pickett, L.M. Quantitative mixing measurements in a vaporizing diesel spray by Rayleigh imaging. *SAE Trans.* **2007**, *116*, 490–504.
194. Pickett, L.M.; Genzale, C.L.; Manin, J.; Malbec, L.-M.; Hermant, L. Measurement uncertainty of liquid penetration in evaporating diesel sprays. In Proceedings of the ILASS Americas, 23rd Annual Conference on Liquid Atomization and Spray Systems, Ventura, CA, USA, 15–18 May 2011.
195. Egermann, J.; Taschek, M.; Leipertz, A. Spray/wall interaction influences on the diesel engine mixture formation process investigated by spontaneous Raman scattering. *Proc. Combust. Inst.* **2002**, *29*, 617–623. [CrossRef]
196. Egermann, J.; Göttler, A.; Leipertz, A. Application of spontaneous Raman scattering for studying the diesel mixture formation process under near-wall conditions. *SAE Trans.* **2001**, *110*, 2182–2188.
197. Emberson, D.; Ihracska, B.; Imran, S.; Diez, A. Optical characterization of Diesel and water emulsion fuel injection sprays using shadowgraphy. *Fuel* **2016**, *172*, 253–262. [CrossRef]
198. Gupta, J.G.; Agarwal, A.K. *Macroscopic and Microscopic Spray Characteristics of Diesel and Karanja Biodiesel Blends*; SAE Technical Paper0148-7191; SAE International: Warrendale, PA, USA, 2016.
199. Payri, R.; Salvador, F.; Bracho, G.; Viera, A. Differences between single and double-pass schlieren imaging on diesel vapor spray characteristics. *Appl. Therm. Eng.* **2017**, *125*, 220–231. [CrossRef]
200. Bruneaux, G. Mixing process in high pressure diesel jets by normalized laser induced exciplex fluorescence: Part i: Free jet. *SAE Trans.* **2005**, *114*, 1444–1461.
201. Adam, A.; Leick, P.; Bittlinger, G.; Schulz, C. Visualization of the evaporation of a diesel spray using combined Mie and Rayleigh scattering techniques. *Exp. Fluids* **2009**, *47*, 439–449. [CrossRef]
202. Liu, R.; Huang, L.; Feng, M.; Ju, D.; Ma, Z.; Lu, X. Schlieren and Mie Scattering Visualization of Liquid and Vapor Phase Behavior for Large Nozzle Diameter Injectors Under Marine Diesel Engine Conditions. *SSRN* **2022**. [CrossRef]
203. Markov, V.; Sa, B.; Devyanin, S.; Grekhov, L.; Neverov, V.; Zhao, J. Numerical analysis of injection and spray characteristics of diesel fuel and rapeseed oil in a diesel engine. *Case Stud. Therm. Eng.* **2022**, *35*, 102129. [CrossRef]
204. Guan, W.; He, Z.; Zhang, L.; El-Seesy, A.I.; Wen, L.; Zhang, Q.; Yang, L. Effect of asymmetric structural characteristics of multi-hole marine diesel injectors on internal cavitation patterns and flow characteristics: A numerical study. *Fuel* **2021**, *283*, 119324. [CrossRef]
205. Zhao, Z.; Zhu, X.; Naber, J.; Lee, S.-Y. Assessment of impinged flame structure in high pressure direct diesel injection. *Int. J. Engine Res.* **2020**, *21*, 391–405. [CrossRef]

206. Jardón-Pérez, L.E.; González-Rivera, C.; Trápaga-Martínez, G.; Amaro-Villeda, A.; Ramírez-Argáez, M.A. Experimental Study of Mass Transfer Mechanisms for Solute Mixing in a Gas-Stirred Ladle Using the Particle Image Velocimetry and Planar Laser-Induced Fluorescence Techniques. *Steel Res. Int.* **2021**, *92*, 2100241. [CrossRef]
207. Bilsky, A.V.; Gobyzov, O.A.; Markovich, D.M. Evolution and recent trends of particle image velocimetry for an aerodynamic experiment (review). *Thermophys. Aeromech.* **2020**, *27*, 1–22. [CrossRef]
208. Jena, A.; Singh, A.P.; Agarwal, A.K. Challenges and Opportunities of Particle Imaging Velocimetry as a Tool for Internal Combustion Engine Diagnostics. In *Novel Internal Combustion Engine Technologies for Performance Improvement and Emission Reduction*; Springer: Singapore, 2021; pp. 43–77.

energies

MDPI

Article

Liquid–Liquid Extraction of Volatile Fatty Acids from Anaerobic Acidification Broth Using Ionic Liquids and Cosolvent

Tao Xing [1,2], Shutai Yu [1,3], Jingliang Tang [1,2], Huiliang Liu [1,2,4], Feng Zhen [1,2], Yongming Sun [1,2] and Xiaoying Kong [1,2,*]

1 Guangzhou Institute of Energy Conversion, Chinese Academy of Sciences, Guangzhou 510640, China
2 Guangdong Provincial Key Laboratory of New and Renewable Energy Research and Development, Guangzhou 510640, China
3 College of Energy and Power Engineering, Lanzhou University of Technology, Lanzhou 730050, China
4 University of Chinese Academy of Sciences, Beijing 100049, China
* Correspondence: kongxy@ms.giec.ac.cn

Abstract: Promoting efficiency of liquid–liquid extraction at a high pH is a main challenge for the recovery of volatile fatty acids (VFAs) from organic wastes. In this study, the extraction efficiency of VFAs from artificial solution and acidification fermentation broth of kitchen wastes using ionic liquids (ILs) was assessed at high pH. The effect of ILs addition ratio in diluent, volumetric solvent to feed ratio (S/F) on extraction efficiency were investigated. The solvent consists of [P666,14][Cl] (IL101) and dodecane was found to be the promising solvent for VFA extraction at pH 6.0, especially for butyric acid. The IL-101 ratio in dodecane and S/F was significant factors for the liquid–liquid extraction of VFAs. In general, a higher IL-101 ratio and S/F can promote the extraction efficiency of single VFAs. As a result, the maximum extraction rate of acetic acid (38.4–49.9%) and butyric acid (66.0–92.1%) from different VFA concentration solutions was observed at 10% IL-101 in dodecane and S/F = 2/1. The solvent was also effective in different types of real fermentation broth of kitchen wastes. The maximum extraction rate and selectivity of butyric acid was 60.2%/70.5% in butyric acid type broth and 74.6%/62.7% in mixture acid type broth.

Keywords: liquid–liquid extraction; volatile fatty acid; ionic liquids; anaerobic fermentation; kitchen wastes

Citation: Xing, T.; Yu, S.; Tang, J.; Liu, H.; Zhen, F.; Sun, Y.; Kong, X. Liquid–Liquid Extraction of Volatile Fatty Acids from Anaerobic Acidification Broth Using Ionic Liquids and Cosolvent. *Energies* **2023**, *16*, 785. https://doi.org/10.3390/en16020785

Academic Editor: Fernando Rubiera González

Received: 28 November 2022
Revised: 30 December 2022
Accepted: 1 January 2023
Published: 10 January 2023

1. Introduction

Volatile fatty acids (VFAs) refer to short chain carboxylic acids, such as acetic acid, propionic acid, butyric acid, and valeric acid. VFAs can not only be directly extracted and used in food, pharmaceutical, and other industries but also used as an important platform compound, which can be used as raw materials to further prepare biofuels and high value-added chemicals [1]. The potential uses of VFAs are: (1) As a carbon source for removing nutrients, such as N and P in sewage [2]; (2) Further fermentation to prepare medium chain fatty acids [3]; (3) Fermentation to produce biodegradable polymer [4]; (4) As a raw material for microbial hydrogen production [5]; and (5) As raw materials to prepare biodiesel and other biofuels [6,7]. Presently, commercial production of VFAs is mainly achieved through chemical pathways [8]. However, the non-renewable characteristics of petrochemical raw materials and their processing have caused some environmental problems [9]. In response to the environmental problems caused by VFA production, there has been renewed interest in biological VFA production [10,11]. This includes the use of organic waste to produce VFAs through anaerobic fermentation, which is recyclable and sustainable.

To produce VFAs from organic wastes, it is important to improve the production of VFAs through the optimization of fermentation conditions and microorganisms [12,13].

Furthermore, the effective recovery of VFAs from the fermentation broth is a main obstacle for producing VFAs from organic wastes [14]. Owing to the complex composition of the fermentation broth, it is more difficult the VFAs recovery process in the fermentation broth than that in the petrochemical industry [15,16]. The main extraction methods used for anaerobic acid-producing fermentation broth include liquid–liquid extraction, membrane separation, adsorption, electrodialysis, ion exchange, and gas extraction [17–20]. Liquid–liquid extraction is a common separation method, which has the advantages of good separation efficiency and can separate VFAs in a low concentration fermentation broth [21–24]. Mostafa used tri-n-octyl phosphine oxide (TOPO, 20%) and kerosene (80%) as extractant and diluent, respectively, to recover VFAs [25]. The extraction rate of VFAs in the fermentation broth reached 75% under the optimum condition. Alkaya used 20% TOPO + kerosene as an organic phase to recover VFAs from the anaerobic acidification solution with beet processing waste as substrate [26]. Owing to less VFAs in the non-dissociated state, a higher pH was unfavorable for extraction. When pH was up to 5.5, the total recovery efficiency drops from 67% (pH = 2.5) to 32%. Reyhanitash et al. used trioctylamine (TOA) and other ionic liquids (ILs) as extractants to recover VFAs [27]. They found that the pH value could significantly affect the extraction rate of VFAs. For liquid–liquid extraction, an undissociated state of VFAs and lower pH than pKa (Acid dissociation constants) of VFAs is required [28]. However, the pH of fermentation broth is often higher than the pKa of VFAs. A main challenge for the recovery of VFAs from fermented broth remains that the pH, which is suitable for acidogenic anaerobic microorganisms function, is typically too high for effective extraction [29]. Finding a method to extract VFAs at relatively high pH values, between 5 and 7, is important [26].

ILs extractant is an environmentally friendly extractant with high thermal stability and low volatilization [30]. Therefore, two kinds of quaternary phosphine ILs ([P666,14][Cl] and [P666,14][Phos]) and three kinds of diluents were selected in this experiment to study the extraction efficiency of VFAs from the anaerobic fermentation broth of kitchen waste. The effects of pH, ratio of ILs addition, volumetric solvent to feed ratio (S/F), and VFA concentration were also studied using artificial solution. The result could find an effective method for liquid–liquid extraction of VFAs from anaerobic fermentation broth at relatively high pH values.

2. Materials and Methods

2.1. Chemicals

Acetic acid (>99.5%), n-butyric acid (>99.0%), trioctylamine (>98.0%), n-octanol (>99.0%), dodecane (>98.0%), mineral oil (25cSt, 40 °C) were purchased from Shanghai Maclin Biochemical Technology Co., Ltd, (Shanghai, China). [P666,14][Cl] (Cyphos® IL101, >95.0%), [P666,14][Phos] (Cyphos® IL104, >90.0%) were purchased from Sigma-Aldrich (Burlington, MA, USA).

2.2. Inoculum and Substrates

Kitchen wastes were obtained from the dining hall of Guangzhou Institute of Energy conversion. The raw materials were crushed by the shredder (DXF-20C, Guangzhou, China) for 1 min after manually removing impurities and then were stored at −20 °C in a refrigerator for future use. The inoculum was obtained from the mesophilic anaerobic reactor which had been acclimated for a long-time using kitchen wastes to produce biogas. The inoculum was heated at 90 °C for 1 h to inhibit methanogenesis activity. The physical and chemical properties of substrate and inoculum are shown in Table 1.

Table 1. Physicochemical properties of substrate and inoculum.

	Kitchen Waste	Inoculum
TS/%	20.85 ± 0.81	2.37 ± 0.06
VS/%	19.58 ± 0.40	1.20 ± 0.03
pH	/	7.5

2.3. Experiment Setup

2.3.1. Extraction of VFAs with Different Extraction Agent

The concentration of acetic acid and butyric acid artificial solution was 10.0 g/L. The pH of the solution was adjusted to 3, 4, 5, and 6 using 4 M KOH before extraction. To extract using different extractants, 20% IL-101 or IL-104 were added to dodecane to be solvents, respectively. In the control group, 20% TOA in n-octanol was used as solvent. In the second experiment, n-octanol, dodecane, and mineral oil were used as different diluents for extraction. No extractant was added to one group, whereas 20% IL-101 was added to the diluent as extractant. Liquid–liquid extraction experiments were performed at the volumetric S/F of 1/1 using 3 mL of the feed. The mixtures were shaken in a constant temperature shaker (30 ± 1 °C, >300 rpm) for 120 min. Then, different phases were separated using a centrifuge (10,000 rpm for 5 min) and sampled for analysis. All treatments were performed in triplicate.

2.3.2. Extraction Efficiencies of Single VFAs at Different Conditions

IL-101 + dodecane was used as solvents to extract acetic acid and butyric acid from the artificial solution, whose concentrations were 2.0, 5.0, and 10.0 g/L. The pH was adjusted to 6.0 before extraction. The addition ratio of IL-101 were 2%, 4%, 6%, 8%, and 10% and S/F were 1/2, 1/1, and 2/1.

2.3.3. Acidification and VFA Production

The fermentation trials were carried out in a 2000 mL glass bottle with an operating volume of 1500 mL at a mesophilic temperature of 37 °C. Substrates of 1200 mL inoculum and 8.0 gVS /100 mL inoculum were added into reactors. The initial pH of the system was adjusted to 5, 7 and 10 using 4 M HCl and 4 M KOH before fermentation. Nitrogen was flushed into flask for 5 min to ensure anaerobic environment. The experiment lasted for 7 days. Before extraction, the broth was separated using a centrifuge (10,000 rpm for 5 min).

2.3.4. Liquid–Liquid Extraction of VFAs from Fermentation Broth

IL101+dodecane was used as solvents to extract VFAs from fermentation broth. The ratio of IL101 in dodecane set as 4% and 8% and S/F was 2/1, respectively. As a comparison, the artificial solution with same acid concentration and pH were also used for extraction experiments. All treatments were performed in triplicate.

2.4. Analytical Methods

TS, VS and pH were determined by previous methods [31]. VFAs and ethanol were determined by gas chromatography (GC9790Plus, FULI INSTRUMENS, Fuli, China) with Agilent DB-FFAP capillary column (30 m × 0.32 mm × 0.25 mm, Santa Clara, CA, USA). The operation conditions of GC were determined by previous methods [32]. VFAs concentrations of solvent phase were determined by mass balance.

Extraction rate (ER) represents the proportion of a certain VFAs extracted from the aqueous phase into the solvent phase when the extraction reaches equilibrium.

$$ER\% = \frac{C_{sol} \times V_{sol}}{C_{in} \times V_{in}} \times 100\% \tag{1}$$

where C_{in} (mg/L) and V_{in} (mL) represent VFAs concentration and volume of initial feed solution; C_{sol} (mg/L) and V_{sol} (mL) represent VFAs concentration and volume of solvent after extraction.

3. Results and Discussion

3.1. Extraction of Single VFAs with Different Extraction Agent

The extraction efficiency of the extractant on acetic acid and butyric acid under different pH conditions is shown in Figure 1. The TOA + n-octanol solvent is a common VFA extractant. As shown in Figure 1, the TOA + n-octanol solvent has good extraction efficiency of acetic acid and butyric acid at pH 3 with the extraction rates reaching 83.5 ± 2.5% and 92.7 ± 0.7%, respectively. With the increase in pH, the extraction rate decreased rapidly. When the pH was 6, the extraction rates of acetic acid and butyric acid were only 11.2 ± 3.8% and 27.8 ± 0.4%, respectively. Similarly, the maximum extraction rates of acetic acid and butyric acid were 89.8 ± 0.2% and 90.6 ± 0.1%, respectively, at pH 3 when using IL-104 + dodecane as the solvent. However, the extraction rates of acetic acid and butyric acid decreased to 14.6 ± 3.8% and 11.5 ± 1.8% when pH increased to 6. For these two solvents, VFAs mainly exist in molecular form, which is conducive to the separation in the solution when pH < pKa (4.76 for acetic acid and 4.82 for butyric acid), while the ionic state is dominant, which is not conducive to the extraction of organic solvents when pH > pKa [33]. Compared with the above mentioned two solvents, the extraction efficiency of VFAs using IL-101 + dodecane was less affected by pH, especially butyric acid. Although the extraction rate of acetic acid decreased from 63.2 ± 0.5% to 46.6 ± 3.6% when pH was up to 6, it is still higher than TOA + n-octanol and IL-104 + dodecane. The extraction rate of butyric acid is always above 95%, which is not influenced by the pH. According to Tamada and King (1990), high hydrophobicity may result in this greater efficiency for liquid–liquid extraction of longer chain VFAs [34]. Ion exchange and intermolecular interactions (e.g., hydrogen bonding) may be responsible for extracting VFAs. However, the extraction of VFAs is affected by other parameters, such as functional groups and steric hindrance [26,35]. Under the combined influence of these factors, IL-101 showed better extraction performance than IL-104 for effective extraction at a higher pH and selective extraction of butyric acid.

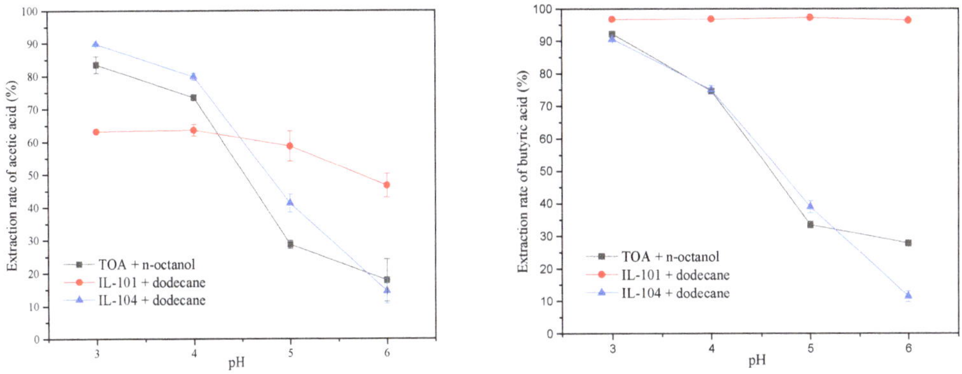

(a) (b)

Figure 1. Extraction rates of acetic acid (**a**) and butyric acid (**b**) using different solvents at different pH.

To determine the influence of different diluents on extraction efficiency, the extraction rate of acetic acid and butyric acid were investigated. As shown in Figure 2a, all diluents could extract acetic acid without IL addition. The extraction rates of acetic acid by n-octanol and dodecane decreased with the increase in pH, and the extraction rate decreased to 27.7 ± 0.3% and 19.8 ± 1.9% at pH 6, respectively. The extraction rate of acetic acid for mineral oil not sensitive to pH is much lower than that of n-octanol and dodecane. After adding IL-101, the extraction rate can be improved at all pH conditions. The extraction using IL-101 + n-octanol became insensitive to the pH. The extraction rate of acetic acid still reached 64.3 ± 2.5% when the pH was 6. The extraction rate for IL-101 + dodecane decreased with the increase in pH. The extraction rate of acetic acid was 46.6 ± 3.6% at pH 6, which was only 70.6% of that at pH 3. The extraction rate of acetic acid by IL-101 + mineral oil was still lower than 40%.

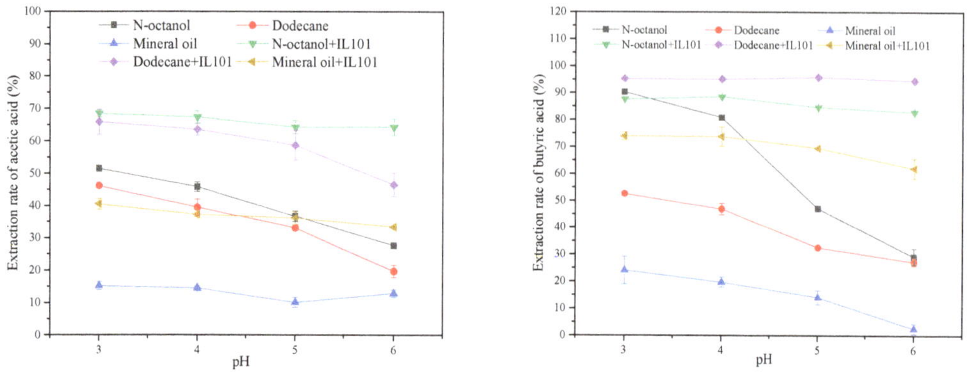

(a) (b)

Figure 2. Extraction rates of acetic acid (**a**) and butyric acid (**b**) using IL 101 with different diluents.

As shown in Figure 2b, all diluents could extract butyric acid without IL addition. However, the extraction rate decreased with the increase in pH. The extraction rates for n-octanol, dodecane, and mineral oil decreased from 90.3 ± 0.2%, 52.7 ± 0.2%, and 24.2 ± 5.1% at pH 3 to 29.2 ± 3.0%, 27.3 ± 1.4%, and 2.6 ± 1.9% at pH 6, respectively. After adding IL-101 to the diluent, the effect of pH on extraction efficiency reduced significantly. Particularly at high pH, higher extraction rates were obtained. The extraction rates for IL-101 + n-octanol, IL-101 + dodecane, and IL-101 + mineral oil reached 82.8 ± 0.3%, 94.6 ± 0.5%, and 62.0 ± 3.8%, respectively. The extraction efficiency of VFAs by IL 101+ dodecane in this study was relatively higher than the previous report which using deep eutectic solvents as extractant to extract VFAs at high pH value [36].

At a higher pH, IL-101 + n-octanol and IL-101 + dodecane have great extraction ability and can be used for acetic acid separation. However, IL-101 + dodecane also presented impressive butyric acid extraction. As the value of butyric acid is usually higher than that of acetic acid, butyric acid extraction should always be preferred in the fermentation broth. Therefore, IL-101 + dodecane may be more favorable for hybrid VFAs owing to its selectivity of butyric acid.

3.2. Extraction Efficiencies of Single VFAs at Different Conditions

The effects of different conditions (ratio of IL addition and volumetric S/F) on extraction efficiencies of different concentrations of acetic acid and butyric acid were studied. The extraction efficiency of acetic acid is shown in Figure 3a. Under all acetic acid concentrations, the extraction rate increased with the increase in IL-101 ratio, and a higher S/F was beneficial to obtain a higher extraction rate. At the concentration of 2 g/L acetic acid, the highest extraction rate was only 38.4 ± 0.7% when S/F was 2/1 and the addition ratio of IL-101 was 10%. With the increase in acetic acid concentration, the extraction rate also increased. When the acetic acid concentration was 5 g/L, the extraction rate improved, i.e., 29.8 ± 1.5%–39.8 ± 1.1% (S/F = 1/2), 34.3 ± 0.7%–46.0 ± 4.2% (S/F =1/1), and 33.4 ± 0.3%–46.6 ± 7.4% (S/F = 2/1). Under this condition, there is little difference in extraction rates between S/F = 1/1 and 2/1, and the highest extraction rates can reach more than 46%, which was more than 20% higher than that of low acetic acid concentration. When the acetic acid concentration was up to 10 g/L, the extraction rates were 29.3 ± 1.0%–37.9 ± 1.6% (S/F = 1/2), 34.2 ± 0.7%–44.6 ± 0.2% (S/F = 1/1) and 31.2 ± 0.9%–49.9 ± 2.1% (S/F = 2/1). Under these conditions, the maximum extraction rate was obtained when the addition ratio of IL-101 was 10% and S/F = 2/1, which was slightly higher than that at S/F = 1/1.

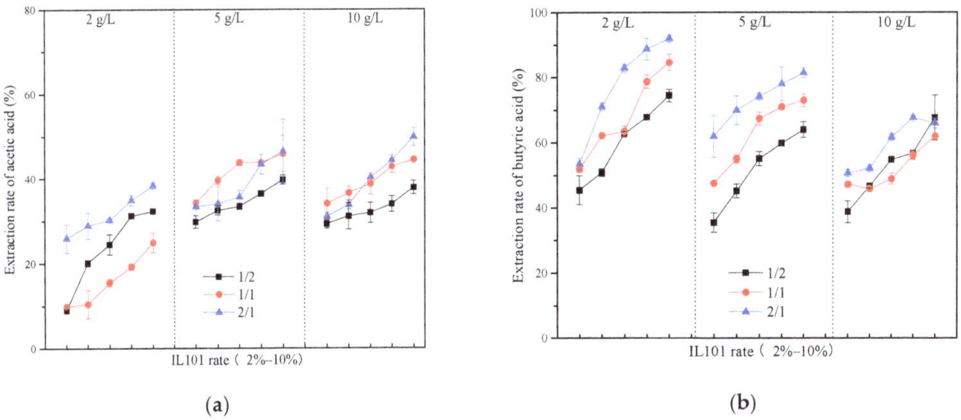

Figure 3. Effect of ratio of IL addition and S/F on extraction efficiency at different VFA concentration (**a**) acetic acid; (**b**) butyric acid.

The extraction efficiency of butyric acid at different concentrations is shown in Figure 3b. Similar to acetic acid, the extraction rate increased with the increase in IL-101 ratio, and a higher S/F was beneficial to obtain a higher extraction rate. The extraction efficiency of butyric acid was obviously better than that of acetic acid. When the concentration of butyric acid was 2 g/L and S/F = 2/1, the extraction rate of butyric acid was 53.6 ± 1.8%–92.1 ± 1.1%, and the maximum extraction rate was 1.85 times that of acetic acid. Different from acetic acid, the extraction rate of butyric acid decreased with the increase in butyric acid concentration. When the concentration of butyric acid was 5 g/L, the extraction rates were 35.4 ± 2.9%–64.0 ± 2.4% (S/F = 1/2), 47.6 ± 0.4%–73.0 ± 1.9% (S/F = 1/1), and 62.0 ± 6.4%–81.5 ± 1.5% (S/F = 2/1). The maximum extraction rate decreased by more than 10% compared to the low concentration butyric acid. When the concentration of butyric acid was 10 g/L, the extraction efficiency further decreased at high S/F values of 1/1 and 2/1, and the maximum extraction rates were 62.0 ± 2.2% and 66.0 ± 1.5%, respectively. For high concentration butyric acid extraction, S/F has little influence on the maximum extraction efficiency.

When extracting low concentration butyric acid, adding 10% IL-101 can achieve more than 90% extraction rate. Although the extraction rate decreased when 5 g/L butyric acid was extracted, adding 10% IL-101 could still achieve more than 80% extraction rate. However, when the extraction concentration of butyric acid was higher, the extraction efficiency significantly decreased, and adding 10% IL-101 was not sufficient to obtain a higher extraction rate.

3.3. Extraction Efficiencies of VFAs from Acidification Broth

Adjusting the initial pH value is an effective method to increase VFA production and regulate composition [37]. pH is also an important factor in VFA extraction. The concentrations of VFAs at different pH are shown in Table 2. At the end of fermentation, the concentration of VFAs and ethanol was 2.95 g/L at pH_{in} = 5, which is much lower than pH_{in} = 7 and 10. For this condition, the proportion of acetic acid was 78.3% and butyric acid was not detected. When the initial pH was 7, the concentration of VFAs and ethanol in fermentation broth increased significantly, reaching 9.28 g/L. Additionally, the VFA composition changed. Butyric and acetic acids accounted for 51.39% and 29.89% in total VFA amount, respectively, which is typical for butyric acid fermentation [38]. Under alkaline conditions (pH_{in} = 10), the concentration of VFAs and ethanol in the fermentation broth continued to increase, and the concentration of fermentation products reached 19.52 g/L. The production of VFAs and ethanol was much higher than that in neutral or acidic fermentation broth. In this broth, the proportion of acetic acid and butyric acid was 49.22% and 29.99%, respectively, which is typical for mixture acid fermentation broth [38].

Table 2. VFAs content in different type fermentation broth.

Initial pH	End pH	Ethanol (mg/L)	Acetic Acid (mg/L)	Propionic Acid (mg/L)	Butyric Acid (mg/L)	VFAs + Ethanol (mg/L)
5.0	3.9	1062	1475	410	0	2948
7.0	5.3	822	2773	915	4769	9279
10.0	5.8	3089	9610	969	5856	19,524

The high butyric acid content was obtained at pH 7 and 10 with different acetic acid proportions. The broth and its corresponding artificial solution, which is called "butyric acid type" (pH_{in} = 7) and "mixture acid type" (pH_{in} = 10), was used for extraction experiments, and the results are shown in Figures 4 and 5. When the butyric acid type artificial solution was extracted using 8% IL-101 solvent in dodecane, the extraction rate of VFAs and ethanol, butyric acid, acetic acid, and propionic acid were 47.5%, 65.3 ± 0.4%, 27.2 ± 0.9%, and 56.9 ± 1.0%, respectively. After extraction, butyric acid, acetic acid, and propionic acid accounted for 70.8%, 17.1%, and 11.8% of total VFA in the solvent phase, respectively. For the butyric acid type fermentation broth, the total extraction rates of VFAs and ethanol, butyric acid, acetic acid, and propionic acid were 43.9%, 60.2 ± 0.8%, 25.0 ± 1.1%, and 50.7 ± 0.9%, respectively. Owing to the complex composition of fermentation broth, which may contain monosaccharides or other acids, the extraction rate of VFAs and butyric acid slightly decreased in the fermentation broth. However, the extraction rate of VFAs was still more than 90% of that in the artificial solution. IL-101 + dodecane could effectively extract VFAs in the real fermentation broth. After extraction. butyric acid, acetic acid, and propionic acid accounted for 70.5%, 17.0%, and 11.4% of total VFA in the solvent phase, respectively. Butyric acid has better extraction selectivity, whereas acetic acid and propionic acid maintain a lower proportion. Emrah Alkaya et al. used TOPO + kerosene to extract Beet-pulp anaerobic acidifying solution at pH 5.5 [26]. They found that butyric acid also has better extraction selectivity, but its extraction rate is lower than 50%, which was lower than this study.

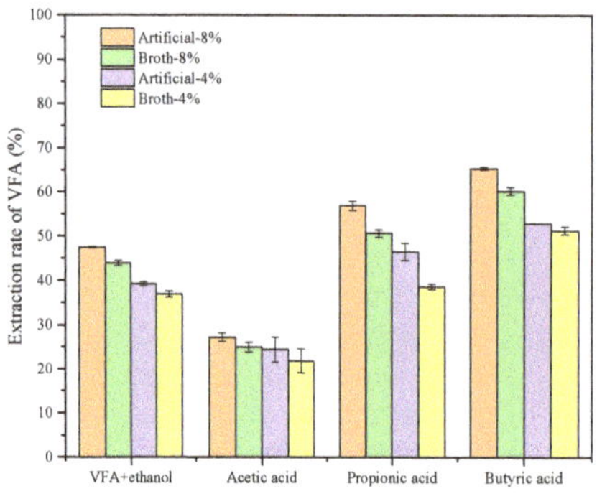

(**a**)

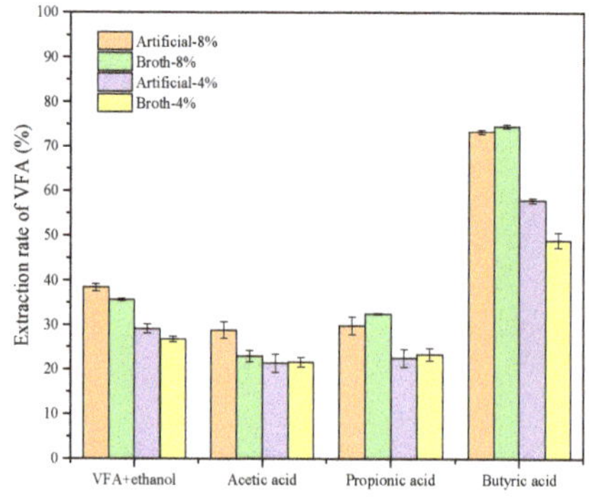

(**b**)

Figure 4. Extraction rates of VFA from different type fermentation broth and artificial solution ((**a**) butyric acid type; (**b**) mixture acid type).

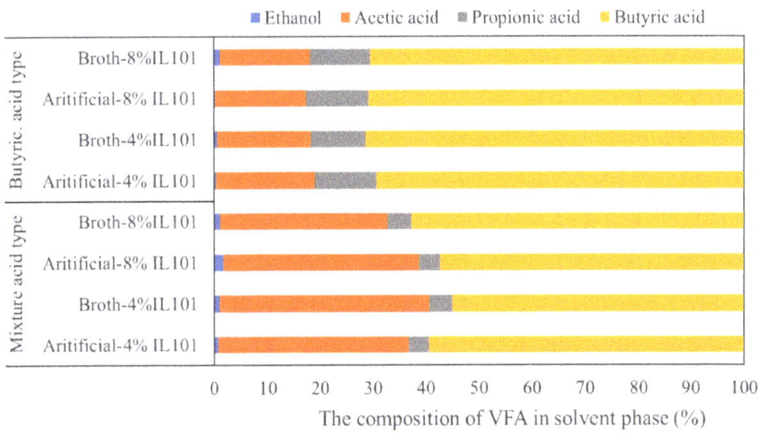

Figure 5. The composition of VFA in solvent phase after extraction.

When the mixture acid type artificial solution was extracted using 8% IL-101 solvent in dodecane, the extraction rate of VFAs and ethanol, butyric acid, acetic acid, and propionic acid were 38.4%, 73.3 ± 0.5%, 28.8 ± 1.9%, and 29.8 ± 2.1%, respectively. After extraction. butyric acid, acetic acid, and propionic acid accounted for 57.3%, 37.0%, and 3.9% of total VFA in the solvent phase, respectively. For the mixture acid type fermentation broth, the total extraction rate of VFAs and ethanol, butyric acid, acetic acid, and propionic acid were 35.7%, 74.6 ± 0.4%, 22.9 ± 1.3%, and 32.5 ± 0.1%, respectively. There was little loss of extraction efficiency in the fermentation broth. The extraction efficiency of butyric acid from real fermented broth was close to that in a previous report which used octanoic acid as extractant to extract single acid from artificial solution at pH 3 [39].

Compared with the butyric acid type fermentation broth, the extraction rate of total VFAs and ethanol from mixture acid type fermentation broth decreased by 17.9%, which may result from the higher concentration of VFAs and higher pH. However, the extraction rate of butyric acid was 1.4 times of that in butyric acid type fermentation broth. However, the proportion of acetic acid in the solvent phase reached 31.6%, whereas that of butyric acid decreased to 62.7%. This may be due to the higher concentration of acetic acid in broth, which was more than three times of that in the butyric acid fermentation broth. This shows that the concentration and type of VFAs in the fermentation broth can affect the extraction efficiency and selectivity. Both "butyric acid type" and "mixture acid type" broth can obtain a higher extraction efficiency for longer chain VFAs, such as butyric acid.

When the ratio of IL-101 in dodecane was reduced to 4%, the extraction rate of VFAs and ethanol in butyric acid and mixture acid type broth decreased to 37.0% and 26.8%, respectively, which was only 84.3% and 75.1% of that when using 8% IL-101 solvent in dodecane. The extraction rate of butyric acid significantly decreased, i.e., by 14.8% in butyric acid type broth and 34.3% in mixture acid type broth. The extraction rate of butyric acid was significantly affected by reducing the addition of IL-101 in the mixture acid type broth. This may be related to high acetic acid concentration, which may compete with butyric acid when IL-101 was limited. The extraction rate of acetic acid decreased by 12.4% in the butyric acid type broth, but had no significant change in mixture acid type broth extraction. Additionally, with the decrease in IL-101 ratio, there was little change in the butyric acid selectivity for extraction in butyric acid type broth. However, the butyric acid selectivity significantly decreased for extraction in mixture acid type broth, which was unfavorable to extraction of butyric acid.

4. Conclusions

To find an effective extractant to recover VFAs from anaerobic fermentation broth at relatively high pH values, several ILs and organic diluents were used to investigate the effect on extraction efficiency in artificial and real fermentation broth. We observed that the solvent consists of IL-101, and dodecane can extract VFAs effectively at pH 6.0. Particularly, IL-101 + dodecane presented a great performance in butyric acid extraction efficiency, which was above 90%. For acetic acid and butyric acid extraction, the effect of IL-101 ratio in dodecane was as crucial as the effect of S/F on the recovery of VFAs. In general, a higher IL-101 ratio and S/F can promote the extraction efficiency of single VFAs. The solvent was also effective in a hybrid VFA system, and different types of real broth were derived from kitchen wastes acidification fermentation. Furthermore, it showed a favorable selectivity for butyric acid. The maximum extraction rate and selectivity of butyric acid was 60.2%/70.5% in butyric acid type broth and 74.6%/62.7% in mixture acid type broth.

Author Contributions: Conceptualization, T.X. and X.K.; methodology, T.X. and S.Y.; software, J.T.; validation, T.X. and S.Y.; formal analysis, J.T.; investigation, H.L.; resources, F.Z.; data curation, S.Y.; writing—original draft preparation, T.X.; writing—review and editing, T.X. and X.K.; visualization, J.T.; supervision, X.K. and Y.S.; project administration, Y.S.; funding acquisition, F.Z. and X.K. All authors have read and agreed to the published version of the manuscript.

Funding: This research was funded by the National Natural Science Foundation of China, grant number 21978289; Key R & D project of Heilongjiang Province, grant number GY2021ZB0253; and the Technological Project of Heilongjiang Province "the open competition mechanism to select the best candidates", grant number 2022ZXJ05C01.

Data Availability Statement: The data presented in this study are available on request from the corresponding author.

Acknowledgments: We thank all our colleagues for their recommendations and support during this study.

Conflicts of Interest: The authors declare no conflict of interest. The funders had no role in the design of the study; in the collection, analyses, or interpretation of data; in the writing of the manuscript; or in the decision to publish the results.

References

1. Atasoy, M.; Owusu-Agyeman, I.; Plaza, E.; Cetecioglu, Z. Bio-based volatile fatty acid production and recovery from waste streams: Current status and future challenges. *Bioresour. Technol.* **2018**, *268*, 773–786. [CrossRef] [PubMed]
2. Zheng, X.; Chen, Y.; Liu, C. Waste Activated Sludge Alkaline Fermentation Liquid as Carbon Source for Biological Nutrients Removal in Anaerobic Followed by Alternating Aerobic-Anoxic Sequencing Batch Reactors. *Chin. J. Chem. Eng.* **2010**, *18*, 478–485. [CrossRef]
3. Spirito, C.M.; Richter, H.; Rabaey, K.; Stams, A.J.M.; Angenent, L.T. Chain elongation in anaerobic reactor microbiomes to recover resources from waste. *Curr. Opin. Biotechnol.* **2014**, *27*, 115–122. [CrossRef] [PubMed]
4. Akaraonye, E.; Keshavarz, T.; Roy, I. Production of polyhydroxyalkanoates: The future green materials of choice. *J. Chem. Technol. Biotechnol.* **2010**, *85*, 732–743. [CrossRef]
5. Uyar, B.; Eroglu, I.; Yücel, M.; Gündüz, U. Photofermentative hydrogen production from volatile fatty acids present in dark fermentation effluents. *Int. J. Hydrog. Energy* **2009**, *34*, 4517–4523. [CrossRef]
6. Bhatt, A.H.; Ren, Z.J.; Tao, L. Value Proposition of Untapped Wet Wastes: Carboxylic Acid Production through Anaerobic Digestion. *iScience* **2020**, *23*, 101221. [CrossRef]
7. Choi, J.D.; Chang, H.N.; Han, J.I. Performance of microbial fuel cell with volatile fatty acids from food wastes. *Biotechnol. Lett.* **2011**, *33*, 705–714. [CrossRef]
8. Huang, Y.L.; Wu, Z.; Zhang, L.; Ming Cheung, C.; Yang, S.-T. Production of carboxylic acids from hydrolyzed corn meal by immobilized cell fermentation in a fibrous-bed bioreactor. *Bioresour. Technol.* **2002**, *82*, 51–59. [CrossRef]
9. Den Boer, E.; Lukaszewska, A.; Kluczkiewicz, W.; Lewandowska, D.; King, K.; Reijonen, T.; Kuhmonen, T.; Suhonen, A.; Jaaskelainen, A.; Heitto, A.; et al. Volatile fatty acids as an added value from biowaste. *Waste Manag.* **2016**, *58*, 62–69. [CrossRef]
10. Lü, F.; Wang, Z.; Zhang, H.; Shao, L.; He, P. Anaerobic digestion of organic waste: Recovery of value-added and inhibitory compounds from liquid fraction of digestate. *Bioresour. Technol.* **2021**, *333*, 125196. [CrossRef]

11. Lee, W.S.; Chua, A.S.M.; Yeoh, H.K.; Ngoh, G.C. A review of the production and applications of waste-derived volatile fatty acids. *Chem. Eng. J.* **2014**, *235*, 83–99. [CrossRef]
12. Bruni, C.; Foglia, A.; Eusebi, A.L.; Frison, N.; Akyol, Ç.; Fatone, F. Targeted Bio-Based Volatile Fatty Acid Production from Waste Streams through Anaerobic Fermentation: Link between Process Parameters and Operating Scale. *ACS Sustain. Chem. Eng.* **2021**, *9*, 9970–9987. [CrossRef]
13. Arslan, D.; Steinbusch, K.J.J.; Diels, L.; Hamelers, H.V.M.; Strik, D.P.B.T.B.; Buisman, C.J.N.; De Wever, H. Selective short-chain carboxylates production: A review of control mechanisms to direct mixed culture fermentations. *Crit. Rev. Environ. Sci. Technol.* **2016**, *46*, 592–634. [CrossRef]
14. Bhatia, S.K.; Yang, Y.H. Microbial production of volatile fatty acids: Current status and future perspectives. *Rev. Environ. Sci. Bio-Technol.* **2017**, *16*, 327–345. [CrossRef]
15. Begum, S.; Arelli, V.; Anupoju, G.R.; Sridhar, S.; Bhargava, S.K.; Eshtiaghi, N. Optimization of feed and extractant concentration for the liquid–liquid extraction of volatile fatty acids from synthetic solution and landfill leachate. *J. Ind. Eng. Chem.* **2020**, *90*, 190–202. [CrossRef]
16. Zhu, X.; Leininger, A.; Jassby, D.; Tsesmetzis, N.; Ren, Z.J. Will Membranes Break Barriers on Volatile Fatty Acid Recovery from Anaerobic Digestion? *ACS ES&T Eng.* **2020**, *1*, 141–153. [CrossRef]
17. Zacharof, M.-P.; Mandale, S.J.; Williams, P.M.; Lovitt, R.W. Nanofiltration of treated digested agricultural wastewater for recovery of carboxylic acids. *J. Clean. Prod.* **2016**, *112*, 4749–4761. [CrossRef]
18. Edmiston, P.L.; Gilbert, A.R.; Harvey, Z.; Mellor, N. Adsorption of short chain carboxylic acids from aqueous solution by swellable organically modified silica materials. *Adsorption* **2017**, *24*, 53–63. [CrossRef]
19. Fargues, C.; Lewandowski, R.; Lameloise, M.-L. Evaluation of Ion-Exchange and Adsorbent Resins for the Detoxification of Beet Distillery Effluents. *Ind. Eng. Chem. Res.* **2010**, *49*, 9248–9257. [CrossRef]
20. Aghapour Aktij, S.; Zirehpour, A.; Mollahosseini, A.; Taherzadeh, M.J.; Tiraferri, A.; Rahimpour, A. Feasibility of membrane processes for the recovery and purification of bio-based volatile fatty acids: A comprehensive review. *J. Ind. Eng. Chem.* **2020**, *81*, 24–40. [CrossRef]
21. Rodríguez-Llorente, D.; Bengoa, A.; Pascual-Muñoz, G.; Navarro, P.; Águeda, V.I.; Delgado, J.A.; Álvarez-Torrellas, S.; García, J.; Larriba, M. Sustainable Recovery of Volatile Fatty Acids from Aqueous Solutions Using Terpenoids and Eutectic Solvents. *ACS Sustain. Chem. Eng.* **2019**, *7*, 16786–16794. [CrossRef]
22. Van den Bruinhorst, A.; Raes, S.; Maesara, S.A.; Kroon, M.C.; Esteves, A.C.C.; Meuldijk, J. Hydrophobic eutectic mixtures as volatile fatty acid extractants. *Sep. Purif. Technol.* **2019**, *216*, 147–157. [CrossRef]
23. Reyhanitash, E.; Brouwer, T.; Kersten, S.R.A.; van der Ham, A.G.J.; Schuur, B. Liquid–liquid extraction-based process concepts for recovery of carboxylic acids from aqueous streams evaluated for dilute streams. *Chem. Eng. Res. Des.* **2018**, *137*, 510–533. [CrossRef]
24. Antony, F.M.; Pal, D.; Wasewar, K. 4-Separation of bio-products by liquid–liquid extraction. *Phys. Sci. Rev.* **2021**, *6*, 20180065. [CrossRef]
25. Mostafa, N.A. Production and recovery of volatile fatty acids from fermentation broth. *Energy Convers. Manag.* **1999**, *40*, 1543–1553. [CrossRef]
26. Alkaya, E.; Kaptan, S.; Ozkan, L.; Uludag-Demirer, S.; Demirer, G.N. Recovery of acids from anaerobic acidification broth by liquid–liquid extraction. *Chemosphere* **2009**, *77*, 1137–1142. [CrossRef]
27. Reyhanitash, E.; Zaalberg, B.; Kersten, S.R.A.; Schuur, B. Extraction of volatile fatty acids from fermented wastewater. *Sep. Purif. Technol.* **2016**, *161*, 61–68. [CrossRef]
28. Yang, S.; White, S.A.; Hsu, S.T. Extraction of carboxylic acids with tertiary and quaternary amines: Effect of pH. *Ind. Eng. Chem. Res.* **1991**, *30*, 1335–1342. [CrossRef]
29. Elhami, V.; Antunes, E.C.; Temmink, H.; Schuur, B. Recovery Techniques Enabling Circular Chemistry from Wastewater. *Molecules* **2022**, *27*, 1389. [CrossRef]
30. Reyhanitash, E.; Fufachev, E.; van Munster, K.D.; van Beek, M.B.M.; Sprakel, L.M.J.; Edelijn, C.N.; Weckhuysen, B.M.; Kersten, S.R.A.; Bruijnincx, P.C.A.; Schuur, B. Recovery and conversion of acetic acid from a phosphonium phosphinate ionic liquid to enable valorization of fermented wastewater. *Green Chem.* **2019**, *21*, 2023–2034. [CrossRef]
31. Xing, T.; Wang, Z.; Zhen, F.; Liu, H.; Wo, D.; Li, L.; Guo, Y.; Kong, X.; Sun, Y. Initial pH-driven production of volatile fatty acid from hybrid Pennisetum. *Bioresour. Technol.* **2021**, *347*, 126426. [CrossRef]
32. Xing, T.; Wang, Z.; Zhen, F.; Liu, H.L.; Wang, F.; Zhang, Y.; Kong, X.Y.; Sun, Y.M. Methane production of rare earth element-rich Dicranopteris dichotoma and effects of La(III) on anaerobic digestion performance of lignocellulose. *J. Chem. Technol. Biotechnol.* **2022**, *97*, 1987–1994. [CrossRef]
33. Malmary, G.; Albet, J.; Putranto, A.M.H.; Molinier, J. Recovery of carboxylic acids from aqueous solutions by liquid-liquid extraction with a triisooctylamine diluent system. *Braz. J. Chem. Eng.* **2001**, *18*, 441–447. [CrossRef]
34. Tamada, J.A.; Kertes, A.S.; King, C.J. Extraction of carboxylic acids with amine extractants. 1. Equilibria and law of mass action modeling. *Ind. Eng. Chem. Res.* **1990**, *29*, 1319–1326. [CrossRef]
35. Liptaj, T.; Marták, J.; Schlosser, Š. 1-NMR study of structural changes of alkyl-phosphonium decanoate ionic liquid induced by water and butyric acid extraction. *J. Mol. Liq.* **2020**, *302*, 112573. [CrossRef]

36. Van Osch, D.J.G.P.; Zubeir, L.F.; van den Bruinhorst, A.; Rocha, M.A.A.; Kroon, M.C. Hydrophobic deep eutectic solvents as water-immiscible extractants. *Green Chem.* **2015**, *17*, 4518–4521. [CrossRef]
37. Li, L.; Li, Z.; Song, K.; Gu, Y.; Gao, X.; Zhao, X. Short-chain fatty acids resource recovery potential from algal sludge via anaerobic fermentation under various pH values. *Chemosphere* **2021**, *275*, 129954. [CrossRef]
38. Zheng, M.; Zheng, M.; Wu, Y.; Ma, H.; Wang, K. Effect of pH on types of acidogenic fermentation of fruit and vegetable wastes. *Biotechnol. Bioprocess Eng.* **2015**, *20*, 298–303. [CrossRef]
39. Rocha, M.A.A.; Raeissi, S.; Hage, P.; Weggemans, W.M.A.; van Spronsen, J.; Peters, C.J.; Kroon, M.C. Recovery of volatile fatty acids from water using medium-chain fatty acids and a cosolvent. *Chem. Eng. Sci.* **2017**, *165*, 74–80. [CrossRef]

sustainability

Article

Impacts of Government Policies on the Adoption of Biomass Power: A System Dynamic Perspective

Zhao Xin-gang [1,2], Wang Wei [1,2,*], Hu Shuran [1,2] and Liu Xuan [1,2]

1 School of Economics and Management, North China Electric Power University, Beijing 102206, China
2 Beijing Key Laboratory of New Energy and Low-Carbon Development, North China Electric Power University, Beijing 102206, China
* Correspondence: 120192106138@ncepu.edu.cn

Abstract: As a kind of renewable energy, biomass power has great development potential in mitigating greenhouse gas emissions. Therefore, under the background of carbon peak and carbon neutrality, the diffusion of biomass power generation technology has practical significance. To address these issues, this paper constructs a system dynamics model to study the impact of different policy effects on the diffusion of biomass power generation technologies. The results show that the feed-in tariff policy can significantly promote the installed capacity growth of biomass power generation projects; on the other hand, carbon emission trading increases the investment value of projects and promotes the growth of the installed capacity of biomass power generation projects, to a certain extent, so relevant policies need to be improved to achieve the promotion of biomass power generation technology in the future.

Keywords: government policies; biomass power; system dynamic; adoption

Citation: Xin-gang, Z.; Wei, W.; Shuran, H.; Xuan, L. Impacts of Government Policies on the Adoption of Biomass Power: A System Dynamic Perspective. *Sustainability* **2023**, *15*, 1723. https://doi.org/10.3390/su15021723

Academic Editors: Dong Li, Fuqiang Wang, Zhonghao Rao and Chao Shen

Received: 30 November 2022
Revised: 10 January 2023
Accepted: 13 January 2023
Published: 16 January 2023

1. Introduction

1.1. Background and Motivation

China's economic transformation and structural adjustment have reached a critical point. The nation has a major strategic plan for its economic and social development, by promoting the revolution in energy production and consumption. In addition, China is dominated by a coal energy consumption structure, which has caused a large number of carbon emission problems. In this regard, China is facing severe pressure to reduce its carbon emissions. In order to meet this challenge, the Chinese government has made a solemn commitment to reach carbon peak emissions by 2030 and achieve carbon neutrality by 2060 (http://www.gov.cn/zhengce/2021-10/24/content_5644613.htm, accessed on 24 October 2021). Therefore, the development of the renewable energy industry can help to achieve this goal. As part of the renewable energy industry, electricity generation from biomass can reduce electricity generation from conventional fossil energy sources. Hence, biomass energy industry promotion can save energy and reduce emissions. The diffusion of biomass energy is a process that is widely adopted over time [1]. Biomass power generation has better economic and ecological benefits. On the one hand, biomass power generation can replace the coal consumption required by traditional coal-fired power generation, to a certain extent, thereby reducing carbon dioxide emissions [2]; on the other hand, biomass power generation can solve the environmental pollution caused by agricultural and forestry wastes, and enterprises can bring economic benefits to places with abundant resource endowments, to a certain extent, through recycling and the utilization of biomass resources [3].

Biomass power generation industry development is dependent on economic factors and social recognition [4]. To promote the development of the biomass power industry, the Chinese government has implemented a series of measures and incentives. These include feed-in tariffs, research and development subsidies and quotas. As a strategic emerging industry, the biomass power generation industry has developed more slowly than other renewable energy technologies, as shown in Figure 1. Therefore, the research questions of this paper are as follows: What are the factors leading to the slow development of the biomass power generation industry? What is the effect? Can government subsidies promote biomass energy investment? The motivation of this paper is to examine the impact of policy effects on the deployment of biomass power generation industries. To answer these questions, this paper selects a system dynamic approach to analyze policy effects on the adoption of biomass power technology. The research results of this paper can provide decision support for biomass energy investors. In addition, for policy makers, the top-level design of the system can be carried out in a scientific and reasonable way when making relevant policies.

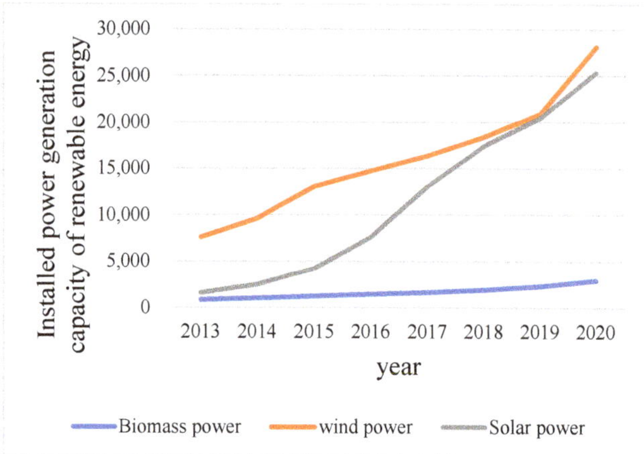

Figure 1. The installed capacity of renewable energy (unit: kW). (Source: China Electricity Council (https://www.cec.org.cn/, accessed on 26 October 2022).

1.2. Literature Reviews

The related research topics of the biomass power generation industry include the development status, cost problems of the biomass power generation industry, policy effects, investment decisions, environmental impacts, etc.

First, a review of the relevant research on the development status of the biomass power generation industry was conducted, since the industry has gradually gained attention. For example, Guan et al. [5] reviewed the problems existing in the development of the biomass industry and believed that in order to ensure the healthy development of the biopower industry, various policy combinations and government policies should be consistent. Lin and He [6] analyzed the feasibility of biomass power generation projects and believed that biomass power generation projects could have a great development potential in bioenergy-rich areas. Ref. [1] studied the diffusion of biomass energy technology and showed that resource potential and economic factors were the main reasons affecting the diffusion of technology. Liu et al. [7] believed that the biomass power generation industry faces the problem of a high production cost during the development process, so the future development of the industry needs to be improved regarding financing, the level of technological progress and the industrial chain. Zhao and Yan [8] believed that the advantages of China's biomass power generation industry lie in its abundant biomass resources, while the disadvantages lie in its high power generation cost and installation cost.

Second, in terms of cost, relevant studies have calculated the investment cost of biomass power using learning curves. This research shows that the learning effect caused by the accumulation of experience in the biomass energy industry can significantly reduce the cost [9]. For example, Nevzorova et al. [10] investigated the source of the driving force of biogas power in mature markets using the technology innovation system, and they found that policy support and technical capacity were the primary driving factors of the biomass power. Lin and He [9] calculated the investment cost of the biomass power generation industry by using a learning curve, and the results showed that cost reduction was the reason for the rapid development of biomass power generation.

Third, there is a focus on policy effect evaluation. Zhang et al. [11] analyzed the development prospect of biomass from the perspective of cost effectiveness and believed that high fuel cost was one of the main factors affecting biomass power generation. Irfan et al. [12] believed that biomass power generation industry development should be improved regarding the aspects of the financial mechanism, environmental awareness, policy incentives, and technology. Borges et al. [13] believed that information dissemination is important in the diffusion of biomass power generation technology and that the lack of relevant incentive production and R&D policies is a major reason for the slow diffusion rate.

Fourth, regarding the investment decision of bio-power projects, government policies generally encourage the development of the biomass power generation industry through financial rewards, thereby raising the enthusiasm of biomass energy investors. For example, Zhang et al. investigated the deployment of the biomass power industry through the lens of renewable energy incentive policies, and their findings revealed that, when compared to the feed-in tariff policy, the renewable energy quota policy could promote the development of the biomass power generation industry in the long run [14]. Biomass power generation project investors' perceptions of investment risk, income and other factors determine their decision-making behavior. Since the investment process has an uncertainty problem, this delays the investors' investment decision. For example, [15] believed that the adoption of renewable energy technology can be studied from the perspective of investors, and the key factors affecting the decision-making behavior of investors are risk and return. Wang et al. [16] used the real option model to investigate the uncertainty in the investment process of the biomass power generation industry. The study showed that government subsidies could stimulate the investment enthusiasm of biomass investors; however, despite full government subsidies, a difference still remained between the investment cost of biomass power and the market price. Liu et al. [17] indicated that the subsidy policy is critical to the development of the biomass power industry, and their results showed that with an increase in the investment subsidy, the scale of biomass grows rapidly.

As for the environmental benefits, some researchers investigating the potential of carbon emission reduction, such as He et al. [3], insisted that biomass energy to generate electricity can not only bring environmental benefits but also control greenhouse gas emissions. Ref. [18] analyzed the biomass power generation industry from the perspective of environmental benefits, and the results revealed that biomass power generation reduced GHG emissions.

1.3. Contributions and Innovations

Over time, the existing research has laid the foundation for the development of the biomass energy industry. However, compared with this related research, the contribution and innovations of this study are summarized as follows.

(1) The existing literature analyzed the technical and economic problems of the biomass power generation industry [19]; however, the problem of the diffusion of the biomass power industry from the perspective of technology adoption is lacking. Thus, this paper studies the adoption behavior of biomass power from the perspective of technology diffusion. At the same time, the related factors affecting the diffusion of biomass power technology are dynamically analyzed in order to clarify the diffusion mechanism.

(2) In order to understand the influence of policy effects on the diffusion of biomass power projects, this paper constructs a system dynamics model to explore the influence of

the interaction of the influencing factors on the diffusion, to provide support for investors' decisions.

2. Materials and Methods

In 2020, the National Development and Reform Commission and other departments issued the "*Implementation Plan on Improving the Construction and Operation of Biomass Power Generation Projects*", which showed that biomass power projects have attracted much attention [20]. This is the reason why this study chose the biomass power projects as a case study. The detailed reasons are as follows: Firstly, from the perspective of resource endowment, biomass energy resources are relatively abundant. Secondly, the biomass power industry has good environmental benefits. Given the targets for carbon peak and carbon neutrality, the development of biomass power technology can help to achieve the dual carbon target. Finally, biomass power generation has good economic benefits, and biomass power investors can bring considerable economic benefits to the resource endowment area by recycling biomass energy resources. Therefore, the large-scale deployment of biomass energy projects has reference significance for biomass energy diffusion in other regions.

2.1. Reasons for System Dynamic Model

Forrester of Massachusetts Institute of Technology first proposed system dynamics in 1956 [21]. System dynamics (SD) aims to analyze an information feedback system and understand its future development trend by imitating the internal structure and development dynamic behavior in the real world [22]. So system dynamics is suitable for the study of renewable energy technology. The advantages of this method are that it has the characteristics of dynamic feedback mechanism, which can well reflect the influence of key variables in the system on the whole system [23]. The diffusion of biomass power is also a complex social system, which involves many influences, and there are interaction behaviors between different factors that make the process show complexity characteristics. On the other hand, system dynamics aims at understanding the relationships among factors in the system and the decision-making behavior of the subject, rather than the pursuit of prediction accuracy. It is also called a "policy laboratory". Due to these advantages, system dynamics is widely used in renewable energy policy evaluation [24]. Moreover, the research topic of this paper conforms to the research paradigm of system dynamics method. Therefore, this paper chooses the SD model to investigate the dynamic diffusion process of biomass projects.

2.2. Model Construction

2.2.1. Model Description

The diffusion of biomass power generation technology is essentially an investment behavior, so investment willingness determines the decision behavior of investors [24]. However, there are many factors influencing investors' decision-making behavior, and this paper only examines the impact of key factors on the diffusion of biomass power generation technology. When measuring the diffusion level, existing studies usually use the installed capacity scale to characterize the diffusion level of renewable energy technologies [25,26]. Therefore, the installed capacity of biomass power is chosen to characterize the diffusion level in this paper.

When making investment decisions, investors usually consider the benefits and risks of investing in biomass power generation projects. Therefore, return on investment is an important factor affecting the promotion of biomass power projects [14]. Accordingly, the diffusion process of biomass power projects is shown in Figure 2. Since biomass power generation projects require high initial investment costs, government incentive policies are needed to support their development. Government policies can increase investment income for investors in biomass power projects. Investment in biomass power projects, driven by considerable income, can increase investors' enthusiasm for investment and making investment decisions.

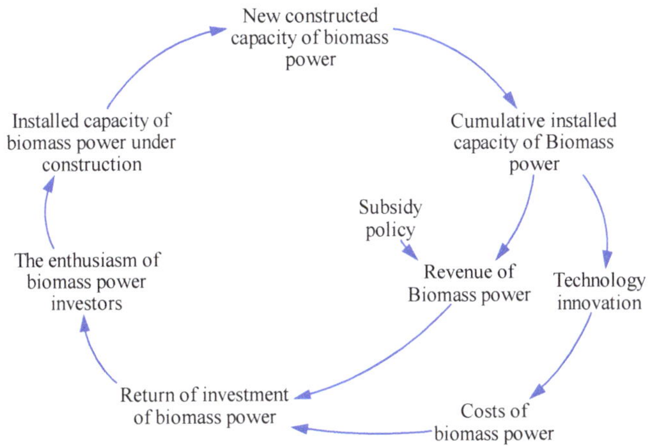

Figure 2. Causal loop diagram.

Finally, with the increase in investment in biomass power projects, the installed capacity of biomass power projects increases, to achieve the diffusion process. On the other hand, the impact of technological progress on the proliferation of biomass power projects reduces investment costs, as knowledge and experience accumulate during the installation of biomass power generation technologies, which is known as the effect of learning by doing. Biomass industry technological progress causes investment cost to reduce, which increases investors' return on investment for improvement of biomass power generation projects, boosts the enthusiasm lever of potential investors and also make more investors participate in biomass power generation projects, thus increasing the installed capacity of biomass power.

2.2.2. Mathematical Model

(1) Installed capacity of biomass power projects

$$CIC_{t,biomass} = CIC_{initial,biomass} + \int (NIC_{t,biomass} - DIC_{t,biomass})dt \quad (1)$$

$$DIC_{biomass} = CIC_{biomass} / LS_{biomass} \quad (2)$$

$$ICC_{biomass} = \int (IR_{biomass} - NIC_{biomass})dt \quad (3)$$

$$IC_{new,biomass} = Delay1(ICC_{biomass}, time) \quad (4)$$

where CIC_t denotes the cumulative constructed capacity of biomass power projects, NIC is new constructed capacity of biomass power projects, DIC is the depreciated capacity of biomass power projects over time, LS is working years during the life cycle of a biomass power project, and ICC denotes the installed capacity under construction of biomass power.

(2) Revenue composition of biomass power projects

For investors of biomass power projects, their income mainly comes from the income of electricity sales and the income generated by reducing carbon emissions. Since biomass power generation technology has great potential to mitigate the greenhouse effect, greenhouse gas emissions can be reduced through biomass power generation. Ahmad et al. [27] believed that renewable energy generation can avoid carbon emissions, so the reduced carbon emissions can be taken into account regarding the investment income of investors.

For the carbon trading market, the Chinese government has set up seven pilot cities. Therefore, reduced greenhouse gas emissions from biomass power projects can bring additional returns to investors in biomass power projects. Therefore, this paper considers the benefits of carbon emission reduction. The main formulas are as follows:

$$G_{power} = CIC \times h \tag{5}$$

$$R = R_{electricity} + R_{carbon} \tag{6}$$

$$PG = (G_{power} - EC) \times (1 - lr) \tag{7}$$

$$R_{electricity} = FIT_{biomass} \times PG_{biomass} \tag{8}$$

$$R_{carbon} = P_{ce} \times C_{effi} \times PG_{biomass} \tag{9}$$

where $R_{electricity}$ denotes the revenue from selling the electricity, R_{carbon} is the income from the sale of the reduced carbon emissions, PG is the amount of electricity that is connected to the grid system, EC is the biomass power plant self-consumption rate, and lr denotes the line loss rate.

(3) Cost composition of biomass power projects

For biomass power projects, cost is mainly composed of biomass equipment investment cost, material cost, and operation and maintenance cost. Therefore, the detailed formula is shown as follows:

$$TC = In_{cost} + Material_{cost} + OM_{cost} \tag{10}$$

where In_{cost} is investment cost of biomass power, $Material_{cost}$ denotes the cost of material, and OM_{cost} is the operation and maintenance cost during the lifetime. In renewable energy technology research, learning curve is often used to measure the impact of technological progress on cost.

The learning curve was first proposed by Wright [28], which means that the unit cost of a certain technology decreases with the accumulation of production experience. The calculation of equipment investment cost conforms to the general rule of learning curve, so it can be calculated by means of learning curve. The detailed formula is as follows:

$$In_{Cost,biomass} = A \times CIC_{t,biomass}^{\alpha} \tag{11}$$

$$LnIn_{Cost,biomass} = LnA + \alpha LnCIC_{t,biomass} \tag{12}$$

$$LR = 1 - 2^{\alpha} \tag{13}$$

where LR is the learning rate of biomass power, and $^{\alpha}$ is a coefficient.

Existing studies show that due to the disclosure of relevant data in the biomass power industry and other related reasons, there are certain difficulties in obtaining relevant data, and there are numerical differences in the learning rate estimated by the learning curve [3]. According to the research results of [9], the value range of coefficient is roughly between 5.6% and 7.8%. According to the research results of [29], the coefficient of biomass power in China is roughly 0.48. Yu-zhuo et al. [14] also assumed the value of coefficient is 0.48. In summary, according to the results of existing studies on the coefficient of biomass power generation industry, this value is roughly between 0.02 and 0.48 [3,14,30,31]. He [3] believed that there were great differences in the learning rate values of different studies, which were mainly explained by the stage of technological progress and the different types of technological maturity. Due to space limitations, only the main equations are listed. The rest of coefficient of this paper can be found in the Supplementary Materials.

3. Results

(1) Effect of different subsidy levels on installed capacity

This study set three scenarios to investigate the effect of different subsidy levels on the installed capacity of the biomass power industry. We evaluated the change in the installed capacity of biomass power generation projects under different subsidy policy scenarios by floating 10% or less under the benchmark scenario as a reference. The feed-in price subsidy in the benchmark scenario is set at 0.75 RMB/kWh. As shown in Figure 3, the feed-in tariff subsidy policy positively correlates with the installed capacity of biomass power generation projects. Higher levels of feed-in tariffs have increased investor interest, resulting in an increase in the installed capacity of physical power projects. The higher the subsidy level is, the greater the installed capacity of biomass power projects is, and vice versa. The reason is that the feed-in tariff subsidy is a direct financial subsidy policy, which provides corresponding capital support to biomass investors according to the power generation and can enhance the investment enthusiasm of investors. In addition, under the subsidy mechanism, investors in biomass power generation projects can obtain stable and expected investment returns. Therefore, the smaller the perceived investment risk is, the more investors are interested in investing in biomass generation projects. Figure 4 shows that the higher the subsidy price level is, the higher the revenue is, from which investors in biomass power generation projects can obtain profit. Driven by expected profits, investors are also more willing to invest in biomass power projects.

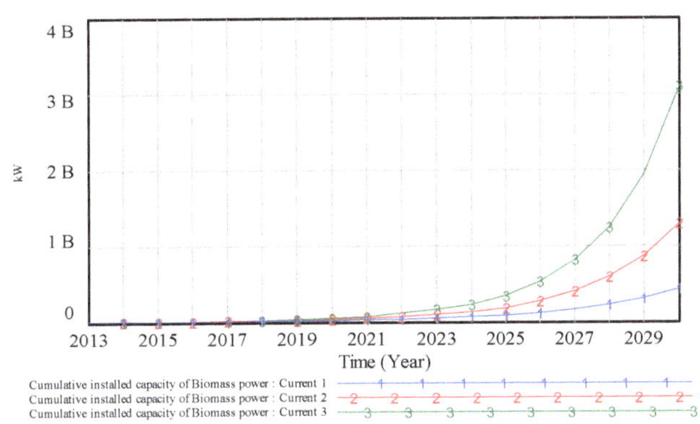

Figure 3. The cumulative installed capacity of biomass power under subsidy scenarios.

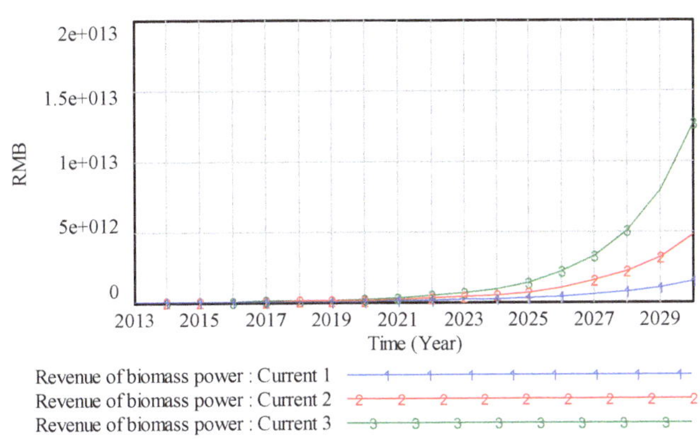

Figure 4. The revenue of biomass power projects under the subsidy scenarios.

(2) Effect of different carbon price levels on installed capacity

To evaluate the impact of carbon emission trading on biomass power project investment, this paper considers the effect of different carbon price levels on the biomass power generation. According to Figures 5 and 6, The carbon price level also has a significant impact on the installed capacity of biomass power generation projects. Specifically, the higher the carbon price is, the installed capacity of biomass power projects also increases. This is mainly because carbon price can increase the additional income of investors in biomass power generation projects. With the increase in profits level, investors' perception of future investment income can reduce investment risk, so they are willing to invest in biomass power generation projects.

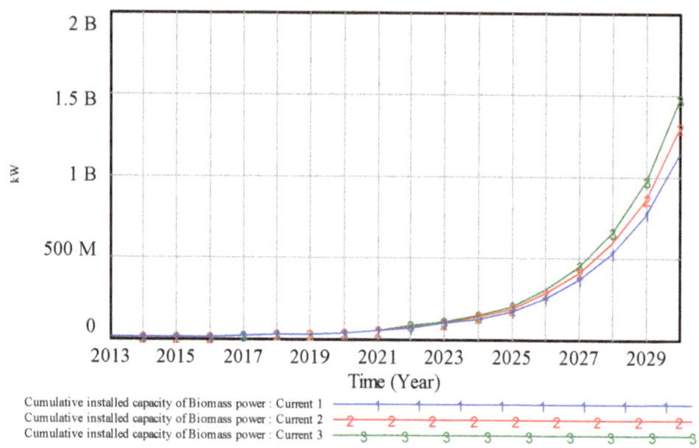

Figure 5. Cumulative constructed capacity of biomass power under the carbon price.

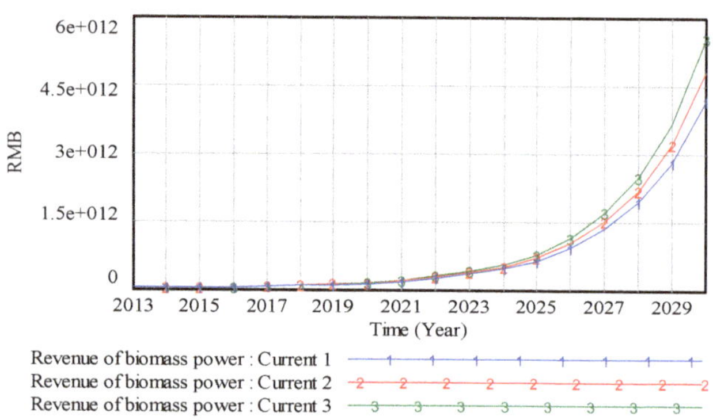

Figure 6. Revenue of biomass power projects under carbon price scenarios.

4. Discussion

To prove the scientific rationality of the research results of this paper, this paper is compared with the relevant research. Consistent with [14], the feed-in tariff subsidy policy boosts the growth of the installed capacity of renewable energy, because the feed-in tariff can compensate for the capital input of investors, thus producing reasonable profit returns. This is also the reason why the feed-in tariff policy significantly promotes the rapid development of the biomass power generation industry. However, with the improvement in the technological innovation level of the biomass power generation industry, the investment cost changes dynamically. Therefore, the level of government subsidy policy should be adjusted in a timely manner, according to the dynamic change of the biomass power generation cost.

In terms of carbon trading policies, Zhang et al. [29] believed that the revenue from carbon emission reduction can improve the investment value of renewable energy power projects. China's carbon trading market is currently in the stage of cultivation and development, so the design of relevant market institutional parameters needs to be constantly improved, which is conducive to carbon trading as a market-based tool to promote the development of the biomass power generation industry. Therefore, it is important for policy makers to improve the relevant institutional norms of the carbon trading market to promote renewable energy technologies. Therefore, the research conclusion of this paper is reasonable, to some extent.

Finally, technological innovation is a source of cost reduction. To improve the technology innovation of the biomass industry, policy makers should strengthen research and development subsidies for the biomass power industry, especially for the technological innovation of core equipment.

5. Conclusions

In order to alleviate the environmental problems caused by greenhouse gas emissions, the Chinese government has put forward ambitious targets for carbon peaking and carbon neutrality. As a kind of clean energy, biomass power generation can improve environmental quality; on the other hand, the development of the biomass power industry has good economic benefits. Therefore, the biomass power generation industry has a wide range of development scenarios under the double carbon targets. In order to study the diffusion of biomass power generation technologies under different policies, a system dynamics model is constructed to study the dynamic effects of policies on the diffusion level. According to the simulation results in this paper, the following conclusions can be drawn. First, the feed-in tariff policy can significantly promote the installed capacity growth of biomass power generation projects. With the increase in the subsidy level, the investment enthusiasm of biomass power generation investors is high, which expands the investment scale of biomass power generation projects. Second, carbon emission trading increases the investment value of the project and promotes the growth of the installed capacity of biomass power generation projects, to a certain extent.

Supplementary Materials: The following supporting information can be downloaded at: https://www.mdpi.com/article/10.3390/su15021723/s1, Table S1: The explanation of relevant variables; Table S2: The key varia-ble's value.

Author Contributions: Conceptualization, methodology, software, formal analysis, writing—original draft, visualization, resources, data curation, W.W.; validation, investigation, funding acquisition, supervision, project administration, Z.X.-g.; formal analysis, writing—original draft, visualization, H.S.; formal analysis, writing, L.X. All authors have read and agreed to the published version of the manuscript.

Funding: This paper is supported by the Beijing Municipal Social Science Foundation (No. 16JDYJB031), the Fundamental Research Funds for the Central Universities (No. 2020YJ008) and the Fundamental Research Funds for the Central Universities (No. 2018ZD14).

Institutional Review Board Statement: Not applicable.

Informed Consent Statement: Not applicable.

Data Availability Statement: Not applicable.

Acknowledgments: Many thanks to the editor for their time and effort in processing this manuscript and also many thanks to the anonymous reviewers for their valuable comments.

Conflicts of Interest: The authors declare no conflict of interest.

References

1. Zhang, M.; Zhou, D.; Ding, H.; Jin, J. Biomass power generation investment in China: A real options evaluation. *Sustainability* **2016**, *8*, 563. [CrossRef]
2. Li, Y.; Lin, J.; Qian, Y.; Li, D. Feed-in tariff policy for biomass power generation: Incorporating the feedstock acquisition process. *Eur. J. Oper. Res.* **2023**, *304*, 1113–1132. [CrossRef]
3. He, J.X. The Efficiency and Prospect of Biomass Power Development in China. Ph.D. Thesis, Xiamen University, Xiamen, China, 2019. (In Chinese).
4. He, J.; Zhu, R.; Lin, B. Prospects, obstacles and solutions of biomass power industry in China. *J. Clean. Prod.* **2019**, *237*, 117783. [CrossRef]
5. Guan, Y.; Tai, L.; Cheng, Z.; Chen, G.; Yan, B. Biomass molded fuel in China: Current status, policies and suggestions. *Sci. Total Environ.* **2020**, *724*, 138345. [CrossRef]
6. Lin, B.; He, J. Is biomass power a good choice for governments in China? *Renew. Sustain. Energy Rev.* **2017**, *73*, 1218–1230. [CrossRef]
7. Liu, J.; Wang, S.; Wei, Q.; Yan, S. Present situation, problems and solutions of China's biomass power generation industry. *Energy Policy* **2014**, *70*, 144–151. [CrossRef]
8. Zhao, Z.Y.; Yan, H. Assessment of the biomass power generation industry in China. *Renew. Energy* **2012**, *37*, 53–60. [CrossRef]
9. Lin, B.; He, J. Learning curves for harnessing biomass power: What could explain the reduction of its cost during the expansion of China? *Renew. Energy* **2016**, *99*, 280–288. [CrossRef]
10. Nevzorova, T.; Karakaya, E. Explaining the drivers of technological innovation systems: The case of biogas technologies in mature markets. *J. Clean. Prod.* **2020**, *259*, 120819. [CrossRef]
11. Zhang, Q.; Zhou, D.; Zhou, P.; Ding, H. Cost analysis of straw-based power generation in Jiangsu Province, China. *Appl. Energy* **2013**, *102*, 785–793. [CrossRef]
12. Irfan, M.; Zhao, Z.-Y.; Panjwani, M.K.; Mangi, F.H.; Li, H.; Jan, A.; Ahmad, M.; Rehman, A. Assessing the energy dynamics of Pakistan: Prospects of biomass energy. *Energy Rep.* **2020**, *6*, 80–93. [CrossRef]
13. Borges, C.P.; Sobczak, J.C.; Silberg, T.R.; Uriona-Maldonado, M.; Vaz, C.R. A systems modeling approach to estimate biogas potential from biomass sources in Brazil. *Renew. Sustain. Energy Rev.* **2021**, *138*, 110518. [CrossRef]
14. Yu-Zhuo, Z.; Xin-Gang, Z.; Ling-Zhi, R.; Ji, L.; Ping-Kuo, L. The development of China's biomass power industry under feed-in tariff and renewable portfolio standard: A system dynamics analysis. *Energy* **2017**, *139*, 947–961. [CrossRef]
15. Dinica, V. Support systems for the diffusion of renewable energy technologies—An investor perspective. *Energy Policy* **2006**, *34*, 461–480. [CrossRef]
16. Wang, X.; Cai, Y.; Dai, C. Evaluating China's biomass power production investment based on a policy benefit real options model. *Energy* **2014**, *73*, 751–761. [CrossRef]
17. Liu, D.; Liu, M.; Xiao, B.; Guo, X.; Niu, D.; Qin, G.; Jia, H. Exploring biomass power generation's development under encouraged policies in China. *J. Clean. Prod.* **2020**, *258*, 120786. [CrossRef]
18. Hiloidhari, M.; Baruah, D.C.; Kumari, M.; Kumari, S.; Thakur, I.S. Prospect and potential of biomass power to mitigate climate change: A case study in India. *J. Clean. Prod.* **2019**, *220*, 931–944. [CrossRef]
19. He, J.; Liu, Y.; Lin, B. Should China support the development of biomass power generation? *Energy* **2018**, *163*, 416–425. [CrossRef]
20. Improve the Implementation Plan of Biomass Power Generation Project Construction and Operation. 2020. Available online: http://www.gov.cn/zhengce/zhengceku/2020-09/16/content_5543821.htm (accessed on 1 October 2022).
21. Forrester, J.W. Counterintuitive behavior of social systems. *Theory Decis.* **1971**, *2*, 109–140. [CrossRef]
22. Selvakkumaran, S.; Ahlgren, E.O. Review of the use of system dynamics (SD) in scrutinizing local energy transitions. *J. Environ. Manag.* **2020**, *272*, 111053. [CrossRef]
23. Zhang, L.; Chen, C.; Wang, Q.; Zhou, D. The impact of feed-in tariff reduction and renewable portfolio standard on the development of distributed photovoltaic generation in China. *Energy* **2021**, *232*, 120933. [CrossRef]
24. Yu, X.; Ge, S.; Zhou, D.; Wang, Q.; Chang, C.T.; Sang, X. Whether feed-in tariff can be effectively replaced or not? An integrated analysis of renewable portfolio standards and green certificate trading. *Energy* **2022**, *245*, 123241. [CrossRef]
25. She, Z.Y.; Cao, R.; Xie, B.C.; Ma, J.J.; Lan, S. An analysis of the wind power development factors by Generalized Bass Model: A case study of China's eight bases. *J. Clean. Prod.* **2019**, *231*, 1503–1514. [CrossRef]
26. Lu, Z.Y.; Li, W.H.; Xie, B.C.; Shang, L.F. Study on China's wind power development path—Based on the target for 2030. *Renew. Sustain. Energy Rev.* **2015**, *51*, 197–208. [CrossRef]
27. Ahmad, S.; Tahar, R.M.; Muhammad-Sukki, F.; Munir, A.B.; Rahim, R.A. Role of feed-in tariff policy in promoting solar photovoltaic investments in Malaysia: A system dynamics approach. *Energy* **2015**, *84*, 808–815. [CrossRef]
28. Wright, T.P. Factors affecting the cost of airplanes. *J. Aeronaut. Sci.* **1936**, *3*, 122–128. [CrossRef]
29. Zhang, M.; Tang, Y.; Liu, L.; Zhou, D. Optimal investment portfolio strategies for power enterprises under multi-policy scenarios of renewable energy. *Renew. Sustain. Energy Rev.* **2022**, *154*, 111879. [CrossRef]

30. Wene, C.O. *Experience Curves for Energy Technology Policy*; International Energy Agency (IEA): Paris, France, 2000.
31. Junginger, M.; Faaij, A.; Björheden, R.; Turkenburg, W. Technological learning and cost reductions in wood fuel supply chains in Sweden. *Biomass Bioenergy* **2005**, *29*, 399–418. [CrossRef]

energies MDPI

Article

Facilitating Investment in Photovoltaic Systems in Iran Considering Time-of-Use Feed-in-Tariff and Carbon Market

Asrin Seyedzahedi [1] and Salah Bahramara [2,*]

[1] Department of Science, Campus of Bijar, University of Kurdistan, Bijar 6651874871, Iran
[2] Department of Electrical Engineering, Sanandaj Branch, Islamic Azad University, Sanandaj 6616935391, Iran
* Correspondence: s_bahramara@yahoo.com

Abstract: Photovoltaic (PV) systems are the leading solutions for reducing carbon dioxide (CO_2) emissions in Iran's energy system. However, there are some challenges to investing in PV systems in Iran, such as the low energy market price and the high investment cost of PV systems. Although the flat feed-in tariff (FiT) is defined to help purchase energy from the PV systems, it is not attractive to investors. In this paper, a mathematical formulation is developed for the planning problem of the PV systems with battery energy storages (BESs) considering two incentive policies: (1) Designing time-of-use FiT to encourage the PV systems to sell energy to the grid at peak hours (2) Participating in the carbon trading energy market. The insolation in Iran is calculated regarding mathematical formulations which divide Iran into eight zones. The results of the base case show high payback periods for all zones. In the presence of the incentive policies, the payback period decreases considerably from 5.46 yrs. to 3.75 yrs. for the best zone. Also, the net present value increases more than 170 percent in some zones compared to the base case.

Keywords: photovoltaic system; carbon trading market; battery energy storage; planning problem; carbon reduction

Citation: Seyedzahedi, A.; Bahramara, S. Facilitating Investment in Photovoltaic Systems in Iran Considering Time-of-Use Feed-in-Tariff and Carbon Market. *Energies* **2023**, *16*, 1067. https://doi.org/10.3390/en16031067

Academic Editors: Fuqiang Wang, Chao Shen, Dong Li and Zhonghao Rao

Received: 20 December 2022
Revised: 5 January 2023
Accepted: 15 January 2023
Published: 18 January 2023

1. Introduction

The main factor of climate change in the world is the huge emission of greenhouse gases (GHGs), especially carbon dioxide (CO_2). Iran has a share of 2.2 percent of the global annual CO_2 emission [1]. Since fossil fuel-based power plants are mainly responsible for producing electrical energy in Iran (80 percent of the power installed capacities), they are the producers of a large portion of this pollution emission [2]. One of the main solutions to decrease pollution emissions in Iran is developing renewable energy sources (RESs), especially photovoltaic (PV) systems. Iran has great potential to use these systems considering their appropriate insolation; however, there are several challenges to developing the PV systems in Iran, which are described in the following subsection.

1.1. Challenges and Solutions of Using PV Systems in Iran

Iran has large fossil fuel resources, such as crude oil and natural gas. Therefore, installing fossil fuel-based power plants has always been the main choice for Iran's governments to meet their electrical demand. The government has therefore considered a significant discount on selling fossil fuels to power plants, leading to the production of electrical energy at a low price. On the other hand, producing electrical energy through the RESs, such as the PV systems, needs high investment costs, and therefore these systems cannot compete with the fossil fuel-based power plants to sell energy to the grid since the energy market price is far lower in Iran. For this purpose, Iran's government defines the flat feed-in-tariff (FiT) scheme as a supporting policy to purchase energy from RES-based power plants. However, these schemes are not attractive to either the system operator or the investors. From the viewpoint of the system operator, the main problem of the system is the peak hours, especially in summer, when it leads to a 10,000 MW gap between generation and consumption. To decrease the demand in peak hours, the system operator has employed some demand response programs and has also shed the industrial, commercial, and residential loads to maintain the reliability of the system. Implementing the demand response programs, especially shedding the loads, has led to huge economic and social damages both for the people and the government. On the other hand, the flat FiT does not encourage the PV's investors to install the BES. Thus, as a result of not installing these BESs, the investors will not participate in the peak periods, and they will consequently not meet the demand of the system. PV-based power plants can act as Iran's main sources of CO_2 reduction. Although there are carbon markets in which carbon credits are traded between sellers and buyers, the PV-based power plant operators in Iran do not access these markets to earn profit from selling their carbon credits.

To address the abovementioned challenges and to encourage investors to invest in PV systems in Iran, two main solutions can be presented:

- Designing the Time of Use (TOU) FiT scheme: As mentioned above, the main challenge of Iran's power system operation is the peak hours. Designing the TOU FiT scheme can encourage investors to use the BESs in their PV systems, which could then provide more energy for the system during peak hours. This policy can be easily implemented in Iran since load shedding causes significant damage to the system in peak hours.
- Participating in the carbon trading market: This market can earn more revenue for PV systems, so investors are encouraged to invest in these systems. Besides this, the price of the carbon market is increasing due to the major concerns of climate change.

One of the main aims of this paper is to consider these solutions for the investment problem of PV systems.

1.2. *Reviewing the Planning Problems of the PV Systems Proposed in the Literature*

The PV systems are used in hybrid energy systems alongside other energy resources and energy storage to meet the electrical, thermal, and cooling energy demands and to sell the extra energy to the grid. Artificial intelligence models are employed in [3,4] to predict the performance of the PV systems, including the PV panels, the thermoelectric air conditioning systems, the thermal collectors, and the electrolytic hydrogen production systems. A new approach is used in [5] to model a system in which heat, electricity, and hydrogen are produced through different energy resources, such as PV thermal collectors. The feasibility of using the PV-thermal system and the ground source of heat pumps to meet the electrical, heating, and cooling energy demands of the buildings is investigated in [6]. Various aspects of using solar dish/Stirling systems in the solar systems are investigated in [7]. Although PV systems can be used for different applications, the focus of this paper is to use these systems in electrical energy systems.

The optimal planning problem of PV systems is addressed in many studies. Hybrid energy systems, including PV/diesel/wind turbine (WT)/BES, are designed for some remote areas in Saudi Arabia using HOMER software in [8]. A mixed integer linear programming (MILP) model is developed to formulate the problem of the optimal sizing of the PV system and the BESs to reduce the annualized total cost of a fast charging station in [9]. The optimal sizing problem of the PV/BES system for different residential consumers in Zurich, Switzerland, is investigated in [10], considering different electrical power system conditions. The results showed that using the BESs along with the PV system leads to obtaining a better net present value (NPV) compared to without a BES system. To obtain the optimal configuration of the hybrid energy system in a rural area in Bangladesh, a technical and economic evaluation is considered using HOMER software [11]. Then, the optimum configurations are evaluated through their economic and environmental indices using the TOPSIS and AHP methods. The optimal sizing problem of the PV/BES system for a residential location in China is formulated as a mixed-integer nonlinear programming (MINLP) model in [12]. HOMER software is employed in [13] to optimize the PV/diesel/BES hybrid energy system for a rural area in Iraq. The main contribution of the study in [13] is using a new dispatch strategy for the energy management system in the proposed system. For this purpose, the proposed dispatch strategy is implemented in MATLAB software, and then it is linked to HOMER. The results showed that the net present cost (NPC) of the system decreases by 3.95 percent using the proposed dispatch strategy in comparison with the case that uses the cycle charging strategy suggested in the HOMER. A two-stage stochastic optimization problem is developed for the optimal sizing problem of a PV system equipped with the electrical and thermal energy storages for the residential areas in Qatar [14]. The behavior of a hybrid energy system, including a PV/BES/diesel generator, to meet the demand of a university in Medellin, Colombia is investigated in [15]. The results showed that in the presence of the diesel generator, the total cost of the system decreases in comparison with the case that uses only the PV and BES to meet the demand. HOMER software is employed in [16] to obtain the best configuration of the PV/diesel/BES system to meet the demand of the electric vehicle-charging station in three cities in Ethiopia. The results show the optimum systems have a cost of energy (COE) of 0.196 USD/kWh, 0.18 USD/kWh, and 0.188 USD/kWh for Addis Ababa, Jijiga, and Bahir Dar, respectively. A PV/thermal energy storage system is designed for a health center in Tigray, Ethiopia using the TRNSYS model in [17]. The aim of this paper is to solve the problem of how to store the excess energy of the PV system in thermal energy storage. The results show that the daily average excess energy in the mentioned system changes from 2070 Wh to 2959 Wh. The best configuration of the hybrid energy systems to meet the electrical load of two residential consumers in Windhoek, Namibia is obtained through minimizing the NPC using HOMER software in [18].

The authors of [19] designed the optimal combined cooling, heating, and power generation (CCHP) system for a residential building in three climate zones in Iran. The results show a huge amount of CO_2 emission reduction in supplying the demand of the buildings using these systems. This is despite the fact that, under the present conditions of Iran's energy market, these systems are not economical to meet the electric, heating, and cooling demand of the buildings in Iran. Therefore, new financial support is required to encourage investors to use the CCHP systems in Iran. The site selection for installing the PV energy system in Markazi Province in Iran is investigated using the multi-criteria decision-making method detailed in [20]. For this purpose, some economic, environmental, and technical indices are defined to obtain the best locations through the fuzzy method. By the end, the results are reported in the geographical information system (GIS) environment. The correlation coefficient and standard deviation (CCSD) method is used as a multi-criteria decision-making approach to evaluate the PV system's power generation potential in the southeast regions of Iran [21]. The optimal planning of the hybrid energy systems, including PVs, diesel generators, and BESs, using HOMER software is investigated by considering some incentive policies in [22]. The simulation results show the significant NPC and CO_2 reduction in the optimized systems, considering the incentive policies in comparison with the base case. The incentive policies include increasing the fixed price of purchasing energy from these systems, reducing the purchased energy from the grid to meet the demand of the system, and decreasing the investment cost in the system. The previous models and approaches proposed in the literature are also reviewed in Tables 1 and 2.

Table 1. Comparing the present paper with previous studies regarding their modeling approaches.

Ref.	Insolation					Modeling Selling Energy to the Grid		Modeling Carbon Reduction Incentive	
	Mathematical Modeling				Using Forecast Data	Flat FiT	TOU FiT	Fixed Price	Carbon Market
	Latitude	Longitude	Day	/Hour					
[۹]		-	-	-	*	-	-	-	-
[9]	-	-	-	-	*	-	-	-	-
[10]	-	-	-	-	*	*	-	-	-
[11]	-	-	-	-	*	-	-	-	-
[12]	-	-	-	-	*	*	*	-	-

Table 1. *Cont.*

Ref.	Insolation					Modeling Selling Energy to the Grid		Modeling Carbon Reduction Incentive	
	Mathematical Modeling				Using Forecast Data	Flat FiT	TOU FiT	Fixed Price	Carbon Market
	Latitude	Longitude	Day	/Hour					
[13]	-	-	-	-	*	-	-	-	-
[14]	-	-	-	-	*	-	-	*	-
[15]	-	-	-	-	*	*	-	-	-
[16]	-	-	-	-	*	-	-	-	-
[17]	-	-	-	-	*	-	-	-	-
[18]	-	-	-	-	*	-	-	-	-
[19]	-	-	-	-	*	*	-	-	-
[22]	-	-	-	-	*	*	-	-	-
This paper	*	*	*	*	-	*	*	-	*

* stands for 'considered' and - stands for 'not considered'.

Table 2. Comparing the present paper w previous studies regarding the optimization approaches and case studies.

Ref.	PV System			Optimization Approach		Case Study	Whole Country
	PV	Inverter	BES	Mathematical Model	Software	Specified Locations	
[8]	*	*	*	-	HOMER	Abha, Jazan, Makkah, Madinah, Hail, and Arar (Saudi Arabia)	-
[9]	*	*	*	*	-	Oak Ridge, Tennessee (USA)	-
[10]	*	*	*	*	-	Zurich (Switzerland)	-
[11]	*	*	*	-	HOMER	Monpura Island (Bangladesh)	-
[12]	*	*	*	*	-	Changsha (China)	-
[13]	*	*	*	-	HOMER	A rural area (Iraq)	-
[14]	*	*	*	*	-	A residential area (Qatar)	-
[15]	*	*	*	*	-	A university (Colombia)	-
[16]	*	*	*	-	HOMER	Three cities (Ethiopia)	-
[17]	*	*	-	-	TRNSYS	A health center (Ethiopia)	-
[18]	*	*	*	-	HOMER	Residential consumers (Namibia)	-
[19]	*	*	*	-	TRNSYS PSO	Hamedan, Tehran, Ahvaz (Iran)	-
[22]	*	*	*	-	HOMER	Educational complex, Sanandaj (Iran)	-
This paper	*	*	*	*	-	-	Iran

* stands for 'considered' and - stands for 'not considered'.

1.3. Proposed Approach and the Contributions

In this paper, a mathematical formulation is developed for the optimal planning of the PV system in the presence of a BES, considering the FiT schemes and the carbon trading market. The proposed model in this paper is compared with previous studies in both Tables 1 and 2. As shown in Table 2, most of the previous studies use different software for the optimal planning problem of the PV systems and the mathematical model is used in a few studies. Although it seems as if this paper uses a similar mathematical model to that considered in the other studies, the differences between the proposed model in this paper and the previous studies are described in detail in Table 1. First, the mathematical formulations are used to calculate the insolation in each location of Iran for each day and each hour in this paper. In contrast, in the previous studies reviewed in Table 1, only the forecast insolation is used to calculate the power generation of the PV systems. Iran is divided into eight zones regarding the calculated insolation. In the next step, the planning problem of the PV/BES systems is modeled, considering the TOU FiT scheme and the carbon trading market as the two main solutions to encourage the investors of the PV/BESs systems. The results are presented in three cases to show the effect of the incentive policies on the NPV and the payback time of the planning problem for different zones in Iran. Therefore, the main contributions of this paper are as follows:

- Calculating the insolation regarding the latitude and longitude of each location for each day and each hour of the year regarding the mathematical models. In contrast, previous studies have used the forecast insolation, which mainly shows the average daily insolation.
- Modeling the planning problem of the PV/BES systems considering the FiT scheme and the carbon trading market.
- Investigating the NPV and the payback time of the PV/BES systems for different zones in Iran. On the other hand, previous studies have only concentrated on specific locations in a country.

1.4. Paper Organization

The problem is described in Section 2. In the third section of the paper, mathematical formulations to calculate the insolation are presented. In the next section, the planning problem of the PV/BES systems is formulated. The results are presented in Section 5. In the last section, conclusions are given.

2. Problem Description

The proposed approach in this paper to model the investment problem of PV systems is shown in Figure 1. In the first step, the insolation is modeled as (1)–(5). Then, this model is coded in Python software to obtain the insolation of each location on every day and hour of the year. The input parameters of this model are the latitude and longitude of the locations. In the second step, the investment problem of the PV systems is mathematically formulated as (6)–(20). It is a linear programming (LP) model which is coded in the GAMS software. The input parameters of this model are described in Figure 1, including the demand of the system, the investment and the maintenance cost of the PV panels, inverter, and BES and the technical specifications of this equipment. Also, the obtained insolation in the previous step is considered as the input parameter in the proposed model in (6)–(20). The decision variables of this step are the capacities of the PV panels, inverter, and BES, as well as the NPV of the system and the payback time. These steps are described in detail in the next sections.

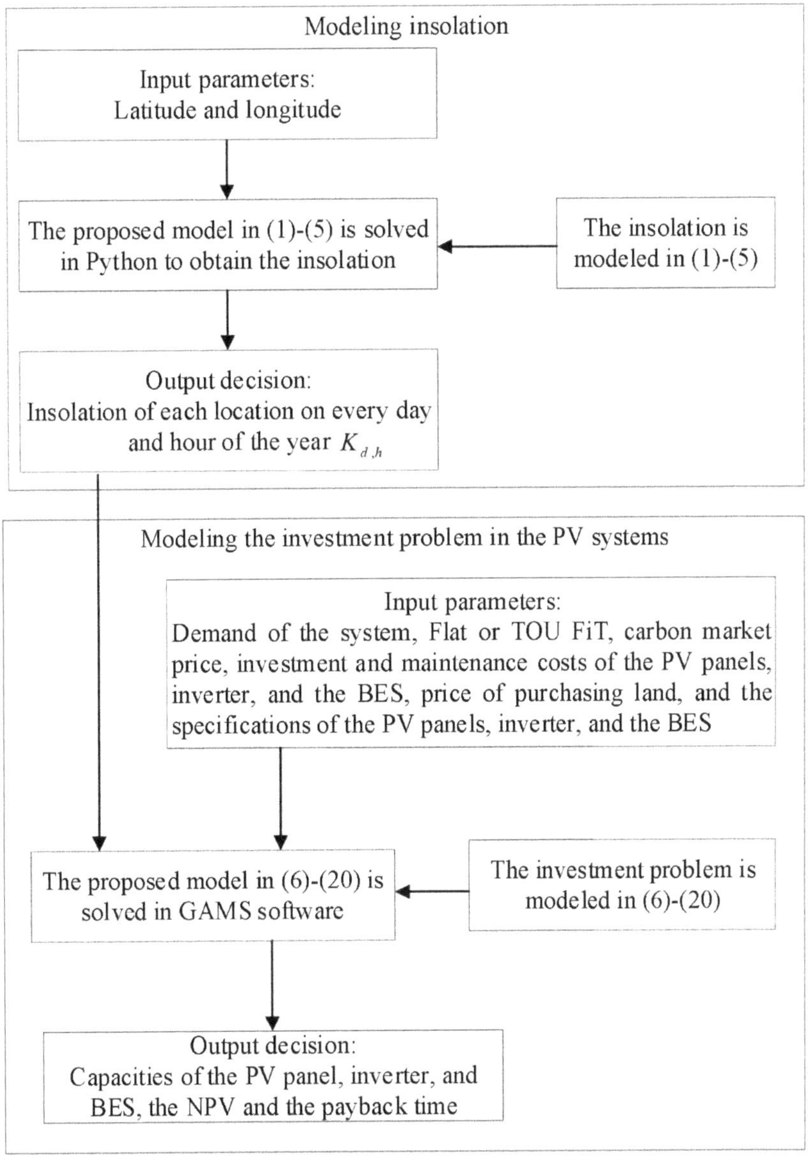

Figure 1. The proposed approach in this paper for modeling the investment problem in PV systems.

3. Modeling Insolation

In this section, the insolation is calculated for various locations (considering their latitudes and longitudes) on every day and hour of the year. For this purpose, the proposed models in [23] are used (The same notations proposed in [23] are used in this section.). The insolation ($K_{\phi,\lambda,d,h}$) is calculated as (1)

$$K_{\phi,\lambda,d,h} = S_d \, T^r_{\phi,\lambda,d,h} \, \sin(\psi) \tag{1}$$

where S_d is the insolation measured on a day (d) of the year (For example, S_5 indicates the insolation measured on the fifth day of the year.), and it is calculated as (2). $T^r_{\phi,\lambda,d,h}$ is the net sky transmissivity, which is modeled as (3)

$$S_d = \mathrm{S}_0\left(\frac{\overline{\mathrm{R}}}{R_d}\right)^2 \tag{2}$$

where S_0 is the average total insolation measured for the average distance between the Sun and the Earth, $\overline{\mathrm{R}}$ is the average distance between the Sun and the Earth, which is 149.6 Giga meters, and R_d is the actual distance of that point with the Sun on a given day (d).

$$T^r_{\phi,\lambda,d,h} = (0.6 + 0.2\sin(\psi)) \tag{3}$$

The local elevation angle (ψ) used in Equations (1) and (3) is calculated as (4)

$$\sin(\psi) = \sin(\phi)\sin(\delta^s) - \cos(\phi)\cos(\delta^s)\cos\left[\frac{\mathrm{C}\,h^{\mathrm{UTC}}}{\mathrm{h^d}} - \lambda\right] \tag{4}$$

where ϕ is the latitude, λ is the longitude, h^{UTC} is the time of the day regarding the UTC, $\mathrm{h^d}$ is 24, which represents the hours in a day, C equals 2π radians, and δ^s is the solar declination angle which is calculated as (5).

$$\delta^{\,s} = \phi^{\mathrm{r}}\cos\left[\frac{\mathrm{C}(d - \mathrm{d^r})}{\mathrm{d^y}}\right] \tag{5}$$

Here ϕ^r is the ecliptic plane, and it is 23.44°, $\mathrm{d^r}$ is the number of the Julian day (It shows the longest day of a year.), which is the 172nd day of the year, $\mathrm{d^y}$ is the number of days in a year, i.e., 365.

To use the proposed formulations in (1)–(5) in the investment problem of the PV systems in the next section, a location with known latitude (ϕ) and longitude (λ) is determined and its insolation ($K_{\phi,\lambda,d,h}$) is calculated for all the days of a year and all the hours of the day (i.e., 365 days and 24 h). This calculated insolation is then reported as $K_{d,h}$.

4. Modeling the Investment Problem in the PV Systems

The NPV of the hybrid PV/BES system modeled in (6) consists of six terms. The system's revenue from selling energy to the grid and participating in the carbon market are modeled as the first and the second terms of (6). These terms are both modeled in (7) (In this section, the index D shows all the days of the year and is the same as $\mathrm{d^y}$ presented in the previous section. Also, index T shows all the hours of the day and is the same as h^d described in the previous section.). The third term of (6) models the salvage of the system. The coefficient (χ) is used in (8) to show the equipment's efficiency reduction after ending the project lifetime. The fourth term of (6) models the investment cost of the system's equipment as modeled in (9). This cost includes the investment cost of the PV panels, the BESs, and the inverter, as described in (10), (11), and (12), respectively. Since the capacity of the inverter should be more than the capacity of the PV panels, the coefficient (α) is used in (12) to model this issue. Also, the coefficient (β) is multiplied by IC in the objective function to model the other investment costs, including the PV panels' structures, wiring, protection devices and so on. The fifth term of (6) is used to model the land price as described in (13). The last term of (6) shows the maintenance cost of the system as modeled in (14).

$$NPV = \sum_{y=1}^{Y}\frac{R_y^{\mathrm{sell}}}{(1+\mathrm{i})^y} + \sum_{y=1}^{Y}\frac{R_y^{\mathrm{carbon}}}{(1+\mathrm{i})^y} + R^{\mathrm{Salvage}} - \beta\, IC - C^{\mathrm{Land}} - \sum_{y=1}^{Y}\frac{MC_y}{(1+\mathrm{i})^y} \tag{6}$$

$$R_y^{\mathrm{sell}} = \sum_{d=1}^{D}\sum_{h=1}^{H} p_{y,d,h}^{\mathrm{sell}}\rho_{y,d,h}^{\mathrm{sell}} \quad , \quad R_y^{\mathrm{carbon}} = \sum_{d=1}^{D}\sum_{h=1}^{H} p_{y,d,h}^{\mathrm{sell}}\,\rho_{y,d,h}^{\mathrm{carbon}} \tag{7}$$

$$R^{\text{Salvage}} = \text{IC}^{\text{PV}}(\frac{\text{LT}^{\text{PV}} - \text{LT}^{\text{Project}}}{\text{LT}^{\text{PV}}})\chi^{\text{PV}} + \text{IC}^{\text{PV}}(\frac{\text{LT}^{\text{BES}} - \text{LT}^{\text{Project}}}{\text{LT}^{\text{BES}}})\chi^{\text{BES}} + \text{IC}^{\text{Inverter}}(\frac{\text{LT}^{\text{Inverter}} - \text{LT}^{\text{Project}}}{\text{LT}^{\text{Inverter}}})\chi^{\text{Inverter}} \quad (8)$$

$$IC = IC^{\text{PV}} + IC^{\text{BES}} + IC^{\text{Inverter}} \quad (9)$$

$$IC^{\text{PV}} = N\,\overline{\text{P}}^{\text{PV}}\rho^{\text{IC\_PV}} \quad (10)$$

$$IC^{\text{BES}} = \overline{\text{E}}^{\text{BES}}\rho^{\text{IC\_BES}} \quad (11)$$

$$IC^{\text{Inverter}} = \alpha N\,\overline{\text{P}}^{\text{PV}}\rho^{\text{IC\_Inverter}} \quad (12)$$

$$C^{\text{Land}} = N\,\text{A}^{\text{Land}}\rho^{\text{Land}} \quad (13)$$

$$MC_y = MC_y^{\text{PV}} + MC_y^{\text{BES}} + MC_y^{\text{Inverter}} = N\,\overline{\text{P}}^{\text{PV}}\rho_y^{\text{MC\_PV}} + \overline{\text{E}}^{\text{BES}}\rho_y^{\text{MC\_BES}} + \alpha N\,\overline{\text{P}}^{\text{PV}}\rho_y^{\text{MC\_Inverter}} \quad (14)$$

The proposed objective function is optimized considering the following constraints:

$$p_{y,d,h}^{\text{PV}}\eta^{\text{Inverter}} + p_{y,d,h}^{\text{discharge}} = p_{y,d,h}^{\text{charge}} + p_{y,d,h}^{\text{sell}} + \text{P}_{y,d,h}^{\text{demand}} \qquad : \forall y,d,h \quad (15)$$

Equation (15) models the power balance of the system where the power generation of the PV system can be used to charge the BES and to sell energy to the grid. Also, the BESs can be discharged to sell energy to the grid. Also, the demand of the system can be met through the PV system and discharging power of the BES.

$$0 \leq p_{y,d,h}^{\text{charge}} \leq \delta\,\overline{\text{E}}^{\text{BES}} \quad , \quad 0 \leq p_{y,d,h}^{\text{discharge}} \leq \delta\,\overline{\text{E}}^{\text{BES}} \qquad : \forall y,d,h \quad (16)$$

The maximum power charging/discharging limitations of the BESs are modeled as (16). In this equation, the coefficient (δ) is used to model the fact that the maximum charging/discharging power of the BESs are coefficients of the BESs' energy maximum capacity.

$$\begin{aligned} E_{y,d,h}^{\text{BES}} &= E_{y,d,h-1}^{\text{BES}} + p_{y,d,h}^{\text{charge}}\eta^{\text{charge}} - \frac{p_{y,d,h}^{\text{discharge}}}{\eta^{\text{discharge}}} && : \forall y,d,h > 1 \\ E_{y,d,h}^{\text{BES}} &= \text{E}_y^{\text{BES\_ini}} + p_{y,d,h}^{\text{charge}}\eta^{\text{charge}} - \frac{p_{y,d,h}^{\text{discharge}}}{\eta^{\text{discharge}}} && : \forall y,d,h = 1 \end{aligned} \quad (17)$$

The stored energy in the BESs depends on both the power charging/discharging of the BESs in that time step and the stored energy in the previous time step, as modeled in (17).

$$\theta\,\overline{\text{E}}^{\text{BES}} \leq E_{y,d,h}^{\text{BES}} \leq \overline{\text{E}}^{\text{BES}} \qquad : \forall y,d,h \quad (18)$$

The minimum and maximum limitations of the stored energy in the BESs are modeled in (18).

$$p_{y,d,h}^{\text{PV}} = \frac{N\,\overline{\text{P}}^{\text{PV}}\text{K}_{y,d,h}}{\text{K}^{\text{Std}}} \quad (19)$$

The power generation of the PV panels in each time step is modeled as (19). For this purpose, the insolation ($K_{y,d,h}$) (It should be noted that $K_{y,d,h}$ is the same as $K_{d,h}$ since the insolation is constant for all years.) is multiplied by the maximum capacity of the PV panel obtained in the standard condition and the number of the panels. The resulting term is then divided by the insolation amount at the standard condition, which is 1000 W/m^2.

$$\sum_{y=1}^{Y}\sum_{d=1}^{D}\sum_{h=1}^{H}(p_{y,t}^{\text{charge}} + p_{y,t}^{\text{discharge}}) \leq \phi\,\overline{\text{E}}^{\text{BES}} \quad (20)$$

Equation (20) models the relationship between the charging/discharging power of the BESs with the BESs' lifetime.

5. Numerical Results

In this simulation, some 0.4kW PV panels are considered. The price of this panel is USD 200. The price of the inverter and the BES for 1 kW and 1 kWh capacities are USD 87.5 and USD 122, respectively. The maintenance cost of the PV panels and the inverter are considered to be 2 percent of their investment cost [24]. This term for the BESs is considered to be 5 percent [25]. The required space for installing the PV system for 1 kW is 12 m^2. The price of purchasing land for installing the PV system is assumed to be 12 USD/m^2. The efficiency of the inverter is 0.98, and the charging/discharging efficiency of the BES is 0.95. The lifetime of the inverter and the PV panel are 10 and 20 years, respectively. For calculating the salvage, it is assumed that the electrical energy production efficiency of the PV panels would reduce by 10 percent after the project lifetime (10 years). It is assumed that the lifetime of each 1 kWh capacity of the battery is 10,000 kWh. The interest rate is 21 percent, and the inflation rate is 18 percent, regarding the average of these values reported by the central bank of Iran. Therefore, the real interest rate would be 2.54 percent. The average hourly value of the demand is 0.14 kW; the peak demand is 0.36 kW, and therefore the load factor is approximately 0.39.

5.1. Insolation in Iran and Defining Different Zones

The amount of insolation in Iran is shown in Figure 2 (This Figure shows the insolation without considering the effect of the temperature and the clearness index.) regarding the mathematical models proposed in Section 3. This data is calculated for 12:00 p.m. in June. Iran is divided into eight zones, as given in Table 3, with regard to the results obtained in Figure 2. These zones are determined by the amount of insolation. For example, the areas with an insolation equal to or greater than 1060 W/m^2 at 12:00 p.m. local time in Iran are labeled as Zone 1. The longitude and the latitude spectrums of each zone are presented, among which one latitude is chosen for each zone for the calculations in the next subsection. The greatest and the least amount of insolation are observed in zones 1 and 8, with more than 1060 W/m^2 and less than 1000 W/m^2 at noon in June, respectively. As shown in this figure, the insolation only depends on the latitude since it is constant in a specified latitude with different longitudes. The average daily insolation in Iran is presented for the different zones in Table 3.

Table 3. Average daily insolation in the eight different zones.

Zone Number	Latitude Spectrum	Longitude Spectrum	The Chosen Latitude for Calculations (φ)	Average Daily Insolation (Wh/m^2)
Zone 1	25° N–26.8° N	44° E–63° E	26° N	6705
Zone 2	26.8° N–29° N	44° E–63° E	28° N	6582
Zone 3	29° N–31° N	44° E–63° E	30° N	6451
Zone 4	31° N–33.7° N	44° E–63° E	32° N	6314
Zone 5	33.7° N–35.8° N	44° E–63° E	34° N	6169
Zone 6	35.8° N–37.7° N	44° E–63° E	36° N	6019
Zone 7	37.7° N–39.4° N	44° E–63° E	38° N	5864
Zone 8	39.4° N–40° N	44° E–63° E	40° N	5703

5.2. Description of Different Cases

In order to investigate the optimal behavior of the PV/BES system in different zones, three cases are defined as described in Table 4. In the base case, the present conditions of Iran's electrical energy system are considered in which the power generation of the PV systems are purchased at the Flat FiT. In this case, the government purchases energy at a fixed price in the year. This price for PV systems with capacities equal to or greater than 20 kW and equal to or smaller than 200 kW is 0.055 USD/kWh, and this increases by 15 percent each year. Therefore, the maximum number of PV panels which can be installed is 500.

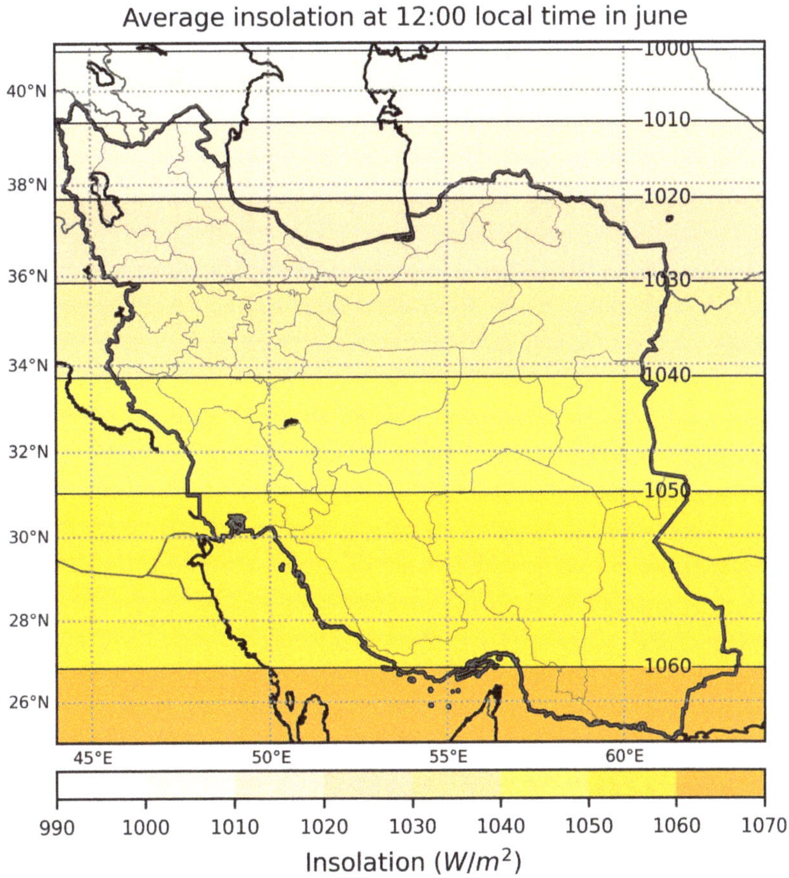

Figure 2. Insolation in Iran at 12:00 p.m. local time in June.

Table 4. Conditions of different cases defined for the simulations.

Cases	Flat FiT	TOU FiT	Carbon Trading Market
Base case	✓	-	-
Case I	-	✓	-
Case II	-	✓	✓

✓ stands for 'considered' and - stands for 'not considered'.

In Case I, it is proposed that the government designs the TOU FiT to encourage the investors to use the BESs in the PV systems. For this purpose, it is assumed that the purchased power from the PV systems increases in the peak periods, as proposed in Figure 3. As shown in this figure, the price increases from 0.055 USD/kWh to 0.11 USD/kWh in hours 21–23 in all seasons and in hours 13–17 only in summer.

In Case II, it is assumed that the PV system's operators can participate in the carbon trading market besides selling their electrical energy under the TOU FiT scheme. The price of the carbon trading market for one year is given in Figure 4 [26]. These prices show the amount of price paid for 1 kg carbon reduction. The emission intensity of electricity production in Iran is 0.492 kgCO_2eq/kWh [27].

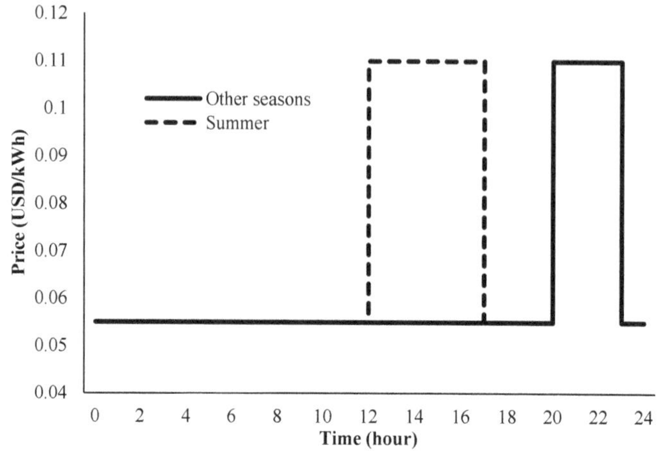

Figure 3. Proposed purchasing energy price from the PV system in Case I.

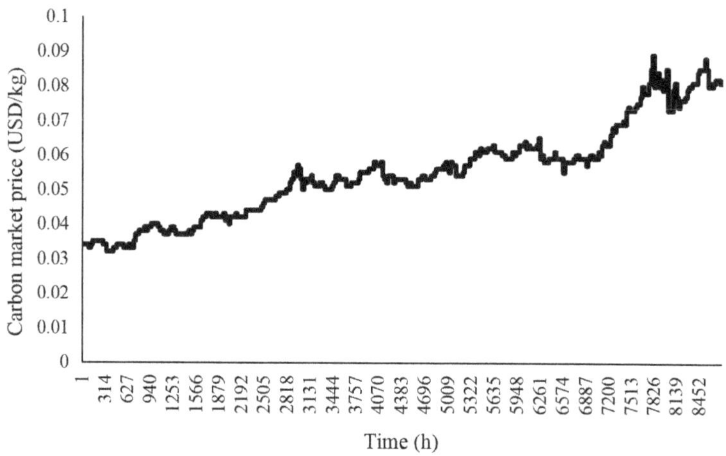

Figure 4. Carbon trading market price used in Case II [26].

5.3. Results

The results are presented for the proposed three cases as follows.

(1) Base case: The results of this case are given in Tables 5 and 6. As shown in these tables, the optimized capacity of the system for all zones is equal. The capacity of the BES is 4 kWh for all zones. This capacity is obtained only to meet the demand of the system in the hours with no power generation of the PV system. It should be noted that since the selling price to the grid is flat, there is no incentive to use the BES in such systems with the aim of selling energy to the main grid. The payback time from the worst zone (Zone 8) to the best zone (Zone 1) decreases by 0.73 yrs. The NPV of the best zone also increases by 32.88 percent more than the worst one. The results presented in Tables 5 and 6 show that investing in the PV system is not attractive for investors in Iran under the present conditions since the payback time in all zones is equal to or greater than 5.46 yrs.

Table 5. The optimized capacity, the NPC, and the payback time of the system in the base case.

Zone's Number	Installed Capacities			NPV (USD)	Payback Time (yr.)
	PV (kW)	Inverter (kW)	BES (kWh)		
Zone 1	200	220	4	227,940.61	5.46
Zone 2	200	220	4	220,992.48	5.54
Zone 3	200	220	4	213,639.02	5.63
Zone 4	200	220	4	205,900.46	5.73
Zone 5	200	220	4	197,798.97	5.83
Zone 6	200	220	4	189,356.58	5.95
Zone 7	200	220	4	180,597.52	6.07
Zone 8	200	220	4	171,547.87	6.19

Table 6. The details of the costs/revenue in the base case.

Zone's Number	Investment Cost (USD)	Maintenance Cost (USD)	Land Price (USD)	Revenue from Selling Energy (USD)	Salvage (USD)
Zone 1	119,738	20,735.45	28,800	376,161.65	45,000
Zone 2	119,738	20,735.45	28,800	369,213.52	45,000
Zone 3	119,738	20,735.45	28,800	361,860.01	45,000
Zone 4	119,738	20,735.45	28,800	354,121.51	45,000
Zone 5	119,738	20,735.45	28,800	346,020.01	45,000
Zone 6	119,738	20,735.45	28,800	337,577.63	45,000
Zone 7	119,738	20,735.45	28,800	328,818.56	45,000
Zone 8	119,738	20,735.45	28,800	319,768.92	45,000

(2) Case I: The results of this case are given in Tables 7 and 8. As shown in these tables, under the TOU FiT scheme, the investors are encouraged to use the BESs in their PV systems to sell energy to the grid. The optimized BESs capacity for the best zone is 94 kWh. In fact, by using the BES with this capacity, the best NPV and payback period are obtained for this zone. For the other zones, when decreasing the power generation of the PV systems, the size of the BES capacity increases to obtain the optimum results for that zone. The optimized BES capacity for zones 4–8 reaches 100 kWh. The payback period from the worst zone (Zone 8) to the best zone (Zone 1) decreases by 0.6 yrs. Also, the NPV of the best zone increases by 23.13 percent in comparison with the worst one.

Table 7. The optimized capacity, the NPC, and the payback time of the system in case I.

Zone's Number	Installed Capacities			NPV (USD)	Payback Time (yr.)
	PV (kW)	Inverter (kW)	BES (kWh)		
Zone 1	200 kW	220 kW	94	373,478.79	5.09
Zone 2	200 kW	220 kW	96	365,084.23	5.16
Zone 3	200 kW	220 kW	99	356,105.87	5.23
Zone 4	200 kW	220 kW	100	346,574.65	5.31
Zone 5	200 kW	220 kW	100	336,523.68	5.39

Table 7. *Cont.*

Zone's Number	Installed Capacities			NPV (USD)	Payback Time (yr.)
	PV (kW)	Inverter (kW)	BES (kWh)		
Zone 6	200 kW	220 kW	100	325,968.27	5.48
Zone 7	200 kW	220 kW	100	314,878.32	5.58
Zone 8	200 kW	220 kW	100	303,323.96	5.69

Table 8. The details of the costs/revenue in case I.

Zone's Number	Investment Cost (USD)	Maintenance Cost (USD)	Land Price (USD)	Revenue from Selling Energy (USD)	Salvage (USD)
Zone 1	130,718	25,529.82	28,800	539,670.21	45,000
Zone 2	130,962	25,636.36	28,800	531,674.99	45,000
Zone 3	131,328	25,796.17	28,800	523,295.64	45,000
Zone 4	131,450	25,849.44	28,800	513,964.09	45,000
Zone 5	131,450	25,849.44	28,800	505,711.57	45,000
Zone 6	131,450	25,849.44	28,800	493,357.72	45,000
Zone 7	131,450	25,849.44	28,800	482,267.76	45,000
Zone 8	131,450	25,849.44	28,800	470,713.41	45,000

The behavior of the PV system on a summer day and a day in other seasons in Case I in Zone 1 is shown in Figures 5–8. The power balance of the system on a summer day is shown in Figure 5. The charging/discharging process of the BES is repeated twice on this day, as shown in Figure 5. The BES is charged in hours 9 and 10, and then it is discharged in hours 15 and 17 to sell energy to the grid at the high-selling energy price. In the second cycle, the BES is charged in hours 18 and 19 and then it is discharged in hours 21 and 22 at the high-selling energy price. The power balance of the system in a day in seasons other than summer is shown in Figure 7. Since the peak price of selling energy to the grid is noticed only in one period in the non-summer days, i.e., 21–23, the BES is only discharged in hours 21 and 22, as shown in Figure 7. For this purpose, the BES is charged in hours 17 and 18. As shown in these two figures, the charging/discharging of the BES takes place regarding the TOU FiT described in Figure 3. The results of supplying the demand of the system are shown in Figures 6 and 8 for a day in summer and a non-summer day, respectively. As shown in these figures, the demand of the system is supplied through the PV system in hours 6–19, and the BES is discharged to meet the demand in the other hours. The results of this case show that this behavior of the PV system in storing the PV power generation in the BES in the off-peak hours, with the aim of discharging the BES in the peak periods, increases the system's revenue from selling energy to the grid.

(3) Case II: The results of this case are given in Tables 9 and 10. As shown in these tables, considering the carbon trading market besides the TOU FiT scheme, investing in PV systems is more attractive for investors since the payback time of the project is significantly shorter compared to the other cases. The payback time for the best zones (Zones 1–4) becomes less than four years, which is very attractive for investors. In the best zone (Zone 1), this payback time is optimized and also the NPV of the project obtained with the BES capacity equals 45 kWh. The BESs' optimized capacities increase in the other zones since the power generation of the PV systems decreases in these zones. The payback period from the worst zone (Zone 8) to the best one (Zone 1)

decreases by 0.6 yrs. The NPV of the best zone also increases by 21.46 percent more than the worst one.

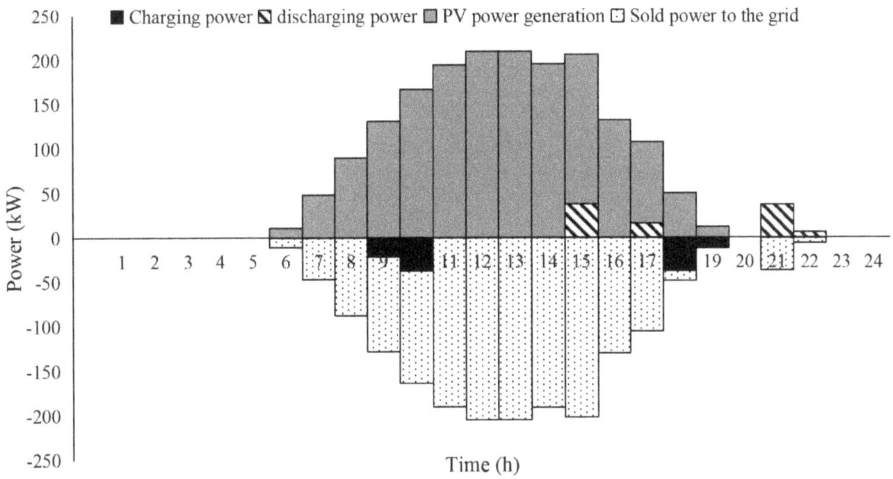

Figure 5. The power balance of the PV system in Case I in Zone 1 on a summer day.

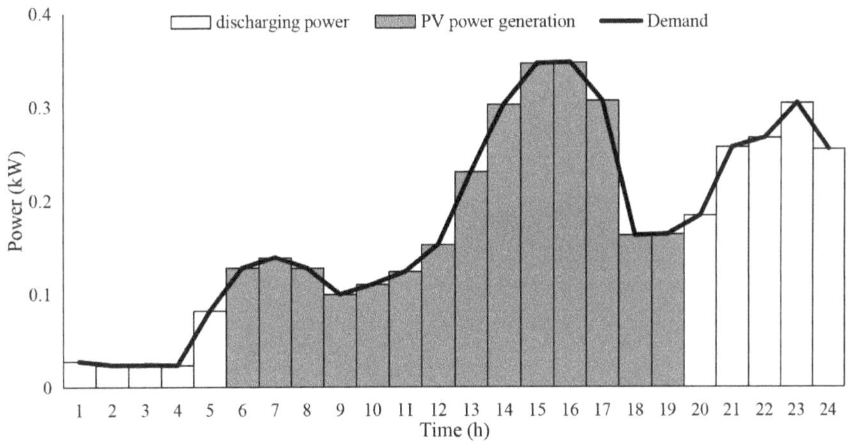

Figure 6. Results of supplying the demand for Case I in Zone 1 on a summer day.

Table 9. The optimized capacity, the NPC, and the payback time of the system in case II.

Zone's Number	Installed Capacities			NPV (USD)	Payback Time (yr.)
	PV (kW)	Inverter (kW)	BES (kWh)		
Zone 1	200	220	45	592,299.54	3.75
Zone 2	200	220	52	579,601.25	3.82
Zone 3	200	220	61	566,082.77	3.9
Zone 4	200	220	69	551,787.52	3.99
Zone 5	200	220	79	536,763.71	4.07
Zone 6	200	220	84	521,038.8	4.16
Zone 7	200	220	88	5046,56.34	4.25
Zone 8	200	220	92	4876,66.17	4.35

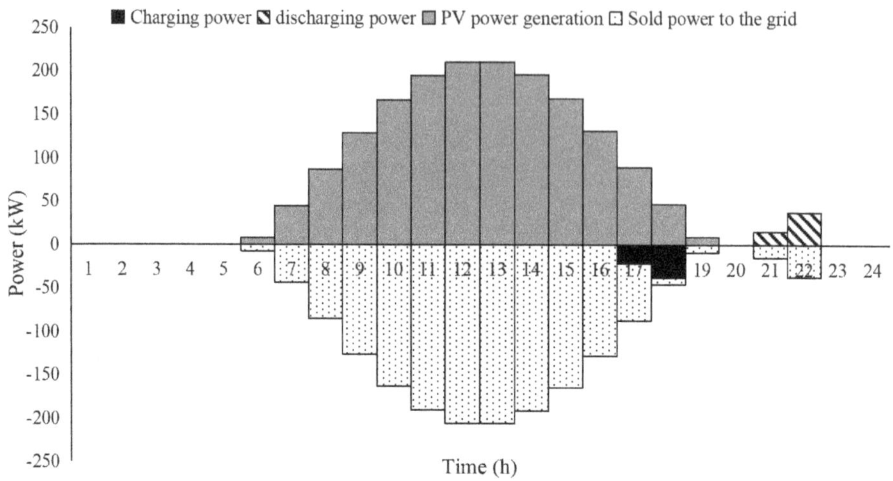

Figure 7. The power balance of the PV system in Case I in Zone 1 on a non-summer day.

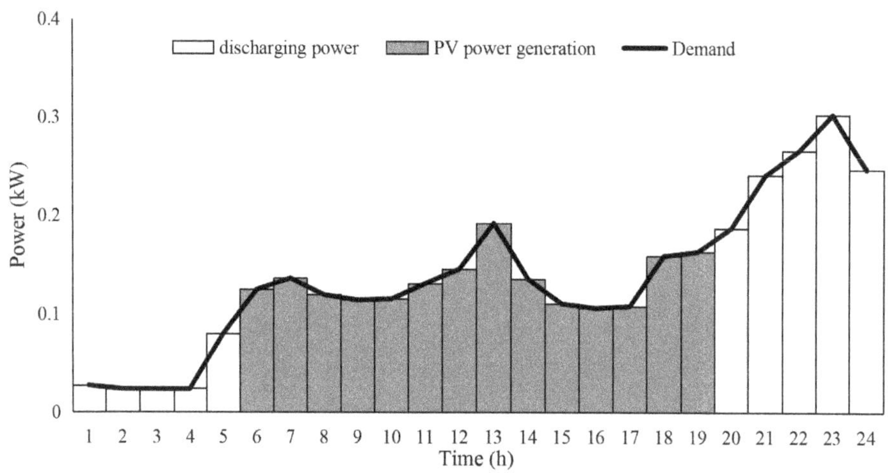

Figure 8. Results of supplying the demand for Case I in Zone 1 on a non-summer day.

Table 10. The details of the costs/revenue in case II.

Zone's Number	Investment Cost (USD)	Maintenance Cost (USD)	Land Cost (USD)	Revenue from Selling Energy (USD)	Revenue from Carbon Reduction (USD)	Salvage (USD)
Zone 1	124,740	22,919.55	28,800	529,651.45	219,055.65	45,000
Zone 2	125,594	23,292.45	28,800	522,671.12	214,735.38	45,000
Zone 3	126,692	23,771.88	28,800	515,525.95	210,159.1	45,000
Zone 4	127,668	24,198.05	28,800	507,611.56	205,375.6	45,000
Zone 5	128,888	24,730.74	28,800	499,599.2	2003,60.86	45,000
Zone 6	129,498	24,997.11	28,800	[illegible]	1952,15.72	45,000
Zone 7	129,986	25,210.19	28,800	479,744.51	1899,05.23	45,000
Zone 8	130,474	25,423.28	28,800	469,025.62	1844,32.62	45,000

The amount of the increment/reduction of the NPV/payback time in Cases I and II in comparison with the base case are given in Table 11. As shown in the results, for the best zone, the NPV increases 63.85 percent in Case I and 159.8 percent in Case II in comparison with the base case. Also, the payback period in this zone in Cases I and II are 6.78 percent and 31.32 percent shorter, respectively, than the base case. The results show that proposing the TOU FiT and participating in the carbon trading market has more effect on the NPV of the zones with lower power generation of the PV systems. From the viewpoint of the payback period, in cases I and II, the payback time reduces more in the zones with low and high power generation of the PV systems, respectively.

Table 11. Comparison of the results of three cases.

Zone's Number	NPV (USD)	NPV Increment (%)		Payback Time (yr.)	Payback Time Reduction (%)	
	Base Case	Case I	Case II	Base case	Case I	Case II
Zone 1	227,940.61	63.85	159.8	5.46	6.78	31.32
Zone 2	220,992.48	65.2	162.3	5.54	6.86	31.05
Zone 3	213,639.02	66.69	165	5.63	7.11	30.73
Zone 4	205,900.46	68.32	168	5.73	7.33	30.34
Zone 5	197,798.97	70.13	171.4	5.83	7.55	30.19
Zone 6	189,356.58	72.15	175.2	5.95	7.9	30.09
Zone 7	180,597.52	74.35	179.4	6.07	8.07	29.99
Zone 8	171,547.87	76.82	184.3	6.19	8.08	29.73

5.4. Discussion

The whole capacity of the PV systems installed in Iran is 484 MW, which is only 0.55 percent of the whole power plant capacity. This capacity of the PV systems produces approximately 848 GWh in a year and prevents 0.417 Mt CO_2 emission per year. The whole CO_2 emission of the electrical energy systems in Iran is approximately 176.8 Mt. Therefore, it can be concluded that with the present trend of investing in the PV system, it is not possible in Iran to reach a low carbon energy system. Therefore, in this paper, two main solutions are proposed to facilitate the investment in PV systems: using the TOU FiT and participating in the carbon trading market. The main findings of the results are as follows:

- The payback period of the PV systems in Iran in the base case (present condition) is not at all attractive to investors.
- In Cases I and II, the payback periods for all zones decrease significantly, especially in Case II, which could encourage investors to invest in PV systems.
- The difference between the payback period in the best and the worst zones in the base case is 0.73 yrs. This difference in both Cases I and II decreases to 0.6 yrs. Therefore, it can be concluded that the economic conditions in all zones in Iran can be more attractive in the presence of these incentive policies.
- The results show that investment in the PV system in all cases is more attractive in the best zones, such as Zones 1 and 2, than the others. Therefore, it can be concluded that defining similar incentive policies for the whole country cannot lead to investing in PV systems in all zones. For this purpose, some solutions should be considered by the government besides the ones proposed in this paper. For example, the price of the purchased energy from the PV systems can increase in the worst zones compared to the best ones.
- The results show that in the base case, the PV systems do not use the BES system to sell energy to the grid, and the BESs are only used to meet the system's demand. Therefore, they cannot have a noticeable participation in helping the system in the peak periods. This is why, in Cases I and II, the BESs are employed in all zones, and they are charged in the off-peak hours and then discharged in the peak hours to sell

energy to the grid. This means that the requirements of Iran's power system could be met during peak hours.

5.5. Comparison of the Proposed Approach in this Paper with the Previous Studies

The aim of this subsection is to prove the main contributions of this paper in comparison with the proposed studies through the reported results.

- Regarding the model used in this paper, the insolation is obtained for all locations in Iran and Iran is then divided into eight zones. This approach of dividing Iran into different zones is used to determine the NPV and the payback period in these zones. This is because, in previous studies, the forecast data related to the insolation used for the investment problem is solved for specific locations.
- As the results show, considering the TOU FiT and participating in the carbon energy market leads to obtaining a better NPV and also a better payback period in comparison with the base case. The results show the different impacts of these incentive policies on the different zones. This issue shows how the proposed model in this paper can be used in Iran to increase investment in PV systems. This is because these incentive policies are not considered in the investment problem in the PV systems in the previous studies.
- The obtained results in this paper can be used both by the investors and the government of Iran. From the viewpoint of the investors, they could then select the best locations for their investment. On the other hand, the government notices the importance of introducing different incentives for different zones to encourage investors to invest in all zones. In fact, the reported results in the previous studies could not be used by investors to select the best location for investing in a country. Also, those models cannot be used by governments to design incentive policies for their entire countries.

6. Conclusions and Future Works

In this paper, the planning problem of the PV system in the presence of the BESs is formulated considering the TOU FiT and the carbon trading market. In order to investigate the behavior of the PV systems in Iran, the insolation is calculated regarding the mathematical formulations by dividing Iran into eight different zones. The average daily insolation for the best and the worst zones are 6705 Wh/m^2 and 5703 Wh/m^2, respectively. The results of the planning problem are reported for three cases. For the base case with the flat FiT and without participating in the carbon market, the payback periods change from 5.46 yrs. to 6.19 yrs. from the best zone to the worst one. In Case I, considering the TOU FiT, the payback periods decrease between 6.78 percent and 8.08 percent for different zones. The results show a significant reduction of the payback period in Case II regarding the carbon trading market, where the payback period in the best zone decreases by 31.32 percent. The results show that investing in PV systems is not attractive for investors considering the present conditions in Iran (base case). The incentive policies proposed in Cases I and II decrease the payback periods of the investment in all zones. Therefore, the investment in PV systems would be attractive for investors under the conditions proposed in Cases I and II.

The results report different effects of the TOU FiT and participating in the carbon market on the NPV and the payback time of the zones. The greatest increments in the NPV in both Cases I and II occur in the zones with low PV power generation, considering the TOU FiT leads to improving the payback time in the worst zones (with low PV power generation). Also, both the TOU FiT and the participation in the carbon market lead to improving the payback time in the best zones (with high PV power generation). Therefore, the main suggestion for Iran's government to increase the investment in the PV system is to define different flat/TOU FiTs for different zones since the NPV and the payback periods are different in each zone. In fact, defining a similar FiT for the country is not generally attractive for investors to invest in PV systems.

Some suggestions are presented as follows, which can be considered for future works:

- Modeling the other incentive policies in the investment problem of the PV systems.
- Developing the proposed model in this paper for the hybrid energy systems with modeling the other energy resources such as WT and biomass energy resources.
- Developing the approach in this paper to model the investment problem of PV systems in the presence of thermal energy resources and thermal demands.

Author Contributions: Conceptualization, A.S. and S.B.; methodology, A.S. and S.B.; software, S.B.; formal analysis, A.S. and S.B.; writing—review and editing, A.S. and S.B.; supervision, S.B. All authors have read and agreed to the published version of the manuscript.

Funding: This research received no external funding.

Data Availability Statement: The data is available upon request.

Conflicts of Interest: The authors declare no conflict of interest.

Nomenclature

Acronyms	
BES	Battery energy storage
CCHP	Combined cooling, heating, and power
CO_2	Carbon Dioxide
FiT	Feed-in-Tariff
GHG	Greenhouse gas
MILP/MINLP	Mixed-integer linear/nonlinear programming
PV	Photovoltaic
RES	Renewable energy sources
TOU	Time of Use
WT	Wind turbine
Indices/sets	
t/T	Index/set of hours of the day
d/D	Index/set of days of the year
y/Y	Year of the project
Variable:	
NPV	Net present value (USD)
Positive variables:	
$E_{y,t}^{\text{BES}}$	Stored energy in the BES (kWh)
IC	Investment cost (USD)
IC^{BES}	Investment cost of the BES (USD)
IC^{Inverter}	Investment cost of the inverter (USD)
C^{Land}	Investment cost of the purchasing the land (USD)
IC^{PV}	Investment cost of the PV panels (USD)
MC_y	Maintenance cost of the system (USD)
MC_y^{PV}	Maintenance cost of the BES (USD)
MC_y^{BES}	Maintenance cost of the BES (USD)
MC_y^{Inverter}	Maintenance cost of the inverter (USD)
N	Number of the PV panels
$p_{y,t}^{\text{charge}}$	Power charging of the BES (kW)

$p_{y,t}^{\text{discharge}}$	Power discharging of the BES (kW)
$p_{y,t}^{\text{sell}}$	Selling energy to the grid (kW)
$p_{y,t}^{\text{PV}}$	Power generation of the PV panels (kW)
R_{y}^{sell}	Revenue from selling energy to the grid (USD)
R_{y}^{carbon}	Revenue from participating in the carbon trading market (USD)
R^{Salvage}	Salvage value (USD)
Parameters:	
A^{Land}	The required space for installing each PV panel (m^2)
$\overline{\text{E}}^{\text{BES}}$	Capacity of the BES (kWh)
$\text{E}_{y}^{\text{BES\_ini}}$	Initial stored energy in the BES (kWh)
i	Real interest rate (%)
$\text{K}_{y,t}$	Insolation (kW/m^2)
K^{Std}	Insolation of the standard condition (kW/m^2)
LT^{BES}	Lifetime of the BES (yr.)
$\text{LT}^{\text{Inverter}}$	Lifetime of the inverter (yr.)
LT^{PV}	Lifetime of the PV panels (yr.)
$\text{LT}^{\text{Project}}$	Lifetime of the project (yr.)
$\text{P}_{y,d,h}^{\text{demand}}$	The demand of the system (kW)
$\overline{\text{P}}^{\text{PV}}$	Capacity of the PV panel (kW)
$\eta^{\text{charge}}/\eta^{\text{discharge}}$	Charging/discharging power efficiency of the BES
$\rho_{y,t}^{\text{carbon}}$	Carbon trading market price (USD/kg)
$\rho^{\text{IC\_PV}}$	Cost of purchasing one PV panel (USD/kW)
$\rho^{\text{IC\_BES}}$	Cost of purchasing the BES (USD/kWh)
$\rho^{\text{IC\_Inverter}}$	Cost of purchasing the inverter (USD/kW)
ρ^{Land}	The price of purchasing land (USD/m^2)
$\rho_{y}^{\text{MC\_PV}}, \rho_{y}^{\text{MC\_BES}}, \rho_{y}^{\text{MC\_Inverter}}$	Maintenance price of the PV panels, BES, and inverter (USD/kW,USD/kWh, and USD/kW)
$\rho_{y,t}^{\text{sell}}$	Fixed/TOU FiT (kW)
$\chi^{\text{PV}}, \chi^{\text{BES}}, \chi^{\text{Inverter}}$	Efficiency reduction of the PV panels, BES, and inverter after ending the project
Coefficients	
α	The coefficient used to show the increment size of the inverter in respect to the PV panels
δ	The coefficient used to show the maximum charging /discharging power of the BES
θ	The coefficient used to show the minimum stored energy in the BES
ϕ	The coefficient used to show the lifetime of the BES

References

1. Our World in Data. Iran: CO_2 Country Profile. Available online: https://ourworldindata.org/co2/country/iran (accessed on 25 December 2021).
2. Iran CO2 Emission. Available online: https://countryeconomy.com/energy-and-environment/co2-emissions/iran, countryeconomic.com (accessed on 25 December 2021).

3. Almodfer, R.; Zayed, M.E.; Elaziz, M.A.; Aboelmaaref, M.M.; Mudhsh, M.; Elsheikh, A.H. Modeling of a solar-powered thermo-electric air-conditioning system using a random vector functional link network integrated with jellyfish search algorithm. *Case Stud. Therm. Eng.* **2022**, *31*, 101797. [CrossRef]
4. Elaziz, M.A.; Senthilraja, S.; Zayed, M.E.; Elsheikh, A.H.; Mostafa, R.R.; Lu, S. A new random vector functional link integrated with mayfly optimization algorithm for performance prediction of solar photovoltaic thermal collector combined with elec-trolytic hydrogen production system. *Appl. Therm. Eng.* **2021**, *193*, 117055. [CrossRef]
5. El-Hadary, M.I.; Senthilraja, S.; Zayed, M.E. A hybrid system coupling spiral type solar photovoltaic thermal collector and electrocatalytic hydrogen production cell: Experimental investigation and numerical modeling. *Process. Saf. Environ. Prot.* **2023**, *170*, 1101–1120. [CrossRef]
6. Bae, S.; Nam, Y. Feasibility analysis for an integrated system using photovoltaic-thermal and ground source heat pump based on real-scale experiment. *Renew. Energy* **2022**, *185*, 1152–1166. [CrossRef]
7. Zayed, M.E.; Zhao, J.; Elsheikh, A.H.; Li, W.; Sadek, S.; Aboelmaaref, M.M. A comprehensive review on Dish/Stirling concentrated solar power systems: Design, optical and geometrical analyses, thermal performance assessment, and applications. *J. Clean. Prod.* **2020**, *283*, 124664. [CrossRef]
8. Al Garni, H.Z.; Mas'Ud, A.A.; Baseer, M.; Ramli, M.A. Techno-economic optimization and sensitivity analysis of a PV/Wind/diesel/battery system in Saudi Arabia using a combined dispatch strategy. *Sustain. Energy Technol. Assess.* **2022**, *53*, 102730. [CrossRef]
9. Rehman, W.U.; Bo, R.; Mehdipourpicha, H.; Kimball, J.W. Sizing battery energy storage and PV system in an extreme fast charging station considering uncertainties and battery degradation. *Appl. Energy* **2022**, *313*, 118745. [CrossRef]
10. Han, X.; Garrison, J.; Hug, G. Techno-economic analysis of PV-battery systems in Switzerland. Renew. *Sustain. Energy Rev.* **2022**, *158*, 112028. [CrossRef]
11. Chowdhury, T.; Hasan, S.; Chowdhury, H.; Hasnat, A.; Rashedi, A.; Asyraf, M.R.M.; Hassan, M.Z.; Sait, S.M. Sizing of an Island Standalone Hybrid System Considering Economic and Environmental Parameters: A Case Study. *Energies* **2022**, *15*, 5940. [CrossRef]
12. Wu, Y.; Liu, Z.; Liu, J.; Xiao, H.; Liu, R.; Zhang, L. Optimal battery capacity of grid-connected PV-battery systems considering battery degradation. *Renew. Energy* **2021**, *181*, 10–23. [CrossRef]
13. Aziz, A.S.; Tajuddin, M.F.N.; Zidane, T.E.K.; Su, C.-L.; Alrubaie, A.J.K.; Alwazzan, M.J. Techno-economic and environmental evaluation of PV/diesel/battery hybrid energy system using improved dispatch strategy. *Energy Rep.* **2022**, *8*, 6794–6814. [CrossRef]
14. Al-Aali, I.; Modi, V. Decarbonizing the electricity sector in Qatar using PV combined with ice thermal and battery storage. *Energy Strat. Rev.* **2022**, *44*, 101014. [CrossRef]
15. Ropero-Castaño, W.; Muñoz-Galeano, N.; Caicedo-Bravo, E.F.; Maya-Duque, P.; López-Lezama, J.M. Sizing Assessment of Islanded Microgrids Considering Total Investment Cost and Tax Benefits in Colombia. *Energies* **2022**, *15*, 5161. [CrossRef]
16. Muna, Y.B.; Kuo, C.-C. Feasibility and Techno-Economic Analysis of Electric Vehicle Charging of PV/Wind/Diesel/Battery Hybrid Energy System with Different Battery Technology. *Energies* **2022**, *15*, 4364. [CrossRef]
17. Kahsay, M.B.; Lauwaert, J. Excess Energy from PV-Battery System Installations: A Case of Rural Health Center in Tigray, Ethiopia. *Energies* **2022**, *15*, 4355. [CrossRef]
18. Amupolo, A.; Nambundunga, S.; Chowdhury, D.S.P.; Grün, G. Techno-Economic Feasibility of Off-Grid Renewable Energy Electrification Schemes: A Case Study of an Informal Settlement in Namibia. *Energies* **2022**, *15*, 4235. [CrossRef]
19. Sigarchian, S.G.; Malmquist, A.; Martin, V. Design Optimization of a Small-Scale Polygeneration Energy System in Different Climate Zones in Iran. *Energies* **2018**, *11*, 1115. [CrossRef]
20. Yousefi, H.; Hafeznia, H.; Yousefi-Sahzabi, A. Spatial Site Selection for Solar Power Plants Using a GIS-Based Boolean-Fuzzy Logic Model: A Case Study of Markazi Province, Iran. *Energies* **2018**, *11*, 1648. [CrossRef]
21. Dahooie, J.H.; Kashan, A.H.; Naeini, Z.S.; Vanaki, A.S.; Zavadskas, E.K.; Turskis, Z. A Hybrid Multi-Criteria-Decision-Making Aggregation Method and Geographic Information System for Selecting Optimal Solar Power Plants in Iran. *Energies* **2022**, *15*, 2801. [CrossRef]
22. Amini, S.; Bahramara, S.; Golpîra, H.; Francois, B.; Soares, J. Techno-Economic Analysis of Renewable-Energy-Based Micro-Grids Considering Incentive Policies. *Energies* **2022**, *15*, 8285. [CrossRef]
23. Stull, R.B.; Ahrens, C.D. *Meteorology for Scientists and Engineers*; Brooks/Cole: Monterey, CA, USA, 2000.
24. Khezri, R.; Mahmoudi, A.; Haque, M.H. A Demand Side Management Approach for Optimal Sizing of Standalone Renewable-Battery Systems. *IEEE Trans. Sustain. Energy* **2021**, *12*, 2184–2194. [CrossRef]
25. Mongird, K.; Viswanathan, V.V.; Balducci, P.J.; Alam, J.E.; Fotedar, V.; Koritarov, V.S.; Hadjerioua, B. *Hadjerioua, Energy Storage Technology and Cost Characterization Report*; Pacific Northwest National Lab.(PNNL): Richland, WA, USA, 2019; p. 1573487. [CrossRef]
26. EMBER. EU Carbon Price Tracker. Available online: https://ember-climate.org/data/data-tools/carbon-price-viewer/ (accessed on 25 December 2021).
27. EMBER. Available online: https://ember-climate.org/countries-and-regions/countries/iran/ (accessed on 25 December 2021).

energies

MDPI

Article

Simulation of the Temperature of a Shielding Induction Motor of the Nuclear Main Pump under Different Turbulence Models

Likun Ai [1,2], Yiping Lu [1,3,*], Jiade Han [3] and Wenxu Suo [4]

1 School of Electrical and Electronic Engineering, Harbin University of Science and Technology, Harbin 150080, China
2 School of Mechanical and Electronic Engineering, Qiqihar University, Qiqihar 161006, China
3 School of Mechanical and Power Engineering, Harbin University of Science and Technology, Harbin 150080, China
4 Harbin Electric and Power Equipment Co., Ltd., Harbin 150040, China
* Correspondence: luyiping@hrbust.edu.cn

Abstract: Thermal design and the choice of turbulence models are crucial for motors. In this project, the geometrical model of the vertical shielding induction motor for a small nuclear main pump was established by SolidWorks software and the finite volume method was adopted to investigate the temperature of the motor, especially the temperatures of bearings lubricated water. To make the numerical simulation of flow and heat transfer in the rotating clearance of the shielding induction motor more accurate, the effects of four types of different two equation turbulence models on the temperature field of the shielding induction motor were studied. The results showed that different choices of turbulence models had little effect on the temperature of the winding insulation but influenced the temperature of the lower guide bearing lubricating water and the secondary cooling water outlet. The SST k-ω model showed the lowest relative error result of the temperature of the winding insulation and the bearing lubricating water in the primary loop system of the shielding induction motor. The temperature of the clearance water, the spiral tube water and the spiral groove water increased approximately linearly along the axial direction.

Keywords: shielding induction motor; turbulence models; prototypical experiment; temperature field; bearing lubricating water; numerical analysis

Citation: Ai, L.; Lu, Y.; Han, J.; Suo, W. Simulation of the Temperature of a Shielding Induction Motor of the Nuclear Main Pump under Different Turbulence Models. *Energies* **2023**, *16*, 2792. https://doi.org/10.3390/en16062792

Academic Editor: Gianluca Brando

Received: 12 February 2023
Revised: 10 March 2023
Accepted: 13 March 2023
Published: 17 March 2023

1. Introduction

As one type of green and low-carbon energy, small-capacity nuclear power has a broad space for development in the current global world. The shielding induction motor of the nuclear main pump is used to drive the reactor cooling pump, and its thermal behavior, which is closely related to its lifespan and efficiency, and is crucial for the primary circuit loop system of a nuclear reactor in the nuclear power plant [1–4]. Thermal investigation of the shielding induction motor of the nuclear main pump is in the ascendant.

Generally, the thermal behavior of the motor can be determined by the losses and the cooling capacity [5–8]. The peak temperature of the stator winding insulation does not exceed the permitted temperature of 200 °C, and the temperature of the bearing lubricating water in the primary cooling water is lower than the alarm temperature of 95 °C, two important factors that affect the reliability and lifetime of the thermal design and the management of the shielding induction motor. In addition to prototype testing [7], many studies have been conducted to predict the temperature distribution of a motor with various motor types and operating conditions adopted by different simulation methods [1–8]. For example, Wang et al. used a thermal equivalent network to calculate the actual increase in temperature of the stator winding of a 5.5 MW shielded induction motor [8]. Ding et al. used the finite volume method to study the temperature distribution characteristics of the solid components of canned pump motors [9]. Lu et al. mainly studied the influence of water friction losses using different calculation methods [10], and the influence of the uncertainty of the thermal conductivity of the winding insulation and core lamination on the characteristics of the temperature field for the shielding motor [11]. There was no comparison of results between the numerical simulation by different types of turbulence models [8–11]. For the water clearance in the primary circuit loop, water lubricated bearings are the key supporting equipment in the rotor system. Xie et al. investigated the influences of the laminar model, the turbulent k-ε model, k-kl-Ω model, and k-Ω model on the bearing lubrication [12]. Furthermore, the influence factors of friction losses induced by clearance flow were also investigated for the flywheel [13,14] and the effect of viscosity-temperature on the dynamic characteristics of rotor of the rotor system was studied, the clearance design method between the stator and the rotor cans was also discussed [15], and many studies focused on obtaining the suitable heat transfer coefficient for the machines [16–20]. In addition, the effect of the turbulence model was also carried out on different types of machines [21,22]. Compared to the quasi-steady state RSM model, Kim et al. found that the SST k-ω model showed the lowest error result of the volumetric flow rate distribution of the ducts in a large-capacity high-speed air-cooled induction motor [22].

Based on the above-mentioned literature, the main focus of research on shielding induction motor of the nuclear main pump is mainly on the peak temperature of the winding, its influencing factors, and the flow and heat transfer behavior in water clearance. In addition, limited studies have been conducted on the effects of the selection of turbulence models on temperature prediction, especially for the clearance water temperature. The geometric structure and parameter range of the shielded induction motor studied in this paper are different from those in the previous literature. The research purpose of this paper is to identify the simulation accuracy of the temperature of a shielding induction motor under different turbulence models. The four types of two-equation turbulence models have been selected for numerical simulation of the flow and heat transfer of the shielding motor for the small reactor vertical nuclear main pump. A numerical analysis was also conducted to examine the effects of different turbulence models on the motor temperature with the prototype test results, the relative error analysis was carried out, and the lowest relative error turbulence model suitable for the rotating clearance flow of the shielding motor was finally determined. Additionally, the temperature distribution characteristics of the motor under this turbulence model were studied.

2. Geometrical Model Description

The geometrical model of this project was a small capacity nuclear main pump shielding induction motor, and it was a vertical arrangement. The main parameters of the analyzed motor are indicated in Table 1. The insulation materials at each part were N grade. Considering the structure of the motor and the layout characteristics of the primary and secondary cooling water, a quarter of the 3-D entity motor was selected as the calculation domain model. To make the calculation boundary correspond to reality, the coolant in the main pump of the reactor under the motor was also included in the calculation domain, as shown in Figure 1a.

Table 1. Main parameters of the small capacity shielding induction motor.

Parameters (Unit)	Value
Rated current (A)	78.3
Rated rotational speed (r/min)	2923
Rated frequency (Hz)	50
Number of stator slots	30
Number of rotor slots	22
Number of poles	2
Stator core length (mm)	660
Water clearance thickness (mm)	2

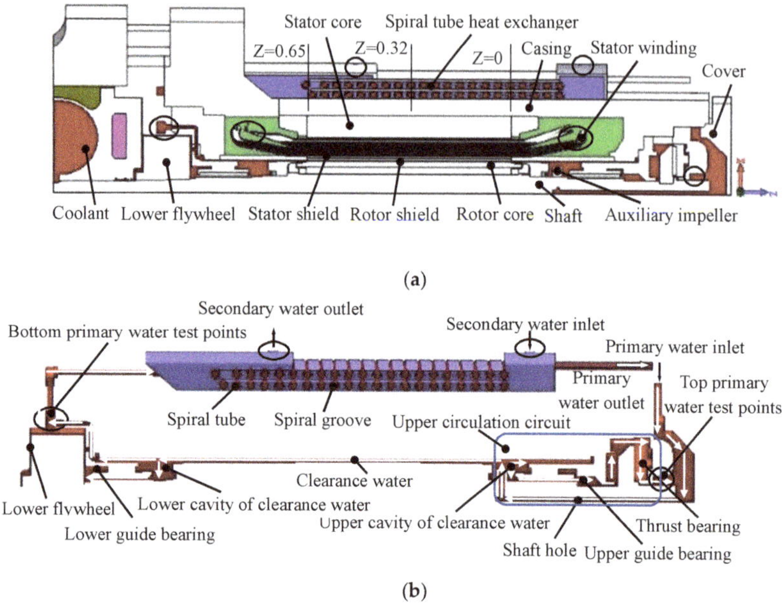

Figure 1. Geometrical model of the solution region. (**a**) Model of shielding induction motor. (**b**) Waterway schematic.

Because the motor cooling capacity is limited, there were primary and secondary cooling water paths within the motor and its spiral heat exchanger. Primary cooling water that flows out of the spiral heat exchanger was sucked into the duct of the upper cover and down through the shaft hole above. The flow rate of primary cooling water depended mainly on the rotation speed of the auxiliary impeller to generate a driving pressure head. The water flow thrown away by the Coriolis force was divided into two parts at the outlet of the auxiliary impeller. A part turned to flow upward, lubricating the upper guide bearing and the thrust bearing, and then returned to the duct of the top cover to form an upper circulation circuit. The other part flowed downward, through the clearance of the stator and rotor shield, the lower guide bearing, and then into the four spiral tubes outside the casing, forming the main cooling loop of the motor. The spiral heat exchanger was directly wrapped outside the motor casing and is arranged in the same height position as the iron core. Secondary cooling water from the upper inlet flowed down into the four spiral grooves within a double layer between spiral tubes and baffled, conducting a countercurrent heat transfer process with the high-temperature primary cooling water in the spiral tubes, and exited at the outlet of the heat exchanger. The waterway structure of the primary and secondary cooling water is shown in Figure 1b.

3. Mathematical Model and Solution Conditions

3.1. Basic Assumptions

To simplify the solution process, during the cooling water flow in the motor, the flow was in a turbulent state, and it can be regarded as an incompressible fluid treatment because the Mach number was much less than 0.7, and the electric losses as a heat source in the components of the motor generated were assumed to be uniformly distributed in the simulation process. The stator winding was composed of multiturn coils. Each side of the turn had multiple flat strands in parallel for the purpose of suppressing the skin effect, which reduced the eddy current losses [23]. For the convenience of calculation, the equivalent simplification of the stator winding model was made without changing the actual physical process [24].

3.2. Mathematical Model

When the shielding induction motor was in operating condition, the water of the shaft hole rotated together with the shaft. Under the action of shear force, Taylor-Couette-Poisson flow would be formed in the water clearance of the bearing lubricating and between the stator and the rotor shield. For the fluid region, the three-dimensional incompressible flow is described based on the Reynolds-averaged Navier–Stokes equation (RANS). The N-S equation combined with different types of turbulence model is the most widely adopted by the CFD numerical simulation approach. Three-dimensional flow and heat transfer coupling equations including mass, momentum, and energy conservation equations can be sketched as follows.

$$\frac{\partial(\rho u_i)}{\partial x_i} = 0 \tag{1}$$

$$\frac{\partial(\rho u_i u_j)}{\partial x_j} = -\frac{\partial P}{\partial x_i} + \frac{\partial}{\partial x_j}\left(\mu \frac{\partial u_i}{\partial x_j} - \rho\overline{u_i' u_j'}\right) \tag{2}$$

$$\frac{\partial(\rho u_i T)}{\partial x_i} = \frac{\partial}{\partial x_i}\left(\frac{\lambda}{c_p}\frac{\partial T}{\partial x_i}\right) + S_T \tag{3}$$

where, $i, j = 1, 2, 3$, x_i, x_j correspond to the three coordinate components x, y, z, u_i, u_j is the velocity component in the direction of x, y, z, p is the pressure, S_T is the heat source term, T is the temperature, ρ is the density, c_p is the specific heat at constant pressure, λ is the thermal conductivity. μ is the dynamic viscosity.

The four types of turbulence models with detailed two transport equations selected in this work are listed in Table 2 [25]. These models represent the turbulent properties of the flow. The first transported variable is the turbulent kinetic energy k, and the second transported variable is ω or ε.

Table 2. Details of four types of turbulence models.

Types of Models	Value
Standard k-ε	$\frac{\partial(\rho k u_i)}{\partial x_i} = \frac{\partial}{\partial x_j}\left[\left(\mu + \frac{\mu_t}{\sigma_k}\right)\frac{\partial k}{\partial x_j}\right] + G_k + G_b - \rho\varepsilon - Y_M + S_k$ $\frac{\partial(\rho \varepsilon u_i)}{\partial x_i} = \frac{\partial}{\partial x_j}\left[\left(\mu + \frac{\mu_t}{\sigma_\varepsilon}\right)\frac{\partial \varepsilon}{\partial x_j}\right] + G_{1\varepsilon}\frac{\varepsilon}{k}(G_k + C_{3\varepsilon}G_b) - C_{2\varepsilon}\rho\frac{\varepsilon^2}{k} + S_\varepsilon$
RNG k-ε	$\frac{\partial(\rho k u_i)}{\partial x_i} = \frac{\partial}{\partial x_j}\left(\alpha_k \mu_{eff}\frac{\partial k}{\partial x_j}\right) + G_k + G_b - \rho\varepsilon - Y_M + S_K$ $\frac{\partial(\rho \varepsilon u_i)}{\partial x_i} = \frac{\partial}{\partial x_j}\left(\alpha_\varepsilon \mu_{eff}\frac{\partial \varepsilon}{\partial x_j}\right) + C_{1\varepsilon}\frac{\varepsilon}{k}(G_k + C_{3\varepsilon}G_b) - C_{2\varepsilon}\rho\frac{\varepsilon^2}{k} - R_\varepsilon + S_\varepsilon$
standard k-ω	$\frac{\partial}{\partial x_i}(\rho k u_i) = \frac{\partial}{\partial x_j}\left(\Gamma_k \frac{\partial k}{\partial x_j}\right) + G_K - Y_K + S_K$ $\frac{\partial}{\partial x_i}(\rho \omega u_i) = \frac{\partial}{\partial x_j}\left(\Gamma_\omega \frac{\partial \omega}{\partial x_j}\right) + G_\omega - Y_\omega + S_\omega$
SST k-ω	$\frac{\partial}{\partial x_i}(\rho k u_i) = \frac{\partial}{\partial x_j}\left(\Gamma_k \frac{\partial k}{\partial x_j}\right) + G_K - Y_K + S_K$ $\frac{\partial}{\partial x_i}(\rho \omega u_i) = \frac{\partial}{\partial x_j}\left(\Gamma_\omega \frac{\partial \omega}{\partial x_j}\right) + G_\omega - Y_\omega + D_\omega + S_\omega$

In the foregoing equations, G_k and G_ω are the generations of k and ω due to the mean velocity gradients. G_b is the k production due to buoyancy, Y_M is the contribution of fluctuating expansion in compressible turbulence at the rate of global dissipation, $G_{1\varepsilon}$, $G_{2\varepsilon}$, and $G_{3\varepsilon}$ are constants, σ_k and σ_ε are the same as in the standard k-epsilon model, σ_k and σ_ε are turbulent Prandtl numbers, S_k, S_ε, and S_ω are user-defined source terms, α_k and α_ε inverse effective Prandtl numbers, μ_{eff} is effective viscosity, Γ_k and Γ_ω are the effective diffusivity of k and ω, respectively, Y_k and Y_ω are the dissipation of k and ω due to turbulence, D_ω is the cross-diffusion term. The units of quantities encountered in this paper were used by the SI system.

3.3. Solution Conditions

3.3.1. Boundary Conditions

In the shielding induction motor test, the corresponding primary and secondary cooling water inlet temperature were set as 32.6/19.7 °C, and the inlet flow rate of primary and secondary cooling water in the internal clearance and heat exchanger grooves of the motor was 10/14.2 m^3/h. The gauge pressure at the primary and secondary cooling water outlet was set at 0 Pa. The outer wall of the physical model was exposed to ambient room air of 25 °C, the combined convection and radiation heat transfer coefficient was set at 10 W/(m^2·K), and the surface emissivity of the steel wall was taken as 0.8. The temperature of the reactor coolant was maintained at 287.3 °C, see Figure 1a. On the left and right boundary surfaces of the geometric model, the primary cooling water in the spiral tube and the motor and the secondary cooling water in the spiral groove are periodically connected one by one to form a continuous flow channel of primary and secondary cooling water. Figure 2 shows the schematic diagram of the periodic connection of the spiral tube, in which the radius difference and the height difference of the spiral line were ignored, as well as the corresponding influence on the temperature and velocity distribution. After periodic connections, the height, pitch, and number of spiral tubes in the model remain unchanged. The yellow, blue, green, and gray spiral tubes were connected in turn to form a connected flow channel.

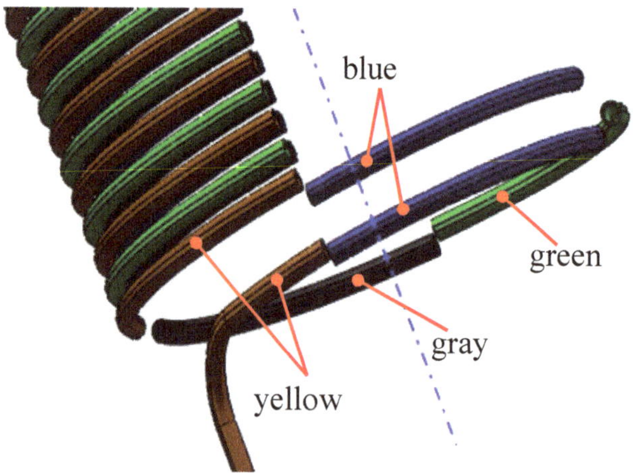

Figure 2. Schematic diagram of the periodic connection of spiral tubes on the left and right sides.

3.3.2. Heat Source and Physical Property Parameters

In addition to the above conditions, the electric losses in the motor were vital heat source data leading to the increase in temperature of solid components, including rotor and stator shield losses, stator winding losses, rotor copper losses, rotor end ring losses, iron losses, etc. The values of electric losses and eddy current losses in the main components could be obtained by the finite element method (FEM) [26] at a rated speed of 2923 r/min, and the thermal conductivity λ of typical components of the motor are shown in Table 3. The heat source was assigned according to the body average. Due to the high rotational speed, the friction losses of the waters were also considered and calculated using a modular program embedded in fluent software [10]. The thermal conductivity of the stator and rotor core lamination was anisotropic, while the others were isotropic. Parameters of the properties of primary and secondary cooling water, such as thermal conductivity, density, specific heat capacity, and dynamic viscosity, were set linearly with temperature according to the fitting curve by UDF under the gauge pressure 15.05/0.41 MPa.

Table 3. Values of electric losses and thermal conductivity of main components of the motor.

Components	Electric Losses (kW)	Thermal Conductivity(W/m·k)
Stator shield	29.27	$\lambda = 0.015T + 9.697$
Rotor shield	8.308	up ↑
Rotor copper bar	10.32	$\lambda = 0.14T + 115.16$
Rotor core	0.49	$\lambda_{radial,\ tangent} = 23.6$, $\lambda_{axial} = 3.2$
Stator core	0.9082	up ↑
Stator winding	6.17	$\lambda = 400.99 - 0.08T$
Rotor end ring	0.454	up ↑
Insulation	—	0.18

The mean of 'up ↑' is the same as the above data.

3.4. Solving Method

The verification of grid independence was detailed, and finally an independent solution was obtained. For parts with a large influence on temperature rise, the mesh was finely divided. Four groups of grids with different numbers were designed to verify independence, and the total number of grids in the calculated domain eventually chosen was approximately 3.54 million [11]. ANSYS FLUENT 19.1 was applied in the numerical simulations. The mass, pressure, momentum, and energy equations were each discretized using a second-order upwind scheme. A separation and implicit solution scheme based on a pressure solver was adopted. The SIMPLE scheme was used for pressure–velocity coupling, the convergence residuals of the discrete equations of flow and temperature field were 1×10^{-3} and 1×10^{-6}, respectively. The convective heat transfer coefficient between the coolant and the motor wall was automatically calculated according to the convective heat transfer and conduction equilibrium equations in the software module program.

4. Results and Discussion

4.1. Prototype Test

To obtain comprehensive performance, a multi-parameters test rig of the 0.24 MW shielding induction motor was established in the factory for the reactor coolant pump for small reactors, as shown in Figure 3. As previously mentioned, the water temperature for bearings lubricating and the insulation temperature were the key testing parameters for performance. Therefore, four temperature test points were installed in the lubricating water above the thrust bearing through the end cover considering that the water clearance was too narrow and the inconvenience of arrangement of the measuring holes and were called top primary cooling water test points for brevity. At the same time, the four test points were installed in the lower annular duct at 90° intervals along the circumferential direction under the lower position of the motor to monitor the water temperature of the lubricated lower guide bearing not exceeding the alarming temperature for real time, and they were called bottom primary cooling water test points (see Figure 1b). For the thermo-safety of stator winding insulation, a total of 6 temperature test points were embedded on the surface of the insulating material at the centers of the winding nose with the worst heat dissipation in the upper and lower end rooms of the stator winding based on previous research and experience, and this means that the insulation temperature measured can be taken as the peak temperature of the motor. Furthermore, two water temperature test points were also installed at the position of the inlet and outlet of secondary cooling water. The specific test point positions are shown in the circles in Figures 1 and 3.

Before the test, all measuring instruments were calibrated and the PT100 platinum thermistor sensor four-wire was used. The error of the test instrument itself is very small. In addition to the inlet and outlet temperature of the secondary cooling water, there were many other test points and the accidental error was relatively small. The position of the sensor installation at the top and bottom of the primary cooling water measuring points is shown in Figure 3b,c. When the sensors were installed, it should ensure that they were closely attached to the ambient wall to avoid thermal contact resistance, which was convenient for heat conduction and reduces the response time of the temperature change.

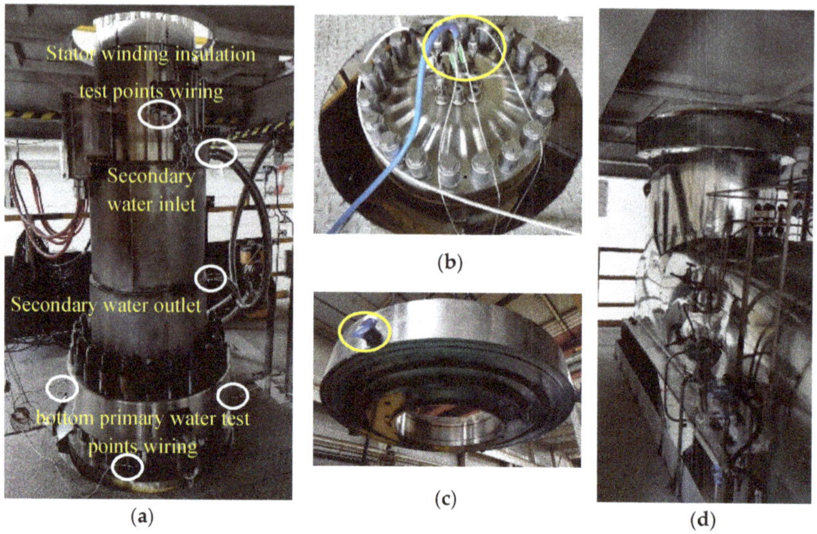

Figure 3. Experimental equipment and locations of the test point. (**a**) Test rig on the second floor. (**b**) Top primary cooling water test points wiring. (**c**) Bottom primary cooling water test points annular duct. (**d**) Test loop on the first floor.

The test system scheme is shown in Figure 4. The main pump coolant circuit, flowmeters, and main control valve system were installed on the first floor, and the accompanying test loop was connected. The motors pressure and temperature measurements module system were installed on the second floor. They can also be seen in the pictures shown in Figure 3d.

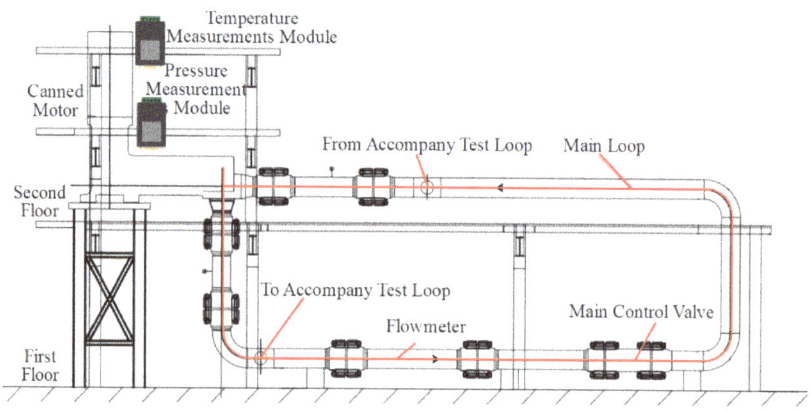

Figure 4. Schematic diagram of the motor test rig.

The motor prototypical experiment has been conducted for 48 h continuously in the steady thermal rated operating condition. Fragments extracted from the test data of some parameters recorded by the instrument in the 300 s interval of the steady state are shown in Figure 5.

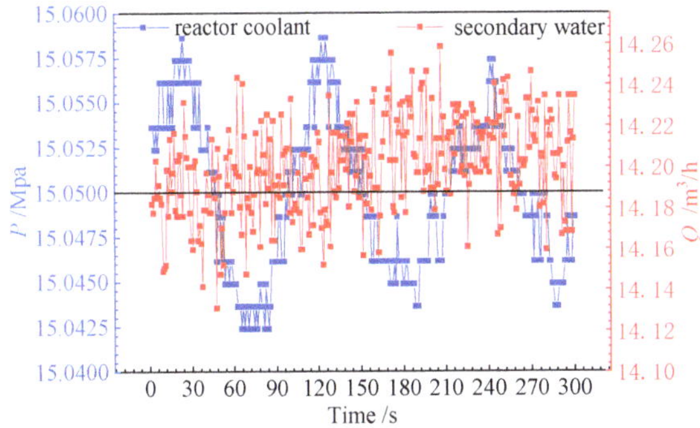

Figure 5. Extracted fragments from experimental data.

Figure 6 shows the average and standard deviation of the test data for the temperature of the upper and lower end rooms of the stator winding insulation test points ($T_{\text{up-ins}}/T_{\text{lo-ins}}$), the temperature of the top primary cooling water test points (T_{top}), and the temperature of the bottom primary cooling water test points (T_{bottom}). The error bar can visually display the uncertainty of the test data. In Figure 6, it can be seen that the standard deviation of the T_{top} measurement point was the smallest (0.072), and the standard deviation of the $T_{\text{low-ins}}$ measuring points was the largest (1.029). This shows that the test data were relatively stable and reliable. The temperature of the secondary cooling water outlet test points (T_{second}) was 28.17 °C. In the subsequent analysis, the results simulated by different types of turbulence models were compared with the test averages of the prototype machine according to the same position in Figure 6.

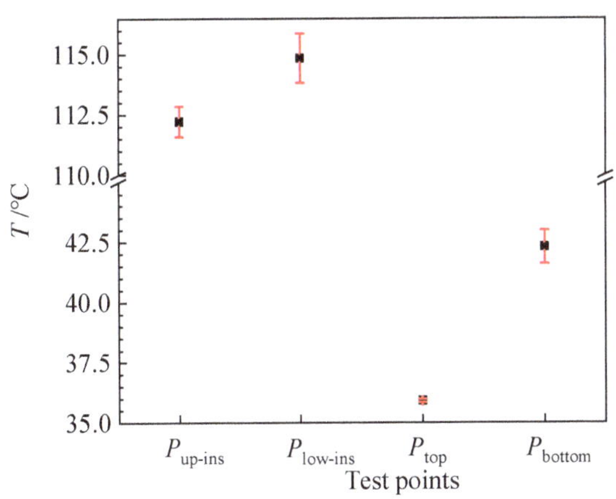

Figure 6. Temperature error bar curve of the test points.

4.2. Effect of Turbulence Models on the Temperature of Stator Winding Insulation

In this project, the primary cooling water was in a highly turbulent state, the rotational Reynolds number and the axial Reynolds number were 9.3×10^4 and 5.6×10^3, respectively, which belong to the Taylor-Couette-Poiseuille flow in the motor. For convenience in subsequent analysis, the relative error δ is defined as the ratio of the absolute error of the simulated value δ to the average tested value $\overline{T_{\text{test}}}$ multiplied by 100%. Generally speaking, the magnitude of relative error can better reflect the credibility of the measurement, as shown in Formula (4):

$$\delta = \frac{T_{\text{simu}} - \overline{T_{\text{test}}}}{\overline{T_{\text{test}}}} \times 100\% \tag{4}$$

As mentioned above, the test temperature of the stator winding insulation could be taken as the peak temperature of the motor. Therefore, it was one of the key parameters for verifying the accuracy of the predictions. Figure 7 showed δ the relative error of the temperature of the stator winding insulation adopted by four types of turbulence models by CFD simulation.

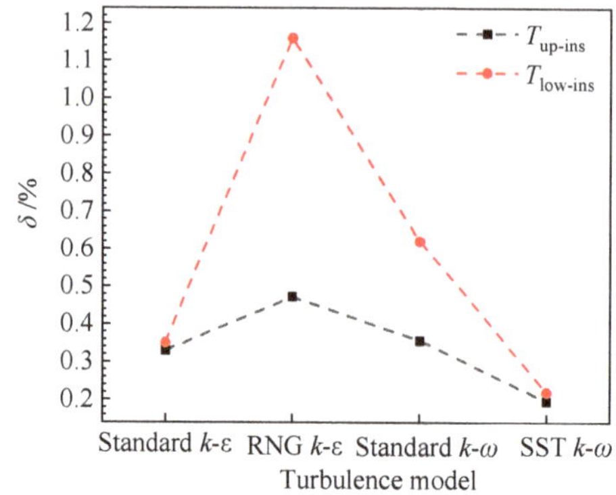

Figure 7. Relative error of the peak temperature of the stator winding insulation with four turbulence models.

It can be seen from Figure 7 that the selection of four types of turbulence models had little effect on the temperature of the stator winding insulation by the finite volume method. The temperature of stator winding insulation was affected by both the convective and thermal conduction resistance. Part of the heat generated by the stator end winding was transferred to its surface insulation by heat conduction. The weak natural convection existed on the surface of the insulation in the end rooms of the stator winding. Another part of the heat was transferred to the stator shield and casing through heat conduction. The heat conduction path was relatively long. The turbulent transport equation mainly affected the convective heat transfer between the rotating clearance water and the stator and rotor shield. The heat source density of the stator shield (157.1 MW/m^3) was much higher than that of the stator winding (0.23 MW/m^3). The above reasons led to the temperature of the winding insulation not being sensitive to change in the turbulence model.

The relative error of the insulation simulation results compared to the experimental results is observed in Figure 7 by the SST k-ω model (0.22%, in the lower end room) followed by the standard k-ε (0.35%, in the lower end room) and the standard k-ω (0.62%, in the lower end room) models, whereas the RNG k-ε model provided the maximum relative error of 1.16% (in the lower end room). The error of the RNG k-ε model was relatively large, includes false diffusion error, discrete error, and rounding error. However, the maximum relative error did not exceed 2%. The relative error of the upper end room of the stator winding insulation was smaller than that of the lower part of the stator. Comparatively speaking, it is more accurate to use the SST k-ω model to calculate the peak temperature of the stator winding insulation of the shielding induction motor.

4.3. *Effect of Turbulence Models on the Temperature of Primary and Secondary Cooling Water*

The relative error of the temperatures of the primary and secondary cooling water with four turbulence models is shown in Figure 8. It can be seen from Figure 8 that the relative error of T_{top} was large, but it basically did not change with the turbulence model. The value of T_{top} calculated by four types of turbulence model was basically the same. The minimum relative error compared to the experimental results was observed in the SST k-ω model (−9.078%) followed by the RNG k-ε (−9.173%) and standard k-ε (−9.179%) models, whereas the standard k-ω model provided the maximum relative error of −9.184%, all of which were lower than the experimental results. The temperature rise of the lubricating water of the thrust bearing was mainly caused by the friction losses of the water. Due to its narrow internal water clearance (see Figure 1b), the four temperature measuring sleeves inserted at this place were ignored when the physical model was established, resulting in the simulated value of flow resistance being less than the prototype test value. The simulated value of the T_{top} was lower than the actual value and the relative error was negative. In addition, the top primary cooling water test points were far away from the heat generating parts of the motor, and their position was near the primary cooling water inlet. The mixing effect of some primary cooling water and the water from the measuring point was obviously stronger than that of turbulent convection heat transfer. The effect of the four turbulence models on the temperature of the top primary cooling water test points was not distinct.

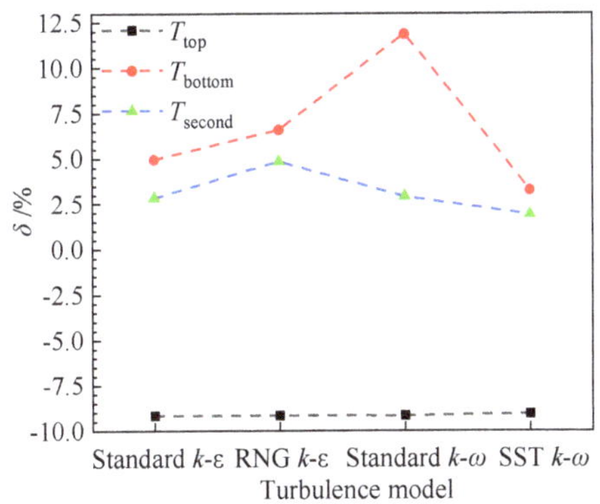

Figure 8. Relative error of primary and secondary cooling water temperatures with four turbulence models.

Furthermore, it can be seen from Figure 8 that the T_{bottom} calculated by four types of turbulence models was different, and the relative error was relatively larger than the T_{second}. The minimum relative error compared to the experimental results was observed in the SST k-ω model (3.26%) followed by the standard k-ε (4.96%) and RNG k-ε (6.59%) models, whereas the standard k-ω model provided the maximum relative error of 11.84%. This was due to the relatively strong sensitivity of the calculation results of the standard k-ω model, which depended on the k and ω values of the free flow inside and outside the shear layer, which must be corrected.

Moreover, the T_{second} calculated by four types of turbulence models was different. The minimum relative error compared to the experimental results was observed in the SST k-ω model (1.92%) followed by the standard k-ε (2.85%) and standard k-ω (2.91%) models, whereas the RNG k-ε model provided the maximum relative error of 4.85%. The relative error of the SST $k-\omega$ model was the smallest.

This result may have been due to the better boundary layer flow analytical ability of the SST k-ω model compared to other turbulence models. Accounting for the transport of the turbulence shear stress in the definition of turbulent viscosity and incorporating a damped cross-diffusion derivative term in the ω equation, these features made the SST k-ω model more accurate and reliable for a narrow clearance water flow with a rotating inner cylinder and the temperature field for the shielding induction motor with spiral tube heat exchanger outside the casing.

From the above analysis, it can be concluded that the SST k-ω model is suitable to simulate the temperature field in a combination of a shielding induction motor and a spiral heat exchanger, followed by the standard k-ε, standard k-ω, and RNG k-ε, which are the worst.

4.4. Temperature Distribution Characteristics of the Motor

The temperature of the solid parts of the motor was mainly determined by the magnitude of the heat source and the cooling capability. In this project, the selection of four types of turbulence models for numerical simulation had little effect on the temperature and distribution characteristics of the solid parts of the shielding induction. Figure 9 shows one of the typical temperature distributions of the computational domain model adopted with the SST k-ω model. It can be seen from Figure 9 that the reactor coolant pump is located below the motor [11]. The lower clearance water of the flywheel was in direct contact with the reactor coolant water leaked from the labyrinth seal. The heat of the high-temperature coolant was transferred to the rotor and stator parts above due to a temperature difference, resulting in the temperature of the winding and its insulation at the lower stator end being relatively higher than that at the upper stator end. In addition, although the electric losses inside the shielding induction motor were mainly generated by components at the middle position of the stator and rotor segment, such as the iron core, winding, stator, rotor shield, and so on, the heat was mainly transferred to the wall near the cooling water by conduction, then it was carried away mainly by rotational convective heat transfer of the primary cooling water in the annular clearance between the stator and rotor shield, the temperature of the winding and its insulation was lower in the middle position of the core than in the end room, but significantly higher than ambient other parts. The temperature of the casing in the middle position of the core was also a key parameter for thermal stress.

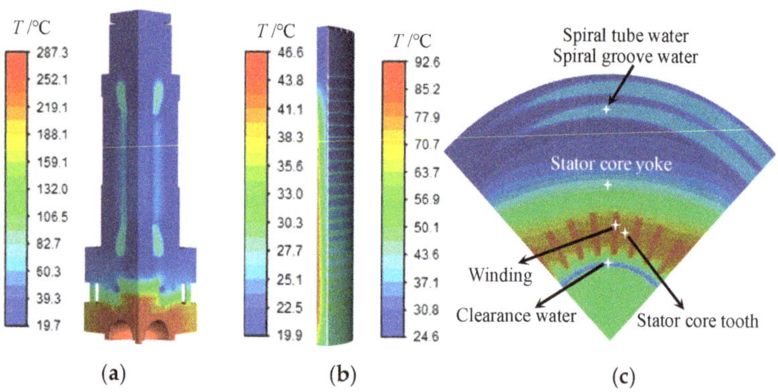

Figure 9. Temperature contour of the shielding induction motor. (**a**) A quarter of the motor. (**b**) Casing. (**c**) Sampling line position and temperature contour of cross section at z = 0.32 m.

Figure 9b shows the temperature contour of the casing extract of Figure 9a. It can be seen from Figure 9b that the inner surface of the casing ambient the stator core was with high temperature, and the temperature gradually decreased along the radius direction with about 20 °C temperature drops. The accurate temperature distribution of the casing provided important basic data for thermal stress calculation.

The spiral heat exchanger was located between the upper and lower stator end room in the middle of the motor, which was the main position of the heat source of the motor and the primary and secondary cooling water heat exchange. Therefore, it was necessary to study the temperature distribution characteristics in the middle of the motor. Figure 9c shows the temperature contour of the cross section at z = 0.32 m. It can be seen from Figure 9c that the temperature of the stator winding was the highest, and the temperature of the solid parts decreased gradually in both directions of increasing and decreasing radius. The temperature of the primary cooling water in the spiral tube of the heat exchanger was higher than the temperature of the secondary cooling water and the casing, the temperature difference between the casing and the secondary cooling water was small, and the heat exchanger had good heat exchange performance. At the same time, a total of six sampling lines for analysis were established along the axial range z = 0–0.65 m of the motor (see

Figure 1), through the radial center position of the clearance water, the stator core tooth, the stator winding, the stator core yoke, the spiral tube water, the spiral groove water (see Figure 9c), and the basic characteristic curve of the temperature distribution along the axial direction is given in Figure 10.

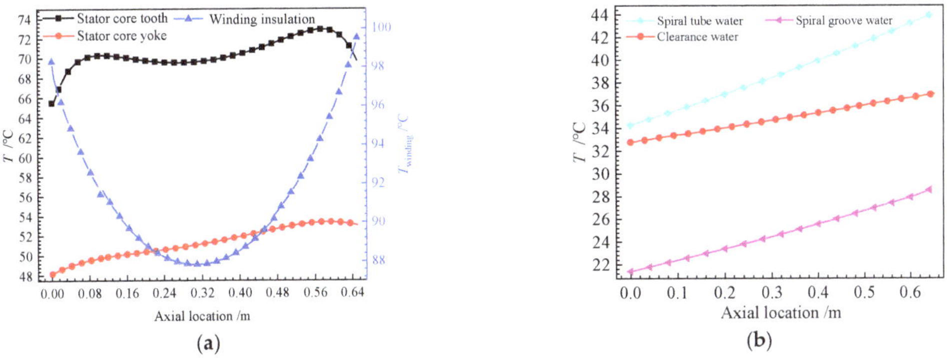

Figure 10. Temperature distribution on the sampling lines. (**a**) Solid components. (**b**) Primary and secondary cooling water.

It can be seen from Figure 10a that the temperature curve of the stator winding was low in the middle and high at both ends, and the lower end was higher than the upper end. The distribution characteristics of the stator winding temperature were the same as those previously studied [10]. The temperature curve of the stator core yoke gradually increased along the axial position and basically changed linearly. Because of its small heat source density, the temperature was low. The temperature curve of the stator core tooth increased non-linearly along the axial direction. The temperature of the lower end of the stator core tooth was higher than that of the upper end and its heat flux density was higher, so the temperature was higher. In addition to the upper and lower end positions, the temperature of the core lamination was lower than that of the stator winding, which absorbed the heat from the stator winding. The axial thermal conductivity of the laminated core was relatively small, and the axial temperature difference was small. The accurate temperature distribution of the stator core could also provide important basic data for the thermal stress calculation of the motor.

It can be seen from Figure 10b that the temperature curves of the clearance water, the spiral tube water, and the spiral groove water curves increased linearly along the axial direction. The temperature of the spiral tube water was the highest, followed by the clearance water, and the spiral groove water was the lowest. Therefore, the low temperature secondary cooling water in the spiral groove could absorb the heat of the high temperature primary cooling water in the spiral tube, and the heating rates of the secondary and primary cooling water were basically the same, which were higher than the heating rate of the clearance water.

Figure 11 presents the temperature contour of primary cooling water in the computational domain and some periodic boundary at the middle position using the SST k-ω model models. It can be seen from Figure 11a that the temperature of the primary cooling water gradually increased from top to bottom in the motor, while, on the contrary, the temperature gradually decreased in the spiral tube of the heat exchanger. The maximum temperature of primary cooling water was located on the coplanar cell of the local adjacent body at the lower annular ducts near the high temperature coolant below, the maximum simulated temperature was 91.8 °C.

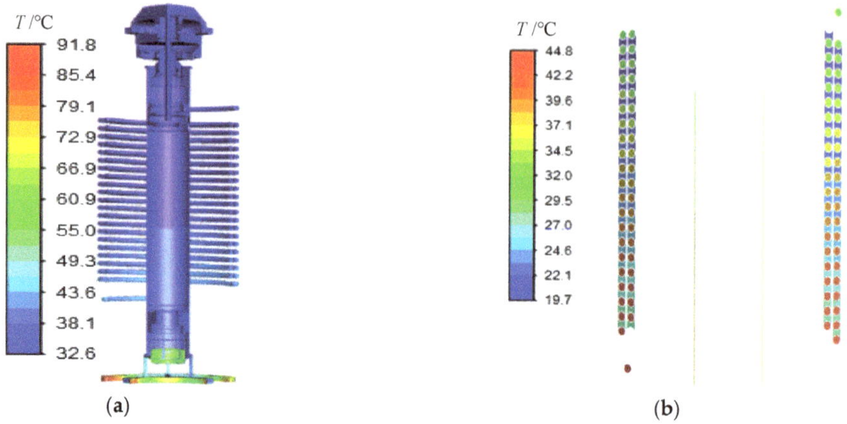

Figure 11. Temperature contour of the water. (**a**) Primary cooling water. (**b**) Primary and secondary cooling water on both sides of the periodic boundary.

Meanwhile, it can be seen from Figure 11b that the maximum temperature of the primary cooling water from the motor to the heat exchanger was 44.8 °C. Secondary cooling water in the spiral groove gradually flowed from the upper inlet section to the bottom, and the maximum temperature was at the bottom of the heat exchanger, the primary cooling water inlet position. The characteristics of the temperature distribution for the primary and secondary cooling water in the spiral tube and groove were consistent with those of the countercurrent heat exchanger in this project.

5. Conclusions

To ensure accurate thermal design of the shielding induction motor and obtain temperature characteristics, the four kinds of two equation turbulence models (namely the standard k-ε model, the RNG k-ε model, the standard k-ω, and the SST k-ω model) were adopted to predict the temperature field results of the shielding induction motor in CFD analyses, and the results were compared with experiments data. Some main conclusions can be obtained as below:

(1) Changes in the four types of turbulence models had little effect on the temperature of the winding insulation. The maximum relative error was 1.16% and the minimum relative error provided by the SST k-ω model was 0.22%.

(2) The selection of the turbulence model influenced the temperature of the lower guide bearing lubricating water and the temperature of the secondary cooling water outlet. The SST k-ω model showed the lowest relative error result of the temperature of the lower guide bearing lubricating water and the secondary cooling water outlet. The minimum relative error provided by the SST k-ω model was 3.26% and 1.92%, respectively. The SST k-ω model is more suitable for simulation of the temperature field of the shielded induction motor.

(3) The temperature in the clearance water, the spiral tube water, the spiral groove water, and the stator core yoke increased approximately linearly along the axial direction in the position of the spiral heat exchanger and the stator core located. Accurate calculation and research of the temperature field will provide basic data for the thermal stress calculation of the shielding induction motor. This paper will lay the foundation for the efficient and stable operation of shielded induction motor.

Author Contributions: L.A. and Y.L. are the main authors of this manuscript. All the authors contributed to this manuscript. Y.L. and L.A. conceived the novel idea; L.A. and J.H. performed the analysis; L.A. analyzed the data and contributed analysis tools; W.S. provided experimental data; L.A. wrote the entire paper. Y.L. checked, reviewed and revised the paper. All authors have read and agreed to the published version of the manuscript.

Funding: This research was funded by the Heilongjiang Education Department special Project in Funda-mental Business, grant number 135509110.

Data Availability Statement: The measurement data presented in this study are not publicly available due to restrictions of privacy.

Conflicts of Interest: The authors declare no conflict of interest.

References

1. Kim, C.; Lee, K. Numerical investigation of the air-gap flow heating phenomena in large-capacity induction motors. *Int. J. Heat Mass Transf.* **2017**, *110*, 746–752. [CrossRef]
2. Janga, J.; Chiub, H.; Yanet, W.; Tsai, M.C.; Wang, P. Numerical study on electromagnetics and thermal cooling of a switched reluctance motor. *Case Stud. Therm. Eng.* **2015**, *6*, 16–27. [CrossRef]
3. Bonnett, A.H. Operating temperature considerations and performance characteristics for IEEE 841 motors. In Proceedings of the Industry Applications Society Forty-Seventh Annual Conference 2002, Houston, TX, USA, 13 September 2002; pp. 1120–1131.
4. Yoon, M.K.; Jeon, C.S.; Kauh, S.K. Efficiency increase of an induction motor by improving cooling performance. *IEEE Trans. Energy Convers.* **2002**, *17*, 1–6. [CrossRef] [PubMed]
5. Rehman, Z.; Seong, K. Three-D Numerical Thermal Analysis of Electric Motor with Cooling Jacket. *Energies* **2018**, *11*, 95. [CrossRef]
6. Gundabattini, E.; Mystkowski, A.; Idzkowski, A. Thermal Mapping of a High-Speed Electric Motor Used for Traction Applications and Analysis of Various Cooling Methods—A Review. *Energies* **2021**, *14*, 1472. [CrossRef]
7. Boglietti, A.; Mandrile, F.; Carpaneto, E.; Popescu, M.; Staton, D. Stator Winding Second-Order Thermal Model including End-Winding Thermal Effects. *Energies* **2021**, *14*, 6578. [CrossRef]
8. Wang, D.; Liang, Y.; Li, C.; Yang, P.; Zhou, C.; Gao, L. Thermal equivalent network method for calculating stator temperature of a shielding induction motor. *Int. J. Therm. Sci.* **2020**, *147*, 106–149. [CrossRef]
9. Ding, S.; Meng, F.; Ge, Y. Temperature field investigation of canned primary pump motors in nuclear power stations. *Proc. CSEE* **2012**, *32*, 149–155.
10. Lu, Y.; Mustafa, A.; Rehan, M.A.; Razzaq, S.; Ali, S.; Shahid, M.; Adnan, A.W. The effects of water friction loss calculation on the thermal field of the canned motor. *Processes* **2019**, *7*, 256. [CrossRef]
11. Lu, Y.; Ai, L.; Han, J.; Wang, B.; Suo, W. Sensitivity Investigation of Uncertainty Factors to Temperature Rise of Nuclear Main Pump Motor for Small Reactor. *Proc. CSEE* **2022**, 1–10.
12. Xie, Z.; Song, P.; Hao, L.; Shen, N.; Zhu, W.; Liu, H.; Shi, J.; Wang, Y.; Tian, W. Investigation on effects of Fluid-Structure-Interaction (FSI) on the lubrication performances of water lubricated bearing in primary circuit loop system of nuclear power plant. *Ann. Nucl. Energy* **2020**, *141*, 107355. [CrossRef]
13. Xu, R.; Song, Y.; Gu, X.; Wang, Y.; Wang, D. Research on the flywheel clearance flow in canned motor RCP. *Ann. Nucl. Energy* **2021**, *150*, 107816. [CrossRef]
14. Wang, Y.J.; Wang, D.Z.; Guo, W.Z.; Yin, J.L.; Hu, Y.Y. The effect of smaller turbulent motions on heat transfer in the annular gap flow of flywheel. *Ann. Nucl. Energy* **2016**, *91*, 1–7. [CrossRef]
15. Xu, R.; Song, Y.; Gu, X.; Linb, B.; Wang, D. Research on the clearance flow between stator and rotor cans in canned motor RCP. *Ann. Nucl. Energy* **2021**, *164*, 108583. [CrossRef]
16. Howey, D.A.; Childs, P.R.N.; Holmes, A.S. Air-Gap Convection in Rotating Electrical Machines. *IEEE Trans. Ind. Electron.* **2012**, *59*, 1367–1375. [CrossRef]
17. Luo, G.; Yao, Z. Decoupling tests on axial heat-transfer in highly turbulent Taylor-Couette flow using thermal waves. *Exp. Therm. Fluid Sci.* **2021**, *128*, 110439. [CrossRef]
18. Leng, X.Y.; Krasnov, D.; Li, B.W.; Zhong, J.Q. Flow structures and heat transport in Taylor–Couette systems with axial temperature gradient. *J. Fluid Mech.* **2021**, *920*, A42. [CrossRef]
19. Staton, D.A.; Cavagnino, A. Convection Heat Transfer and Flow Calculations Suitable for Electric Machines Thermal Models. *IEEE Trans. Ind. Electron.* **2008**, *55*, 3509–3516. [CrossRef]
20. Sun, P.; Lu, Y.; Tong, J.; Lu, Y.; Liang, T.; Zhu, L. Study on the Convective Heat Transfer and Fluid Flow of Mini-Channel with High Aspect Ratio of Neutron Production Target. *Energies* **2021**, *14*, 4020. [CrossRef]
21. Lu, Y.; Ge, Y.; Wang, H.; Wang, F. Verification and analysis on turbulence model of flow field simulation for multiple ventilation stator ducts of air- cooled turbo generators. *Electr. Mach. Control.* **2018**, *22*, 63–69.
22. Kim, C.; Lee, K.; Yook, S. Effect of air-gap fans on cooling of windings in a large-capacity, high-speed induction motor. *Appl. Therm. Eng.* **2016**, *100*, 658–667. [CrossRef]

23. Khang, H.V.; Arkkio, A. Eddy-current loss modeling for a form-wound induction motor using circuit model. *IEEE Trans. Magn.* **2012**, *48*, 1059–1062. [CrossRef]
24. Simpson, N.; Wrobel, R.; Mellor, P.H. An accurate mesh-based equivalent circuit approach to thermal modeling. *IEEE Trans. Magn.* **2014**, *50*, 269–272. [CrossRef]
25. Bazgir, A.; Nikou, M.K.; Heydari, A. Numerical CFD analysis and experimental investigation of the geometric performance parameter influences on the counter-flow Ranque-Hilsch vortex tube (C-RHVT) by using optimized turbulence model. *Heat Mass Transf.* **2019**, *55*, 2559–2591. [CrossRef]
26. Gao, L.; Wang, D.; Liang, Y.; Wang, C. Evaluation of can eddy current losses and can circulating current losses between the can and end structures in double canned induction motors. *IET Electr. Power Appl.* **2019**, *13*, 635–642. [CrossRef]

energies

MDPI

Article

Power Output Optimisation via Arranging Gas Flow Channels for Low-Temperature Polymer Electrolyte Membrane Fuel Cell (PEMFC) for Hydrogen-Powered Vehicles

James Chilver-Stainer [1], Anas F. A. Elbarghthi [1,2], Chuang Wen [1,*] and Mi Tian [1,*]

[1] Faculty of Environment, Science and Economy, University of Exeter, Exeter EX4 4QF, UK; jc1119@exeter.ac.uk (J.C.-S.); anas.elbarghthi@tul.cz (A.F.A.E.)

[2] Department of Applied Mechanics, Faculty of Mechanical Engineering, Technical University of Liberec, Studentská 1402/2, 46117 Liberec, Czech Republic

* Correspondence: c.wen@exeter.ac.uk (C.W.); m.tian@exeter.ac.uk (M.T.)

Abstract: As we move away from internal combustion engines to tackle climate change, the importance of hydrogen-powered vehicles and polymer electrolyte membrane fuel cell (PEMFC) technology has dramatically increased. In the present study, we aimed to determine the optimal configuration for the power output of a PEMFC system using computational fluid dynamics (CFD) modelling to analyse variations of the primary serpentine design of gas flow channels. This helps improve efficiency and save on valuable materials used, reducing potential carbon emissions from the production of hydrogen vehicles. Different numbers of serpentine gas channels were represented with various spacing between them, within the defined CFD model, to optimise the gas channel geometry. The results show that the optimum configuration was found to have 11 serpentine channels with a spacing of 3.25 mm. In this optimum configuration, the ratio between the channel width, channel spacing, and serpentine channel length was found to be 1:2.6:38 for PEMFCs. Furthermore, the inclusion of fillets to the bends of the serpentine gas channels was found to have a negative effect on the overall power output of the fuel cell. Moreover, the optimisation procedures with respect to the number of gas channels and the spacing revealed an optimal power density exceeding 0.65 W/cm^2.

Keywords: hydrogen-powered vehicle; polymer electrolyte membrane; fuel cell; hydrogen; gas flow channel; micro porous layer; optimal configuration; computational fluid dynamics; power output

Citation: Chilver-Stainer, J.; Elbarghthi, A.F.A.; Wen, C.; Tian, M. Power Output Optimisation via Arranging Gas Flow Channels for Low-Temperature Polymer Electrolyte Membrane Fuel Cell (PEMFC) for Hydrogen-Powered Vehicles. *Energies* **2023**, *16*, 3722. https://doi.org/10.3390/en16093722

Academic Editor: Antonino S. Aricò

Received: 20 March 2023
Revised: 13 April 2023
Accepted: 23 April 2023
Published: 26 April 2023

1. Introduction

As governments set "net zero" carbon emission targets, hydrogen power and technology are becoming increasingly relevant, competing with other sources of clean and renewable energy, such as solar, wind, and biofuels. Batteries and hydrogen are a significant energy storage technology that have the potential to mitigate and lessen the harm being inflicted on the environment, especially for hydrogen and electric car applications [1]. Hydrogen fuel cell technology was invented in 1932 by Francis Thomas Bacon, but has rarely been commercialised in the motor industry due to the lack of hydrogen fuel stations and infrastructure. However, the future of hydrogen-powered vehicles (HPVs) is looking bright, with governments setting low carbon targets and the banning of sale of internal combustion vehicles in the UK by 2030. Additionally, there will be more renewable electricity sources as we move towards a greener economy. There will be more excess electricity produced during peak output periods, and this excess electricity can be stored as hydrogen through electrolysis, ready to be used in HPVs. With the emphasis moving away from hydrocarbon fuel for vehicles, hydrogen is an attractive alternative fuel source, with much higher energy density than hydrocarbons and faster re-fuelling times than electricity and batteries. However, one of the problems with producing hydrogen-fuelled vehicles is that they do not compare favourably with the power output and acceleration of electric vehicles.

CFD analysis of low-temperature PEMFCs used in hydrogen-powered vehicles can be used to optimise the gas channel configuration of the PEMFCs that are stacked within HPVs to increase the power output of the vehicle. Hydrogen fuel cell (HFC) technology is and has been thought to be a viable green low-carbon alternative to the internal combustion engine. HPVs are 64.7% fuel efficient compared to 20.8% for internal combustion engines over one drive cycle [2].

PEMFCs use hydrogen compressed gas and oxygen from the air to generate electricity and water. HFCs have two sides, a hydrogen gas side with the anode and an oxygen side with the cathode. The electricity is generated by the polymer electrolyte membrane (PEM) blocking the movement of electrons from the hydrogen side to the oxygen side, or from one gas diffusion layer electrode to the other. This means that the hydrogen ions can pass through the membrane, leaving the electrons to react with the oxygen from the air. With nowhere else to go, the electrons run up the anode and down the cathode, generating electricity. The oxygen then reacts with the electrons from the cathode and the hydrogen ions, producing water as a waste product. Catalysts like platinum allow hydrogen molecules to be split into protons and electrons on the anode side. Furthermore, on the cathode side, the platinum catalyst enables oxygen reduction by reacting with the protons generated by the anode. These enhanced dynamics decrease the reaction temperature in the PEMFC, to a lower temperature for use in the HPVs below 80 °C [3].

The PEMFC working principle is based on the gas that flows into the cell through gas channels that are connected to bipolar conductive plates or 'flow plates' for the flow of electrons. The gasses then diffuse into a layer called the gas diffusion layer (GDL). This is where the hydrogen-oxygen reaction occurs as the gasses react with the catalyst's assistance. There can be a layer between the catalyst and GDL called the micro porous layer (MPL) used to reduce water content in the reactive area up against the catalyst [4]. Each fuel cell has a certain standard flow channel configuration within HPVs. These fuel cells are stacked in arrays to increase output. Therefore, any optimised design needs to be stackable. They can be stacked in series to increase voltage or in parallel to increase the current, according to the needs of the vehicle. Usually, in HPVs, wired connections are inefficient. Therefore, a bipolar plate connection is generally used [3].

PEMs are known as the 'heart' of an HFC and are one of the most critical areas to understand. One of the main causes of failure in an HFC is the degradation of the PEM. These polymers, with ions attached to them, have a negative charge, repel electrons, and attract hydrogen ions [5]. Early PEMs only had a lifetime of 200 h at 60 °C and were made using a sulfonated polystyrene backbone. Currently, the most commonly used PEM is Nafion 120 [5]. A Nafion PEM has a lifespan of 60,000 h, operating at a temperature of 22–82 °C at one atmospheric pressure [5]. The lifetime of a PEM determines the lifetime of a PEMFC and is critically important. Mechanical degradation can occur due to the fragility of the material, especially when using thin membranes. Non-uniform pressure between the membrane electrode assembly and the bipolar plates in the system can accelerate this degradation [5]. Degradation is significantly increased when low humidity, high pressure and temperature are applied simultaneously. Penetration of catalyst particles into the membrane can cause local areas of high stress, resulting in an increase in the acceleration of degradation [5]. Nafion 120 is different from most PEMs, with a higher mechanical stability due to its polytetrafluroethylene backbone [5]. This allows a thinner PEM layer and therefore reduced ohmic internal resistance.

Another way a fuel cell can fail is cell reversal. This occurs when oxygen forms at the anode due to localised water electrolysis. This cell reversal causes the current to flow outside the channels due to the high water content. This can be prevented by adding MPLs into the PEMFC [6]. This layer is situated between the catalyst and the gas diffusion layer (GDL). It uses large pores in its material composition to separate and transport water produced in the hydrogen-oxygen reaction away from the reactive area up against the catalyst, allowing more reactants to collide with the catalyst, thereby increasing power output and efficiency of the PEMFC [4].

Chemical degradation is another area that is a cause of failure of PEMs. During use, there is a small chance that radicals (a particle with one unpaired valence electron) can be formed and these radicals attack carbon-hydrogen bonds, causing degradation of the PEM [7]. A platinum catalyst was found to negate the production of most of these radicals due to the occurrence of other catalytic reactions that can occur in its presence [5].

Humidity within the GDL and MPL impacts the power output of an HFC. One study used a CFD model of one gas channel within a PEMFC to measure how humidity affects power output [8]. It was discovered that the cell's performance at low humidities was influenced by the ohmic polarisation within the cell. Excessive humidity on the cathode side causes the catalyst to flood with water, resulting in a loss of performance. The water forms between the GDL and the catalyst interface [9]. Higher bipolar plate temperatures can be used to prevent this, as less water is able to condense. Alternatively, a dehumidifier can be used before the inlet [8]. However, this is not possible for low temperature PEMFCs used in HPVs, so other solutions are needed. An MPL layer helps prevent flooding by managing the water content in the GDL and reducing the effect of water flooding on the fuel cell's performance by allowing reactant gasses to reach the catalyst layer [6]. Another way of increasing the cell's performance by removing water content is by inserting porous sponges into the serpentine flow field [9].

An area of contention In PEMFC gas channel configuration research is co-current flow vs. counter-flow between the anode and cathode sides of the fuel cell. One model, using co-current flow, found that serpentine channels have better performance than parallel flow channels [10]. Multiple serpentine flow models either use co-current flow with gas channels directly above each other or counter-flow with the gas channels running between each other from the cathode to anode sides [11]. Most models seem to do one or the other, and both models appear to be almost as effective as each other [12]. When channels are lined directly above each other, co-current flow has been found to have the highest power output [13]. There has been much discussion over which model is best. A recent study by Yuan et al. compared these models using current density data and found that the counter-flow model was better for creating the highest power output [12]. However, this study employed more hydrogen, so it can be argued that the co-current model is more efficient, as hydrogen fuel lasts longer when the cell is running. Due to the higher power output in the counter-flow model, the authors only consider this model in this research. Based on the earlier study, HPVs and PEMFCs are more relevant in this field, as the world will move to more carbon-neutral technology and increase the engine's power and performance. This study implemented a CFD model of low-temperature PEMFC, using data and equations from the Ansys PEMFC module manual [14] to create an initial model of a PEMFC. The data and the parameters—including the boundary conditions from the literature—were used to validate the model, together with a set of potential differences and power density comparison. Moreover, this study illustrates the optimum voltage for the highest power output and uses CFD software to optimise the number of air and hydrogen serpentine gas channels within a single PEMFC. In addition, this research aimed to utilise a new model to design different numbers of channels to find the configuration with the maximum power density output across the PEM of the HFC. However, the geometry of the highest power density number of channels will be further optimised by comparing standard 90-degree bend gas channel bends and filleted curved gas channel bends of different radii to find the highest power density and to find the best gas channel configuration overall.

The novelty and research work in this study includes:

- The development of a new CFD model for low temperature PEMFCs—using data and equations from the ANSYS PEMFC module—validated with a set of potential difference and power density plots against current density, to compare with existing literature.
- CFD modelling is used to optimise the number of air and hydrogen serpentine gas channels within a single PEMFC, where we use the new model to design different

numbers of channels to find the maximum power density output across the PEM of the HFC.

- We use potential difference and power density to work out and illustrate the optimum voltage for the highest power. The PEMFC geometry is optimised by comparing standard 90-degree bend gas channel bends with filleted curved gas channel bends of different radii to find the highest power density and the best overall gas channel configuration.

2. Theory and Methodology

A similar experiment to this research has been conducted by Ferng et al. [14]. However, the previous investigation solely focused on the most common serpentine bend channel shape and parallel channel shape. An area of contention at that time was whether a serpentine channel shape was better than a parallel flow shape for the reactant gas flow channels. The purpose of this research is to address this question. For the validation, the d'Adamo et al. modelling case study will be used with a 25 cm^2 active area of the PEMFC and a set 0.7 V potential at relative humidity inlets of 100% with an operating temperature of 300.15 K [15]. In this study, a 'sweep' function was used on the membrane electrode assembly to increase mesh density, and therefore the accuracy of the results. The higher the mesh density around boundaries, the more accurate the results relating to the linear flow boundary layer thickness where the wall affects the velocity of the linear flow. The first layer of all inflations should be the same as the thickness of the boundary layer [15]. This meshing method, using sweeps and inflations, can increase the PEMFC model's accuracy. Furthermore, this study used similar boundary conditions of d'Adamo et al. to compare the PEMFC model [15]. Regarding the boundary conditions, it has been found that it is best to have a constant pressure from the inlet to the outlet in order to achieve the highest electrical power output [13]. This also means that improved gas channel configurations will have a low pressure drop. Table 1 shows all the channel configuration parameters used in the analysis and validation.

Table 1. Channel and Membrane Electrode Assembly Parameters.

Part	Direction	Dimension (mm)
Domain	[X,Y,Z]	[50,50,2.68]
Gas Channels	Section Width	1.25
	Section Height	1.25
	Length of serpentine branches	48
	distance between serpentine branches	1.25
GDL	[X,Y,Z]	[50,50,0.1273]
Membrane	[X,Y,Z]	[50,50,0.0254]

2.1. PEMFC CFD Modelling

This research uses Ansys fluent 2021R1 to model PEMFC channel configurations. The models were formulated using the Ansys Fluent PEMFC add-on module manual, listing all of the modelling software's governing equations [16–19], including the information on the many processes and energy losses that are less known within HFCs. The main reasons for choosing this software lie in the ability to solve complex fluid flow problems numerically and to simulate coupled models with multiple variables needed for a PEMFC as a laminar flow model [20–23]. This research is split into three different model sections, including an initial model that set out and ran a simple model to gain insight into polarisation (potential difference) and power density curves for the default parameters embedded in the Ansys fluent module [14]. The next set of models validated a polarisation curve against the literature and their experimental results using the same geometry. Finally, the third set of models used different geometry to optimise the design of the reactant gas channels within that same model.

In the simulation, Navier stokes equations are used to model fluid flow applied across a mesh of cells, in which individual momentum vectors are calculated using numerical methods. These are then used to convert a set of partial differential equations over iterations into solvable equations. The model follows the conservation of mass from the inlet and outlet boundary conditions, and the total energy is used to model static and dynamic temperature throughout the model from the hydrogen-oxygen reaction.

The current flux equation was activated through modelling the PEMFC because of the existence of electrochemistry, and to model the movement of current [14]. Electrochemistry modelling equations can be used to model the flow of current through the materials within the PEMFC by setting a voltage potential at both ends of the PEMFC on the anode and cathode side as fixed boundary conditions; the initial current flux R at the anode and cathode can be found. The calculations are also explained by a current flux boundary condition that is ready to be identified numerically towards the centre of the mesh, and therefore at the PEM. The current flux total can be found across the membrane by an integral method across a plane area at the centre of the PEM, giving the model results [14]. This process is repeated using different gas channel configurations for optimisation. One of the primary energy losses in PEMFCs is the activation loss due to the slowness of the hydrogen-oxygen reaction taking place near the membrane surface [24]. This is usually found by subtracting the potential difference at the membrane from the potential at the flow plate on the anode side [14].

In the post-processing of CFD results, several numerical calculations need to be carried out to gather relevant outcomes, such as polarisation and power density curves. The fundamental equation for calculating electrical power can be stated as follows:

$$P = VI \tag{1}$$

where P represents the power (W), V is the potential difference (V), and I is the current (A). This equation characterises the fundamental to the measurement of power output of fuel cells from a fixed voltage and a measured current density. Therefore, this equation needs to be modified to suit the current density measurement as follows:

$$P_d = VI_d \tag{2}$$

where P_d is the power density (W/m^2), and I_d represents the current density (A/m^2). The power and current densities are calculated over the active area of the PEMFC. Current density is gathered from the CFD results across the centre plane of the PEM. This is then converted into the power density curve of the PEMFC at different fixed voltages.

2.2. Defining Material Properties

There were two sets of material properties used in the experiment. The first set was the default Ansys values for the material for the initial model, and the validation properties as seen in Table 2. The default Ansys properties are described in Table 3.

Table 2. Properties of Graphite and Nafion.

PEMFC Part	PEM	Flow Plate Electrodes
Material	Nafion	Graphite
Density (kg/m^3)	1970	2250
Electrical conductivity (S/cm)	1×10^{-16}	125,000
Thermal Conductivity (W/m K)	0.445	20
Specific Heat (J/kg K)	903	707.68

Table 3 shows the material properties that arise automatically when Ansys fluent is released. This was used as the initial models' materials. Although the GDL layer is assigned material for electrical and thermal conduction purposes, in reality, the layer is just space for the reactant gasses to diffuse into. The material properties for the validation and optimisation models used the properties shown in Table 2 for the flow plate and PEM layers. The other layers were modelled with the same density as the flow plate in Table 2. This was so that the model could be simplified to represent any set of materials for any PEMFC. The thermal conductivity, specific heat, and electrical conductivity were kept the same as the default properties to decrease the chance of error in the model due to the change in properties from the initial model to the validation model. The material properties for the validation model are shown in Table 4.

Table 3. Ansys default material properties.

Material Represented	PEMFC Part	Density (kg/m^3)	Specific Heat J/kg·K	Thermal Conductivity W/m·K	Electrical Conductivity (S/m)
graphite	flow plate	2719	871	100	1,000,000
	GDL	2719	871	10	5000
Epoxy	MPL	2719	871	10	5000
Platinum	catalyst	2719	871	10	5000
Nafion 120	PEM	1980	2000	2	1×10^{-16}

Table 4. Validation Model Material Properties [15].

Material Represented	PEMFC Part	Density (kg/m^3)	Specific Heat /kg·K	Thermal Conductivity W/m·K	Electrical Conductivity (S/m)
graphite	flow plate	2250	707.68	20	1.25×10^7
	GDL	2250	871	10	5000
Epoxy	MPL	2250	871	10	5000
Platinum	catalyst	2250	871	10	5000
Nafion 120	PEM	1970	903	0.445	1×10^{-16}

2.3. *Biconjugate Gradient Stabilisation Method (BCGSTAB)*

The BCGSTAB method is used to solve non-symmetrical linear systems while helping avoid irregular convergence patterns. This helps stabilise variables that may have irregular fluctuations in value while the model runs. In the context of an HPV, these variables would be controlled and regulated by subsystems within the HPV. When the variable starts fluctuating, usually at the beginning of a set of iterations, the residual of the variable will return to a stable set of values and begin to converge, thanks to the BCGSTAB method. During the running of the models, the species molar counts and the electric and protonic potential variables use BCGSTAB. The species molar counts used the BCGSTAB method because the reactant content in the early iterations of the model can drop very suddenly due to the reaction taking place, and the water content can increase significantly as a waste product. The electric and protonic potential would likely be controlled by a set of diodes to prevent current in the opposite direction and circuit breakers. The BCGSTAB method assisted with getting convergence in the model and helped prevent floating point exemption errors. These errors are caused by a large gradient in the residuals of the model, meaning that the governing Navier stokes equations divide by zero resulting in an invalid result.

2.4. *F-Cycle Method and Coupled 2nd Order Modelling*

The F-cycle multigrid method was used for all calculated variables within the models. This technique is used to speed up the solution and convergence of an iterative model. It reduces the number of iterations needed for convergence, utilising a set of simultaneous correction equations. This method adds neighbouring discretised equation coefficients and generates these correction equations. These coefficient additions allow the correction equations to be solved using fewer iterations. The F-cycle has a faster solution time than a V-cycle due to more coefficients making up correction equations. This saves computational power and time, and because of these advantages, the F-cycle was used in the initial and validation models.

The model used a coupled solver that solves for continuity, momentum, energy, and species simultaneously, due to it saving computational power compared to other methods. The models use a second-order scheme which is second-order accurate in terms of the Taylor series as a model. All models had a courant number of 200, which is the speed of sound multiplied by the ratio of the time step length to the cell length. Throughout the F-cycle modelling process, there are several errors that can occur, including data errors, syntax errors, logic errors, and parameter errors. These can be avoided by double checking input parameters before running the model, thoroughly assessing the CFD, F-cycle, and BCGSTAB code before use, and ensuring that the digital inputs are within their acceptable range [25].

All additional parameters used in the simulation which support the verification based on the boundary conditions are listed in Tables 5 and 6.

Table 5. Further Boundary conditions for the Ansys fluent PEMFC add-on module.

Boundary Condition Input	Value
W-diff Model	Wu
Liquid vapour source relaxation factor	0.2
Devised vapour/liquid relaxation factor	0.2
Osmotic drag relaxation factor	1
Gas diffusion layer liquid removal	0.5

Table 6. Anode and Cathode Inlet and Outlet Boundary Conditions.

Region	Surface	Property	Value
Anode	Inlet	Mole Fraction H_2	0.9764
		Mass Flow Rate (kg/s)	3.93×10^{-7}
		Pressure (bar)	1.5
		Temperature (K)	300.15
		Mole Fraction H_2O	0.0236
	Outlet	Mole fraction H_2, H_2O	0.9764, 0.0236
		Pressure (bar)	1.5
		Temperature (K)	300.15
Cathode	Inlet	Mole Fraction O_2	0.2075
		Mass Flow Rate (kg/s)	2.07×10^{-5}
		Pressure (bar)	2
		Temperature (K)	300.15
		Mole Fraction H_2O	0.0119
	Outlet	Mole fraction O_2, H_2O	0.2075, 0.0119
		Pressure (bar)	2
		Temperature (K)	300.15

3. Model Descriptions

3.1. Physical Model

First, a geometry was needed to carry out the modelling and optimisation. This was found by using the example model case study as a guide to create a similar geometry to validate the Ansys model [13]. One difference intended between this model and the PEMFC model is the inclusion of an MPL layer. As noted above, this is an extra layer in the fuel cell between the Catalyst and GDL layers that removes water content. This means that the liquid phase of the model can be removed, saving computing power, and this causes better convergence in models. The model used the 25 cm^2 active area from the case study model, as shown in Figure 1. To simplify, the channels were assumed to have 90-degree corners, to get a standardised set of results that could be used for any fuel cell. The model used the same materials as in the literature.

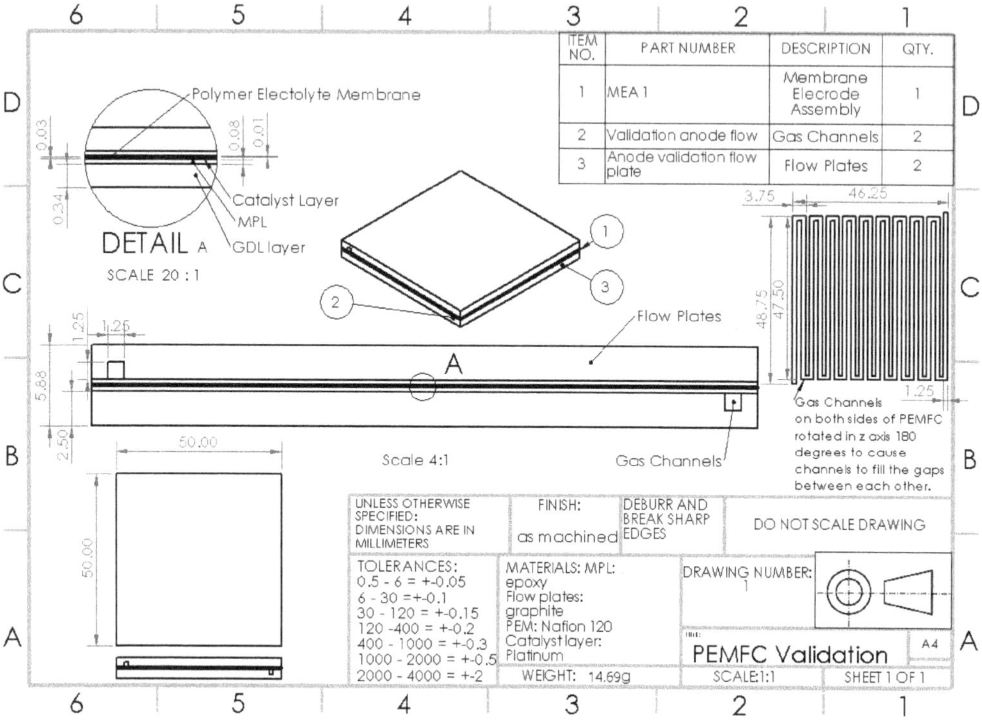

Figure 1. Initial Model Geometry.

3.2. Meshing and Mesh Dependency Analysis

For the PEM meshing, a sweep method was used with one division to save computational power, because it is a very thin layer. The catalyst and MPL layers had three division sweep methods because that is where most of the reaction is taking place, and increasing the mesh density captures the change in temperature and species gradients more accurately. The GDL would also be ideal for sweeping, but due to the irregular surface of the gas channels, it is not possible to compute. Therefore, the GDL was computed with trapezoidal cells, ensuring that they are pyramidal cells close to the sweep method of the MPL layer, to increase accuracy. The gas channels have an inflation layer around them to increase the accuracy of the flow results around the walls of the channels. This inflation was three layers thick and had a growth ratio of 1.2. The flow plates were left with a standard trapezoidal mesh with a standard cell size from meshing software (3.55 mm). This was to save on computational power and had the benefit of the mesh being adaptable for any configuration of gas channels.

To find the best cell size for the MEA, GDL, and gas channels, the initial Model was run with different cell sizes for each to find the set that saved the most computational power while getting the most accurate results. To get the pyramidal cells in the GDL layer of the mesh, the cell size needs to be half of the MEA (PEM, MPL, and catalyst layers) cell size. In addition, due to the inflation giving high accuracy in the gas channels to save on computational power, the cell size in the gas channels is set to double that of the MEA. Each cell size set from every layer gave a different number of cells to run in the model. The results of this are shown in Table 7 and Figure 2.

Figure 2 below shows how current density across the PEM from the initial model exponentially decays with cell number to an asymptote around 7525 A/m^2. This is the ideal point to choose, as the cell size of the layers would be at the bottom of the exponential curve, as it flattens out. However, to save computational power and reduce the models' run-time, 1 mm for flow/gas channels was chosen, with 0.5 mm for the MEA and 0.25 mm for the GDL layers. Figure 3 below shows the final mesh used in the models with the sweep methods and inflations.

Table 7. Layer cell size to cell number.

Cell Size Flow Channels [mm]	Cell Size GDL [mm]	Cell Size MEA [mm]	Number of Cells
1.5	0.375	0.75	2,000,968
1.25	0.3125	0.625	2,109,151
1	0.25	0.5	2,270,056
0.75	0.1875	0.375	3,333,827
0.6	0.15	0.3	4,820,839
0.625	0.25	0.5	2,403,259

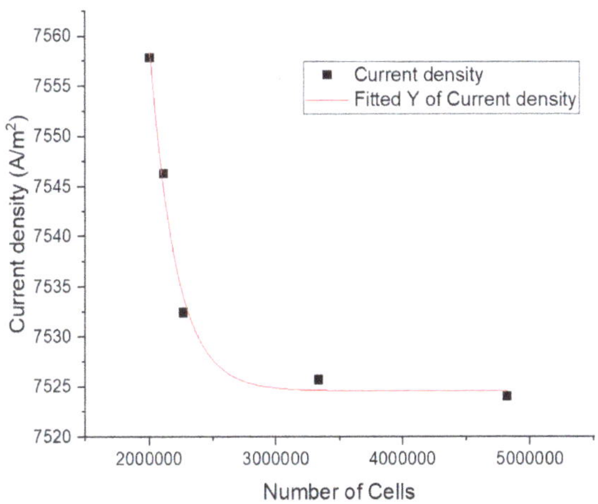

Figure 2. Mesh dependency analysis: Number of cells in mesh plotted against current density.

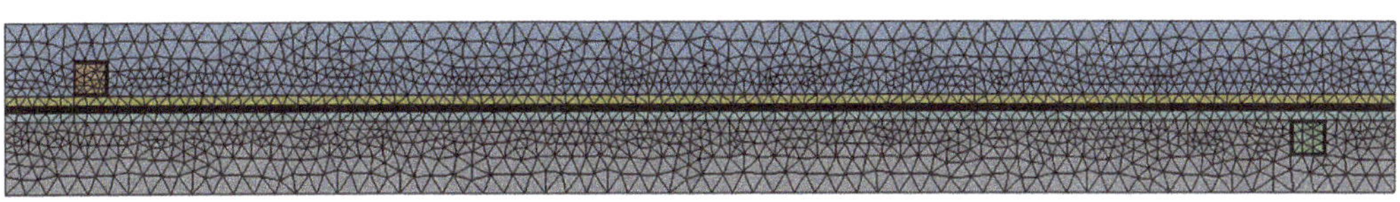

Figure 3. Final model mesh with the scale inflation and sweep methods.

3.3. Initial Model Setup and Boundary Conditions

To get the first model setup, named sections for each layer body were created, following the Ansys PEMFC module manual [14]. Similarly, for the cathode's and anode's inlets and outlets, it was important to ensure that the inlets of the anode and cathode are above and below each other for counter-flow to occur. Furthermore, to control voltage, named sections on the top and bottom surfaces were created for the cathode and anode terminals. Within the PEMFC add-on module, each named section body layer needed to be added to their respective chosen layer within the software for the anode and cathode (flow/gas channels, flow plates, GDLs, MPLs, catalyst layers, and the PEM). This model was run for over 200 iterations.

First, a contact resistivity was added to model an imperfect contact between flow plates and the GDL layer of the PEMFC. This contact resistivity was 1×10^{-6} Ωm^2 as a general value used [14]. The simulation temperature was set to 333.15 K as an operating temperature.

The anode and cathode inlets were mass flow inlets with flow rates of 1×10^{-7} kg/s and 1.4×10^{-6} kg/s, or inlet velocities of 0.064 m/s and 0.896 m/s, respectively. The cathode needed a higher mass flow rate to account for the lower reactant content, as oxygen only makes up roughly 20% of the air [14]. Therefore, for this model, the species content was set up as mass fractions, the same for the inlet and outlet. For the cathode, oxygen was a mass fraction of 0.2 and a water content of 0.17. Additionally, for the anode, the hydrogen mass fraction was 0.5, and 0.5 for water content. To best model, simplify, and optimise the gas channel configuration, a constant liner flow was used with constant pressure and species content for the anode and cathode sides of the PEMFC model. The Reynolds number for the anode and cathode were 11.3 and 91.8, respectively. For the initial model, one atmospheric pressure was used as the operating pressure. The anode terminal was set at 0 V potential difference, and the cathode terminal was set to different voltages for the polarisation curve. This was from 0.8 V to 0.2 V, repeating the model each time. The initial model material properties were implemented in this model to get a set of polarisation and power density curves.

3.4. Validation Model Setup and Boundary Conditions

The contact resistivity from the validation model was kept the same as in the initial model. The boundary conditions were taken from the literature [15]. All other boundary conditions were kept the same as in the initial model except for the values shown in Table 8.

Table 8. Validation model boundary conditions.

Boundary Condition	Set Value
Temperature (K)	300.15
Anode Mole Fraction H_2	0.9764
Cathode Mole Fraction O_2	0.2075
Anode Mole Fraction H_2O	0.0236
Cathode Mole Fraction H_2O	0.0119
Pressure (Pa)	175,000

The case study model runs two different pressures for the anode and cathode. This method is not possible using Ansys fluent because the solid-fluid modelling of the PEMFC MEA causes the model to fail with floating point exemption errors due to the pressure gradient in the system. Therefore, the pressure chosen for the validation model of 175,000 Pa was between the two anode and cathode pressures from the literature [15]. The mass flow rates of the inlets were kept the same as the initial model, and for the same reason, they were set up due to the different pressures. This may lead to different results from the literature. The temperature chosen was the same as the literature's model [15]. The validation model gives a set of polarization and power density curves relating to the literature. This will provide the fuel cell's optimum power output, which can then be used in geometry variation and optimisation models. Additionally, a contour of the current density was taken across the PEM to show where the most electrical power is being generated in the cell through the hydrogen-oxygen reaction. This model was run for over 500 iterations.

3.5. Geometry Variation

During the research, many sets of similar geometry were used for different serpentine flow channels. Figure 1 shows the initial 19-channel configuration. The outer two channels with the inlets and outlets were fixed. To optimise the PEMFC serpentine gas channel configuration, the number of gas channels for both the anode and cathode was reduced by two each time from nineteen down to three channels. The spacing between the channels was found using Equation (3), with 46.35 mm being the distance between the two outer channels:

$$Channel\ Spacing(\text{mm}) = \frac{(46.35 - 1.25 \times N)}{(N-1)} \quad (3)$$

where N represents the number of serpentine gas channels. The spacing was implemented so the channels were evenly spaced as if the set of gas channels were continuous and could be implemented into a stack of cells. These new PEMFCs can be modelled to find the power density of the new configurations. Therefore, it was possible to find the best power output configuration, which can be used to test different fillet radii. Additionally, a contour of the current density was taken across the PEM, to show where in the cell the most electrical power was being generated through the hydrogen-oxygen reaction for the optimum number of channels.

3.6. Fillet Radii Testing

The serpentine channels for each configuration have used 90-degree squared-off bends. This configuration can now be compared to filleted channels (with rounded bends) by testing whether filleting these bends can increase the power output of the PEMFC. The original 19-channel configuration and the best geometry from the geometry variation were tested. Bends in the serpentine channels were modelled with different radii for comparison purposes. These were then put through the validation model at the optimum power output voltage to find the highest power outputting fillet radius. This was performed from a fillet radius of 0 mm to 1.625 mm for the optimum number of channels. Furthermore, for the original 19-channel configuration, filleted and 90-degree corners were tested from a radius of 0.625 mm. These results made it possible to find the optimal geometry for a PEMFC.

4. Results and Discussion

This CFD study compared the modelled results to experimental results and discovered an underestimation of the current density of the membrane at lower fixed voltages [15]. The authors thought that this was due to the uncertainty in material characterisation. The model and results were plotted on a polarisation/power density curve shown in Figure 4.

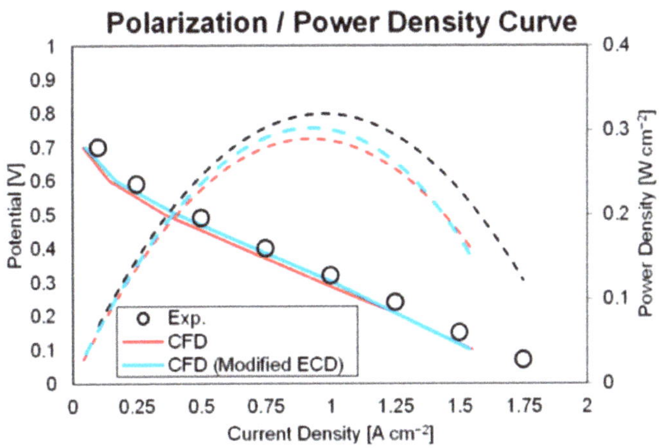

Figure 4. Polarization (solid lines)/Power density curve (dashed lines) compared with [15].

Figure 4 will help compare the validation model of this research against the case study's experimental and model results. The validation model is intended to validate the CFD model by getting similar characteristic results to Figure 4 and getting as close as possible numerical values to this study's experimental results. The potential difference against current density can be plotted for different fixed voltages, and the power density can be calculated from each current density.

4.1. Initial and Validation Models Polarization and Power Density Graphs

The figures in this section present the data collected from the experimental work. Figures 5 and 6 present how voltage and power density vary with current density for the initial model and validation model, respectively. The behaviour of the validation model results in Figure 5 are very similar to that of the experimental results in the literature shown in Figure 4. The validation model also shows a peak power density output at 0.3 V. Therefore, this is the optimum power output voltage ready to be used as the voltage for the geometry variation. That is 83.3 cm^2 of active area for every volt of potential difference. For the initial model shown in Figure 5, there is a slightly irregular shape in the power density curve. This is likely to be due to the higher operating temperature and lower operating pressure. The higher temperature means that it is easier for the hydrogen-oxygen reaction to occur, resulting in slightly higher power output at lower current densities. The validation model's numerical values are similar but not similar enough to be counted as error. Therefore, the difference is likely to be due to the different mass flow rates and pressure, as it cannot be modelled differently from the anode to cathode side.

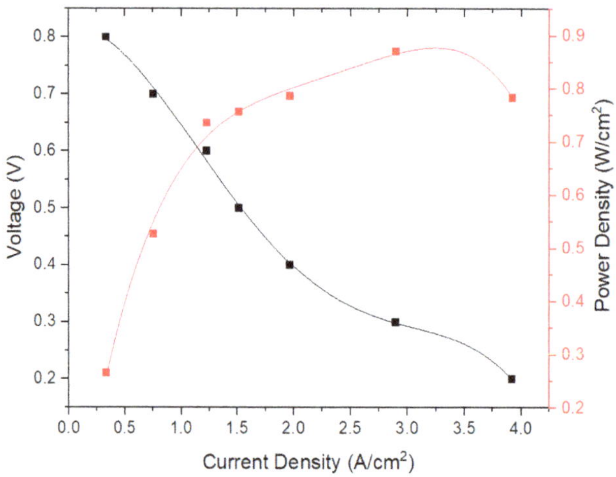

Figure 5. Initial model potential difference from anode to cathode and power density curves plotted against current density.

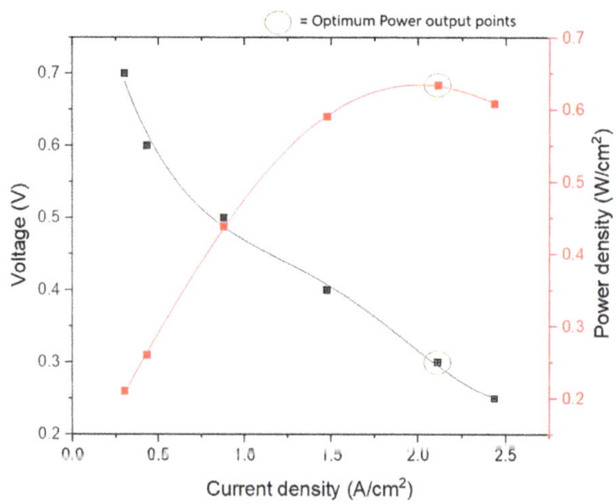

Figure 6. Validation model potential difference from anode to cathode and power density curves plotted against current density with optimum power output points labelled.

4.2. Geometry Variation Optimization Results

Figures 7–9 demonstrate the optimisation of how different geometries affect the power density of the PEMFC. Moreover, they illustrate a clear optimum number of gas channels and spacing between those gas channels. The optimum number of channels was found to be 11, giving the cell the highest power density and, consequently, the highest power output. The channel spacing for this set of channels was 3.25 mm.

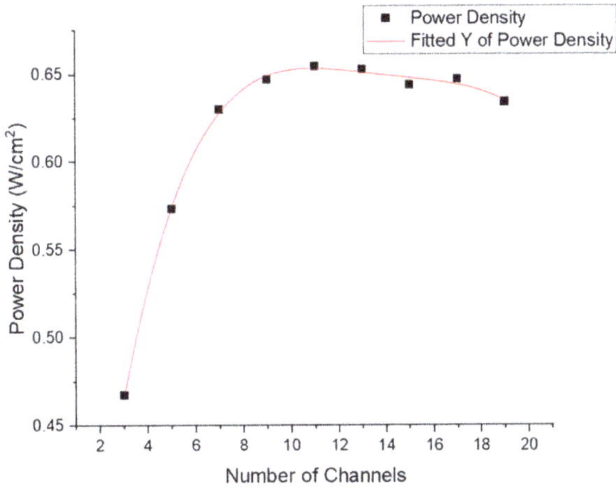

Figure 7. Optimisation of the number of gas channels by recording the power density of each geometry run.

Figure 8. Current flux density contour across optimum 11 channel configuration PEM.

In Figure 8, the console shows most of the reaction taking place towards the inlet side of the serpentine channels. The higher the current density, the more hydrogen-oxygen reaction takes place at that location on the PEM surface. This is likely to be due to the oxygen reacting and running out as it is depleted due to the PEMFC being cathode dependent, as discussed previously. The oxygen depletion for the 11-channel configuration is shown in Figure 10. The current density is higher where the cathode gas channels are located, especially close to the inlet of the cathode. Figure 9 shows the power density increasing slightly with channel spacing at the beginning, reaching a maximum point and then decreasing with higher channel spacing. This is likely due to the higher current density close to the inlet of the cathode, where more hydrogen-oxygen reactions can occur.

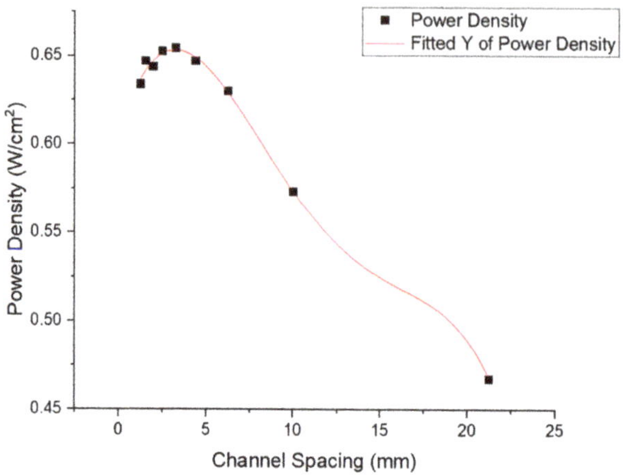

Figure 9. Optimisation of channel spacing from the different number of gas channel geometry configurations by recording power density.

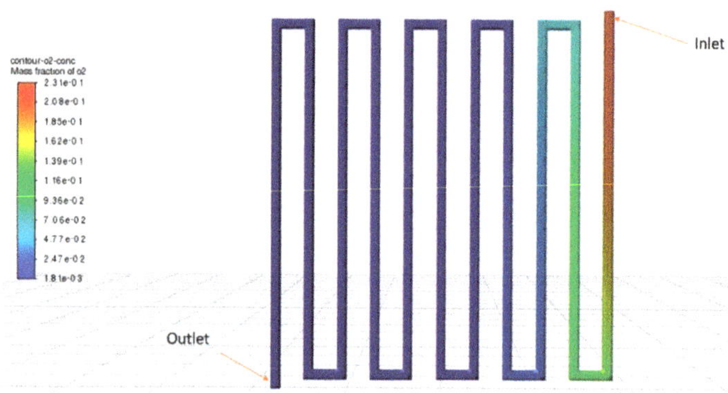

Figure 10. Oxygen mass fraction cathode 11 gas channel configuration.

At a lower spacing of the channels, this initial high level of reaction can spread further across the cell. This means a higher overall current density, as there is more oxygen to react before being depleted, close to the cathode inlet. In higher channel spacing geometry configurations, after the maximum point, the power density decreases are likely to be due to fewer gas channels, which means that there is less reaction occurring, as there is less area for the reactants to come into contact with the catalyst and react for both the anode and cathode reactants. To apply this to any PEMFC, the ratio between channel width, serpentine channel height (in this case, 47.5 mm as shown in Figure 1 and in the final model), and channel spacing can likely be used. The value is 1:2.6:38, respectively, to maximize the power output of PEMFC with the same control variables, including an operation pressure of 175,000 Pa, an operation temperature of 300.15 K, and inlet velocities of 0.064 m/s and 0.896 m/s for the anode and cathode, respectively.

4.3. Filleting Results and Final Design

Figure 11 shows the power density plotted against the radius of the fillet. It shows that the higher the fillet radius, the lower the power density of the fuel cell. The graph shows a small increase in power density after the initial decrease, and this could be due to the 90-degree angle turns in the geometry slowing down the flow of oxygen around the cell. However, this is outweighed by the potential of covering more cell area with a lower to no fillet radius, increasing the overall current flux and power output of the cell at 0.3 V. Figure 12 shows how, at the cathode inlet, oxygen gas can flow further into the cell than without a fillet. This is further supported by Table 9, where the less powerful 19-channel configuration is given a fillet of 0.75 mm, which increases the cell's power output. Figure 11 shows that the higher the fillet radius in general, the lower the cell's power output due to the less active area covered by the cathode channels. Therefore, the best gas channel configuration and the final design of this research is the 11-channel configuration, as shown in Figure 13.

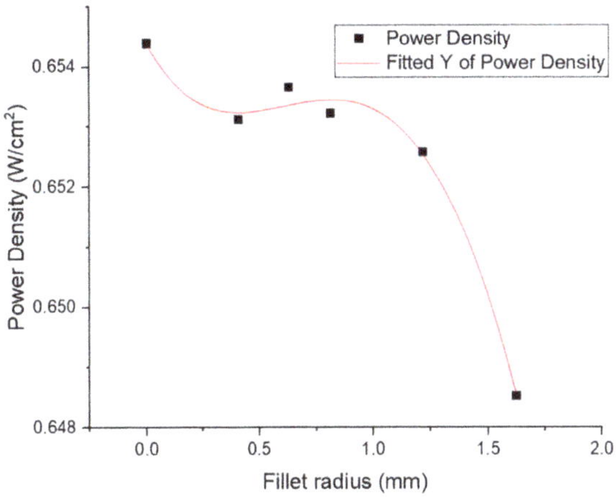

Figure 11. Optimisation of fillet radius of 11-channel PEMFC recording power density.

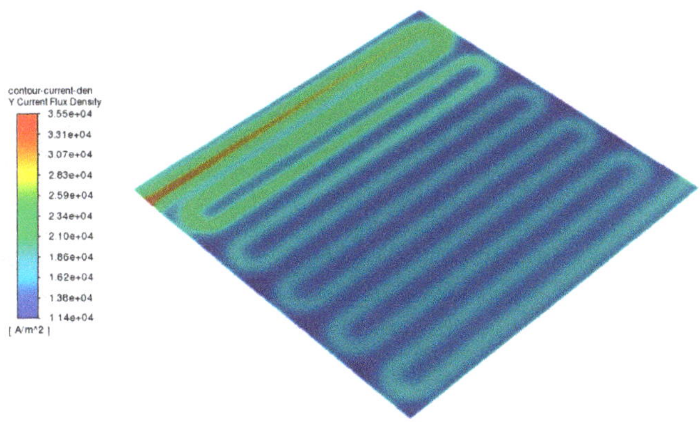

Figure 12. Current flux density contour for 1.625 mm fillets.

Table 9. Power density of 19-channel configuration filleted vs. not filleted.

Fillet Radius (mm)	Power Density (W/cm^2)
0	0.6338817
0.75	0.6341631

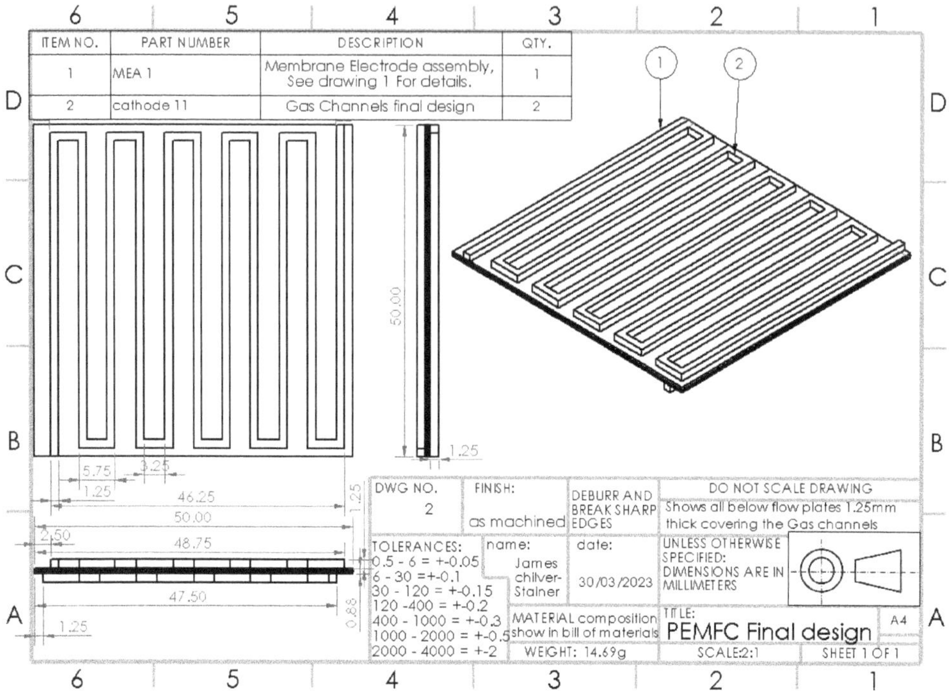

Figure 13. Final 11-channel configuration of the research.

5. Conclusions

In this research, the PEMFC model results follow the trends of the experimental results from the literature closely. The numerical values of the validation model were similar but not the same as the experimental results from the case study model, even though the model used the same boundary conditions. This is likely due to the many conditions of the model not being specified in the study literature, due to using different software to model. This includes flow courant number, which variables used BCGSTAB, and the order of the method used. In addition, as discussed in validation model boundary conditions and Setup Section 3.3, the mass flow rates and pressure boundary conditions differed from the case study and can affect the results' validity. However, the trends between the models and experimental results are the same, and the model is sufficiently valid to satisfy the purposes of this research. The most relevant findings can be drawn as follows:

1. The optimum power output voltage was found to be 0.3 V for a 25 cm^2 active area. This can be applied to any PEMFC by extrapolating a 1 volt of potential difference for every 83.3 cm^2 active area.
2. The optimum configuration was determined to have 11 gas channels with a spacing of 3.25 mm, giving a channel width to channel spacing to serpentine channel length ratio of 1:2.6:38 for any PEMFC with a pressure of 175,000 Pa, an operation temperature of 300.15 K, and inlet velocities of 0.064 m/s and 0.896 m/s for the anode and cathode, respectively.
3. Further research is required both to validate this ratio over a range of PEMFC sizes with different active areas, and experiments need to be conducted using a PEMFC to properly validate these results with real-life data so that they can be used in fuel cells in the future.

4. The optimum 3.25 mm channel spacing is likely due to the combination of a high initial oxygen spread before being depleted, producing current flux and power. Additionally, it could be due to the increased area covered by having more channels.
5. There is scope for further improvement in the modelling. There is a possibility of having an increased space close to the inlet of the cathode and a lower spacing further away from the inlet close to the outlet. This would mean that the initial oxygen can spread across the cell faster and have the increased area covered close to the outlet. According to the results of this study, there is potential to increase the power output of the cell further.
6. According to the results, fillets to the bends of the serpentine channels decrease the power output for the 11-gas channel configuration due to the less active area covered by the serpentine channels.

These accurate models of 3D PEMFC Serpentine Gas Channel Fluid Flow results could be used to find the optimum serpentine gas channel configuration for any PEMFC with the same control variables, and with further research, for any set of control variables. This would be especially useful for increasing the acceleration of HPV's and their fuel use capabilities. It could also potentially mean less fuel cell stacks would be needed to power an HPV, saving material costs and reducing carbon emissions when manufacturing HPVs.

Future research in this area should focus on validating the results of this study through experimentation, particularly with different PEMFC sizes and active areas. Additionally, there is scope for further optimisation of the gas flow channel configuration to improve power output even further. In particular, exploring the potential benefits of fillets at the inlet and outlet of the cathode channels could lead to significant improvements. Furthermore, the findings of this study can be extended to explore the impact of gas channel configuration on other fuel cell technologies, such as solid oxide fuel cells (SOFCs) and alkaline fuel cells (AFCs). Such investigations could pave the way for even more efficient and cost-effective fuel cell technologies, making them more accessible and widespread in various applications.

Author Contributions: Conceptualization, J.C.-S.; Methodology, J.C.-S., C.W. and M.T.; Formal analysis, J.C.-S.; Investigation, J.C.-S. and M.T.; Writing—original draft, J.C.-S.; Writing—review & editing, A.F.A.E., C.W. and M.T.; Supervision, C.W. and M.T. All authors have read and agreed to the published version of the manuscript.

Funding: This research was funded by Royal Society Grants RGS\R1\231093 and IEC\NSFC\211452.

Data Availability Statement: The research data supporting this publication are provided within this paper.

Acknowledgments: The second author acknowledges the support of the Student Grant Competition of the Technical University of Liberec under project No. SGS-2023-5323. M.T. acknowledges the support from the Royal Society Grants RGS\R1\231093 and IEC\NSFC\211452.

Conflicts of Interest: The authors declare no conflict of interest.

References

1. Balasankar, A.; Arthiya, S.E.; Ramasundaram, S.; Sumathi, P.; Arokiyaraj, S.; Oh, T.; Aruchamy, K.; Sriram, G.; Kurkuri, M.D. Recent Advances in the Preparation and Performance of Porous Titanium-Based Anode Materials for Sodium-Ion Batteries. *Energies* **2022**, *15*, 9495. [CrossRef]
2. Liu, X.; Reddi, K.; Elgowainy, A.; Lohse-Busch, H.; Wang, M.; Rustagi, N. Comparison of well-to-wheels energy use and emissions of a hydrogen fuel cell electric vehicle relative to a conventional gasoline-powered internal combustion engine vehicle. *Int. J. Hydrog. Energy* **2020**, *45*, 972–983. [CrossRef]
3. Shusheng, X.; Qiujie, S.; Baosheng, G.; Encong, Z.; Zhankuan, W. Research and development of on-board hydrogen-producing fuel cell vehicles. *Int. J. Hydrog. Energy* **2020**, *45*, 17844–17857. [CrossRef]
4. Zhang, X.; Ma, X.; Shuai, S.; Qin, Y.; Yang, J. Effect of micro-porous layer on PEM fuel cells performance: Considering the spatially variable properties. *Int. J. Heat Mass Transf.* **2021**, *178*, 121592. [CrossRef]
5. Collier, A.; Wang, H.; Yuan, X.Z.; Zhang, J.; Wilkinson, D.P. Degradation of polymer electrolyte membranes. *Int. J. Hydrog. Energy* **2006**, *31*, 1838–1854. [CrossRef]

6. Tang, H.; Wang, S.; Pan, M.; Yuan, R. Porosity-graded micro-porous layers for polymer electrolyte membrane fuel cells. *J. Power Sources* **2007**, *166*, 41–46. [CrossRef]
7. Trogadas, P.; Parrondo, J.; Ramani, V. Degradation Mitigation in Polymer Electrolyte Membranes Using Cerium Oxide as a Regenerative Free-Radical Scavenger. *Electrochem. Solid State Lett.* **2008**, *11*, B113–B116. [CrossRef]
8. Kim, H.-Y.; Kim, K. Numerical study on the effects of gas humidity on proton-exchange membrane fuel cell performance. *Int. J. Hydrog. Energy* **2016**, *41*, 11776–11783. [CrossRef]
9. Marappan, M.; Narayanan, R.; Manoharan, K.; Vijayakrishnan, M.K.; Palaniswamy, K.; Karazhanov, S.; Sundaram, S. Scaling up Studies on PEMFC Using a Modified Serpentine Flow Field Incorporating Porous Sponge Inserts to Observe Water Molecules. *Molecules* **2021**, *26*, 286. [CrossRef]
10. Hashemi, F.; Rowshanzamir, S.; Rezakazemi, M. CFD simulation of PEM fuel cell performance: Effect of straight and serpentine flow fields. *Math. Comput. Model.* **2012**, *55*, 1540–1557. [CrossRef]
11. Krastev, V.; Falcucci, G.; Jannelli, E.; Minutillo, M.; Cozzolino, R. 3D CFD modeling and experimental characterization of HT PEM fuel cells at different anode gas compositions. *Int. J. Hydrog. Energy* **2014**, *39*, 21663–21672. [CrossRef]
12. Yuan, L.; Jin, Z.; Yang, P.; Yang, Y.; Wang, D.; Chen, X. Numerical Analysis of the Influence of Different Flow Patterns on Power and Reactant Transmission in Tubular-Shaped PEMFC. *Energies* **2021**, *14*, 2127. [CrossRef]
13. Kahraman, H.; Orhan, M.F. Flow field bipolar plates in a proton exchange membrane fuel cell: Analysis & modeling. *Energy Convers. Manag.* **2017**, *133*, 363–384.
14. Canonsburg, T.D. Ansys Fluent Fuel Cell Modules Manual, Knowl. *Creat. Diffus. Util.* **2012**, *15317*, 724–746.
15. D'adamo, A.; Riccardi, M.; Borghi, M.; Fontanesi, S. CFD Modelling of a Hydrogen/Air PEM Fuel Cell with a Serpentine Gas Distributor. *Processes* **2021**, *9*, 564. [CrossRef]
16. Tian, P.; Liu, X.; Luo, K.; Li, H.; Wang, Y. Deep learning from three-dimensional multiphysics simulation in operational optimization and control of polymer electrolyte membrane fuel cell for maximum power. *Appl. Energy* **2021**, *288*, 116632. [CrossRef]
17. Varghese, G.; Venkatesh Babu, K.P.; Joseph, T.V.; Chippar, P. Combined effect of channel to rib width ratio and gas diffusion layer deformation on high temperature—Polymer electrolyte membrane fuel cell performance. *Int. J. Hydrog. Energy* **2022**, *47*, 33014–33026. [CrossRef]
18. Sapkota, P.; Brockbank, P.; Aguey-Zinsou, K.-F. Development of self-breathing polymer electrolyte membrane fuel cell stack with cylindrical cells. *Int. J. Hydrog. Energy* **2022**, *47*, 23833–23844. [CrossRef]
19. Wang, Y.; Wang, X.; Fan, Y.; He, W.; Guan, J.; Wang, X. Numerical Investigation of Tapered Flow Field Configurations for Enhanced Polymer Electrolyte Membrane Fuel Cell Performance. *Appl. Energy* **2022**, *306*, 118021. [CrossRef]
20. Su, G.; Yang, D.; Xiao, Q.; Dai, H.; Zhang, C. Effects of vortexes in feed header on air flow distribution of PEMFC stack: CFD simulation and optimization for better uniformity. *Renew. Energy* **2021**, *173*, 498–506. [CrossRef]
21. Cao, Y.; El-Shorbagy, M.; Dahari, M.; Cao, D.N.; El Din, E.M.T.; Huynh, P.H.; Wae-Hayee, M. Examining the relationship between gas channel dimensions of a polymer electrolyte membrane fuel cell with two-phase flow dynamics in a flooding situation using the volume of fluid method. *Energy Rep.* **2022**, *8*, 9420–9430. [CrossRef]
22. Yuan, H.; Dai, Y.; Li, H.; Wang, Y. Modeling of high-temperature polymer electrolyte membrane fuel cell for reaction spatial variation. *Int. J. Heat Mass Transf.* **2022**, *195*, 23209. [CrossRef]
23. Huang, F.; Qiu, D.; Peng, L.; Lai, X. Optimization of entrance geometry and analysis of fluid distribution in manifold for high-power proton exchange membrane fuel cell stacks. *Int. J. Hydrog. Energy* **2022**, *47*, 22180–22191. [CrossRef]
24. Yun, S.-H.; Woo, J.-J.; Seo, S.-J.; Yang, T.-H.; Moon, S.-H. Estimation of approximate activation energy loss and mass transfer coefficient from a polarization curve of a polymer electrolyte fuel cell. *Korean J. Chem. Eng.* **2012**, *29*, 1158–1162. [CrossRef]
25. Versteeg, H.K.; Malalasekera, W. An Introduction to Computational Fluid Dynamics: The Finite Volume Method. Available online: http://ftp.demec.ufpr.br/disciplinas/TM702/Versteeg_Malalasekera_2ed.pdf (accessed on 22 April 2023).

energies

MDPI

Article

Detection Stability Improvement of Near-Infrared Laser Telemetry for Methane Emission from Oil/Gas Station Using a Catadioptric Optical Receiver

Dai Geng [1], Di Wang [2,*], Yushuang Li [2,*], Wei Zhou [2] and Hanbing Qi [3]

[1] School of Mechanical Science and Engineering, Northeast Petroleum University, Daqing 163318, China
[2] School of Physics and Electronic Engineering, Northeast Petroleum University, Daqing 163318, China
[3] School of Architecture and Civil Engineering, Northeast Petroleum University, Daqing 163318, China
* Correspondence: wangdinepu@stu.nepu.edu.cn (D.W.); dqpilys@nepu.edu.cn (Y.L.)

Abstract: Open-path laser telemetry of methane leakage yields security guarantees of energy storage and transportation for oil/gas station production operation. In order to further improve the long-term detection stability under the condition of long-distance non-cooperative targets, a catadioptric optical receiver (COR) consisting of a Fresnel lens, cone reflector and parabolic reflector is proposed to focus the laser echo light that deviates gradually with the increase in atmospheric turbulence. The geometric configuration parameters of COR are optimized by the ray-tracing method, and the condensing performance of COR is further verified. The self-developed methane laser telemetry system coupled with COR is calibrated in the laboratory and then moved to the field for a signal receiving stability experiment under turbulence interference. The results show that the receiving angle of COR increases 3.8 times compared with the Fresnel lens optical receiver (FOR). The RMSE and IS of the COR system are 0.00173 V and 84.79%, respectively. For comparison, these two evaluating indicators of the FOR system are 0.00288 V and 76.23%. This self-developed methane laser telemetry system coupled with COR is feasible for improving the long-term detection stability of remote leakage monitoring in oil/gas stations.

Keywords: methane emission; energy security; optical telemetry; laser absorption spectroscopy; optical receiver; detection stability

Citation: Geng, D.; Wang, D.; Li, Y.; Zhou, W.; Qi, H. Detection Stability Improvement of Near-Infrared Laser Telemetry for Methane Emission from Oil/Gas Station Using a Catadioptric Optical Receiver. *Energies* **2023**, *16*, 3854. https://doi.org/10.3390/en16093854

Academic Editor: Zissis Samaras

Received: 17 March 2023
Revised: 20 April 2023
Accepted: 27 April 2023
Published: 30 April 2023

1. Introduction

Methane emissions have 80 times the climate-change effects of carbon dioxide over a 20-year timeframe and are responsible for a quarter of today's global warming [1,2]. An increase in the energy intensity technology is correlated with a decrease in greenhouse gas emissions [3]. Currently, natural gas is considered to be an important link in the transition from the fossil energy system to a renewable energy system. In other words, natural gas use has risen dramatically and can replace ordinary fuels and electricity power for both environmental and economic reasons.

As a result, new storage and transport units are being constructed, and larger amounts of petroleum and especially natural gas are distributed worldwide. This change presents both opportunities and challenges. One of these challenges is how to best control methane emissions while embracing this new energy economy. The most significant segment in the oil and gas production and supply chain for methane emissions is natural gas field production (over 50%), followed by petroleum systems as a whole (over one-third) [4,5].

Methane emission inventories are undercounted as reported by the U.S. EPA (Environmental Protection Agency) [6,7]. Recent research has shown that methane emissions from the oil and gas sectors are also underestimated in Canada [8]. Methane fugitive emissions (leaks) are a major issue in both the oil and gas sectors, not only in environmental and economic terms because of the wasting of important natural energy but also importantly from a safety perspective [9]. For these and other reasons, it is important to develop excellent technologies to monitor natural gas and specifically methane emissions. Laser absorption spectroscopy sensing technology is making it easier to monitor leaks and control emissions, putting the industry in a position to recover the lost revenue that each year's methane emissions represent [10–14].

The monostatic open-path tunable laser absorption spectroscopy (MOP-TLAS) monitoring system is widely established for the protection of standoff facilities in the oil and gas sector [15], such as for leakage detection and warnings in the natural gas depot. In a MOP-TLAS monitoring system, the laser transmitter and receiver are located in the same fixed position. The effective path length is doubled, resulting in increased sensitivity. Depending on the desired distance to cover, simple reflective surfaces to highly precise retro-reflectors are used. He et al. [16] presented a highly effective method to measure methane emissions from landfills based on MOP-TLAS. The methane concentrations in six different directions were measured by placing six flat mirrors coated with aluminum films around a testing distance of 10 m.

Xia et al. [17] developed a movable MOP-TLAS platform combined with a retro-reflector array to continuously measure the atmospheric variations of the concentrations of methane and carbon dioxide over paths of up to 2.6 km. Zhu et al. [18] presented field deployment results of a portable MOP-TLAS for the localization and quantification of fugitive methane emissions, and the employment of a retro-reflector achieved the flexible measuring of different area sources under distances ranging from 1 to 1000 m.

The MOP-TLAS monitoring technology has made great breakthroughs in the telemetry range and concentration limit for methane detection in the oil/gas industry, but any deviation introduced by the laser transmission or the optical transceiver system will deteriorate the detectability and reliability of this technique [19]. Although the collimator can compress the divergence angle of the laser beam to a small range, the existence of the divergence angle will also lead to a decrease in parallelism of laser transmission with the increase in the detection distance.

In the MOP-TLAS monitoring system, a retro-reflector typically consists of three mirrors or reflective prism faces, which return an incident laser beam in the opposite direction [20]. It is obvious that the reflected laser echo is not a strictly parallel beam when it reaches the surface of the receiving lens. Since there is a certain off-axis angle between the laser echo and the optical axis of the receiving lens, the optical signal finally focused on the photodetector will inevitably have light intensity loss, which will intensify with the extension of the detection distance.

When laser absorption spectroscopy is used for gas remote detection in open space, weak signals carrying effective absorption information are extremely sensitive to optical noise, uncontrollable environmental noise, etc. Therefore, the more laser echo intensity received by the photodetector will be beneficial to improve the signal-to-noise ratio (SNR) of the system and further ameliorate the measurement accuracy [21–24]. To this end, several scholars have provided valuable resolves into this issue. In terms of concentration signal processing, Li et al. [25] applied normalized WMS-$2f/1f$ to improve the sensitivity and robustness of methane continuous monitoring.

Some scholars also optimized from the perspective of the optical transmitter and receiver. Li et al. [26] introduced a focus-tunable lens into a TLAS telemetry collimation system to dynamically adjust the divergent performance of the laser beam to maximize the received optical signal in methane remote detection. To receive more signal light and enhance the signal-to-noise ratio, Xiao and Hu [27] used a lens and its focal length, the spot size of the reflecting surface and a filter plate to optimizing the optical path.

However, there are still some defects in the actual application of the above methods. Turning the focal length of collimator lens by controlling the current will increase the complexity of the system circuit and introduce more random noise. Optimizing the lens size and focal length is only a compromise improvement. The immutable parameters of optical receiving elements will be difficult to adapt to the change of telemetry distance.

In this work, a catadioptric optical receiver (COR), which consists of a Fresnel lens, cone reflector and parabolic reflector is proposed to improve the remote detection stability performance of the methane laser telemetry sensor in the non-cooperative target scenario. The relation between the convergence point offsets of the Fresnel lens and the incidence angle of laser echo is analyzed. The geometric configuration parameters of COR are optimized by the TracePro 7.4.3 optical software, and the condensing performance of COR is further verified. The COR model with optimal parameters is customized and deployed into the independently developed methane laser telemetry system and a signal receiving stability experiment under turbulence interference is carried out to evaluate the reliability of COR.

2. Methods

2.1. Optical Model of Convergence Point Offsets for Fresnel Lens

A Fresnel lens has a significant advantage in enhancing the receiver performance of light [28]. However, when the incident angle of echo light rays is not 0, it will affect the echo receiving power of the detector. The optical path of echo light rays with different incident angle is shown in Figure 1. u is the incident angle of echo light rays. u' is the exiting angle of echo light rays. μ_1 is the incident angle on the incident plane of the Fresnel lens. $\mu_{1'}$ is the refraction angle on the incident plane. μ_2 is the incident angle on the exiting plane of the Fresnel lens. $\mu_{2'}$ is the refraction angle on the exiting plane. α is the vertex angle of the serrate prism. f is the focal length of the Fresnel lens. L is the horizontal distance between the exiting point and the incident plane. h is the vertical distance between the exiting point and the optical axis. δ is the deviation of the echo light ray convergence point. Additionally, n is the refractive index of the Fresnel lens.

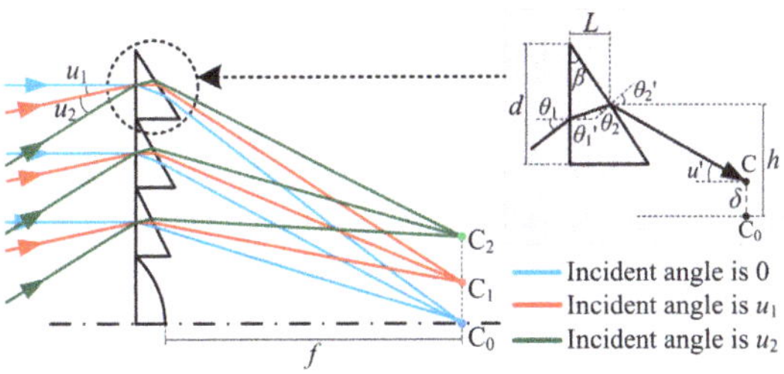

Figure 1. Deviation optical path of the Fresnel lens convergence point.

According to the refraction law and the geometric relationships, the refraction angle of each interface can be derived as follows:

$$u = \theta_1 \tag{1}$$

$$\theta_1' = \arcsin(\sin\theta_1 / n) \tag{2}$$

$$u' = \arcsin\left[n \sin\left(\beta_i + \theta_1'\right)\right] - \beta_i \tag{3}$$

$$\tan u' = \frac{h - \delta}{f - (d/2)\tan\beta} \tag{4}$$

By combining the above equations, it can be obtained that the convergence point offset of a Fresnel lens is

$$\begin{gathered}\delta = h - f\tan\{\arcsin[n\sin(\alpha + \arcsin(\sin u / n))] - \alpha\} + \\ \tfrac{d}{2}\tan\alpha\tan\{\arcsin[n\sin(\alpha + \arcsin(\sin u / n))] - \alpha\}\end{gathered} \tag{5}$$

2.2. Optical Model of Segmentation Cell of a Parabolic Reflector for COR

When the echo light rays are parallel to the optical axis of the Fresnel lens, the signal reception performance of the methane laser telemetry sensor can be greatly improved, but with a poor focusing ability to the echo light rays with a large off-axis angle. A composite parabolic concentrator (CPC) can be designed with a large field angle, but its light gathering performance is low. In the fields of solar photovoltaic and visible light communication, the Fresnel lens and composite parabolic reflector are usually combined to form a secondary focusing system [29,30].

Therefore, based on the design idea of multi-layer structure decomposition and multi-level light gathering, a novel catadioptric optical receiver that consists of a Fresnel lens, cone reflector and parabolic reflector is proposed, and its parameters are optimized using TracePro optical software. The parabolic reflector can be divided into multiple annular condenser cells. The reflected optical path of the echo light ray in each cell is shown in Figure 2.

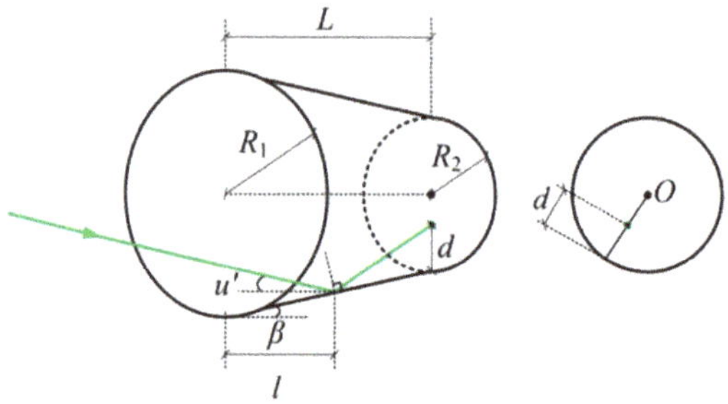

Figure 2. Optical model of a segmentation cell of a parabolic reflector for COR.

As shown in Figure 2, the incident aperture of the annular condenser cell is R_1, the exit aperture is R_2, the length is L, the reflecting surface inclination is $\tan\beta$, the angle between the echo light ray and the optical axis is u', the horizontal distance between the reflection point and the incident port is l, and the distance from the falling point of the outgoing light to the photosensitive surface of detector along the radial direction to the edge of photosensitive surface is d, which can be obtained from the geometric relationship:

$$d = \left[\tan(2\beta + u') - \tan\alpha\right] \cdot (L - 1) \tag{6}$$

Only when $0 < d < 2R_2$ is met can the incident light can be received by the photosensitive surface of the detector.

3. Results and Discussion

3.1. Structural Parameters for Optimization Analysis of COR

The modeling and simulation of the Fresnel lens was implemented in TracePro software, and its specific parameters are shown in Table 1. As shown in Figure 3, the incident light rays from 0° to 5° were traced, and the deviation of the convergence point is marked. As the incident light gradually deviates from the optical axis of the Fresnel lens, the focal shift along the X-axis direction (i.e., the radial direction of the Fresnel lens) increases proportionally with the change of the incident angle, and the focal shift is about three times the incident angle, which is consistent with the analytical result of Equation (5).

Table 1. Optical parameters of the Fresnel lens in this work.

Diameter [mm]	Focal Length [mm]	Thickness [mm]	Pitch [°]	Groove Spacing [mm]	Refractive Index
50	93	2	1	10	1.516

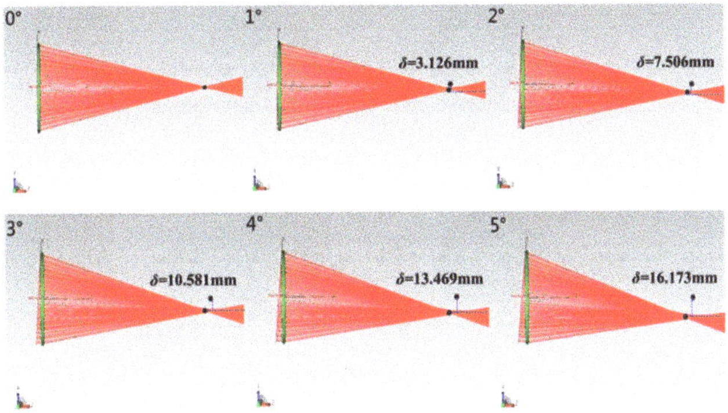

Figure 3. Focusing deviation of the Fresnel lens at different incident angles.

The simulation analysis of the Fresnel lens optical reception in TracePro software was performed, and a receiving plane with a size of 5 mm at the focus of the Fresnel lens as the photosensitive surface of the detector was added. As shown in Figure 4, the optical efficiency was 82.5% at the incident angle with 1.3°. According to the definition of the receiving angle of the condenser system, the incident angle corresponding to 90% of the optical efficiency under the vertical incidence was the receiving angle. Due to the absorption loss when the light passes through the semi-transparent medium, the maximum optical efficiency of the Fresnel lens in this paper was 92.2%. Therefore, the receiving angle was 1.3° for the single Fresnel lens used as the condenser.

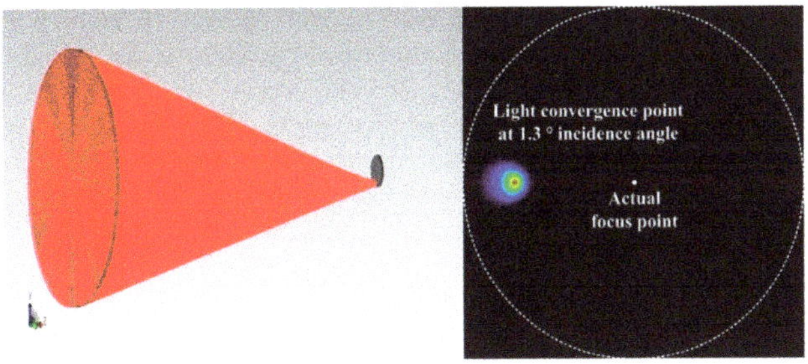

Figure 4. Light convergence point at 1.3° incident angles.

When the incident angle of the echo light ray was less than 1.3°, the detector could achieve the maximum optical efficiency only by using a Fresnel lens. Therefore, the functional contribution of COR was started from an incident angle greater than 1.3°. When the incident angle was greater than 1.3°, the parabolic reflector part of COR was optimized and analyzed according to the edge-ray principle.

The principle is as follows: the optical receiving efficiency of the detector under angle-by-angle light ray is maximized, and a light ray that has entered the reflection cell cannot escape through secondary reflection. Considering the complexity and cost of the actual preparation process, the parabolic reflector part of COR is divided into the annular condenser cells with a length L_1 of 2 mm, and the exit aperture is $R_{2,1}$ of the first-stage annular condenser cell is set to be 5 mm. When the incident angle of the echo light ray is 1.4°, the first-stage annular condenser cell model with incident apertures of 5.2, 5.4, 5.6, 5.8, 6.0, 6.2 and 6.4 mm is ray-traced, and the simulation results are shown in Figure 5.

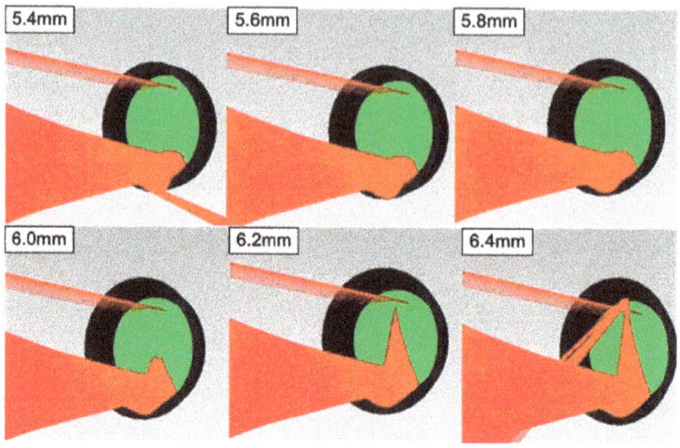

Figure 5. The ray tracing results of the first-stage annular condenser cell under different incident apertures.

The amount of light received on the detector can be increased by adjusting the incident aperture. However, when the radius of the incident plane reaches 6.4 mm, some light rays have been reflected. Further analysis of the 6.3 mm aperture model shows that the incident light is also reflected. Therefore, the incident aperture corresponding to the best focusing effect of 1.4° incident light is 6.2 mm. However, when the incident aperture reaches 6.4 mm, some light has been reflected from the model.

Further analysis of the 6.3 mm incident aperture model shows that a small amount of the echo light ray also escapes. Therefore, for the echo light ray with an incident angle of 1.4°, the incident aperture corresponding to the best focusing effect is 6.2 mm. In this case, the optical receiving efficiency of detector does not reach the maximum value, because a small amount of light rays are projected on the incident aperture edge. When the second-stage cell is added, the extended overall length will improve this problem.

Similarly, the exit aperture of the previous stage is used as the incident aperture of the next stage. By changing the reflecting surface inclination, the structural parameters of the annular condenser cell at all stages can be determined in turn, as shown in Table 2.

Table 2. Structural parameters of the annular condenser cell from the first stage to the twelfth stage.

Stage Number	1	2	3	4	5	6	7	8	9	10	11	12
Incident aperture [mm]	6.20	7.00	7.80	8.50	9.10	9.60	10.10	10.50	10.84	11.14	11.34	11.44
Exit aperture [mm]	5.00	6.20	7.00	7.80	8.50	9.10	9.60	10.10	10.50	10.84	11.14	11.34
Length [mm]	2.00	2.00	2.00	2.00	2.00	2.00	2.00	2.00	2.00	2.00	2.00	2.00

The structural parameters were input into MATLAB (R2018b) software for curve fitting to obtain the parabolic equation. The incident aperture of parabolic reflector part was 11.44 mm, the exit aperture was 5 mm, and the length was 24 mm. The optical receiving efficiency of the parabolic reflector part of COR is shown in Figure 6. The parabolic reflector part had good focusing performance for the echo light in the range of 0° to 2.9° of the incident angle. With the continuous increase in the incident angle, its focusing performance dropped sharply.

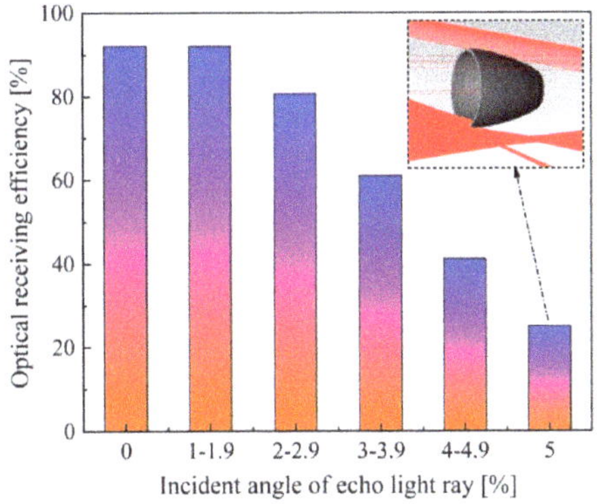

Figure 6. The optical receiving efficiency of the parabolic reflector part of COR under different incident angles of echo light rays.

When the incidence angle was 5°, its optical receiving efficiency was only 27.2%. The main reason is that the length of the parabolic reflector is short, and the shift of the convergence point of the echo light rays is about three times of the incident angle, which means that most of the light rays cannot enter the parabolic reflector. Therefore, further optimization and analysis can be carried out by adding the cone reflector at the front of the parabolic reflector part and changing the reflecting surface inclination and length.

The length of the first-stage cone reflector was set to 2 mm, the exit aperture is known to be the same as the incident aperture of the parabolic reflector, and the reflecting surface inclination can be changed by changing the incident aperture of cone reflector. The cone reflector model with incident apertures of 11.44, 11.6, 11.8, 12, 12.2, 12.4 and 12.6 mm was built, and the concentrating performance simulation of the echo light rays under different incident angles was implemented. The number of echo light rays for each incident angle condition was 7351, and the number of echo light rays received by the detector is shown in Figure 7.

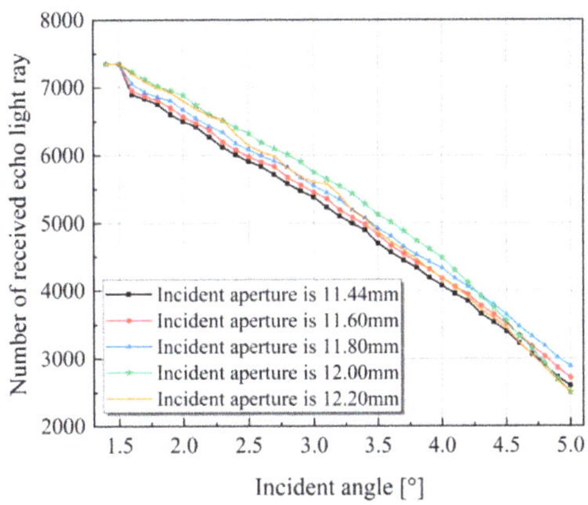

Figure 7. The relationship between the number of received echo light rays and the incident aperture of the first-stage cone reflector.

The echo light rays can be fully received by the detector regardless of the incident aperture when the incident angle is less than 1.5°. When the incident angle changed from 1.6° to 4.2°, the number of echo light rays received by the detector first decreased and then increased with increasing incident aperture. The maximum value was achieved when the incident aperture was 12 mm. When the incident angle ranged from 4.2° to 5°, the change trend of the number of echo light rays received by the detector was consistent with that of the former.

However, the number of received echo light rays reached a maximum at the incident aperture of 11.8 mm. Considering the design principles and practical manufacturing difficulties, the incident aperture of the first-stage cone reflector was set as 12 mm. Thus, when the reflecting surface inclination was 0.28, there was an optimal solution for the concentrating performance under the incident angel changed from −5° to 5°.

The reflecting surface inclination and incident aperture were set and maintained as constant for the first-stage cone reflector. It should be noted that the first-stage cone reflector was given a length change only. To ensure that the light passing through the Fresnel lens was not blocked when entering the COR, a Fresnel lens installation section with an inner diameter of 50 mm and a length of 4 mm was designed at the COR front end. In addition, the second-stage cone reflector was added, and its incident aperture was consistent with the size of the Fresnel lens.

The ray-tracing simulation was performed, and the number of received echo light rays by the detector is shown in Figure 8. The different length combinations of the first-stage and second-stage cone reflectors determine the concentrating performance for the echo light rays with an incident angle greater than 3.9°. With the increase in the incident angle, the increase in the length of the first-stage cone reflector led to increasingly less echo light rays on the detector. When the length was 2 mm, the echo light rays were better received by the detector.

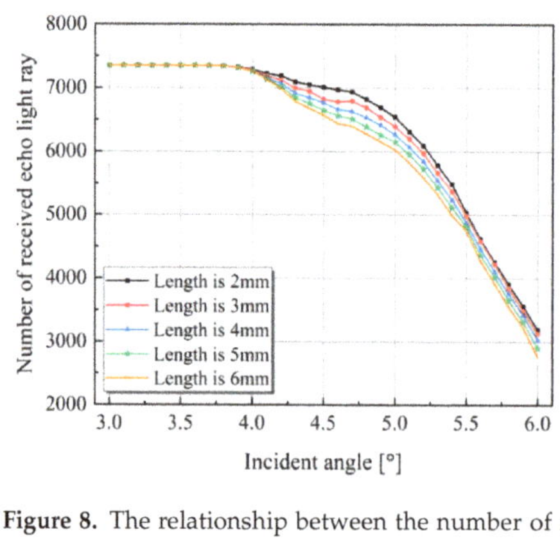

Figure 8. The relationship between the number of received echo light rays and the length of the first-stage cone reflector.

The complete optical structure of COR was obtained through optimization design as shown in Figure 9. COR was successively connected by the Fresnel lens installation section, the second-stage cone reflector, the first-stage cone reflector and the parabolic reflector. The Fresnel lens was located at the incident end. When the incident angle of the echo light rays were less 1.3°, the echo light rays were directly focused on the detector by the refraction of the Fresnel lens.

For echo light rays with incident angle greater than 1.3°, the echo light rays first pass through the refraction of the Fresnel lens, some of which can be directly incident to the detector, and the rest enter the detector through multiple reflections of the cone reflector and parabolic reflector. In order to evaluate the optical receiving performance of COR, two optical models, the single Fresnel lens optical receiver (FOR) and the Fresnel lens and parabolic reflector combining optical receiver (POR) were established in the TracePro software as the contrast reference.

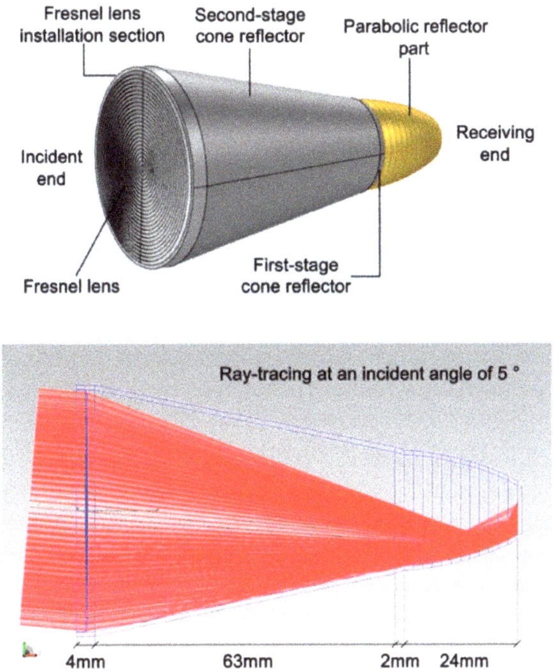

Figure 9. The relationship between the number of received echo light rays and the length of the first-stage cone reflector.

The simulation analysis of the echo light ray receiving performance was conducted, and the results are shown in Figure 10. When the incident angle of echo light ray was 0°, the optical receiving efficiencies of all three optical receivers were 92.2%. Thus, the Fresnel lens of these three optical receivers plays a direct focusing role when the echo light rays are parallel to the optical axis. As the incident angle increases gradually, the optical efficiency of these three optical receivers declines. COR was compared with FOR: the optical receiving efficiency of FOR was reduced to 0 when the incident angle was greater than 2°. Thus, the detector could not receive the echo light ray.

In comparison, the concentrating performance of COR with an incident angle greater than 2° was significantly improved. According to the definition of the receiving angle, the receiving angle of FOR was only 1.3°, while the receiving angle of COR was 4.9°, which is 3.8 times that of FOR. At this time, the optical efficiency of COR was 83.5%. Compared with POR, COR and POR had basically the same optical receiving efficiency in the incident angle variant range of 0° to 2°. When the incident angle was between 2° and 5°, the optical efficiency of COR was 30% higher than that of POR. When the incident angle was greater than 5°, although the optical efficiency of both decreased rapidly with the increase in the incident angle, the optical efficiency of COR was still palpably higher than that of POR.

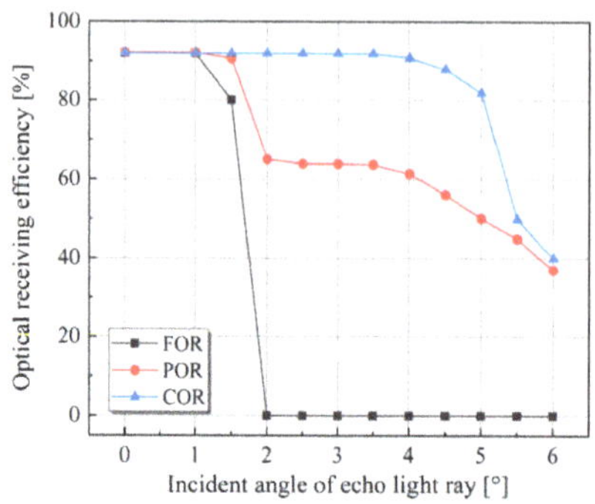

Figure 10. The variation of the optical receiving efficiency of three optical receivers over the incident angle of echo light rays.

3.2. Performance Analysis of Methane Laser Telemetry Coupled with the COR System

The schematic of the independently developed methane laser telemetry coupled with the COR system is shown in Figure 11. The sensor architecture mainly includes an optical subsystem and electrical subsystem. In the optical subsystem part, a single-mode continuous distributed feedback laser diode (EP-1653-DM, Eblana Photonics, Dublin, Ireland) using a TO39 package structure was employed as the laser sources with the wavelength ~1653.7 nm (corresponding to the absorption features of methane), output power of 10 mW and optical linewidth of 2 MHz. A K9 Plano convex lens (100006, United Optical Technology (Beijing) Co., Ltd., Beijing, China) was involved to control the divergence angle of the laser beam and was installed on the optical lens mounts in front of the laser diode.

The collimated emission laser beam was diffusely reflected by the non-cooperative target (white latex paint wall) after passing through the methane cloud. The echo light was collected by the COR to ensure the reception strength of the laser echo signal. A Fresnel lens (Φ = 50 mm, f = 93 mm) functioned as a refraction tool of COR to take the lead in receiving as many echo signals as possible. A stereolithography-600 3D printer (Shenzhen Sogaworks Technology Co., Ltd., Shenzhen, China) was used to fabricate the reflective cavity part of COR, and the inner wall of the reflective cavity was plated with a silver film. The reflectivity reached more than 95%.

The COR was embedded in the stacked lens tubes to maintain the mechanical stability. An InGaAs PIN photodiode (LSIPD-L2.5, Beijing Lightsensing Technologies Ltd., Beijing, China) was placed at the exit end of COR to transform the optical signal into an electrical signal. The active diameter was 2.5 mm, and the spectral responsivity was 0.83 mA/mW. Since the spectral reception range of the photodetector was optical signals of all wavelengths from 800 to 1700 nm, a visible light band-pass filter was configured to filtrate out the ambient background stray light. The STM32F407 Microcontroller unit was the main controller for the electrical module.

Excellent linearity and smoothness of the saw tooth wave shape was achieved using a circuit that generated waves with numerical control of the baseline and slope. Combined saw tooth and sine waves were used as laser driver signals. The operational amplifiers were adopted in the photoelectric conversion circuit for obtaining high-speed and ultra-low noise signals. With the utilization of a multistage band-pass filter circuit and programmable gain amplification circuit, methane information was converted from analog to digital. The updated LabVIEW program was adopted to control the system parameters and measure laser spectra signals. The program featured laser scanning and modulation, data acquisition, harmonic signal demodulation and display and methane concentration data output.

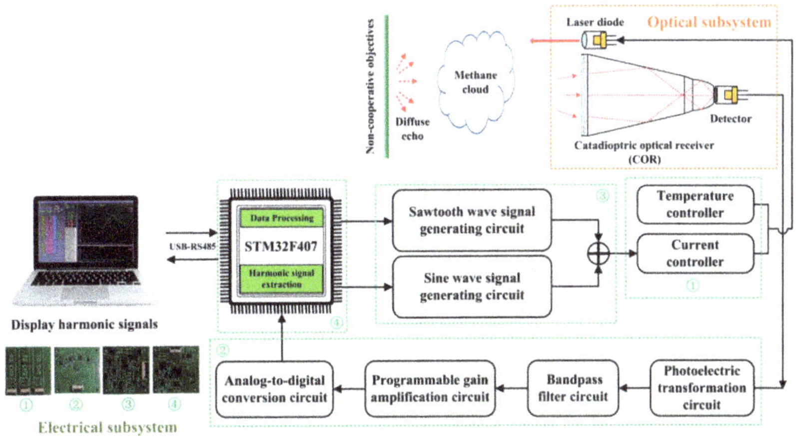

Figure 11. Schematic of the independently developed methane laser telemetry coupled with the COR system.

The previously verified methane concentration inversion method based on the novel wavelength modulation spectroscopy was adopted in this work [21], that is, the methane concentration was characterized using the signal amplitude obtained by normalizing the first harmonic (1*f*) signal to the second harmonic (2*f*), and the reliable telemetry stability was secured using the distorted harmonic waveform recognition algorithm. The signal receiving experiment of the methane laser telemetry system was first performed in the hallway of NEPU Chemical Laboratory building. In order to visually show the feasibility of COR, we replaced the COR in the methane laser telemetry system with a single Fresnel lens as a reference object for comparison.

The detection system was mounted on the tripod to reduce the interference of mechanical vibration on the results during the test. The tripod was positioned at forty different detection distances within a 10 m linear range with a variation interval of 0.25 m. To simulate the scene of natural gas leakage clouds, a gas sampling bag made of thermoplastic polyurethanes was placed in the optical path, which was filled with methane of 1%vol. For each distance, the methane $2f/1f$ signal amplitude was collected for one minute, and the mean values in the time interval were plotted as a function of the distance as shown in Figure 12.

The measured signals from the methane laser telemetry system coupled with COR are shown in Figure 12a. Due to the nonlinear behavior of the laser intensity modulation, the residual amplitude modulation produced a background in the 2*f* signal. It can be compressed by using the the first harmonic (1*f*) normalized second harmonic (2*f*) signal (i.e., $2f/1f$ siganl).

As shown in Figure 12a, the change rule of the second harmonic signal amplitude of these two systems with the detection distance is consistent, and both of them show a trend of increasing first and then decreasing. However, within the detection distance of less than 1.25 m, the second harmonic amplitude of the COR system is significantly higher than that of the FOR system, which will also provide a better detection signal-to-noise ratio.

The reason for this phenomenon can be explained as follows: when the distance between the optical receiving system and the target surface is relatively short, the incident angle of the diffuse reflection echo light is relatively large relative to the optical axis of the optical receiving system, and the Fresnel lens of the FOR system causes the echo light to not be fully focused on the detector. The COR system not only relies on the refraction focusing function of the Fresnel lens but also uses the secondary reflection function to further focus the echo light with too large incidence angle so that its detector can receive more echo light with the absorption signal.

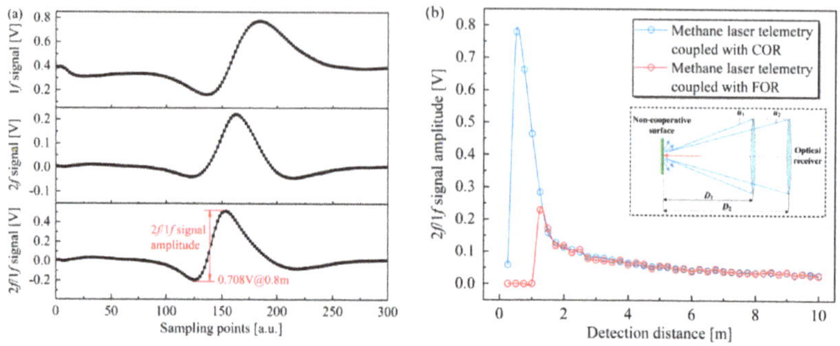

Figure 12. (**a**) Observed signals at a detection distance of 0.8 m from the methane laser telemetry system coupled with COR. (**b**) The $2f/1f$ signal amplitude of the methane laser telemetry system coupled with COR or Fresnel lens variations according to detection distance.

With the extension of the detection distance, the incident angle of the echo light decreases gradually. When it is less than the effective receiving angle of the Fresnel lens, both systems can concentrate all the echo light to the detector only by relying on the Fresnel lens. However, the light intensity of the echo light decreases rapidly with the increase in the detection distance, which was confirmed in our previous work [31], resulting in the continuous reduction of the amplitude of the second harmonic signal, and the signal intensity of these two systems is basically the same. It is obvious that the signal reception effect of the COR system shows more satisfactory results when measured in close range.

To evaluate the detection limit of the methane laser telemetry system coupled with COR, the 300 ppm·m methane was continuously monitored for more than 1200 s with the scan frequency of 5 Hz. Figure 13 shows the Allan deviation analysis for the continuous measurement. The Allan deviation reflects the fluctuation of the measurement result with the passage of the integration time. It indicates that the optimum integration time was 319.6 s, and the corresponding detection limit was 5.13 ppm·m. When the integration time was within 320 s, the trend of the Allan deviation decreasing was basically the same as ~$1/\text{sqrt}(\tau)$, which indicates that thermal noise of passive device was dominant in the measurement results at this time. When the integration time was over 320 s, increasing the integration time could not further improve the detection limit.

The transmission of the laser beam through the atmospheric air beside the absorbing gaseous components and atmospheric aerosols is affected by air flow turbulence, which results in the amplitude and beam deflection of the received echo light rays [17]. Therefore, atmospheric turbulence will cause the incident angle of the echo light to fluctuate randomly when it enters the optical receiver of the methane laser telemetry system. The weakening of received laser signal becomes more pronounced with increasing incident angles. In order to investigate the stability performance of the methane laser telemetry sensor under the effect of atmospheric turbulence, time series measurements of methane gasbag were performed in an empty outdoor space.

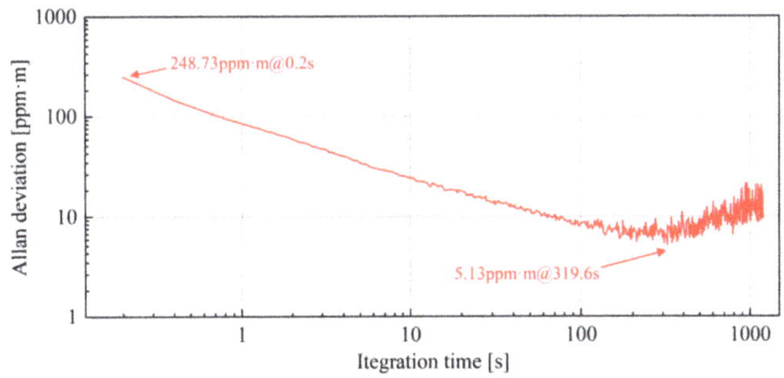

Figure 13. Allan deviation analysis for the 300 ppm·m methane measurement.

A comparative experiment was implemented by using a methane laser telemetry system with COR and the system with FOR. For both systems, the sampling period of the methane laser telemetry sensor was 0.5 s. The methane gasbag with an integrated concentration of 300 ppm·m was monitored at a detection distance of 10 m within 900 s. The calibration experiment of the signal amplitude of the methane gasbag was completed in the laboratory indoor in advance, and the signal amplitude at this integrated concentration was 0.0652 V. Two industrial axial flow fans were placed on both sides of the laser light path, and the fan wind speed was set at 12 m/s to simulate the atmospheric turbulence area. The schematic diagram of outdoor experiment and the $2f/1f$ signal amplitude data of these two systems are shown in Figure 14.

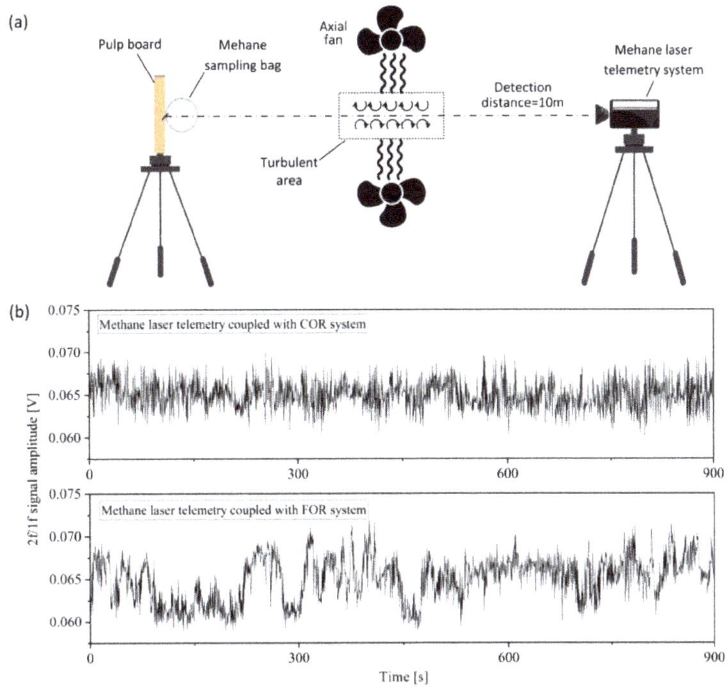

Figure 14. A signal-receiving stability experiment under turbulence interference. (**a**) Schematic diagram of experimental configuration. (**b**) $2f/1f$ signal amplitude measurement data during the monitoring period.

The signal amplitude markedly fluctuated during the measurement cycle for these two systems, which is partly due to the background noise generated by the internal electronic components of the sensor system but in large part because of the atmospheric turbulence. However, the fluctuation range (~0.00989 V) of the measurement data obtained by the COR system was significantly smaller than that of the FOR system (~0.0131 V). The root mean square error (RMSE) and indication stability (IS) were introduced to quantitatively assess the stability performance of these system.

$$\mathrm{RMSE} = \sqrt{\frac{1}{N}\sum_{i=1}^{N}(X_i - X_0)^2} \tag{7}$$

$$\mathrm{IS} = \left(1 - \frac{X_{\max} - X_{\min}}{X_{\mathrm{mean}}}\right) \times 100\% \tag{8}$$

where N is the total number of data samples; X_i is the measured signal amplitude data; X_0 is the real signal amplitude; $X_{\max}$ is the maximum data; $X_{\min}$ is the minimum data; and X_{mean} is the average data. The RMSE and IS of the COR system were 0.00173 V and 84.79%, respectively. These two evaluating indicators of the FOR system were 0.00288 V and 76.23%, which indicates that the COR system had better detection stability performance under the effects of atmospheric turbulence.

4. Conclusions

In this paper, we explored methane emission remote detection stability by a catadioptric optical receiver on the basis of the near-infrared laser absorption spectroscopy telemetry technique. The relation between the convergence point offsets of the Fresnel lens and the incidence angle of laser echo was analyzed. A catadioptric optical receiver consisting of a Fresnel lens, a second-stage cone reflector, a first-stage cone reflector and a parabolic reflector was proposed, and the geometric configuration parameters were optimized using the ray-tracing method. Then, the COR module was manufactured and installed in the self-developed methane laser telemetry system. The signal receiving stability experiment under turbulence interference was performed to evaluate the reliability of COR. The main conclusions of this work are as follows:

(1) Compared with FOR, the receiving angle of COR increased 3.8 times. Compared with POR, their optical efficiency decreases rapidly when the incident angle was greater than 5°, but the optical efficiency of COR was still significantly higher than that of POR.

(2) The RMSE and IS of the COR system were 0.00173 V and 84.79%, respectively. For comparison, these two evaluating indicators of the FOR system were 0.00288 V and 76.23%. The developed catadioptric optical receiver provided superior stability for laser telemetry of methane.

Based on the results obtained, both the simulation and experimental results demonstrated that the self-developed methane laser telemetry system coupled with COR can be arranged as a leakage long-term monitoring tool for methane gas in an oil/gas station. At present, this method has only been tested in excellent indoor and outdoor environments. In practice, the environment of an oil/gas station site can be complicated.

In the process of using lasers for telemetry, unfavorable weather conditions have an impact on laser detection sensitivity, such as the attenuation of laser power due to the absorption and scattering effects of fog, rain and snow in the atmosphere. For future work, we are working hard on the optimization design of the signal filtering and waveform reconstruction to improve the detection limit of the methane laser telemetry system. In addition, we envision the combining methane optical telemetry, scanning tomography and gas diffusion modeling techniques for the acquisition for flux measurements of fugitive methane emissions.

Author Contributions: Conceptualization, D.G.; methodology, D.W.; software, W.Z.; validation, D.G. and Y.L.; formal analysis, D.W.; investigation, D.G.; resources, H.Q.; data curation, Y.L.; writing—original draft preparation, D.W.; writing—review and editing, H.Q.; visualization, D.W.; supervision, D.G.; project administration, H.Q.; funding acquisition, D.W. and H.Q. All authors have read and agreed to the published version of the manuscript.

Funding: This research was funded by Scientific and Technological Achievements Industrialization Cultivation Funding, grant number 15011210801 and Postdoctoral Support Project of Heilongjiang Province, grant number LBH-Q21084.

Data Availability Statement: Data are available on request from the authors. The data that support the findings of this study are available from the corresponding author upon reasonable request.

Conflicts of Interest: The authors declare that they have no conflict of interest.

Nomenclature

d	Deviation of the echo light ray convergence point for COR, cm
f	Focal length of the Fresnel lens, mm
h	Vertical distance between the exiting point and the optical axis, cm
L	Horizontal distance between the exiting point and the incident plane, cm
n	Refractive index of the Fresnel lens, -
N	Total number of data samples, -
R_1	Incident aperture of the annular condenser cell, cm
R_2	Exit aperture of the annular condenser cell, cm
u	Incident angle of echo light ray, °
u'	Exiting angle of echo light ray, °
X_i	Measured signal amplitude data, -
X_0	Real signal amplitude, -
X_{max}	Maximum measurement data, -
X_{min}	Minimum measurement data, -
X_{mean}	Average measurement data, -

Greek letters	
α	Vertex angle of serrate prism, °
$\tan\beta$	Reflecting surface inclination, °
δ	Deviation of the echo light ray convergence point for the Fresnel lens, cm
μ_1	Incident angle on the incident plane of the Fresnel lens, °
$\mu_{1'}$	Refraction angle on the incident plane, °
μ_2	Incident angle on the exiting plane of the Fresnel lens, °
$\mu_{2'}$	Refraction angle on the exiting plane, °

References

1. Allen, D.T.; Cardoso-Saldaña, F.J.; Kimura, Y.; Chen, Q.; Xiang, Z.; Zimmerle, D.; Bell, C.; Lute, C.; Duggan, J.; Harrison, M. A Methane Emission Estimation Tool (MEET) for predictions of emissions from upstream oil and gas well sites with fine scale temporal and spatial resolution: Model structure and applications. *Sci. Total Environ.* **2022**, *829*, 154277. [CrossRef] [PubMed]
2. Floerchinger, C.; McKain, K.; Bonin, T.; Peischl, J.; Biraud, S.C.; Miller, C.; Ryerson, T.B.; Wofsy, S.C.; Sweeney, C. Methane emissions from oil and gas production on the North Slope of Alaska. *Atmos. Environ.* **2019**, *218*, 116985. [CrossRef]
3. Konieczna, A.; Roman, K.; Borek, K.; Grzegorzewska, E. GHG and NH_3 Emissions vs. Energy Efficiency of Maize Production-Technology: Evidence from PolishFarms; a Further Study. *Energies* **2021**, *14*, 5574. [CrossRef]
4. Sun, S.; Ma, L.; Li, Z. A Source-Level Estimation and Uncertainty Analysis of Methane Emission in China's Oil and Natural Gas Sector. *Energies* **2022**, *15*, 3684. [CrossRef]
5. Heydarzadeh, Z.; Kinnon, M.M.; Thai, C.; Reed, J.; Brouwer, J. Marginal methane emission estimation from the natural gas system. *Appl. Energy* **2020**, *277*, 11557. [CrossRef]
6. Brandt, A.R.; Heath, G.; Kort, E.; O'sullivan, F.; Petron, G.; Jordaan, S.; Tans, P.; Wilcox, J.; Gopstein, A.; Arent, D.; et al. Methane leaks from north american natural gas systems. *Science* **2014**, *343*, 733–735. [CrossRef]
7. Wang, J.; Ji, J.; Ravikumar, A.P.; Savarese, S.; Brandt, A.R. VideoGasNet: Deep learning for natural gas methane leak classification using an infrared camera. *Energy* **2022**, *238*, 121516. [CrossRef]
8. MacKay, K.; Lavoie, M.; Bourlon, E.; Atherton, E.; O'Connell, E.; Baillie, J.; Fougere, C.; Risk, D. Methane emissions from upstream oil and gas production in Canada are underestimated. *Sci. Rep.* **2021**, *11*, 8041. [CrossRef]
9. Liu, C.; Wang, Y.; Li, Y.; Xu, M. Experimental study on new leak location methods for natural gas pipelines based on dynamic pressure waves. *J. Nat. Gas Sci. Eng.* **2018**, *54*, 83–91. [CrossRef]
10. Bakkaloglu, S.; Lowry, D.; Fisher, R.E.; France, J.L.; Brunner, D.; Chen, H.; Nisbet, E.G. Quantification of methane emissions from UK biogas plants. *Waste Manag.* **2021**, *124*, 82–93. [CrossRef]

11. Kordella, S.; Ciotoli, G.; Dimas, X.; Papatheodorou, G.; Etiope, G. Increased methane emission from natural gas seepage at Katakolo Harbour (Western Greece). *Appl. Geochem.* **2020**, *116*, 104578. [CrossRef]
12. Schimmelmann, A.; Ensminger, S.A.; Drobniak, A.; Mastalerz, M.; Etiope, G.; Jacobi, R.D.; Frankenberg, C. Natural geological seepage of hydrocarbon gas in the Appalachian Basin and Midwest USA in relation to shale tectonic fracturing and past industrial hydrocarbon production. *Sci. Total Environ.* **2018**, *644*, 982–993. [CrossRef] [PubMed]
13. Hollenbeck, D.; Zulevic, D.; Chen, Y. Advanced Leak Detection and Quantification of Methane Emissions Using sUAS. *Drones* **2021**, *5*, 117. [CrossRef]
14. Zhang, E.J.; Teng, C.C.; van Kessel, T.G.; Klein, L.; Muralidhar, R.; Wysocki, G.; Green, W.M.J. Field Deployment of a Portable Optical Spectrometer for Methane Fugitive Emissions Monitoring on Oil and Gas Well Pads. *Sensors* **2019**, *19*, 2707. [CrossRef]
15. Ro, K.S.; Johnson, M.H.; Varma, R.M.; Hashmonay, R.A.; Hunt, P. Measurement of greenhouse gas emissions from agricultural sites using open-path optical remote sensing method. *J. Environ. Sci. Health A* **2009**, *44*, 1011–1018. [CrossRef] [PubMed]
16. He, H.; Gao, S.; Hu, J.; Zhang, T.; Wu, T.; Qiu, Z.; Zhang, C.; Sun, Y.; He, S. In-situ testing of methane emissions from landfills using laser absorption spectroscopy. *Appl. Sci.* **2021**, *11*, 2117. [CrossRef]
17. Xia, J.; Zhu, F.; Zhang, S.; Kolomenskii, A.; Dong, J.; Okada, K.; Strohaber, J.; Schuessler, H.A. Probing greenhouse gases in turbulent atmosphere by long-range open-path wavelength modulation spectroscopy. *Opt. Laser Eng.* **2019**, *117*, 21–28. [CrossRef]
18. Zhu, H.; Letzel, M.O.; Reiser, M.; Kranert, M.; Bächlin, W.; Flassak, T. A new approach to estimation of methane emission rates from landfills. *Waste Manag.* **2013**, *33*, 2713–2719. [CrossRef]
19. Li, J.; Du, Z.; Liu, C. Standoff Chemical Detection Using Laser Absorption Spectroscopy: A Review. *Remote Sens.* **2020**, *12*, 2771. [CrossRef]
20. Lewellen, J.W.; Harris, J.R. Performance estimates for a multicube retroreflector design. *Opt. Commun.* **2019**, *441*, 26–32. [CrossRef]
21. Shen, S.; Li, W.; Wang, M.; Wang, D.; Li, Y.; Li, D. Methane near-infrared laser remote detection under non-cooperative target condition based on harmonic waveform recognition. *Infrared Phys. Technol.* **2022**, *120*, 103977. [CrossRef]
22. Barrass, S.; Gérard, Y.; Holdsworth, R.J.; Martin, P.A. Near-infrared tunable diode laser spectrometer for the remote sensing of vehicle emissions. *Spectrochim. Acta A* **2004**, *60*, 3353–3360. [CrossRef] [PubMed]
23. Yang, H.; Bu, X.; Cao, Y.; Song, Y. A methane telemetry sensor based on near-infrared laser absorption spectroscopy. *Infrared Phys. Technol.* **2021**, *114*, 103670. [CrossRef]
24. Zhang, Y.; Ding, J.; Zhang, X.; Fang, J.; Zhao, Y. Open-path sensor based on QCL for atmospheric N_2O measurement. *Results Phys.* **2021**, *31*, 104909. [CrossRef]
25. Shao, L.; Fang, B.; Zheng, F.; Qiu, X.; He, Q.; Wei, J.; Li, C.; Zhao, W. Simultaneous detection of atmospheric CO and CH_4 based on TDLAS using a single 2.3 μm DFB laser. *Spectrochim. Acta A* **2019**, *222*, 117118. [CrossRef]
26. Li, G.; Ma, K.; Jiao, Y.; Jiang, Q.; Zhang, X.; Zhang, Z.; Wu, Y.; Song, D. Performance enhancement of DFBL based near-infrared CH_4 telemetry system using a focus tunable lens. *Microw. Opt. Technol. Lett.* **2021**, *63*, 1147–1151. [CrossRef]
27. Xiao, C.; Hu, S. Optimization of the Optical Path of the Methane Telemetry System Based on TDLAS Technology. In Proceedings of the International Conference on Optical Instruments and Technology: Optical Systems and Modern Optoelectronic Instruments, Beijing, China, 2–4 November 2020; p. 114341B.
28. Xu, N.; Ji, J.; Sun, W.; Huang, W.; Li, J.; Jin, Z. Numerical simulation and experimental validation of a high concentration photovoltaic/thermal module based on point-focus Fresnel lens. *Appl. Energy* **2016**, *168*, 269–281. [CrossRef]
29. Cabral, D. Development and performance comparison of a modified glazed CPC hybrid solar collector coupled with a bifacial PVT receiver. *Appl. Energy* **2022**, *325*, 119653. [CrossRef]
30. Peng, X.; Kong, L. Design and optimization of optical receiving antenna based on compound parabolic concentrator for indoor visible light communication. *Opt. Commun.* **2020**, *464*, 125447. [CrossRef]
31. Wang, D.; Wu, G.; Li, D.; Qi, H.; Lv, Y. Research on Echo Power Detection Based on Laser Radiation Transmission Characteristics of Non-Cooperative Target Surface. In Proceedings of the 16th International Heat Transfer Conference, Beijing, China, 13 August 2018.

sustainability

Article

An Investigation of the Effect of Propylene Gas Flame on Emissions and Temperature Distribution of a Preheated Metal Plate

Y. Nhu Quach and Ocktaeck Lim *

School of Mechanical Engineering, University of Ulsan, San 29, Mugeo2-dong, Nam-gu, Ulsan 44610, Republic of Korea; quachnhuyhcmute@gmail.com
* Correspondence: otlim@ulsan.ac.kr; Tel.: +82-10-7151-8218

Abstract: This study investigates the effect of the propylene gas flame on the emissions and temperature distribution of the metal plate during the preheating process. Experimental tests were carried out using a preheating system with a cylindrical chamber for emissions measurement and a metal plate placed near the torch head. Emissions were measured using a gas analyzer, while the temperature distribution of the metal plate was measured using an infrared thermal camera and thermocouples. The findings reveal that the emissions decrease as the equivalence ratio is increased as it approaches a ratio of 1. However, when the appropriate equivalence ratio is reached, NO_x emissions will rise and then gradually fall. The peak temperature of propane fuel is higher than that of other fuels because of the concentrated flame. Propane fuel can achieve a peak temperature of 347.65 °C, surpassing both propylene fuel (275.45 °C) and acetylene fuel (335.45 °C). Using a propylene gas flame results in a reduction in emissions of carbon monoxide and nitrogen oxides compared to a propane flame. But, acetylene fuel produces the most NO_x emissions, reaching 450.79 ppm for the experimental conditions. Additionally, the temperature distribution of the preheated metal plate was more uniform with the propylene gas flame, indicating improved heat transfer. However, the peak temperature of the metal plate was slightly lower when using the propylene gas flame.

Keywords: propylene gas flame; equivalence ratio; preheated metal plate; emissions; temperature distribution

Citation: Quach, Y.N.; Lim, O. An Investigation of the Effect of Propylene Gas Flame on Emissions and Temperature Distribution of a Preheated Metal Plate. *Sustainability* **2023**, *15*, 12306. https://doi.org/10.3390/su151612306

Academic Editors: Fuqiang Wang, Chao Shen, Dong Li and Zhonghao Rao

Received: 14 April 2023
Revised: 23 July 2023
Accepted: 7 August 2023
Published: 12 August 2023

1. Introduction

Preheating is a technique that involves providing heat to a metal plate and is frequently used in industrial processes, including preheating metal before welding and glass manufacturing [1,2]. The use of gas flames for the heating and processing of metals is a widely adopted industrial practice. Propylene gas, in particular, is known for its high energy density, which makes it a popular choice for many industrial applications. However, the use of propylene gas flames in metal processing can have significant environmental impacts, such as the release of harmful emissions into the atmosphere. To address this issue, there is a growing need to better understand the effects of propylene gas flames on the temperature distribution and emissions when heating metal plates [3]. The ratio of the distance between the burner head and the metal plate to the nozzle diameter, the Reynolds number, and the equivalence ratio are a few factors that have a substantial impact on the heat transfer properties of flame. Out of these factors, the equivalence ratio has a very significant effect on the heat transfer of the flame [4].

Numerous studies have been carried out on the heat transfer properties of different flames using analytical and numerical simulation techniques, including computational fluid dynamics (CFD) [4–6]. Liu et al. [7] and Zhen et al. [8] examined how a premixed hydrogen–liquefied petroleum gas (LPG) flame's heat transmission properties changed with hydrogen concentration. According to their findings, relatively high hydrogen concentrations caused a rise in combustion temperature and NO_x production but a decrease in

CO emission. Additionally, they evaluated that the LPG-H_2 and CH_4-H_2 mixes showed improved flame heat transfer with the addition of hydrogen. For a particular hydrogen concentration, the CH_4-H_2 mixture had a higher rate of heat transfer than LPG-H_2.

Several studies have investigated the impact of gas flames on emissions and temperature distributions in metal processing. For instance, Kandilli et al. [9] investigated the effect of natural gas flames on the thermal and environmental performance of a metallic honeycomb monolith. The study found that the use of natural gas flames led to significant emissions of CO, NO_x, and PM. Another study by Wang et al. [10] investigated the effect of propane gas flames on the thermal and environmental performance of a rotary kiln. The study found that the use of propane gas flames resulted in high emissions of CO and NO_x. A study by Zulkefli et al. [11] investigated the effects of LPG flames on the emission of NO_x and CO from a stainless-steel plate. The study found that the emission of NO_x and CO increased with increasing flame temperature, and the emissions were more significant at the edge of the flame than in the center. Another study by Yao-Yao Wang et al. [12] investigated the impact of preheating on the surface quality and corrosion resistance of 316L stainless steel plates that were cut by laser. The authors conducted experiments where the stainless-steel plates were preheated to different temperatures before being cut with a laser. They then analyzed the surface qualities and corrosion resistances of the plates. The study found that preheating the plates to a specific temperature range resulted in improved surface quality and corrosion resistance. The study conducted by Bader A. Alfarraj et al. [13] investigated the emissions and performance of conventional liquefied petroleum gas (LPG) cookstove burners. The results showed that the emissions of carbon monoxide (CO), nitrogen oxides (NO_x), and particulate matter (PM) were found to be higher than the limits set by regulatory agencies. The study also found that the performance of the burners was affected by multiple factors, including the LPG pressure, air–fuel ratio, and burner diameter.

In the context of propylene gas flames, several studies have investigated their impact on emissions and temperature distribution. For example, A.T. Hartlieb et al. [14] investigated the impact of a quartz nozzle on the structure and temperature of a propene flame. Their results indicate that the nozzle can enhance mixing and improve the homogeneity of the flame, leading to a reduction in the required flame temperature. Specifically, the use of the nozzle results in a shift towards fuel-lean combustion, which reduces the temperature in the flame front and promotes complete combustion. The findings suggest that the use of a sampling quartz nozzle could be a viable strategy for controlling the temperature and improving the efficiency of low-pressure propylene (propene) flames. Krishna C. Kalvakala et al. [15] investigated the effects of oxygen enrichment and fuel unsaturation on soot and NO_x emissions in different flames, including propene. The study found that increasing the oxygen concentration in the combustion air led to a decrease in soot emissions in propene flames. However, the increase in oxygen concentration also led to an increase in NO_x emissions in propene flames. Additionally, the study found that fuel unsaturation, such as in propene, led to higher soot emissions compared to saturated fuels like propane. Overall, the results suggest that the combustion of propene can lead to significant emissions of both soot and NO_x, which should be considered in developing effective emission reduction strategies.

While these studies provide insights into the effects of gas flames on metal surfaces [16–20], further research is needed to investigate the specific effects of propylene gas flames on the temperature distribution and emissions of metal plates. Moreover, the impact of the heat transfer characteristics on temperature distribution and thermal efficiency during combustion with a specific focus on NO_x emissions have been extensively studied to date. However, none of the studies have highlighted the effect of the equivalence ratio on the temperature distribution and NO_x emissions.

The current study aims to address this gap by investigating the interaction between propylene gas flames and metal plates and exploring the effects of the flame on temperature distribution and associated emissions. In summary, previous research has investigated the effects of gas flames on metal surfaces, including heat transfer characteristics, emissions, and surface quality [21–25]. However, there is a need for further research to investigate the specific effects of propylene gas flames on the temperature distribution and emissions of flame in heating metal plates, which is the focus of the current study. The findings of this research could contribute to the widespread adoption and use of propylene gas flames.

2. Methodology

2.1. Experiment Setup

The schematic designs for the exhaust gas measurement system and the experimental setup are shown in Figure 1a,b, respectively. The experimental system consists of 7 components. The feed tanks supply the air and fuel to the torch, and the airflow meters are used to manage the flow rate of the fuel mixture. The torch is employed to burn fuel inside the main chamber, and an exhaust gas chamber is added to maintain the homogeneity of the exhaust gas and enhance the measurement accuracy. The signal from the exhaust gas temperature sensor is analyzed using an exhaust gas analyzer (Horiba MEXA-7100 DEGR). The experiments were performed in a well-ventilated laboratory environment with the torch system placed on a laboratory bench. The gas pressure, flow rate, and torch-to-workpiece distance were adjusted as required. The torch was connected to a regulator, which controlled the pressure of the fuel mixture gas, and it was mounted on a stand to ensure stability during the experiments. The fuel was stored in a feed tank and delivered to the torch system through a flexible hose. The air and fuel pressure were measured using a pressure gauge installed on the regulator, and the gas flow rate was measured using a flow meter installed on the flexible hose. The gas pressure and flow rate were adjusted using the regulator to achieve the desired operating conditions. The tests were conducted under steady-state conditions at near-room temperature conditions of approximately 27 °C.

The experiment setup and schematic design for the preheating procedure are shown in Figure 1c,d, respectively. The metal plate's total width, length, and thickness were 0.5 m, 0.5 m, and 0.03 m, respectively. The distance (d) between the torch outlet and the metal plate was 0.06 m. The gas torch combined fuel and air to facilitate combustion. After leaving the exits of the gas torch, the mixture of fuel and air was ignited, generating a combustion flame for preheating the metal plate. The operating conditions were optimized to achieve the best performance of the torch system using LPG as fuel. The optimal gas pressure and flow rate were determined based on the statistical analysis of the data. The torch-to-metal plate distance was also optimized for optimal performance. The temperature distribution on the reverse side of the metal plate was measured using a TVS-200EX infrared camera, as shown in Figure 1d. To compensate for the lower sensitivity of the infrared camera, an additional thermocouple sensor connected to a Midi logger 840 was employed for temperature measurements. To measure the temperature distribution of a metal plate during the preheating process, 9 thermocouples were positioned on the rear of the plate in three lines. The upper line's temperature was measured using Ch1, 2, and 3, the middle line's temperature was obtained using Ch4, 5, and 6, and the lower line's temperature was measured using Ch7, 8, and 9. The calibration of the equipment is shown in Table 1.

Table 1. Calibration of equipment.

No	Equipment	Producer	Specification
1	Horiba MEXA-7100 DEGR	HORRIBA	CO (0-50 ppm)
			HC (0-10 ppm)
			CO_2 (0-10 ppm)
			NO_x (0-10 ppm)
2	TVS-200EX infrared camera	HANDY	±2%
3	Midi logger GL840	GRAPHTEC	±1.5%

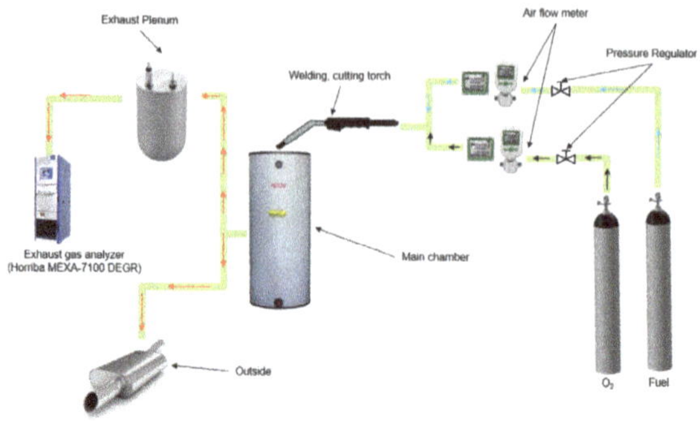

(a)

(b)

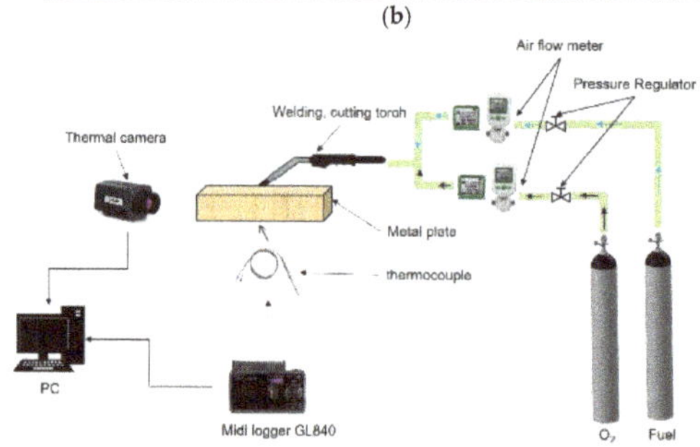

(c)

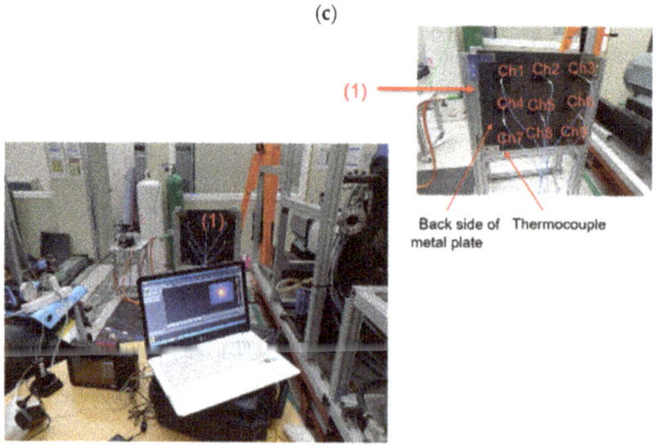

(d)

Figure 1. Experimental setup for the torch system. Schematic (**a**) and detail (**b**) setup for exhaust gas measurement, schematic (**c**) and detail (**d**) setup for preheating process.

2.2. Fuel Properties

Propylene gas, a hydrocarbon gas with the chemical formula C_3H_6, is a colorless and flammable gas. It has a high energy density and burns cleanly, making it a popular choice for heating, cutting, and welding. Propylene gas has a lower heating value than natural gas, but it can be used as a substitute for natural gas in many applications. Propylene gas has a high flash point and low volatility, which makes it relatively safe to handle and store.

Propane gas with the chemical formula C_3H_8, lacking the carbon double bond of propene, is a hydrocarbon gas that is commonly used as a fuel for heating and powering vehicles. It is a colorless, odorless gas that is typically stored in pressurized tanks as a liquid. Propane gas has a higher vapor pressure than propylene gas, making it easier to store and transport. Additionally, propane gas has a low flammability range and can be safely used in enclosed spaces with adequate ventilation. It produces relatively low emissions of pollutants.

Acetylene gas is a hydrocarbon with the chemical formula C_2H_2. It has a high energy density and burns with a high-temperature flame, making it suitable for applications that require high heat. However, acetylene gas also has high flammability, which requires special handling and storage precautions. It also has a narrow flammability range and is sensitive to shock and friction. Acetylene gas produces high emissions of pollutants.

In summary, propylene gas, propane gas, and acetylene gas are all useful hydrocarbon fuels with different fuel properties. Propane gas has the highest heating value. Propylene gas has high-energy fuels, while acetylene gas has the highest flame temperature. Detailed information on the properties of these fuels is presented in Table 2.

Table 2. Fuel properties.

Property	Propane	Propylene	Acetylene
Chemical formula	C_3H_8	C_3H_6	C_2H_2
Liquid density (kg/m^3)	509	609	1097
Calorific value (MJ/kg)	46.34	43.9	49.9
Boiling poing (°C)	−42	−47.7	−84.7
Autoignition temperature (°C)	510	438	305
Flame temperature (°C)	1980	2924	3160
Flame speed (m/s)	0.4	0.5	0.15
Stoichiometric air/fuel	15.8	14.5	13.3
Equivalent AFR	15.5	14.7	10.3

The air–fuel equivalence ratio is the ratio of actual air–fuel ratio (AFR) to stoichiometric air–fuel ratio (AFR). An equivalence ratio of 1.0 corresponds to the stoichiometric air–fuel ratio, while rich air–fuel mixtures have an equivalence ratio of <1.0, and lean mixtures have an equivalence ratio of >1.0. There is a direct relationship between equivalence ratio and air–fuel ratio (AFR).

$$AFR = \frac{m_{air}}{m_{fuel}} \tag{1}$$

where

m_{air}: mass of air.
m_{fuel}: mass of fuel.

$$\varnothing = \frac{AFR}{AFR_{stoich}} \tag{2}$$

where

AFR: actual AFR.

AFR_{stoich}: stoichiometric AFR.

3. Result and Discussion

3.1. The Effect of the Equivalence Ratio on Emission Exhaust Gas

The graph presented in Figure 2 provides a visual representation of the relationship between equivalence ratio and total hydrocarbon (THC) emissions. The data indicate that as the equivalence ratio increases, there is a noticeable reduction in THC emissions when the ratio is below 1. This can be attributed to the presence of excess oxygen in the combustion chamber, which facilitates the combustion process and promotes the oxidation of unburned hydrocarbon molecules.

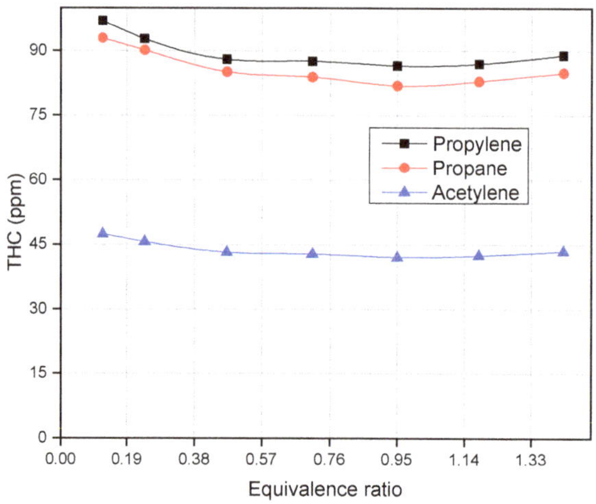

Figure 2. The effect of the equivalence ratio on THC emissions.

However, an interesting observation is made when the equivalence ratio surpasses 1. In this scenario, the THC emissions show a slight increase. This phenomenon can be attributed to the conditions of lean combustion, where the mixture becomes fuel-lean, and there is an insufficient amount of oxygen available for complete combustion. As a result, the combustion process becomes sluggish, leading to an extended combustion time. These factors contribute to the rise in THC emissions.

Furthermore, it is noteworthy that propylene fuel exhibits higher THC emissions compared to propane and acetylene fuels. This can be attributed to the unique combustion characteristics of propylene. The flame rate of propylene is relatively higher, resulting in a shorter burning time. However, this shorter duration may lead to incomplete combustion, where some hydrocarbon molecules are not fully oxidized. As a consequence, propylene fuel emits a greater amount of THCs. These findings align with the fuel properties discussed in Section 2.2, which highlight the combustion behavior and characteristics of the different fuels.

Figure 3 depicts the effect of the equivalence ratio on CO_2 emissions. The findings demonstrate that the carbon dioxide (CO_2) emission decreases as the equivalence ratio values increase. The impact of the equivalence ratio on CO_2 emissions is relatively minimal compared to other emissions. CO_2 is primarily determined by the carbon content in the fuel rather than the equivalence ratio. However, extremely high equivalence ratios can lead to incomplete combustion and increased CO_2 emissions. Aside from that, propane emits significantly more CO_2 than other gases. The fuel is completely burned, leaving behind only carbon dioxide (CO_2) and water. It is evident that propane fuel burns more completely than propylene and acetylene fuel because of its molecular structure and combustion characteristics. Propane gas has a relatively simple chemical structure, which makes it easier to burn completely in the presence of oxygen. Propane gas also has a narrower flammability range, enabling better control and optimization of conditions for achieving complete combustion.

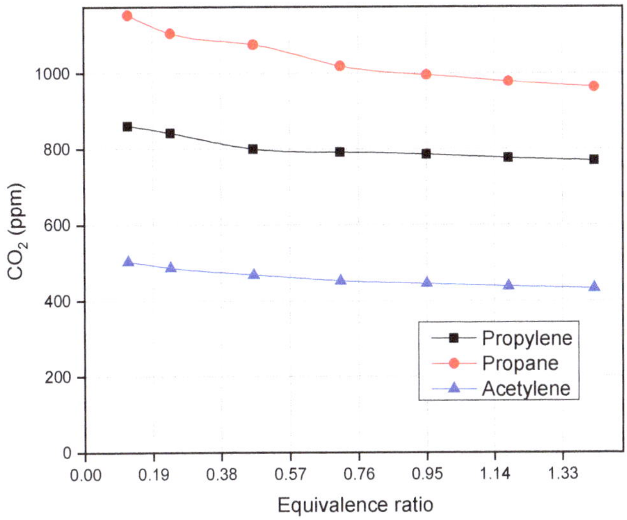

Figure 3. The effect of the equivalence ratio on CO_2 emissions.

Figure 4 depicts the influence of the equivalence ratio on CO emission. The results demonstrate that lowering the equivalence ratio increases CO emissions. It is understandable that increases in CO emission with a decrease in the equivalence ratio were brought on by a drop in the oxygen concentration. Furthermore, because THC emissions rise with lambda (air/fuel) when the relative air–fuel ratio is larger than 1, the presence of unburned hydrocarbons in the reaction zone slows CO oxidation, as seen in Figure 4. Therefore, CO emissions increase during times of oxygen scarcity, implied by the single oxygen atom in the carbon monoxide structure. Furthermore, acetylene creates far less CO than propylene and propane fuel because of its unique combustion properties and the stoichiometry of its combustion reaction. The stoichiometric ratio for acetylene combustion is much lower than for propylene and propane. This means that a smaller amount of air is needed to combust a given amount of acetylene relative to the other fuels. Furthermore, the combustion reaction of acetylene is highly exothermic, meaning that it releases a large amount of heat when it reacts with oxygen. This high heat release helps to ensure that complete combustion occurs, reducing the formation of harmful byproducts like CO.

Figure 5 shows the impact of the equivalence ratio on NO_x emissions. According to the figure, a drop in the equivalence ratio resulted in a sharp decrease in NO_x in most of these experiments. The main contributor to NO_x generation is the reduction in NO_x at an equivalent oxygen concentration. Additionally, at fuel-lean conditions, the availability of oxygen is relatively higher compared to the fuel, resulting in lower peak flame temperatures. This leads to a reduction in the formation of NO_x, as lower temperatures inhibit the reaction between nitrogen and oxygen. Conversely, under fuel-rich conditions, the excess fuel generates higher peak flame temperatures, thereby promoting the formation of NO_x.

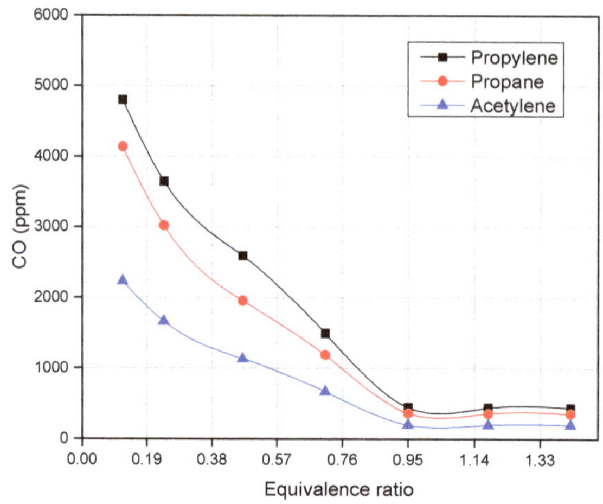

Figure 4. The effect of the equivalence ratio on CO emissions.

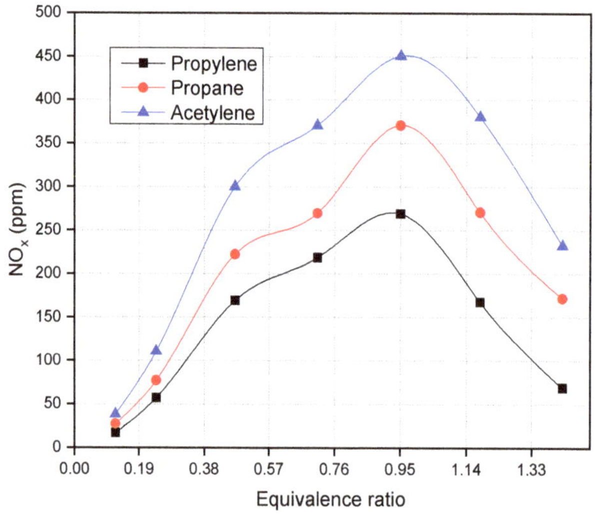

Figure 5. The effect of the equivalence ratio on NO_x emissions.

Furthermore, it is noteworthy that acetylene exhibits the highest NO_x emission value. This can be attributed to the significant heat generated during acetylene combustion, resulting in an increase in chamber temperature, which in turn promotes the formation of NO_x emissions. This is predictable because the N_2 bond is stronger than the O_2 bond, and the N_2 bond requires more energy to break. One factor that contributes to the high NO_x emissions of acetylene is its combustion temperature. Acetylene has a relatively low ignition temperature and a high flame temperature, which leads to rapid combustion and high temperatures. Moreover, acetylene has a triple bond between its carbon atoms, which makes it highly reactive, leading to the creation of an oxygen-rich flame zone, which favors the formation of NO_x emissions.

3.2. Comparison of Temperature Distributions on a Metal Plate

Figure 6a–c depict the temperature contours from the gas torch outlets to the metal plate of propane, propylene, and acetylene, respectively. As shown in Figure 6a–c, the heat transfer rate of propylene fuel is greater than that of propane fuel. However, the central point of propane fuel has a higher temperature, indicating the flame of the propane fuel is more focused.

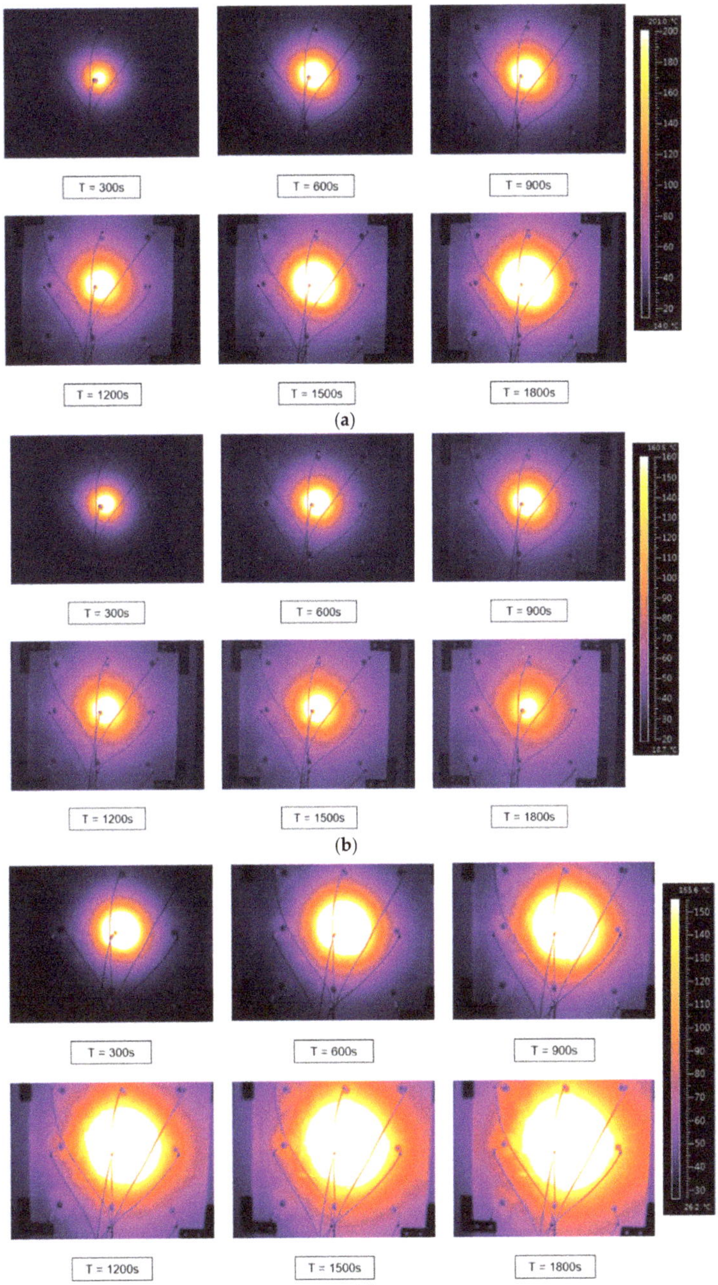

Figure 6. The temperature contours from the gas torch outlets to the metal plate: (**a**) propane flame, (**b**) propylene fuel, and (**c**) acetylene fuel.

Additionally, propylene fuel is considered safer due to its more uniform temperature distribution compared to propane fuel. Propane fuel has a higher flame temperature than propylene fuel, which can lead to localized hotspots during combustion. These hotspots can result in uneven heating of the material being heated, causing thermal stresses and deformation of the material. Additionally, the localized hotspots can increase the risk of ignition or fire if the hotspots exceed the ignition temperature of the material or surrounding environment. In contrast, propylene fuel has a lower flame temperature than propane fuel, which leads to a more uniform temperature distribution during combustion. This uniform temperature distribution reduces the risk of localized hotspots and thermal stresses on the heated material. Furthermore, a more uniform temperature distribution also means that the overall temperature of the heated material can be kept lower, which can reduce the risk of ignition or fire.

Figure 7a–c present the temperature distributions of propylene, propane, and acetylene fuels, respectively, as a function of time. These distributions provide valuable insights into the thermal behavior and characteristics of each fuel throughout the experimental duration.

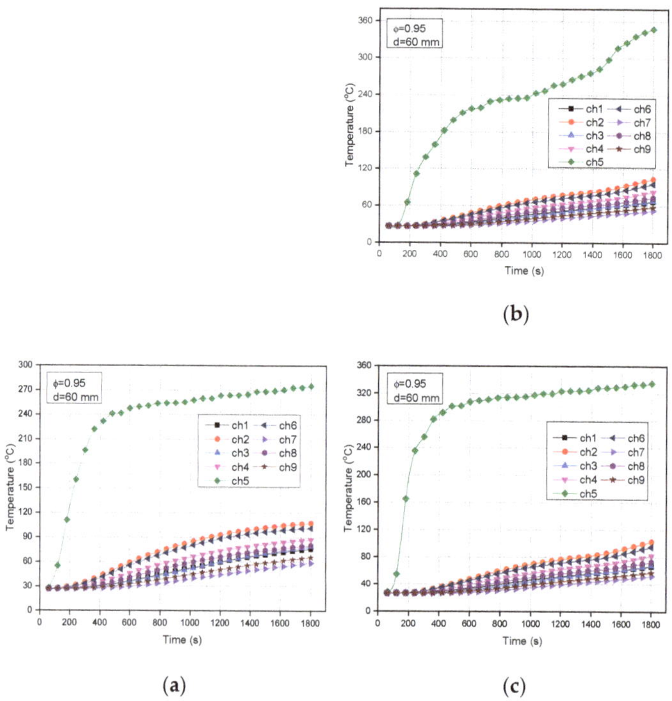

Figure 7. The temperature distribution as a function of time: (**a**) propylene fuel, (**b**) propane fuel, and (**c**) acetylene fuel.

The temperature distribution of propylene fuel, as depicted in Figure 7a, exhibits an interesting pattern. At the central point, the temperature experiences a rapid rise from its initial value, reaching 222.15 °C within 360 s. As time progresses, the temperature gradually approaches a steady-state value, indicating a more stable thermal condition. At the end of the experiment (1800 s), the maximum temperature recorded at the central point of the propylene fuel is 275.45 °C. This finding suggests that propylene fuel has a relatively fast response in terms of temperature increase and achieves a moderate maximum temperature.

In Figure 7b, the temperature distribution of propane fuel is showcased. The central point temperature of the propane fuel gradually increases from the starting temperature and reaches 211.05 °C after 540 s. However, unlike propylene fuel, the temperature profile of propane fuel does not stabilize and continues to exhibit fluctuations beyond the experimental timeframe. This indicates the potential for a further rise in temperature or variability. Impressively, the central point of the propane fuel records a maximum temperature of 347.65 °C at 1800 s, indicating a higher peak temperature compared to propylene fuel.

Figure 7c illustrates the temperature distribution of acetylene fuel. Similar to propylene fuel, the central point temperature of acetylene fuel experiences a rapid initial increase. Within 480 s, the temperature rises quickly from the starting temperature to 301.45 °C. As the experiment progresses, the temperature of the acetylene fuel gradually stabilizes and approaches a steady-state value. At 1800 s, the central point of the acetylene fuel reaches a maximum temperature of 335.45 °C, indicating a relatively high peak temperature.

The observed temperature distributions highlight the distinct characteristics of each fuel gas. Propylene fuel demonstrates a rapid but stable temperature increase, propane fuel exhibits a gradually rising temperature with potential fluctuations, and acetylene fuel showcases a rapid initial increase followed by a relatively stable temperature profile.

Figures 8 and 9 compare the temperatures and temperature increase rates of the three fuels at the middle point. The graphs show that the temperature of acetylene fuel increases at a higher rate than for propene, but the temperature of propane fuel increases faster than other fuels. After a period of burning, the temperature of propylene and acetylene fuel stabilizes, while the temperature of propane continues to rise. Furthermore, the peak temperatures of propane and acetylene fuels are higher than that of propylene fuel due to the concentrated flame. Moreover, acetylene's atomic structure has a triple bond, and propylene's atomic structure has a double bond, which allows it to easily react with oxygen and create heat rapidly, but propane generates more heat due to its high latent heat of vaporization. On the other hand, propane has a lower flame temperature compared to acetylene. Despite this, propane's combustion process is more complete, resulting in a higher energy output per unit mass of fuel. This leads to a more rapid increase in temperature relative to propylene and acetylene.

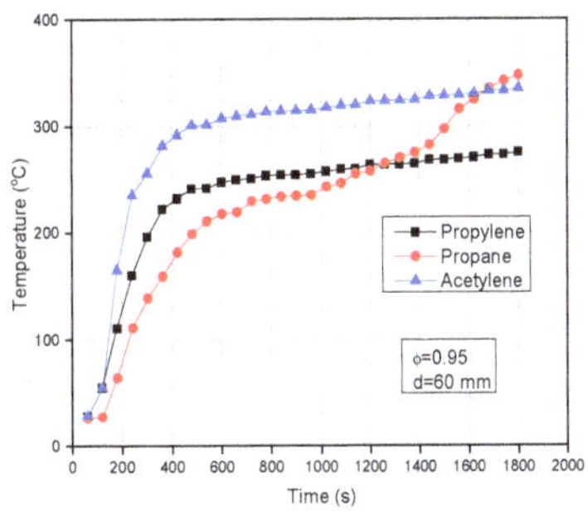

Figure 8. A comparison of temperatures at center points using three fuels.

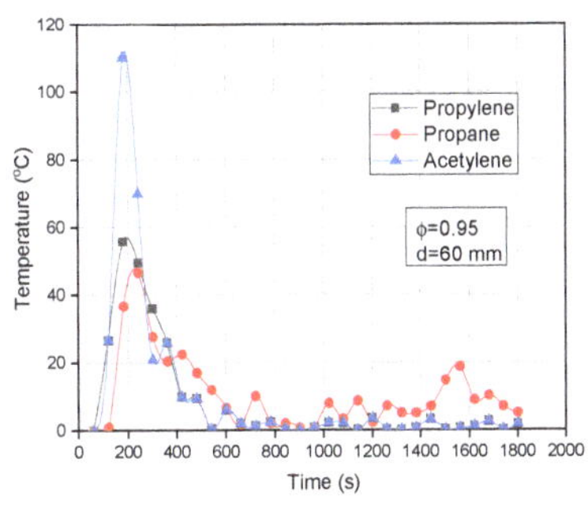

Figure 9. A comparison of the change in temperature rates at the center points using the three fuels.

4. Conclusion

In this study, we used an experimental strategy to overcome some of the shortcomings of previous experimental optimization approaches. We carefully researched the equivalence ratio, which has sensitive impacts on exhaust gases such as NO_x, CO, CO_2, and THC, as well as the temperature distribution when heating a metal plate. The ideal equivalence ratio was also found. When the equivalence ratio is at its ideal value, the torch system performs better. While this is not true for NO_x emissions, as the equivalence ratio of three fuels increases and reaches 1.0, the CO, CO_2, and THC emissions decrease. The combustion parameters of three fuels are optimized at an equivalence ratio of 0.95.

Analyzing the temperature profiles of the different fuels, distinct patterns emerged. Acetylene fuel demonstrated a rapid increase in temperature, surpassing the other fuels. In contrast, the temperature increase for the other fuels was gradual and fluctuating, with the potential for further escalation. Notably, the central point of propane fuel recorded the highest temperature of 347.65 °C at 1800 s, exceeding both propylene fuel (275.45 °C) and acetylene fuel (335.45 °C). Furthermore, except for the center point of the propylene fuel, the temperature was consistently higher compared to the other analyzed fuels. This denotes that, as the flame of propane and acetylene fuels are more concentrated, the flame of propylene fuel spreads over a wider area.

Using a propylene gas flame can lead to a reduction in emissions of carbon monoxide and nitrogen oxides compared to the propane flame. Additionally, the temperature distribution of the preheated metal plate was more uniform with the propylene gas flame, indicating improved heat transfer. These findings highlight the potential benefits of employing propylene gas as a fuel source in preheating systems. Its adoption could enhance energy efficiency, promote environmental sustainability through reduced emissions, and facilitate the production of sustainable chemicals. Moreover, the utilization of propylene gas has the potential to optimize energy consumption in various industrial processes.

Author Contributions: Conceptualization, Y.N.Q.; Methodology, Y.N.Q.; Writing—original draft, Y.N.Q.; Writing—review & editing, Y.N.Q.; Supervision, O.L. All authors have read and agreed to the published version of the manuscript.

Funding: This results was supported by "Regional Innovation Strategy (RIS)" through the National Research Foundation of Korea (NRF) funded by the Ministry of Education (MOE) (2021RIS-003). This research is financially supported by the individual basic research project by the National Research Foundation of Korea (NRF-2021R1F1A1048238, Reliability Improvement of Ammonia-Diesel Dual-Fuel Combustion Model regarding Optimized Combustion Strategy for Improved Combustion Efficiency and Emission Characteristics).

Conflicts of Interest: The authors declare no conflict of interest.

Nomenclature

Ch	Channel
ppm	Parts per million
THC	Total hydrocarbon
Lambda	Air–fuel ratio
Ø	Equivalence ratio

References

1. Bunaziv, I.; Langelandsvik, G.; Ren, X.; Westermann, I.; Rørvik, G.; Dørum, C.; Danielsen, M.H.; Eriksson, M. Effect of preheating and preplaced filler wire on microstructure and toughness in laser-arc hybrid welding of thick steel. *J. Manuf. Process.* **2022**, *82*, 829–847. [CrossRef]
2. Mohan, D.G.; Gopi, S. Induction assisted friction stir welding: A review. *Aust. J. Mech. Eng.* **2020**, *18*, 119–123. [CrossRef]
3. Khalid, A.H.; Milcarek, R.; Dahm, W.; Kim, J. *Low Temperature Soot Regime of Propane/Air in a Micro Flow Reactor with Controlled Temperature Profile*. 2019. Available online: https://keep.lib.asu.edu/_flysystem/fedora/c7/218074/Khalid_asu_0010N_19305.pdf (accessed on 9 April 2023).
4. Sahu, S.; Dewangan, M. Analysis of Heat Transfer Characteristics of Flame Impinging to a Plane Surface Perpendicular to Flame Jet Axis. *Int. J. Eng. Res. Technol.* **2013**, *2*, 578–593.

5. Tajik, A.R.; Kuntikana, P.; Prabhu, S.V.; Hindasageri, V. Effect of preheated mixture on heat transfer characteristics of impinging methane–air premixed flame jet. *Int. J. Heat Mass Transf.* **2015**, *86*, 550–562. [CrossRef]
6. Remie, M.J.; Cremers, M.F.G.; Schreel, K.R.A.M.; de Goey, L.P.H. Analysis of the heat transfer of an impinging laminar flame jet. *Int. J. Heat Mass Transf.* **2007**, *50*, 2816–2827. [CrossRef]
7. Liu, Y.; Xiang, Q.; Li, Z.; Yao, S.; Liang, X.; Wang, F. Experiment and Simulation Investigation on the Characteristics of Diesel Spray Impingement Based on Droplet Impact Phenomenon. *Appl. Sci.* **2018**, *8*, 384. [CrossRef]
8. Zhen, H.S.; Cheung, C.S.; Leung, C.W.; Choy, Y.S. Effects of hydrogen concentration on the emission and heat transfer of a premixed LPG-hydrogen flame. *Int. J. Hydrogen Energy* **2012**, *37*, 6097–6105. [CrossRef]
9. Kandilli, N. Development of a Three Way Catalytic Converter for Elimination of Hydrocarbons, Carbon Monoxide and Nitric Oxide in Automotive Exhaust. 2010. Available online: https://open.metu.edu.tr/handle/11511/20172 (accessed on 9 April 2023).
10. Jiang, Y.; Chen, H.; Zhang, X.; Zhou, Y.; Wang, L. Combustion Condition Recognition of Coal-Fired Kiln Based on Chaotic Characteristics Analysis of Flame Video. *IEEE Trans. Ind. Inform.* **2022**, *18*, 3843–3852. [CrossRef]
11. Cui, Y.; Li, G.; Xu, W.; Zhu, J. Experimental Investigation of NOx and CO Emissions from Fuel Rich-Lean Flame of Natural Gas. *Adv. Mater. Res.* **2012**, *347–353*, 3821–3825. [CrossRef]
12. Wang, Y.-Y.; Yang, X.-N.; Shi, S.-Y.; Wu, Z.; Han, R.-H.; Qi, H.-B. Laser welding 6061 aluminum alloy with laser cladding powder. *J. Laser Appl.* **2021**, *33*, 022006. [CrossRef]
13. Alfarraj, B.A.; Al-Harbi, A.A.; Binjuwair, S.A.; Alkhedhair, A. The Characterization of Liquefied Petroleum Gas (LPG) Using a Modified Bunsen Burner. *J. Combust.* **2022**, *2022*, 1–9. [CrossRef]
14. Hartlieb, A.T.; Atakan, B.; Kohse-Höinghaus, K. Effects of a sampling quartz nozzle on the flame structure of a fuel-rich low-pressure propene flame. *Combust. Flame* **2000**, *121*, 610–624. [CrossRef]
15. Kalvakala, K.C.; Katta, V.R.; Aggarwal, S.K. Effects of oxygen-enrichment and fuel unsaturation on soot and NOx emissions in ethylene, propane, and propene flames. *Combust. Flame* **2018**, *187*, 217–229. [CrossRef]
16. Ritrievi, K.E.; Longwell, J.P.; Sarofim, A.F. Influence of Combustion Duration on the Performance and Emission Characteristics of a Spark-Ignition Engine Fueled with Pure Methanol and Ethanol. *ACS Omega* **2022**, *17*, 14505–14515. [CrossRef]
17. Height, M.J.; Howard, J.B.; Tester, J.W.; Sande, J.B.V. Flame synthesis of single-walled carbon nanotubes. *Carbon* **2004**, *42*, 2295–2307. [CrossRef]
18. Nghia, N.T.; Khoa, N.X.; Cho, W.; Lim, O. A Study the Effect of Biodiesel Blends and the Injection Timing on Performance and Emissions of Common Rail Diesel Engines. *Energies* **2021**, *15*, 242. [CrossRef]
19. Tan, Y.; Croiset, E.; Douglas, M.A.; Thambimuthu, K.V. Combustion characteristics of coal in a mixture of oxygen and recycled flue gas. *Fuel* **2006**, *85*, 507–512. [CrossRef]
20. Khoa, N.X.; Kang, Y.; Lim, O. The effects of combustion duration on residual gas, effective release energy and engine power of motorcycle engine at full load. *Energy Procedia* **2019**, *158*, 1835–1841. [CrossRef]
21. Chander, S.; Ray, A. Flame impingement heat transfer: A review. *Energy Convers. Manag.* **2005**, *46*, 2803–2837. [CrossRef]
22. Khoa, N.X.; Lim, O. The effects of bore-stroke ratio on effective release energy, residual gas, peak pressure rise and combustion duration of a V-twin engine. *J. Mech. Sci. Technol.* **2020**, *34*, 2657–2666. [CrossRef]
23. Zhao, Z.; Wong, T.T.; Leung, C.W. Impinging premixed butane/air circular laminar flame jet—Influence of impingement plate on heat transfer characteristics. *Int. J. Heat Mass Transf.* **2004**, *47*, 5021–5031. [CrossRef]
24. Basu, S.; Miglani, A. Combustion and heat transfer characteristics of nanofluid fuel droplets: A short review. *Int. J. Heat Mass Transf.* **2006**, *85*, 507–512. [CrossRef]
25. Saiki, Y.; Fan, Y.; Suzuki, Y. Radical quenching of metal wall surface in a methane-air premixed flame. *Combust. Flame* **2015**, *162*, 4036–4045. [CrossRef]

sustainability

MDPI

Article

Does Carbon Emissions Trading Policy Improve Inclusive Green Resilience in Cities? Evidence from China

Bin Xiong and Qi Sui *

School of Public Policy and Management, Guangxi University, Nanning 530004, China; xb9873@gxu.edu.cn
* Correspondence: 2201401008@st.gxu.edu.cn

Abstract: With the impact of external globalization uncertainties and the pressure of internal national environmental policies and markets, maintaining inclusive green resilience while coordinating economic, environmental, and social systems is critical for achieving green and sustainable urban development. We define inclusive green resilience for cities in this research and build a system of inclusive green resilience indicators. The DID model and entropy approach were used to examine the impact of carbon trading policies on inclusive green resilience in 184 representative Chinese cities from 2008 to 2018, and PSM-DID was utilized for further validation. According to the findings of the study, carbon emissions pricing policies can considerably increase inclusive green resilience in cities. Mechanism verification demonstrates that carbon trading programs improve inclusive green resilience in cities through industrial restructuring, technical innovation capability, and employment benefits. According to the city heterogeneity study, the implementation of the carbon emissions trading system has a scale effect and significant urban functional differences, and its impact on inclusive green resilience of cities is greater in large and medium-sized cities and non- resource-based cities. This research offers a new way of thinking about inclusive green resilience as well as empirical data for future sustainable policy development.

Keywords: carbon emissions trading; PSN-DID model; inclusive green resilience; social equity; low carbon economy

Citation: Xiong, B.; Sui, Q. Does Carbon Emissions Trading Policy Improve Inclusive Green Resilience in Cities? Evidence from China. *Sustainability* **2023**, *15*, 12989. https://doi.org/10.3390/su151712989

Academic Editors: Fuqiang Wang, Chao Shen, Dong Li and Zhonghao Rao

Received: 7 July 2023
Revised: 2 August 2023
Accepted: 26 August 2023
Published: 29 August 2023

1. Introduction

Under the background of global warming, the world economy is facing a contradiction between high energy consumption and low-carbon economic transition [1]. As the largest developing country, China is also under enormous pressure to achieve the goals of carbon emissions reduction, economization, and environmental protection in its efforts to promote rapid economic growth and improve people's livelihoods. The outline of China's 14th Five-Year Plan clearly states that it aims to achieve a "carbon peak" by 2030 and the "3060 Goal Vision" of carbon neutrality by 2060. The "3060 Goal Vision" emphasizes the importance of cities reducing carbon emissions, managing the relationship between environmental and economic systems while maintaining social equity and justice, improving the affordability of urban economic and social development, and ultimately promoting inclusive green resilience [2,3]. With the legislation and execution of a large number of environmental regulatory regulations, China's carbon emissions intensity will have decreased by 48.4% from 2005 levels by the end of 2020. Nonetheless, despite domestic policy support, China confronts huge obstacles in terms of inclusive green resilience. The key barriers to China's inclusive green resilience are a lack of "inclusive resilience" and the difficulties in harmonizing it with "green development resilience [4,5]. Therefore, how to promote inclusive green resilience in cities while improving low-carbon economic efficiency has become an issue demanding an urgent solution for China.

The notion of inclusive green resilience in cities is taken primarily from the concept of resilience in ecology, with a focus on dynamic adaptive adjustment pathways for the long-term sustained expansion of urban low-carbon economic and social systems [3,6,7]. Rizzo believes that so-called "green resilience" is the integration of technological solutions and resilience thinking that emerged to solve urban ecological problems [8]. Shokry et al. propose that green resilience aims to reduce the vulnerability of socially vulnerable groups to climate risks and impacts [9]. The article argues that urban inclusive green resilience refers to the fact that, in the face of current internal and external shocks, such as resource and environmental constraints, a city recovers by upgrading the level of economic and social inclusiveness, as well as environmental development, and then transitions to a new path of growth and a state of sustainable development. The academic community currently uses the core variable method and the indicator system method to measure inclusive green resilience capacity. When compared to the core variable method, the indicator system method can comprehensively and systematically measure inclusive green resilience capacity. The indicator system method separates green resilience into numerous aspects and selects indicators to quantify each dimension. Shokry et al. constructed an index system from the three dimensions of exposure, sensitivity, and adaptability to measure urban green resilience [10]. However, investigating the measurement of the level of inclusive green resilience by creating indicators based solely on the operational process of green resilience, neglecting the interaction between economic, social, and green development, still poses a problem for scholars. As a result, this essay develops an evaluation index system based on three factors: economic resilience, social inclusive resilience, and ecological environment resilience. The existing literature on the factors influencing urban inclusive green resilience focuses primarily on urban development activities [11,12], and there is still potential to investigate other aspects impacting inclusive green resilience capability. Furthermore, although a large number of scholars have confirmed that environmental policies have a positive influence on the level of urban inclusive green development through qualitative and quantitative analyses, there is still a research gap in exploring the impact of environmental policies on inclusive green resilience capacity.

Carbon emissions trading policy has garnered substantial attention from academics in recent years due to its operating mechanism and low-carbon development as a market incentive-based environmental strategy. Carbon emissions trading policies, according to relevant researchers, can greatly improve regional economic efficiency [13], improve energy efficiency [14], and play a vital role in stimulating low-carbon technology innovation in businesses [15]. In terms of research methods, some researchers have used qualitative and theoretical research approaches to examine the impact of carbon emissions trading policies, such as developing an impact evaluation index system and analyzing and evaluating the consequences of policy implementation [16]. Some researchers analyze the influence of carbon emissions trading programs using quantitative methods such as the instrumental variable method, comprehensive control method, dynamic computable general equilibrium model (CGE), DID model, PSM model, PSM-DID model, and so on [17–19]. The policy assessment impact is disturbed when the process of assessing policy outcomes is influenced by some observable or unobservable confounding factors. To address this type of issue, the DID model may reduce the influence of confounding factors by inserting dummy variables to increase policy assessment accuracy; hence, the DID model has been widely employed in the policy evaluation process. The PM-DID model is also valued in the policy analysis process because it combines the benefits of the DID and the PSM while effectively avoiding the difficulties of selection bias and endogeneity [20,21].

To sum up, although there is a considerable amount of literature on carbon emissions trading policy and inclusive green resilience, some limitations still exist. First of all, although previous studies on carbon emissions trading policy have discussed the impacts of carbon emissions trading policy on enterprises, economy, and ecological environment from different perspectives, few studies have discussed the impacts of carbon emissions trading policy on urban economic, social, and environmental coordination, especially on urban inclusiveness, from the perspective of urban integrity. Secondly, in terms of measuring inclusive green resilience, most scholars mainly focus on the level of green resilience, exploring the level of urban green development space and green infrastructure construction, and few studies have deeply captured and explored the development status of urban economic, social, and ecological comprehensive resilience capacity. In addition, there is an urgent need for empirical studies on whether urban environmental policies have a significant impact on improving inclusive green resilience.

In order to bridge the research gap, based on the current research content, this paper uses the PSM-DID method to thoroughly investigate the response of inclusive green resilience to the carbon emissions trading policy and conducts an in-depth analysis of its influence mechanism. Based on China's actual national conditions, the heterogeneous impact of carbon emissions trading policy on cities in different categories is analyzed in depth.

2. Theoretical Analysis and Research Hypothesis

2.1. Mechanism of Analysis of Carbon Emissions Trading Rights Policy for Inclusive Green Resilience

In this section, we describe the theoretical mechanism by which emissions trading rights policy impacts inclusive green resilience. We extend the theoretical mechanism based on the previous carbon emissions trading rights policy, including three main aspects:

First is urban industrial structure adjustment. The secondary industry has long accounted for a relatively large part in China's economic aggregate. Due to long-term "path dependence" and uncontrolled exploitation of natural resources, there are prominent urban environmental problems and human settlement problems, and it is difficult to achieve sustainable economic and social development. As the key to inclusive green urban growth, reasonable industrial structure adjustment can help reduce energy consumption, reduce the energy "dependence path," promote the development of clean energy, and thus reduce carbon emissions. Through reasonable control of carbon emissions and strengthening of carbon emissions constraints on the energy industry, carbon emissions trading policy will help promote the transformation and upgrading of the original energy-intensive industries to the tertiary sector, forming an industrial restructuring and a change in the mode of economic development and ultimately achieving inclusive green resilience [17,22]. To this end, the following hypothesis is proposed:

Hypothesis 1. *Carbon emissions trading policy promotes the adjustment of urban industrial structure and enables inclusive green resilience.*

Second is urban technological innovation. On the one hand, under the impact of carbon emissions trading policy, technological innovation will accelerate R&D and the application of production technology and environmental protection technology of enterprises. By applying clean technology into energy and production systems, enterprises can fundamentally improve resource utilization efficiency, implement prevention from the front end, and reduce environmental pollution. On the other hand, the development of technological innovation meets the demands of production and life, improves the productivity of social labor, and provides more human, material, and financial resources for industrial enterprises. Hence, the implementation of carbon emissions trading policy can effectively stimulate the innovation capacity of enterprises and create more new jobs, provide more employment opportunities for urban labor, and promote inclusive green resilience. To this end, this paper proposes the following hypothesis:

Hypothesis 2. *Carbon emissions trading policy stimulates the technological innovation capacity of cities and thus enables inclusive green resilience.*

Third, the employment effect is remarkable. On the one hand, the carbon emissions trading policy encourages enterprises with high energy consumption and low efficiency to transform into enterprises with high efficiency and low energy consumption. For enterprises with successful transformation, they can take advantage of the opportunities brought by green transformation to reduce carbon emissions and save carbon quotas, trade and make profits in the carbon trading market, improve their capital stock, and thus promote the expansion of enterprise scale and technological improvement, enabling green and upgraded development. In this process, in order to meet the needs of expanding scale and improving technology, enterprises increase employment and employment income, which meets the public's demand for employment and narrows the income distribution gap [23]. On the other hand, under the influence of carbon emissions trading policy, natural resources and environmental standards inevitably rise. In order to meet their own needs, enterprises reduce the use of natural resources and turn to labor with a relatively low cost instead of natural resources in development. Therefore, the substitution of resource elements by labor promotes the increasing labor demand of industrial enterprises, thus effectively adjusting the balance between social labor supply and demand, enabling coordination and unity between the economic system, social system, and environmental system and promoting inclusive green resilience. To this end, this paper proposes the following hypothesis:

Hypothesis 3. *Carbon emissions trading policy affects urban employment and thus enables inclusive green resilience.*

2.2. Research on the Heterogeneity of Carbon Emissions Trading Policy to Inclusive Green Resilience

The role of carbon emissions trading policy has different significance in cities of different development types. This paper mainly analyzes the difference from two parts: urban development scale and urban function.

First is urban development scale. Urban scale is a primary factor affecting the economic and social development of a city. The economic, social, and environmental benefits vary for cities of different scales, and a city of a certain scale will form economic aggregation and population aggregation to varying degrees. Economic agglomeration and population agglomeration promote the economic and social development of cities but also lead to the consumption of more energy and more carbon dioxide emissions, thus hindering the green development ability of cities. In addition, larger cities face more social problems and need to deal with more problems in their efforts to promote inclusive green resilience. Therefore, the impact of carbon emissions trading policy on the inclusive green resilience of cities is somewhat different due to the effect of urban development scale. Larger cities have higher economic agglomeration and population agglomeration, with great energy consumption and carbon emissions, but industrial enterprises have a stronger influence on the external environment and greater inter-industry competition reaction. When exposed to external conditions such as environmental policy constraints, they can make rapid adjustments to enhance the production and supply capacity of enterprises, thus improving the economic benefits of the whole city. However, due to the low degree of economic development and market activity in small-scale cities, carbon emissions trading policies finds it challenging to effectively foster inclusive green resilience. To this end, the following hypothesis is proposed:

Hypothesis 4. *In the development of large cities, carbon emissions trading policy has a better impact on inclusive green resilience.*

Second is urban function difference. One important reason for China's rapid economic development since the reform and opening up is the driving role of large-scale resource-based cities. For cities that have relied on heavy industry for a long time and excessively exploit natural resources to develop their economy, there is serious environmental pollution, backward infrastructure, a large proportion of labor force emigration, and significant vulnerability in economic development. As a result, these cities have significantly higher energy consumption and carbon emissions intensity than other non-resource-based cities in China. Due to the emphasis on the development of the secondary industry, serious industrial structure imbalance occurs in resource-based cities, resulting in overcapacity, weak urban innovation and development capacity, and serious urban pollution. Carbon emissions trading policy can control the carbon emissions of resource-based cities, adjust the industrial structure, inject new vitality into the development of industrial enterprises, stimulate the internal production and technological innovation of industrial enterprises, reasonably adjust the allocation of the urban labor force, improve the green development ability of enterprises and the well-being of social residents, and ultimately improve the inclusive green resilience level of cities. To this end, the following hypothesis is proposed:

Hypothesis 5. *In resource-based cities, carbon emissions trading policy has a more significant impact on inclusive green resilience.*

3. Research Model and Variables

3.1. Study Area

In order to put China's carbon trading system requirements into action, the National Development and Reform Commission (NDRC) designated Beijing, Tianjin, Shanghai, Chongqing, Hubei Province, Guangdong Province, and Shenzhen City as pilot cities for the carbon emissions trading policy in 2011. In 2013, Shenzhen City led the way in launching the nationwide carbon emissions trading market, and other cities and provinces followed suit. As a result, this paper views the 2013 pilot carbon emissions trading policy as a quasi-natural experiment. After deleting the cities with the most missing data on green inclusive resilience, the panel data of the remaining 184 cities from 2008 to 2018 are used as research samples, of which 139 are non-resource cities, 37 are mature resource cities, and 8 are declining resource cities. Given the policy implementation time lag, 2014–2018 is designated as the implementation year of the carbon emissions trading policy, and 2008–2013 is designated as the year preceding the system's implementation in both the treatment and the control group. The 37 prefectural-level cities taking part in the pilot carbon emissions trading strategy are considered the treatment group, whereas the remaining 147 cities are considered the control group. A double-difference model is built to analyze the influence of carbon emissions trading policy on the level of green inclusion resilience.

3.2. Research Model

PSM-DID model. The DID was proposed by Ashenfelier. The model can deduce differences between the experimental group impacted by the policy and the control group not affected by the policy before and after the policy's implementation, mostly through the use of dummy variables. The DID model, as one of the most widely used tools for assessing policy success, has also been used to investigate the process of the influence of carbon emissions trading policies on the level of economic, social, and other development [24,25]. As a result, this article employs the DID model to assess the impact of carbon emissions pricing policies on the level of green inclusive resilience. The model is as follows:

$$\mathrm{green}_{it} = \alpha_0 + \alpha_1 \mathrm{du} * \mathrm{dt} + \sum_{i=1}^{N} b_j X_{it} + \varepsilon_{it} \tag{1}$$

where i represents a city, t represents the year, green_{it} represents the interpreted variable, du represents the city grouping dummy variable, and dt represents the time grouping dummy variable. X_{it} represents the control variable, including the economic development level (sgdp), innovative research and development ability (echol), urbanization level (city), industrial structure (ind), and population density (pop). ε_{it} is the disturbance term. Since the policy is implemented after 2013 and there is a certain time lag effect, 2014–2018 is defined as the implementation year of the carbon emissions trading policy, which is defined as 1. The period 2007–2013 is defined as 0 as the early stage of system introduction. In the division of treatment group and control group, pilot cities in two provinces and five cities in 2014 are taken as the experimental group and defined as 1. Other cities serve as the control group and are defined as 0.

However, due to the great heterogeneity in the development of different cities in China, it is difficult for different cities to meet the condition of time effect consistency [26,27]. Thus, a control group with characteristics as similar as possible to the experimental group should be selected before adopting the DID method. The PSM-DID model is an estimation method that combines the PSM method and the DID method. The basic concept of propensity score matching is that for some control variables, the assumption that the mean value can be ignored must be met. Compared with the DID method, PSM-DID exhibits the advantage of controlling unobservable differences between groups that do not change over time, so this paper adopts the PSM-DID model for robustness analysis [27]. Specifically, PSM is used to reduce sample selection bias, and radius matching is conducted between the experimental group and the control group samples. On this basis, DID analysis is performed. The PSM-DID model is as follows:

$$\mathrm{green}_{it}^{\ PSM} = \alpha_0 + \alpha_1 du * dt + \sum_{i=1}^{N} b_j X_{it} + \varepsilon_{it} \tag{2}$$

3.3. *Measure the Level of Inclusive Green Resilience in Cities*

Index system construction. This paper comprehensively refers to the inclusive growth evaluation index published by the Asian Development Bank and the description of inclusive green resilience index system by Li et al. and constructs three subsystems, including economic development, social inclusion, and environmental sustainable development [7,28]. Economic resilience is mainly measured by the basis of urban economic development and economic stability. In the selection of indexes, six indexes are used to describe the subsystem of economic development. The social inclusion resilience is mainly measured by eight indexes, including education, employment, health, and infrastructure. In terms of environmental resilience, the analysis mainly focuses on green production and consumption and environmental protection, including seven evaluation indexes. See Table 1 for details.

Table 1. Indicators of inclusive green resilience in cities.

Goal System	Subsystem	Name of Index	Unit	Character
Inclusive green resilience	Economic resilience	GDP growth rate	%	+
		Proportion of tertiary industry	%	+
		Proportion of fiscal revenue	%	+
		Per capita disposable income of urban residents	CNY/person	+
		Per capita disposable income of rural residents	CNY/person	+
		Urban and rural per capita income ratio	%	−
	Social inclusion resilience	Educational resources for every 10,000 people	%	−
		Educational fund input	%	+
		Urban registered unemployment rate	%	−
		Employment training rate	%	+
		Health professionals per 1000 population	Person/1000	+
		Number of beds in medical and health institutions per 1000 population	Person/1000	+
		Length of transportation lines per 10,000 people	Km/10,000 people	+
		The number of buses per 10,000 people	Car/1000 people	+
	Ecological environment resilience	Energy consumption per unit of gross regional product	Ton/1000 yuan	−
		Amount of discharge wastewater	Ton/1000 yuan	−
		Carbon dioxide emissions	Ton/1000 yuan	−
		Sulfur dioxide emissions	Ton/1000 yuan	−
		Urban per capita public green space	Person/square meter	+
		Comprehensive utilization rate of solid waste	%	+
		Intensity of investment in environmental pollution control	%	+

Note: "+", "−" represent positive indicators and negative indicators, respectively.

Research methods. On the basis of constructing an evaluation index system, this paper refers to the research method of Wang et al. (2022) and uses the entropy method to analyze the level of inclusive green resilience [29]. The specific steps are as follows:

First, according to the index attributes, the evaluation data are standardized with the help of the entropy method:

Forward standardization formula:

$$X_{ij} = \left(X_{ij} - X_{minij}\right) / \left(X_{maxij} - X_{minij}\right) \times 0.99 + 0.001 \tag{3}$$

Negative normalization formula:

$$X_{ij} = \left(X_{maxij} - X_{ij}\right) / \left(X_{maxij} - X_{minij}\right) \times 0.99 + 0.001 \tag{4}$$

where X_{ij} represents the evaluation standard value, X_{minij} represents the minimum value of the j-th item in the i-th city, and X_{maxij} represents the maximum value of the j-th item in the i-th city.

Second, calculate the characteristic proportion of the j-th index:

$$P_{ij} = X_{ij} / \sum_{i=1}^{m} X_{ij} \tag{5}$$

Third, calculate the index of the information entropy:

The information entropy calculation formula of the j-th index of the i-th city can be expressed as:

$$e_j = -k \sum_{i=1}^{mT} \left(X_{ij} / \sum_{i=1}^{mT} X_{ij} \right) \ln \left(X_{ij} / \sum_{i=1}^{mT} X_{ij} \right) \tag{6}$$

where $k = 1/\ln mT$, $1 \le i \le m$, and $1 \le j \le 21$.

Fourth, calculate the weight of the evaluation index:

$$w_j = (1 - e_j) / \sum_{j=1}^{21} (1 - e_j) \tag{7}$$

Fifth, calculate the comprehensive score of the weight of the j-th index:

$$S_j = \sum w_j P_{ij} \tag{8}$$

Sixth, according to the obtained comprehensive score of the index weight, the comprehensive index of inclusive green resilience is calculated:

$$green_i = \sum_{i}^{N} S_{ij} \tag{9}$$

where $green_i$ represents the comprehensive index, and N is the number of indexes under all levels in the evaluation system.

3.4. Variable Selection and Data Source

The DID model is used to analyze the impact of carbon emissions trading policy on the level of inclusive green resilience in cities, and its explained variable is the level of inclusive green resilience in cities (green). The data used to calculate the level of inclusive green resilience in cities are mainly collected from the "China Urban Statistical Yearbook," the "China Environmental Statistical Yearbook," and the statistical yearbook and bulletin of each province from 2007 to 2018.

In terms of control variables, urbanization(city) is mainly expressed by the proportion of non-agricultural population in the total population [30]. The level of economic development (sgdp) is expressed as a logarithm of real GDP per capita. Innovative research and development capability (echol) is expressed by the number of urban patent grants. Industrial structure (ind) is expressed as the proportion of added value of secondary industry in GDP. Population density (pop) is expressed by the number of people per unit administrative area.

4. Empirical Results and Robustness Test

4.1. Empirical Results

The descriptive statistical results of the main variables in this paper are shown in Table 2. Figure 1 shows the distribution of inclusive green resilience capacity across cities in 2008, 2013, and 2018. In Table 3, model (1) is the model without adding control variables, and model (2) is based on model (1) with the addition of the time effect. Model (3) is the regression result after adding control variables on the basis of model (1), and model (4) is the regression result after adding control variables on the basis of model (2). All regression results show that carbon emissions trading policy has a significant positive impact on the inclusive green resilience level of cities with or without the inclusion of control variables, indicating that carbon emissions trading policy significantly improves the inclusive development level of cities. In terms of control variables, high population density and technical innovation level can significantly improve the level of inclusive green resilience in cities. The proportion of urbanization has a significant negative impact on the level of inclusive green resilience in cities, indicating that a higher proportion of urbanization reduces the level of inclusive green resilience in cities, whereas the coefficients of other control variables are insignificant, indicating that they are not the core factors affecting the level of inclusive green resilience in cities.

Table 2. Descriptive statistics.

Variable	Mean	SD	Min.	Max.
Green	0.357	0.172	0.010	0.998
sgdp	4.563	0.282	2.670	5.280
echol	5.613	8.962	0.000	64.580
ind	0.497	0.120	0.000	3.739
pop	5.986	0.750	0.177	3.739
city	0.516	0.146	0.189	1.000

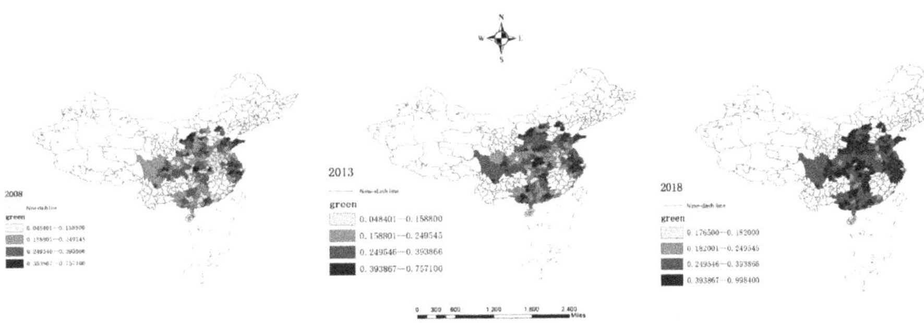

Figure 1. Level of inclusive green resilience of cities in different years.

Table 3. Carbon emissions trading policies and inclusive green resilience in cities: DID regression results.

Variable	(1)	(2)	(3)	(4)
	Green	Green	Green	Green
DID	0.138 *** (20.130)	0.029 *** (7.799)	0.053 *** (11.354)	0.022 *** (5.588)
city			0.358 *** (12.822)	−0.111 *** (−4.181)
sgdp			0.230 *** (25.214)	0.024 ** (2.481)
echol			0.004 *** (6.530)	0.002 *** (5.051)
ind			−0.050 *** (−3.927)	−0.004 (−0.411)
pop			0.042 *** (6.621)	0.025 *** (4.888)
_cons	0.344 *** (205.542)	−39.287 *** (−72.530)	−1.129 *** (−22.535)	−39.564 *** (−33.537)
N	2024	2024	2024	2024
R^2	0.099	0.770	0.650	0.778

Note: ** and *** indicate significance at the levels of 10%. In parentheses is the robust standard error.

4.2. Robustness Test

Parallel trend test. For the DID, a quasi-natural experimental method, an important prerequisite for the establishment of its results is to meet the parallel trend hypothesis—that is, for the inclusive green resilience level of the experimental group and the control group to have the same time variation trend before the impact of carbon emissions trading policy [31]. The research completed a parallel trend test on the level of inclusive green resilience in cities, using 2014 as the base year and data from the first four years and the last four years as samples. The test results are shown in Figure 2. As can be seen from Figure 1, in the years before the pilot of the carbon emissions trading policy, the inclusive green resilience level (green) of cities fluctuates around 0 and does not reject 0 at the 95% confidence interval, which conforms to the common trend hypothesis.

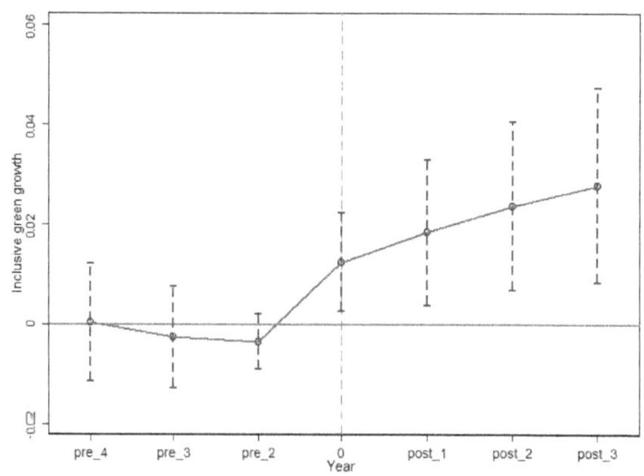

Figure 2. Parallel trend test.

PSM-DID analysis. In order to overcome the differences in the change trends between the pilot cities of the carbon emissions trading policy and other cities and to reduce the bias of the DID analysis, further robustness tests were conducted in this paper using PSM-DID's nearest neighbor matching, kernel matching, and radius matching. As can be seen from Table 4, carbon trading policies still significantly increase the level of inclusive green resilience of cities after using the three PSM-DID tests. The estimated results of the PSM-DID do not differ significantly from those of the above DIDs, further validating the empirical results of this paper. Carbon trading policies play an important role in inclusive green resilience.

Table 4. The average treatment effect.

Stage	Type	Nearest Neighbor Matching	Kernel Matching	Radius Matching
		Green	Green	Green
Before	Treated	1.010	1.010	1.010
	Control	0.051	0.051	0.051
	T-C	0.959	0.960	0.960
After	Treated	0.706	0.917	0.706
	Control	0.050	0.050	0.050
	T-C	0.656	0.868	0.656
S.E. (ATT)		0.131	0.146	0.131
T (ATT)		5.01	5.95	5.01
Variable		Yes	Yes	Yes
N		2024	2024	2024

5. Verification of Influence Mechanism

5.1. Verification of Influence Mechanism

Through DID analysis and the robustness test, it can be seen that the carbon emissions trading policy can promote the inclusive green resilience level of cities. In order to further explore the internal influence mechanism of carbon emissions trading policy on the inclusive green resilience level of cities, based on Formula (1), this paper sets the mechanism variables from the three parts of industrial structure adjustment, technological innovation ability, and employment effect, and, drawing on the study of Wen and Ye (2004), a three-step approach is used to test the model to further investigate the action path of carbon emissions trading policy on inclusive green resilience [31]. The model is as follows:

$$\text{green}_{it} = \beta_0 + \beta_1 du * dt + \sum_{i=1}^{N} b_j X_{it} + \varepsilon_{it} \quad (10)$$

$$M_{it} = \alpha_0 + \alpha_1 du * dt + \sum_{i=1}^{N} b_j X_{it} + \varepsilon_{it} \quad (11)$$

$$\text{green}_{it} = \gamma_0 + \gamma_1 du * dt + \gamma_2 M_{it} + \sum_{i=1}^{N} b_j X_{it} + \varepsilon_{it} \quad (12)$$

where M_{it} is the mechanism variable. If the coefficients α_1 and γ_1 are significant, then there is an effect. In the index selection of variables, we set mechanism variables for industrial structure adjustment, technological innovation capacity, and employment effect in accordance with the requirements of scientificity, representativeness, and comprehensiveness. Industrial structure adjustment is measured by the proportion of added value of the secondary industry and tertiary industry (upgrade). Technological innovation capacity is measured by the number of effective invention patents of industrial enterprises (patents).

Employment effect is measured by the average wage of working employees (wage). The data for each index derive from the statistical yearbook of each city.

5.2. Influence Mechanism Analysis

Table 5 shows the analysis of the internal mechanism by which carbon emissions trading policy affects the level of inclusive green resilience in cities. As can be seen from the table, the three influencing mechanisms can significantly promote the impact of carbon emissions trading policies on inclusive green resilience. First of all, it can be seen from the table that, as the mechanism variable of industrial structure adjustment, the proportion of added value of the secondary and tertiary industries has a significant negative impact, indicating that the economic structure dominated by the secondary industry does not use carbon emissions trading policy to promote inclusive green resilience. This is because, compared with the tertiary industry, the secondary industry has a single industrial structure, high energy consumption, and carbon emission intensity, which leads to serious urban environmental pollution and low economic and social development, and finally results in the weak green sustainable development level of cities. As a result, if we can raise the entry threshold of the secondary industry, reduce the market share of the secondary industry and transform it in a green and clean direction, and increase the proportion of the tertiary industry, we can reduce the negative impact of industrial structure imbalance in the process of enhancing cities' green inclusive resilience through the optimization and upgrading of the industrial structure so as to better play the impact of carbon emission trading policy on the urban inclusive green resilience capacity. Therefore, Hypothesis 1 is verified.

Table 5. A mechanistic test of the impact of carbon trading policies on green inclusive resilience.

Variable	Industrial Structure Adjustment	Technological Innovation Capacity	Employment Effect
upgrade	−0.023 ***		
patents		0.157 ***	
wage			0.028 ***
did	0.052 ***	0.048 **	0.046 ***
city	0.302 ***	0.288 ***	0.272 ***
sgdp	0.242 ***	0.237 ***	−0.219 ***
pop	0.019 ***	0.018 ***	0.018 ***
ind	0.059 **	−0.029 **	−0.037 **
echol	0.003 ***	0.003 ***	0.003
_cons	−1.037 ***	−1.046 ***	−1.047 ***
R^2	0.685	0.689	0.701

Note: ** and *** indicate significance at the levels of 5%, and 10%, respectively. In parentheses is the robust standard error.

Secondly, in terms of the level of technological innovation, the number of patentable inventions of industrial enterprises has a significant impact on the impact of carbon trading policies on inclusive green resilience. Therefore, to enhance inclusive green resilience through technological innovation, more attention should be paid to innovation, invention, scientific research, and technology. Overall, carbon emissions trading policy can promote inclusive green resilience in cities with the help of technological innovation. Therefore, Hypothesis 2 is verified.

Finally, in terms of employment effect, the total number of jobs in a city has a significant impact on the inclusive green resilience of a city due to carbon emissions trading policies. This suggests that, in order to better improve the inclusive green resilience of a city, we should strengthen the improvement of urban employment capacity and employment opportunities. Overall, carbon emissions trading policy can promote inclusive green resilience in cities through employment effects. Therefore, Hypothesis 3 is verified.

6. Heterogeneity Analysis

6.1. Heterogeneity Analysis of City Development Scale

Large-scale cities can lead to economic aggregation and population aggregation, which in turn promotes regional economic development and technological innovation. However, excessive aggregation leads to high energy consumption, increased carbon emissions, and more complex social problems, which is unfavorable to inclusive green urban growth. Therefore, whether carbon emissions trading policy can play an equal role in cities of different scales needs to be further studied. Based on the "Notice on Adjusting the Standards of City Scale Division" issued by the State Council, the size of the resident population in an urban area is used as a criterion in this article, and cities with a resident population of 500,000 or fewer are classified as small cities. Cities with a population of more than 500,000 but fewer than 1 million people are classed as medium-sized. Large cities are classified as having a resident population of more than one million people [32]. The paper divides 184 representative cities into large and medium cities and small cities for the regression test.

Table 6 shows the impact of different types of cities on the level of inclusive green resilience. As can be seen from the table, the carbon emissions trading policy significantly improves the inclusive green resilience level in large and medium-sized cities, but there is an insignificant effect on small cities. This shows that the carbon emissions trading policy can better fit the economic aggregation and population aggregation of large and medium-sized cities to jointly achieve inclusive green resilience. As small cities have lagging development, inferior market factor allocation ability, and weak information response ability compared to large and medium-sized cities, carbon emissions trading policy exerts an insignificant effect on them. In addition, in the process of policy implementation, due to the insufficient advantages of such cities, a large amount of capital and labor need to be consumed after policy implementation, resulting in greater output than income, which is ultimately unfavorable to the sustainable development of cities. Thus, Hypothesis 4 is verified.

Table 6. Heterogeneity analysis of urban development scale.

	Small Cities	Large and Medium Cities
	Green	Green
DID	0.001	0.029 ***
	(0.09)	(6.24)
Control variable	Yes	Yes
Year fixed effect	Yes	Yes
_cons	−0.756 ***	−0.383 ***
	(−4.78)	(−6.62)
N	451	1573
R^2	0.806	0.816

Note: *** indicate significance at the levels of 10%. In parentheses is the robust standard error.

6.2. Heterogeneity Analysis of Urban Function

According to statistics, China has 262 resource-based cities that can be classified into four types: growing, mature, declining, and regenerative. Compared with growth- and regenerative-type resource-based cities, the mature and declining types have advantages in transformation and development and can quickly get rid of resource dependence. As a result of the examination of resource-based and non-resource-based cities, this article separates resource-based cities into typical mature resource-based cities and declining resource-based cities, taking into account the sample cities chosen previously in this research. As shown in Table 7.

Table 7. Heterogeneity analysis of ecological resilience.

	Non-Resource-Based Cities	Resource-Based Cities Group		Resource-Based Cities
		Mature Resource-Based Cities	Declining Resource-Based Cities	
DID	0.054 ***	0.053 ***	0.133 ***	0.046 ***
	(9.42)	(4.04)	(5.88)	(4.70)
_cons	−1.135 ***	−0.780 ***	−2.782 ***	−0.907 ***
	(−16.52)	(−7.52)	(−2.10)	(−10.85)
Control variable	YES	YES	YES	YES
N	1430	385	99	594
F	449.75	85.62	28.64	162.53
R^2	0.685	0.708	0.930	0.705

Note: *** indicate significance at the levels of 10%. In parentheses is the robust standard error.

Carbon trading policies have a significant influence on the green inclusive resilience of both declining and mature resource-based cities at the 1% level, green inclusive resilience improves significantly after the implementation of carbon trading policies, and the increase is greater for declining resource-based cities. On the one hand, declining resource-based cities have more space for improvement in green inclusive resilience than mature resource-based cities, and they are more responsive to environmental policies aimed at carbon reduction and green development. On the other hand, declining resource-based cities experience more severe urban development concerns, such as slow economic growth and significant issues with means of subsistence. As a market-based policy, carbon emissions trading can effectively promote the sustainable development capacity of cities by taking into account both equity and efficiency. It is therefore easier to inject new impetus into the reform and development of cities and to achieve an increase in their green, inclusiveness, and resilience capacity. Thus, Hypothesis 5 is verified.

7. Conclusions and Policy Implications

Based on 184 representative cities from 2008 to 2018 as a study sample, this study analyzes the impact of carbon trading policies on the inclusive green resilience of pilot cities with the help of the DID model. The results show that carbon trading policies can significantly promote inclusive green resilience in cities, at about 2.2% and pass parallel trend and PSM-DID model tests, confirming the reliability of the findings. In order to gain insight into the mechanism of the impact of carbon trading policies on inclusive green resilience, this paper provides a further in-depth analysis from three components: industrial structure, technological innovation capacity, and employment effects. The results show that carbon trading policies can achieve inclusive green resilience by adjusting industrial structure, improving the capacity for technological innovation, and promoting employment effects. Carbon emissions trading programs, at 15.7%, have the largest influence on inclusive green resilience through technological innovation capacity. Carbon emissions pricing policies have a negative and large influence on inclusive green resilience through industrial restructuring, with a value of −2.3%. The results of the heterogeneous analysis show that, on the one hand, carbon trading policies affect cities of scale, and the policy can have a significant positive impact on the inclusive green resilience of large and medium-sized cities. On the other hand, there are significant differences in the impact of carbon emissions trading policies on inclusive green resilience in urban functions. For declining resource-based cities, carbon trading policies can most rapidly improve the inclusive green resilience of cities, at 13.3%, followed by non-resource-based cities, at about 5.4%. Based on the above findings, the paper proposes the following:

First, focus should be on the adjustment of the industrial structure of a city. The secondary industrial economy, which is based on high energy consumption and high emissions, still accounts for a large proportion of China's economic development. In this regard, on the one hand, on the basis of carbon emissions trading policies, the coordination of carbon emission trading policies should be promoted with other mandatory policies the low-carbon incentive and constraint mechanism should be improved while raising the entry threshold of the secondary industry through carbon emissions trading policies to guide and encourage the development of emerging low-carbon industries and promote the development of modern service industries. On the other hand, urban environmental supervision should be strengthened to reduce the intensity of urban carbon emissions with the help of market guidance and government control. With regard to the secondary industry, which is high in energy consumption and pollution, the industrial energy consumption structure should be gradually adjusted, clean technology of resources should be vigorously developed, and the flow of industries with high energy consumption and pollution should be reasonably guided to those with low emissions and high efficiency.

Second, urban technological innovation and development should be promoted. From the analysis of the influence mechanism, we can see that the number of patents of industrial enterprises in cities has a significant positive effect on the development of technological innovation in cities. Therefore, while implementing the carbon emissions trading policy, attention should be paid to enhancing the technological input of cities, actively guiding enterprises to carry out green technological innovation, guaranteeing the technological innovation of non-high-tech enterprises and private enterprises from the policy level, and fundamentally reducing the carbon dioxide emissions intensity of cities.

Third, attention should be paid to the effect of urban employment. The influence mechanism shows that urban employment effects can improve inclusive green resilience. Therefore, the government should increase training in employment skills of the workforce and improve the level of employment skills and the quality of the workforce. In addition, the function of public employment services should be fully utilized to alleviate structural contradictions and employment inequalities in the urban workforce, and the social impact of the workforce should be recognized in enhancing the inclusive green resilience of a city.

Finally, measures should be adjusted to local conditions. Different city sizes have different impacts on carbon trading policies. On the one hand, large and medium-sized cities should implement appropriate tightening of carbon emissions quotas to stimulate market competition, improve industrial enterprises' technological innovation ability, promote employment, and improve inclusive green resilience. Small cities, on the other hand, should establish appropriate relaxation policies, as well as matching auxiliary policies, to provide financial and technical support for technological innovation and green development to alleviate the negative effects of the lack of impetus for urban development and promote inclusive green resilience. In addition, for resource-based cities, government departments at all levels should avoid the "one-size-fits-all" policy. For mature resource-based cities, reasonable carbon emissions targets should be set to encourage cities to develop new pillar industries and green industries, strengthen workforce absorption, and achieve inclusive green urban growth. For declining resource-based cities, we should not only scientifically set reasonable total carbon emissions targets, relax carbon emissions indicators, and stimulate energy-intensive and high-emissions enterprises to reduce carbon emissions by means of technological innovation but also take into account the market economy vulnerability of declining resource-based cities and introduce corresponding carbon emissions-supporting measures in a timely manner. The relationship between industrial structure and the allocation of labor should be properly adjusted to reduce structural unemployment and help gradually achieve sustainable urban transformation.

Our research has several limitations, which may also be the future directions for our follow-up research. First, our analysis exclusively takes into account data from Chinese cities and excludes data from other countries. With the substantial trend of globalization, the network of economic and social links between countries is becoming increasingly intertwined. The study could be expanded in the future to investigate the dynamic impact of national interaction networks on the capacity for inclusive green resilience. Second, due to the scarcity of data on inclusive green resilience indicators for some provinces, the article's study period is limited to 2008–2018. If the research data years could be expanded, the article could more precisely depict variations in impacts over time.

Author Contributions: Conceptualization, B.X. and Q.S.; methodology, B.X.; software, Q.S.; validation, B.X. and Q.S.; formal analysis, Q.S.; resources, B.X.; data curation, Q.S.; writing—original draft preparation, Q.S.; writing—review and editing, B.X.; supervision, B.X. All authors have read and agreed to the published version of the manuscript.

Funding: This research was funded by the National Science Foundation of China (72061001) and the Research Fund Project of Guangxi University (XBS1642).

Institutional Review Board Statement: Not applicable (this study does not involve humans or animals).

Informed Consent Statement: Not applicable (this study does not involve humans).

Data Availability Statement: Not applicable.

Conflicts of Interest: The authors declare no conflict of interest.

References

1. Song, Q.; Liu, T.; Qi, Y. Policy innovation in low carbon pilot cities: Lessons learned from China. *Urban Clim.* **2021**, *39*, 100936. [CrossRef]
2. Duan, H.; Mo, J.; Fan, Y.; Wang, S. Achieving China's energy and climate policy targets in 2030 under multiple uncertainties. *Energy Econ.* **2018**, *70*, 45–60. [CrossRef]
3. Wang, K.L.; Pang, S.Q.; Zhang, F.Q.; Miao, Z.; Sun, H.P. The impact assessment of smart city policy on urban green total-factor productivity: Evidence from China. *Environ. Impact Assess. Rev.* **2022**, *94*, 106756. [CrossRef]
4. He, Y.; Song, W. Analysis of the impact of carbon trading policies on carbon emission and carbon emission efficiency. *Sustainability* **2022**, *14*, 10216. [CrossRef]
5. Yu, X.; Shi, J.; Wan, K.; Chang, T. Carbon trading market policies and corporate environmental performance in China. *J. Clean. Prod.* **2022**, *371*, 133683. [CrossRef]
6. Liu, W.; Qiu, Y.; Jia, L.; Zhou, H. Carbon emissions trading and green technology innovation—A quasi-natural experiment based on a carbon trading market pilot. *Int. J. Environ. Res. Public Health* **2022**, *19*, 16700. [CrossRef]
7. Yu, H.T.; Wang, Y.T.; Li, X.; Wang, C.D.; Sun, M.X.; Du, A.S. Measuring Ecological Capital: State of the Art, Trends, and Challenges. *J. Clean. Prod.* **2019**, *219*, 833–845. [CrossRef]
8. Rizzo, A. Megaprojects and the limits of 'green resilience' in the global South: Two cases from Malaysia and Qatar. *Urban Stud.* **2020**, *57*, 1520–1535. [CrossRef]
9. Shokry, G.; Anguelovski, I.; Connolly, J.J.; Maroko, A.; Pearsall, H. "They didn't see it coming": Green resilience planning and vulnerability to future climate gentrification. *Hous. Policy Debate* **2022**, *32*, 211–245. [CrossRef]
10. Shokry, G.; Connolly, J.J.; Anguelovski, I. Understanding climate gentrification and shifting landscapes of protection and vulnerability in green resilient Philadelphia. *Urban Clim.* **2020**, *31*, 100539. [CrossRef]
11. Starczewski, T.; Rogatka, K.; Kukulska-Kozieł, A.; Noszczyk, T.; Cegielska, K. Urban green resilience: Experience from post-industrial cities in Poland. *Geosci. Front.* **2023**, *14*, 101560. [CrossRef]
12. Bănică, A.; Istrate, M.; Muntele, I. Towards green resilient cities in eastern European union countries. *J. Urban Reg. Anal.* **2020**, *12*. [CrossRef]
13. Mo, J.L.; Agnolucci, P.; Jiang, M.R.; Fan, Y. The impact of Chinese carbon emission trading scheme (ETS) on low carbon energy (LCE) investment. *Energy Pol.* **2016**, *89*, 271–283. [CrossRef]
14. Chen, J.; Huang, S.; Shen, Z.; Song, M.; Zhu, Z. Impact of sulfur dioxide emissions trading pilot scheme on pollution emissions intensity: A study based on the synthetic control method. *Energy Pol.* **2021**, *161*, 112730. [CrossRef]
15. Song, M.; Liu, X.Y. Research on the mechanism of enabling agricultural resilience in digital economy: An analysis of the mediating effect based on human capital. *Jiangsu Soc. Sci.* **2023**, *1*, 103–112.
16. Wang, H.; Li, Y.; Bu, G. How carbon trading policy should be integrated with carbon tax policy—Laboratory evidence from a model of the current state of carbon pricing policy in China. *Environ. Sci. Pollut. Res.* **2023**, *30*, 23851–23869. [CrossRef] [PubMed]
17. Hailemariam, A.; Dzhumashev, R.; Shahbaz, M. Carbon emissions, income inequality and economic development. *Empir. Econ.* **2020**, *59*, 1139–1159. [CrossRef]
18. Huang, H.; Roland-Holst, D.; Springer, C.; Lin, J.; Cai, W.; Wang, C. Emissions trading systems and social equity: A CGE assessment for China. *Appl. Energy* **2019**, *235*, 1254–1265. [CrossRef]
19. Gree, J.F. Does carbon pricing reduce emissions? A review of ex-post analyses. *Environ. Res. Lett.* **2021**, *16*, 43004. [CrossRef]
20. Shi, D.; Ming, X. Whether the "double points" policy can promote substantial innovation of new energy vehicles. *J. Beijing Inst. Technol.* **2023**, *25*, 40–51.
21. Feng, R.; Lin, P.; Hou, C.; Jia, S. Study of the effect of China's emissions trading scheme on promoting regional industrial carbon emission reduction. *Front. Environ. Sci.* **2022**, *10*, 947925. [CrossRef]

22. Fang, Z.; Wolski, M. Human capital, energy and economic growth in China: Evidence from multivariate nonlinear Granger causality tests. *Empir. Econ.* **2021**, *60*, 607–632. [CrossRef]
23. Yang, H.; Gan, T.; Liang, W.; Liao, X. Can policies aimed at reducing carbon dioxide emissions help mitigate haze pollution? An empirical analysis of the emissions trading system. *Environ. Dev. Sustain.* **2022**, *24*, 1959–1980. [CrossRef]
24. Liu, H.Y.; Guo, W.Q. Carbon Emission Trading Policy Pilot and Energy-Environmental Efficiency: An Empirical Study of 287 Prefecture-Level Cities in China. *J. Xi'an Jiaotong Univ. (Soc. Sci.)* **2022**, *42*, 72–86.
25. Shi, D.Q.; Ding, H.; Wei, P.; Liu, J.J. Whether the construction of smart cities can reduce environmental pollution. *Chin. Ind. Econ.* **2018**, *6*, 117–135.
26. Qi, S.; Cheng, S.; Cui, J. Environmental and economic effects of China's carbon market pilots: Empirical evidence based on a DID model. *J. Clean. Prod.* **2021**, *279*, 123720. [CrossRef]
27. Li, H.; Don, Y.L. The Source of China's high-quality economic development and its differences: Based on the perspective of inclusive green Total factor productivity. *Financ. Econ. Res.* **2021**, *47*, 4–18.
28. Wang, L.; Li, S.R.; Zhang, G. Research on the safety measurement of traditional fossil energy in China based on TOPSIS entropy method. *Ind. Technol. Econ.* **2022**, *41*, 124–129.
29. Desalegn, G.; Tangl, A. Enhancing green finance for inclusive green resilience: A systematic approach. *Sustainability* **2022**, *14*, 7416. [CrossRef]
30. Li, L.; Lei, Y.; Wu, S.; He, C.Y.; Yan, D. Study on the coordinated development of economy, environment and resource in coal-based areas in Shanxi Province in China: Based on the multi-objective optimization model. *Resour. Policy* **2018**, *55*, 80–86. [CrossRef]
31. Wen, Z.L.; Zhang, L.; Hou, J.T.; Liu, H.Y. Mediation effects test procedure and its application. *Psychol. Dly.* **2004**, *5*, 614–620.
32. Notification of the Adjustment of the Criteria for Municipal Division. Available online: https://www.gov.cn/gongbao/content/2014/content_2779012.htm (accessed on 6 April 2023).

sustainability MDPI

Article

Research on Provincial Carbon Emission Reduction Path Based on LMDI-SD-Tapio Decoupling Model: The Case of Guizhou, China

Hongqiang Wang [1], Wenyi Xu [1,*] and Yingjie Zhang [2]

1 School of Management, Shanghai University, Shanghai 200444, China; whqsl@shu.edu.cn
2 School of Management, University of Shanghai for Science and Technology, Shanghai 200093, China; alina_sh@163.com
* Correspondence: wyxsharin@163.com

Abstract: The successful implementation of the national carbon emissions reduction work necessitates the collaboration of various regions. Carbon emission reduction strategies need to be adjusted according to local circumstances due to the differences in regional development levels. From 2005 to 2020, carbon emissions were measured in Guizhou Province, and the contribution degree and action direction of various influencing factors were analyzed using the LMDI model. Using an SD model, we performed dynamic simulations of carbon emission trends under eight scenarios and calculated the Tapio decoupling relationship between economic growth and CO_2 emissions. According to the study, carbon emissions in Guizhou Province increased from 2005 to 2020, emphasizing the high pressure for carbon emission reduction. The industry sector ranked first in contribution, contributing 62.71% in 2020. Furthermore, this study found a weak decoupling relationship between economic growth and carbon emissions. The economic scale was the key driver driving the increase in carbon emissions, whereas the industrial fossil energy intensity was the main factor inhibiting the growth of carbon emissions. Additionally, it was predicted that carbon emissions would only peak at 277.71 million tons before 2030 if all three measures were implemented simultaneously, and a strong decoupling relationship with economic growth could be achieved as early as possible. These findings provided Guizhou Province with an effective path for reducing carbon emissions.

Keywords: LMDI; system dynamics; scenario simulation; tapio decoupling analysis; the carbon emission reduction path

Citation: Wang, H.; Xu, W.; Zhang, Y. Research on Provincial Carbon Emission Reduction Path Based on LMDI-SD-Tapio Decoupling Model: The Case of Guizhou, China. *Sustainability* **2023**, *15*, 13215. https://doi.org/10.3390/su151713215

Academic Editors: Fuqiang Wang, Chao Shen, Dong Li and Zhonghao Rao

Received: 24 June 2023
Revised: 24 August 2023
Accepted: 1 September 2023
Published: 3 September 2023

1. Introduction

The greenhouse gas emissions have exacerbated the global greenhouse effect, primarily CO_2, and the resulting climate change issues severely threaten the economy, society, and ecosystem. Reducing carbon emissions has developed into a mutual issue confronting the global community. China, which has the most significant total carbon emissions globally, is still seeing a rise in its carbon emissions, resulting in increased pressure to decrease it. China proposed "dual carbon" goals in September 2020. That is, carbon dioxide emissions aim to peak by 2030, and strive to achieve carbon neutrality by 2060, known as the carbon peak, carbon neutrality '3060' goal. How to implement effective carbon emission reduction measures is an urgent challenge for China today. Due to the variation in the driving mechanisms and peak times of different provinces of CO_2 [1], each region should adopt corresponding emission reduction measures according to its conditions. Thus, exploring the provincial carbon emission reduction path is significant.

To explore the scientific approach to lowering carbon emissions, scholars have continuously proposed new perspectives. Some researchers have investigated from a spatial perspective, such as a regional perspective [2–4] and an industry perspective [5–7], which provides a scientific foundation for formulating effective carbon emission reduction strategies. Some researchers have studied from the perspective of carbon emission responsibility allocation [8,9] and quota allocation [10,11]. Through these studies, governments and policymakers can consider different regions and industries' actual situations and responsibilities more scientifically and comprehensively when formulating carbon emission reduction strategies. Among these, the discussion of carbon emission impact mechanism and peak time is particularly intense in the research of regional carbon emissions. For example, to drive China to achieve its "dual carbon" goals, Guo et al. took the Yangtze River Delta region as an example. They found that the key influencing factors for carbon emissions differ among provinces [12]. Huang et al. analyzed the factors influencing terminal energy consumption in typical urban clusters. The research found that energy intensity and energy consumption structure significantly impact carbon reduction [13]. Yue et al. anticipated China's carbon emission trends from 2020 to 2035 by combining scenario analysis with predictive algorithms, which assisted China in meeting its target for carbon peaks [14].

Several methods are commonly used to study the factors influencing carbon emissions, including the structural decomposition approach (SDA) [15], the logarithmic mean Divisia index (LMDI) model [16,17], the grey relational analysis [18], as well as Stochastic Impacts by Regression on Population, Affluence, and Technology STIRPAT model [19–21]. For its ability to handle zero values with ease and provide an independent analysis path, the LMDI model is frequently employed by academics [22]. For example, using the LMDI model, Xu et al. successfully decomposed the contributing factors of carbon emission. They elucidated the influences of energy intensity, energy structure, industrial structure, and economic growth [23]. Yang et al. found that carbon emissions were predominantly influenced by the significant effects of economic growth and energy intensity [24]. In addition, carbon emission trend prediction is also one of the critical concerns in the academic sphere. In order to investigate the possibility of achieving carbon emission peaks and predict the timing for their occurrence, scholars have used various methods, including the long-term energy alternative planning (LEAP) model [4,25], artificial neural network model [26], STIRPAT model [27,28], system dynamics (SD) model [29,30] and Kuznets curve (EKC) [7] to predict and simulate the future carbon emission scenarios. Among them, system dynamics, as a theory and method to explore the overall system structure, has been widely used in recent years to solve carbon emission reduction problems. Additionally, the model is utilized to explore ways of emission reduction of carbon in both the industrial [31] and residential sectors [32]. For example, a multi-level SD model was developed by Yang et al. to predict China's future carbon emissions levels while also providing suggestions to promote energy low-carbon transformation [33]. Similarly, Du et al. employed a combination of system dynamics modeling and scenario analysis to simulate the trend of carbon emissions [34].

Despite the insightful contributions the LMDI model can offer, it has limitations in directly observing the effectiveness of corresponding measures in reducing emissions. System dynamics modeling proves highly beneficial in dealing with complex issues, but most studies are more subjective in constructing models. Therefore, to avoid the limitations of a single model, some scholars have tried to combine these two models to analyze the interrelationships among various relevant factors and predict carbon emission trends under different scenarios. For example, some scholars combined LMDI models with system dynamics models to explore the crucial factors behind alterations in carbon emissions [35] or predict future carbon emission trends [36]. In summary, national and city levels of carbon emission reduction research are the main subjects of the literature that is now available. However, more attention should be paid to reducing carbon emissions in underdeveloped provinces. Moreover, few studies have simultaneously considered the decoupling relationship between economic growth and carbon emissions across various scenarios. Considering this, this paper introduces the Tapio decoupling index based on previous studies, combines the characteristics of LMDI and SD models, takes Guizhou province as an example, subdivides emission sectors, and explores its carbon emission reduction problem from a more comprehensive perspective.

Guizhou Province possesses the most significant amount of coal resources among the regions in southern China. Over the past decade, its economic growth rate has continuously ranked among the highest in the country. However, as an underdeveloped area in China, Guizhou Province commenced its industrial development comparatively late. Currently, the secondary sector still dominates, and its industrial development highly relies on coal, making it challenging to achieve a carbon emission peak. Nevertheless, existing literature lacks attention to carbon emissions reduction in such regions. Therefore, we chose Guizhou Province as an example in the hope of conducting a more thorough analysis of the carbon emission reduction path at the provincial level.

The paper makes several key contributions. Firstly, the carbon emissions in Guizhou Province are calculated, and the LMDI model is employed to examine the contribution degree and action direction of different influencing factors, thereby identifying crucial driving factors that influence carbon emissions. Secondly, based on the analysis results from the LMDI model, an SD model is constructed to dynamically simulate future carbon emission trends and examine the direct impact of corresponding measures on carbon emissions. Thirdly, by calculating the Tapio decoupling index, this research observes the changing trends in the decoupling relationship between economic growth and carbon emissions under different scenarios. The study's findings provide insights for Guizhou Province in developing strategies to reduce carbon emissions and hold crucial practical significance for achieving national carbon emission reduction goals.

The rest of this paper is structured as follows. The Section 2 is the methodology and data source. The Section 3 presents the results. The Section 4 includes the discussion and policy recommendations. The Section 5 is the conclusions.

2. Methodology and Data Source

2.1. Study Area

Guizhou Province is located in the southeast of southwestern China, between 103°36′ to 109°35′ east longitude and 24°37′ to 29°13′ north latitude. In 2020, the province's gross domestic product (GDP) increased by 4.34 % year-on-year. The primary, secondary, and tertiary industries accounted for 8.70%, 44.16%, and 47.14% of the total output value, respectively. As of 2020, the province's population stood at 35.85 million, with an urban population accounting for 53.16%, an increase of 1.75 percentage points compared to 2019. The total energy consumption amounted to 114.14 million tons of standard coal, indicating an annual decrease of 2.05%. Cement production reached 108.21 million tons, with an annual decrease of 2.17%. Moreover, coal remains the primary energy source in the current energy structure.

2.2. IPCC Carbon Emission Calculation Method

According to data from the International Energy Agency (IEA), carbon emissions from coal in China accounted for 79.61% of the total carbon emissions in 2019. At the same time, global cement production contributes to over 7% of the total carbon emissions annually [37]. Both coal and cement production are core components of economic activities in Guizhou Province, making significant contributions to carbon emissions. Furthermore, Guizhou Province has an annual coal production exceeding 100 million tons and has the richest coal resources in southern China. Currently, it is in a phase of rapid economic growth, with the secondary industry as the dominant sector, heavily reliant on coal for its development. Therefore, considering the scale of carbon emission contributions, regional characteristics, and development conditions of Guizhou Province, as well as the data availability, this study primarily focuses on carbon emissions from energy consumption and cement production.

Carbon emissions from fossil energy consumption were calculated using the Intergovernmental Panel on Climate Change (IPCC) emission factor methodology [38]. Since the carbon emission factors provided in the Guidelines for Provincial Greenhouse Gas Inventory better suit China's national conditions, the carbon content and carbon oxidation rate figures provided by these guidelines were adopted. At the same time, the calorific value was determined using the average low-level heat generation data from the China Energy Statistical Yearbook. The calculation method is shown in Equation (1).

$$C_e = \sum E_k \times NCV_k \times CC_k \times O_k \times \frac{44}{12} \tag{1}$$

where C_e was the energy consumption CO_2 emissions, the subscript k was the subdivided energy type, E_k was the fossil energy consumption, NCV_k was the average low-level heating value of energy source k, CC_k was the carbon content of energy k, O_k was the carbon oxidation rate of energy k, and 44/12 was the conversion coefficient of CO_2 and C. The carbon emission produced by the cement production process can be calculated using Equation (2).

$$C_c = AD \times EF \tag{2}$$

where C_c was the carbon dioxide emissions resulting from cement production, AD was the volume of cement production, and EF was the carbon emission factor associated with cement production, referring to the 0.2906 provided by Shan et al. [39].

2.3. Logarithmic Mean Divisia Index (LMDI)

The logarithmic mean Divisia index decomposition is commonly used to study energy, resources, and the environment [22]. In this paper, we extended the Kaya identity [40] by introducing economic scale, industrial structure, energy intensity, and population structure to analyze the contributions and directions of different influencing factors to CO_2 emissions. We utilized the LMDI-I additive [41] decomposition analysis to decompose carbon emissions from energy consumption and cement production.

We used ES (=1, 2, . . ., 8) to represent the eight major emission sectors, including the primary sector, such as agriculture, forestry, animal husbandry, and fishery, abbreviated as AFAHF; the secondary industry, such as industry and construction; the tertiary sector such as transport, storage, and post, abbreviated as TSP, wholesale, retail trade, hotel, and restaurants, abbreviated as WRTHR, as well as other service industries, abbreviated as OSI, in addition to the above, it also included urban and rural sectors. J (=1, 2, 3) represented coal, petroleum, and natural gas. Equation (3) demonstrates the total CO_2 emissions, Equations (4) and (5) respectively show the carbon emissions from the industrial and residential sectors, and Equation (6) displays the emissions from cement production, while Table 1 defines the variables.

$$C = C_e + C_c = (C_1 + C_2) + C_c = \left(\sum_{ES=1}^{6} \sum_{j=1}^{3} C_{ES,j} + \sum_{ES=7}^{8} \sum_{j=1}^{3} C_{ES,j} \right) + C_c \tag{3}$$

$$C_1 = \sum_{ES=1}^{6} \sum_{ES=1}^{3} \frac{C_{ES,j}}{E_{ES,j}} \times \frac{E_{ES,j}}{E_{ES}} \times \frac{E_{ES}}{G_{ES}} \times \frac{G_{ES}}{G} \times G = \sum_{ES=1}^{6} \sum_{j=1}^{3} b_{ES,j} \times d_{ES,j} \times f_{ES} \times h_{ES} \times G \tag{4}$$

$$C_2 = \sum_{ES=7}^{8} \sum_{j=1}^{3} \frac{C_{ES,j}}{E_{ES,j}} \times \frac{E_{ES,j}}{E_{ES}} \times \frac{E_{ES}}{R_{ES}} \times \frac{R_{ES}}{P_{ES}} \times \frac{P_{ES}}{P} \times P = \sum_{ES=7}^{8} \sum_{j=1}^{3} b_{ES,j} \times d_{ES,j} \times q_{ES} \times r_{ES} \times s_{ES} \times P \tag{5}$$

$$C_c = \frac{C_c}{AD} \times \frac{AD}{G_2} \times \frac{G_2}{P} \times P \tag{6}$$

Table 1. The definition of variables in the equation.

Variable	Meaning	Variable	Meaning
$C_{ES,j}$	Carbon emissions from the consumption of fossil energy j in sector ES	$E_{ES,j}$	Consumption of fossil energy j in sector ES
E_{ES}	The fossil energy consumption of sector ES	G_{ES}	The gross domestic product of sector ES
G	Gross Domestic Product (GDP)	R_{ES}	Household consumption expenditure in department ES
P_{ES}	The number of people in sector ES	P	Total population of the region
$b_{ES,j} = C_{ES,j}/E_{ES,j}$	Carbon emission coefficient of sector ES fossil energy j	$d_{ES,j} = E_{ES,j}/E_{ES}$	Fossil energy structure of sector ES
$f_{ES} = E_{ES}/G_{ES}$	The fossil energy intensity of industrial sector	$h_{ES} = G_{ES}/G$	Industrial structure
$q_{ES} = E_{ES}/R_{ES}$	The fossil energy intensity of residential sector	$r_{ES} = R_{ES}/P_{ES}$	Per capita consumption level
$s_{ES} = P_{ES}/P$	Urban and rural population structure	$m = C_c/AD$	Carbon emission coefficient of cement
$n = AD/G_2$	Cement production intensity	$u = G_2/P$	Industrial sector per capita output value

We assumed that C^t and C^{t-1} denoted the CO_2 emissions for the year t and t − 1, respectively, and ΔC denoted the year-to-year change influence of carbon emissions. Equation (7) illustrates the calculation method.

$$\Delta C = C^t - C^{t-1} = (\Delta b + \Delta d + \Delta f + \Delta h + \Delta G) + (\Delta b + \Delta d + \Delta q + \Delta r + \Delta s + \Delta P) + (\Delta m + \Delta n + \Delta u + \Delta P) \quad (7)$$

where Δb was the fossil energy carbon emission coefficient influence, Δd was the fossil energy structure influence, Δf was the industrial energy intensity influence, Δh was the industrial structure influence, ΔG was the economic scale influence, Δq was the residential energy intensity influence, Δr was the per capita consumption level influence, Δs was the population structure influence, Δm was the cement carbon emission coefficient influence, Δn was the cement production intensity influence, Δu was the industrial sector per capita output value influence, and ΔP was the population scale influence. Since the cement emission factor is constant, the contribution of Δm is zero. The carbon emission influence of each variable can be calculated by Equation (8).

$$\Delta C_x = \sum_{ES,j} \frac{C^t_{ES,j} \times C^{t-1}_{ES,j}}{\ln\left(C^t_{ES,j}\right) - \ln\left(C^{t-1}_{ES,j}\right)} \times \ln\left(\frac{x^t}{x^{t-1}}\right) \quad (8)$$

where x was any of the 11 influences, we introduced the relative contribution degree δ to represent the proportion of the influence of a specific impact factor at a certain stage in the total influence of all impact factors at that stage. The calculation method is shown in Equation (9).

$$\delta = \frac{\Delta C_x}{\left|\sum_x \Delta C_x\right|} \quad (9)$$

where, if $\delta > 0$, the factor had a positive impact on carbon emissions; otherwise, if $\delta < 0$, the factor had an adverse effect, with the absolute value of δ reflecting the degree of contribution.

2.4. Construction of Carbon Emission System Dynamics Model

2.4.1. System Boundary and Research Hypothesis

System dynamics (SD) possesses inherent advantages in addressing complex nonlinear problems. Combining system dynamics and scenario analysis methods can dynamically simulate future trends. The study took the administrative boundary of Guizhou province as the spatial boundary of the system, and we used the overall interaction of the influencing factors obtained through the LMDI model decomposition as the system's behavioral boundary. The time span of analysis was set as 2005–2040, with forecast years ranging from 2021 to 2040, and the model was run using a time step of 1 year for the simulation.

2.4.2. Subsystem Division and System Model Establishment

Based on the decomposition results of the LMDI model, the carbon emission system in Guizhou Province was divided into four subsystems: population, economy, energy, and environment. By analyzing the interaction between variables, we have successfully established a CO_2 emission system for Guizhou province using the Vensim PLE software. Figure 1 displays the stock-flow diagram.

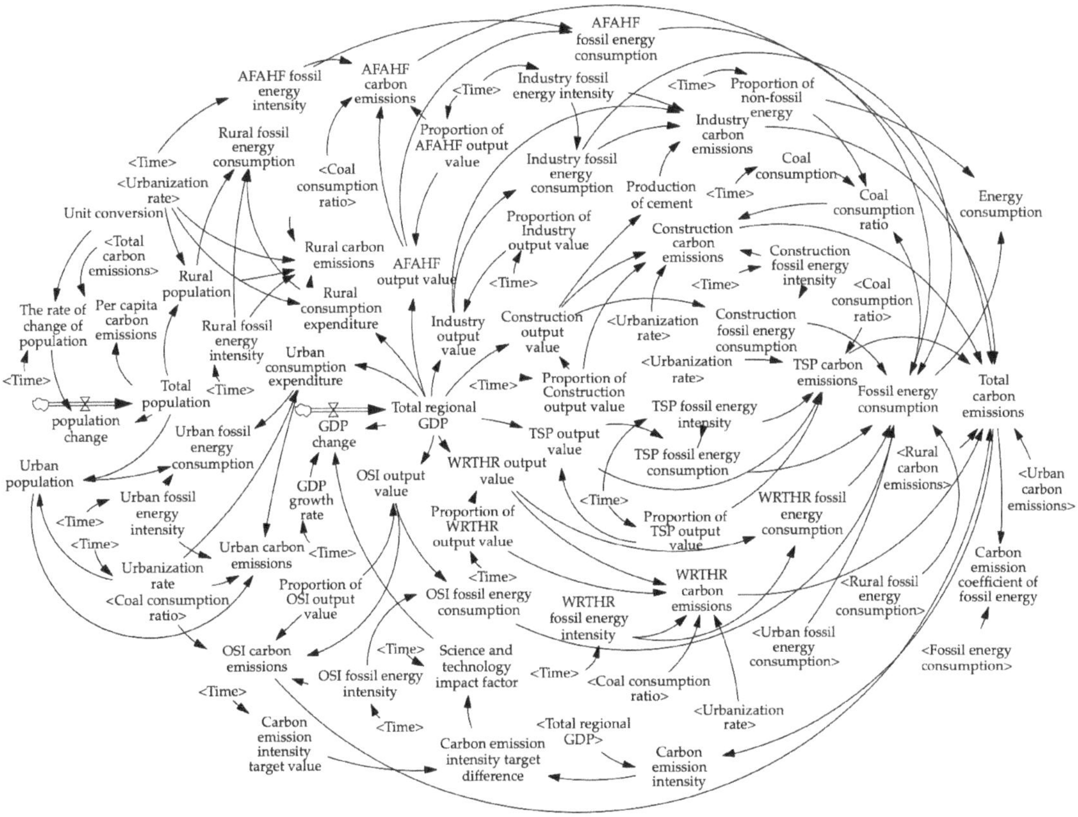

Figure 1. Stock-flow diagram of the carbon emission system.

The population subsystem primarily influenced the other three subsystems through the total population, household consumption expenditure, and urbanization. As urbanization levels increased, residential consumption expenditure and carbon emissions tended to change. The intensity of fossil energy, residential consumption expenditure, and population size determined the amount of residential energy consumption, affecting the total carbon emissions. The economic subsystem mainly influenced the other three through total regional GDP and industrial structure. Changes in GDP played a significant role in influencing consumption expenditure levels, while GDP growth was closely linked to the development and use of fossil energy. Industrial structure changes directly affected fossil energy utilization across different industrial sectors, influencing the environmental subsystem. The energy subsystem is the core system that influences carbon emissions, and most carbon emissions from human activities come from fossil fuel combustion [42]. By improving technology, decreasing the intensity of fossil fuel usage and increasing the percentage of non-fossil energy sources could lead to lowered fossil fuel consumption and optimized energy consumption structure, ultimately affecting carbon emissions in Guizhou Province. The environmental subsystem included carbon emissions from energy consumption and cement production, influencing carbon emission intensity and the total regional GDP.

The relationship functions among the variables were constructed by employing the direct assignment method, regression analysis method, mean value method, ratio analysis method, and table function method. Table 2 lists the relationship equations between some critical variables.

Table 2. The specific relationship equations between variables.

Variables	Equation
Total carbon emissions	TSP carbon emissions + OSI carbon emissions + Rural carbon emissions + AFAHF carbon emissions + Urban carbon emissions + Industry carbon emissions + Construction carbon emissions + WRTHR carbon emissions
AFAHF carbon emissions	EXP (LN (Coal consumption ratio × 100) × (−0.365647) + LN (AFAHF fossil energy intensity) × 0.955531 + LN (AFAHF output value) × 0.50745 + LN (Proportion of AFAHF output value) × (−0.174425) + 5.15538)
Construction carbon emissions	EXP (LN (Coal consumption ratio × 100) × 0.502325 + LN (Construction fossil energy intensity) × 1.11191 + LN (Construction output value) × 0.59081 + LN (Proportion of Construction output value) × 0.33328 + LN (Urbanization rate × 100) × 1.08194 − 1.82582)
Industry carbon emissions	EXP (LN (Industry fossil energy intensity) × 0.0237222 + 0.0136943 × LN (Industry output value) +0.0552457 × LN (Production of cement) +0.00684715 × LN (Industry output value × Industry output value) + 0.894751 × LN (Industry fossil energy consumption) + 1.25979)
OSI carbon emissions	EXP (LN (Coal consumption ratio × 100) × 0.649347 + LN (OSI fossil energy intensity) × 1.02937 + LN (OSI output value) × 1.02964 + LN (Proportion of OSI output value) × 0.262082 − 1.76103)
Rural carbon emissions	EXP (LN (Coal consumption ratio × 100) × 0.234948 + 1.01362 × LN (Rural fossil energy intensity × 100) +LN (Rural consumption expenditure) × 1.10321 + 0.292281 × LN ((1 − Urbanization rate) × 100) + LN (Rural population) × 0.77769−14.1891)
TSP carbon emissions	EXP (LN (Coal consumption ratio × 100) × 0.154314 + LN (TSP fossil energy intensity) × 1.02946 + LN (TSP output value) × 1.12309 + LN (Proportion of TSP output value) × (−0.279472) + LN (Urbanization rate × 100) × (−0.212805) − 0.633044)
Urban carbon emissions	EXP (LN (Urban consumption expenditure) × 0.844486 + LN (Urbanization rate × 100) × 0.470785 + LN (Urban population) × 0.22218 + LN (Urban fossil energy intensity) × 1.01239 − 3.05799)
WRTHR carbon emissions	EXP (LN (Coal consumption ratio × 100) × (0.497733) + LN (WRTHR fossil energy intensity) × 1.06022 + LN (WRTHR output value) × 0.410425 + LN (WRTHR output value × WRTHR output value) × 0.205212 + LN (Proportion of WRTHR output value) × 0.553747 + (Urbanization rate × 100) × 0.0119902 + 0.690354)
Rural consumption expenditure	EXP (−0.692313 × LN (Urbanization rate) + LN (Total regional GDP) × 0.543471 + 2.24237)
Urban consumption expenditure	EXP (−0.29377 × LN (Urbanization rate) + 0.372021 × LN (Total regional GDP) + 5.43239)
Production of cement	1.024/(1 + 19.531 × EXP (−9.754 × Construction output value/866.365)) × 10820.9
Proportion of non-fossil energy	IF THEN ELSE (Time < 2021, -8×10^{-6} × (Time-2004) × (Time-2004) × (Time-2004) × (Time-2004) + 0.0003 × (Time-2004) × (Time-2004) × (Time-2004) − 0.0035 × (Time-2004) × (Time-2004) + 0.0187 × (Time-2004) + 0.0861, 0.0002 × (Time-2004) × (Time-2004) + 0.0007 × (Time-2004) + 0.1125)

2.4.3. Validity Test of SD Model

Historical data testing and sensitivity analysis are essential evaluation methods to validate the SD model. Generally, a model has good prediction accuracy and can be used for simulation and model tuning when the simulation results compared with the historical data have an error of no more than 15%. The model test years were from 2005 to 2020. Following persistent debugging of the model, the relative discrepancy between the simulation outcomes and historical data did not exceed 9%, so the model passed the historical data test and can accurately simulate the future carbon emission trend of Guizhou province.

Sensitivity analysis examines the impact of important parameters on system output results by varying the magnitude of their values [43]. In general, the greater the system's sensitivity to the parameter, the more likely the parameter is the critical factor influencing the system. The study selected six parameters: urbanization rate, GDP growth rate, energy intensity, non-fossil energy proportion, industrial structure, and coal consumption, to test the sensitivity of the system variables. The absolute values of the sensitivity of the model parameters range from 0% to 13.51%, indicating that the model was robust and sensitive. At the same time, there was no over-sensitivity observed. Therefore, the model passed the sensitivity test.

2.5. Scenario Settings

By setting different scenarios, using a system dynamics model can further test the influence of relevant factors' changes on carbon emission reduction. This paper combined LMDI decomposition results and SD model sensitivity analysis while also considering the structural characteristics of the system dynamics model to select the regulatory variables for scenario analysis. These variables were the average annual GDP growth rate, the proportion of tertiary industry output value, and the cumulative decline in fossil energy intensity.

2.5.1. Variable Assumptions

We assumed two scenarios for three variables. Table 3 displays the assumptions made for specific variables. The first variable was the average annual GDP growth rate. We assumed that under the baseline scenario, the average annual GDP growth rate was set at 7% during 2021–2025 with reference to the Guizhou Province's 14th Five-Year Plan. Then the average annual growth rate gradually slowed down over time, which was consistent with the actual economic development of Guizhou Province, and the variable was at a high level of development in this scenario. Under the scenario of emission reduction measures, we assumed that Guizhou Province focused on high-quality economic development and avoided the crude economic development model, and the economic growth rate slowed down. From 2021 to 2025, the average annual GDP growth rate maintained the growth rate of the last five years, which was set at 5.69%, then decreased year by year, and the variable under this scenario was at a medium level of development.

The second variable was the proportion of tertiary industry output value. In the baseline scenario, we assumed that Guizhou Province gradually increased the importance of the tertiary sector's development in accelerating economic restructuring. Still, this transformation took time, so the proportion of tertiary industry output value was expected to increase slowly. At that time, according to the curve-fitting prediction, it was set that the proportion of tertiary industry output value reached 48.37%, 48.84%, 49.23%, and 49.56% in 2025, 2030, 2035, and 2040, respectively, and the variable was at a medium level of development under this scenario. Under the scenario of emission reduction measures, Guizhou Province accelerated the development of the tertiary sector, especially the development of the service industry and the enhancement of technological innovation capacity, to promote the optimization, transformation, and upgrading of the economic structure. It was promoting emission reduction measures provided good opportunities for the development of the tertiary sector, which was expected to increase its share of output value gradually. The proportion of tertiary industry output value in 2025, 2030, 2035, and 2040 was set to be 49.02%, 51.35%, 54.51%, and 57.86%, respectively, and the variable was at a high development level in this scenario.

The third variable was the cumulative decline in fossil energy intensity. Under the baseline scenario, with reference to the Guizhou "14th Five-Year Plan" Comprehensive

Work Program for Energy Conservation and Emission Reduction, the cumulative decline in fossil energy intensity was set to be 13% from 2021 to 2025, and considering the gradual improvement of environmental protection requirements, it increased year by year. The variable under this scenario was at a medium development level. Under the emission reduction measures scenario, to cope with climate change and environmental pollution, we assumed that Guizhou Province would accelerate the reduction of fossil energy intensity and improve the efficiency of energy utilization, and the reduction rate of fossil energy intensity was expected to increase year by year under the impetus of emission reduction measures. The cumulative decline in fossil energy intensity from 2020 to 2040 was set to be 15%, 17%, 19%, and 21%, respectively, and the variables in this scenario were at a high level of development.

Table 3. Variables assumptions.

Regulatory Variables	Years	Average Annual GDP Growth Rate	The Proportion of Tertiary Industry Output Value	Cumulative Decline in Fossil Energy Intensity
High	2021–2025	7.00%	49.02%	15%
	2026–2030	6.00%	51.35%	17%
	2031–2035	5.00%	54.51%	19%
	2036–2040	4.50%	57.86%	21%
Medium	2021–2025	5.69%	48.37%	13%
	2026–2030	5.29%	48.84%	15%
	2031–2035	4.49%	49.23%	17%
	2036–2040	3.99%	49.56%	19%

2.5.2. Scenario Assumptions

Three variables, with six states, were set to the following eight scenarios. Table 4 displays the simulation assumptions.

1. Baseline Scenario (Scenario I) assumed a continuation of the current development trend without implementing additional mitigation measures. The average annual GDP growth rate was set at a high level of development. The proportion of tertiary industry output value and the cumulative decline in fossil energy intensity was set at a medium level of development, which provided a reference for setting the other scenarios.
2. High-quality Economic Development Scenario (Scenario II) assumed that economic development was able to avoid "high consumption" and "high emissions" and focus on high-quality economic development and that the average annual GDP growth rate slowed down to a medium development level. Other indicators were consistent with the baseline scenario.
3. Industrial Structure Optimization Scenario (Scenario III) assumed that Guizhou Province would accelerate the transformation from a "2-3-1" industrial structure, where the secondary sector dominated, to a "3-2-1" structure, where the tertiary sector took the lead, aiming to achieve a reduction in carbon emissions by adjusting the ratio of outputs in Guizhou Province. The proportion of tertiary industry output value was set to be at a high level of development, and other indicators were consistent with the baseline scenario.
4. Fossil Energy Intensity Adjustment Scenario (Scenario IV) assumed that Guizhou Province accelerated the formation of an energy-efficient society and consumed the same amount of fossil energy to generate more GDP than in the baseline scenario. This scenario aimed to reduce carbon emissions by adjusting the cumulative decline in fossil energy intensity in eight sectors, setting it at a high level of development while keeping the other indicators consistent with the baseline scenario.
5. Take Any Two Measures Combination Scenario (Scenarios V, VI, and VII) assumed that the average annual GDP growth rate slowed down under Scenario V while

accelerating the increase in the proportion of tertiary industry output value. Under Scenario VI, the average annual GDP growth rate slowed while the cumulative decline in fossil energy intensity increased. Under Scenario VII, the increase in the proportion of tertiary industry output value accelerated while increasing the cumulative decline in fossil energy intensity.

6. The Combined Scenario (Scenario VIII) assumed that the government took three measures simultaneously to achieve carbon emission reductions. Specifically, this included maintaining the current GDP growth rate while emphasizing high-quality economic development, accelerating the development of the tertiary sector, optimizing the industrial structure, reducing fossil energy intensity, and improving energy use efficiency. The average annual GDP growth rate was set at a medium development level, and the proportion of tertiary industry output value and cumulative decline in fossil energy intensity was set to be at a high development level.

Table 4. Scenario assumptions.

Scenario	Average Annual GDP Growth Rate	The Proportion of Tertiary Industry Output Value	Cumulative Decline in Fossil Energy Intensity
Scenario I	High	Medium	Medium
Scenario II	Medium	Medium	Medium
Scenario III	High	High	Medium
Scenario IV	High	Medium	High
Scenario V	Medium	High	Medium
Scenario VI	Medium	Medium	High
Scenario VII	High	High	High
Scenario VIII	Medium	High	High

2.6. *Tapio Decoupling Index*

The Tapio decoupling model is an extension of the model proposed by the Organization for Economic Cooperation and Development (OECD). The decoupling index constructed by Tapio can describe the status of the association between economic growth and carbon emissions. The calculation method is shown in Equation (10).

$$\varepsilon = \frac{(\Delta C/C_0)}{(\Delta G/G_0)} \tag{10}$$

where $\Delta C/C_0$ was the carbon emission growth rate, $\Delta G/G_0$ was the GDP growth rate. The ε was the decoupling index and the decoupling index status classification criterion is shown in Figure 2 [44].

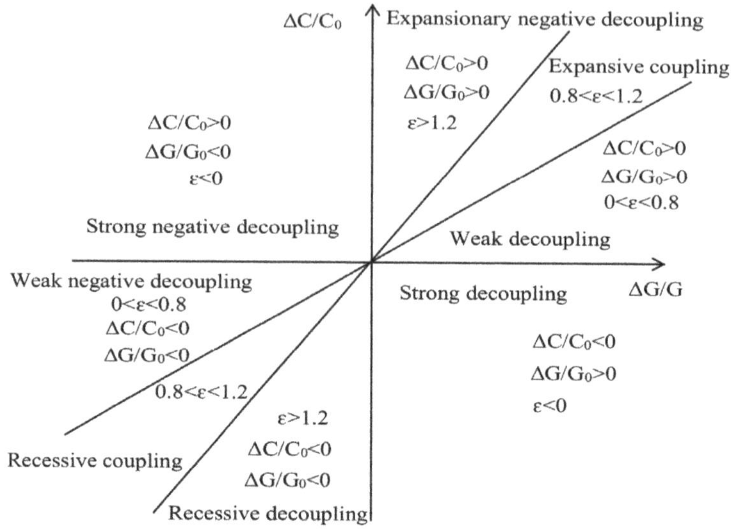

Figure 2. Decoupling index status classification and evaluation criterion.

2.7. Data Source

Considering the data's reliability and accessibility, the study selected 2005–2020 as the study interval and 2021–2040 as the forecast interval. To conduct price deflation on the price-related data using constant prices based on the year 2004 to eliminate the impact of inflation. The energy consumption and standard coal coefficients for the industrial and residential sectors were determined by consulting the China Energy Statistical Yearbook. Cement production, regional GDP, per capita consumption level, consumer price index, regional GDP index, and population numbers were obtained from the annual data of Guizhou Province from China's National Bureau of Statistics.

3. Results

3.1. Analysis of Carbon Emission Trend and Historical Decoupling Status

3.1.1. Carbon Emissions Trend Analysis

The carbon emission status of Guizhou Province from 2005 to 2020 is shown in Figure 3. Generally, carbon emissions in Guizhou province have been rising since 2005, with a faster growth rate from 2005 to 2012 and a slower growth rate after that, but still no downward trend. At that time, carbon dioxide was mainly emitted through coal consumption, but the proportion of carbon emissions from petroleum and natural gas gradually increased. This indicated that the energy consumption structure of Guizhou Province had been gradually improving.

In terms of carbon emission sectors, the industrial structure of Guizhou province was dominated by the secondary sector from 2005 to 2020. The industry has consistently ranked first in carbon emissions since 2005, accounting for 78.68% of 2007 and 62.71% in 2020. Therefore, this industry is crucial for Guizhou Province to achieve low carbon transformation. Over the study period, it became clear that CO_2 emissions from various tertiary industry sectors, including transportation, storage and post, wholesale, retail trade, hotel, restaurants, and other service industries, rapidly increased due to the rising share of tertiary industry output value. In addition, despite the continuous improvement in urbanization and living standards in Guizhou province, the carbon emission proportion of urban residents' lifestyles remained low and unchanged due to the ongoing reduction in energy intensity, resulting in limited room for carbon reduction. Furthermore, the rural sector's carbon emissions have also been on a downward trend due to the decreasing population, improving energy structure, and decreasing energy intensity.

Even though the fossil energy intensity in Guizhou Province was declining, most sectors still heavily relied on coal, resulting in a continuous increase in carbon emissions. Economic growth accompanied by massive CO_2 emissions makes it challenging to achieve decarbonization, and the current carbon emission situation is far from optimistic. Therefore, it is imperative to promptly explore effective carbon emission reduction strategies to advance the national "dual carbon" goal and enable Guizhou Province to achieve a carbon emission peak as soon as possible.

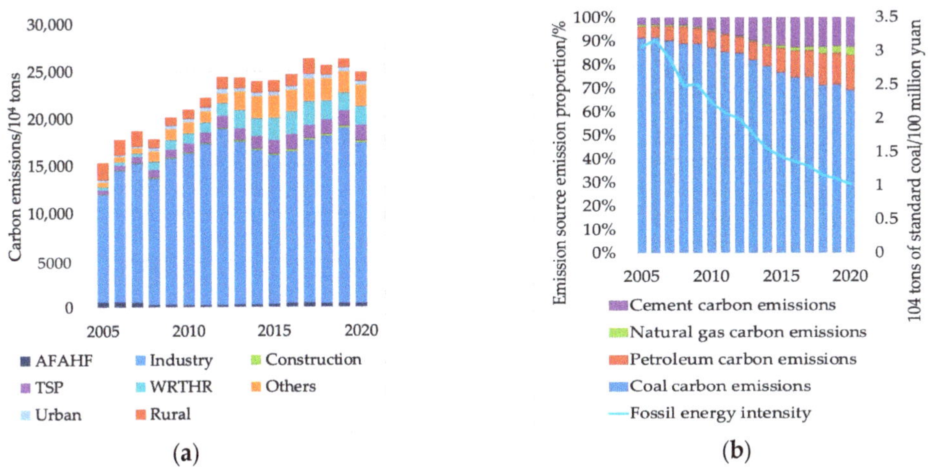

Figure 3. Carbon emission status of Guizhou province from 2005 to 2020: (**a**) Carbon emissions in various sectors; (**b**) Carbon emission proportion and energy intensity, illustrating the relative contributions of carbon emissions from coal, petroleum, natural gas, and cement production.

3.1.2. Historical Decoupling Status Analysis

The decoupling status of various sectors in Guizhou Province from 2005 to 2020 is shown in Table 5. Overall, the decoupling index of economic growth and carbon emissions was getting smaller and smaller. This phenomenon indicated that the green development model is gradually gaining importance as society develops. However, a strong decoupling of these two variables was not achieved during the study period. This illustrated the continued dependence of economic growth on fossil energy consumption. In terms of sectoral classification, the decoupling index gradually decreased in all sectors except agriculture, forestry, animal husbandry, and fishery, as well as industry. Especially in the wholesale, retail trade, hotel, restaurants, and other service industries, they achieved strong decoupling during the study period. Since the industrial sector plays a critical role in reducing carbon emissions, it should strive to achieve decoupling of economic growth and carbon emissions at the earliest.

Table 5. Decoupling of historical relationships.

Sector	2005–2010		2010–2015		2015–2020	
	ε	Decoupling State	ε	Decoupling State	ε	Decoupling State
Total carbon emissions	0.45	WD	0.19	WD	0.07	WD
AFAHF	−2.14	SD	0.70	WD	0.72	WD
Industry	0.51	WD	−0.02	SD	0.17	WD

Table 5. *Cont.*

Sector	2005–2010		2010–2015		2015–2020	
	ε	Decoupling State	ε	Decoupling State	ε	Decoupling State
Construction	2.09	END	0.37	WD	0.25	WD
TSP	1.18	EC	0.64	WD	0.25	WD
WRTHR	1.90	END	1.61	END	−0.29	SD
OSI	1.07	EC	1.31	END	−0.08	SD

Note: The WD was the weak decoupling, the SD was the strong decoupling, the END was the expansionary negative decoupling, and the EC was the expansive coupling.

3.2. LMDI Factor Decomposition Results Analysis

The LMDI decomposition findings are illustrated in Figure 4. The research indicated two key factors contribute to the growth of carbon emissions in Guizhou Province: economic scale and fossil energy intensity in industrial sectors. The dominant factor that curbed carbon emissions growth was the intensity of fossil energy usage, while the chief driver behind their growth was the economic scale. They were also pivotal in designing carbon emission reduction paths. Therefore, in formulating policies to reduce carbon emissions, it is crucial to focus on promoting clean energy, optimizing industrial production processes, and improving energy efficiency. These measures aim to achieve an effective reduction in carbon emissions. In addition, the per capita output value of the industrial sector promoted the growth of carbon emissions in Guizhou Province. In contrast, cement production intensity, fossil energy structure, energy carbon emission coefficients, and industrial structure inhibited carbon emissions.

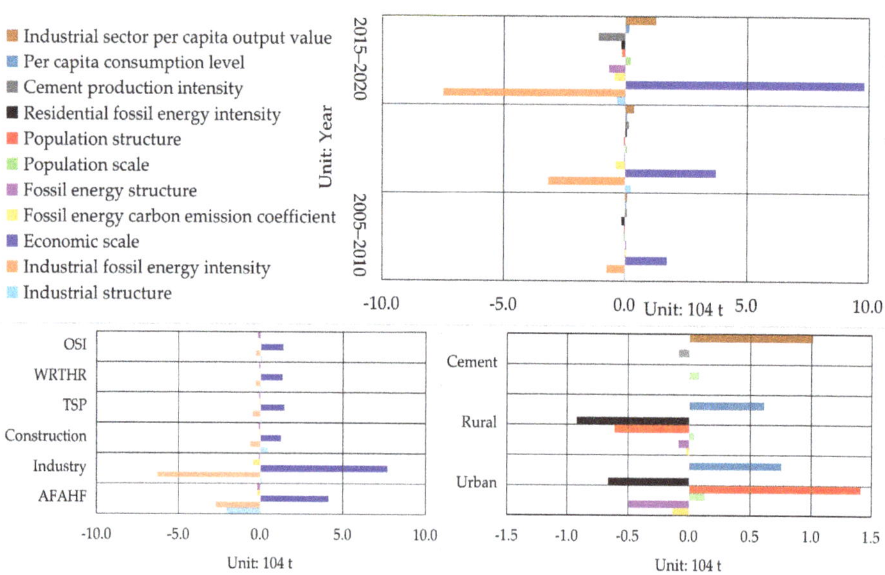

Figure 4. The contribution of influencing factors.

Looking at it from an industrial sector perspective, the economic scale and the fossil energy intensity of the industrial sector were the most significant influencing factors. Different industrial sectors were affected differently by the industrial structure. It promoted carbon emissions in other sectors except for the primary industrial output value. This is because the share of the output value of the primary industry in Guizhou province continues to decline while secondary and tertiary industries flourish. In terms of residential sectors, the three main factors influencing carbon emissions were per capita consumption level, the fossil energy intensity of residential sectors, and population structure. Per capita consumption level led to increased carbon emissions, whereas residential sectors' energy intensity helped reduce them. In addition, the population structure was also a critical contributing factor, which promoted carbon emissions from urban residents and suppressed carbon emissions from rural residents, a phenomenon consistent with the characteristics of urban-rural population mobility in Guizhou Province.

3.3. Simulation Analysis

Eight scenarios were simulated separately using the debugged model. The simulation results are shown in Figure 5.

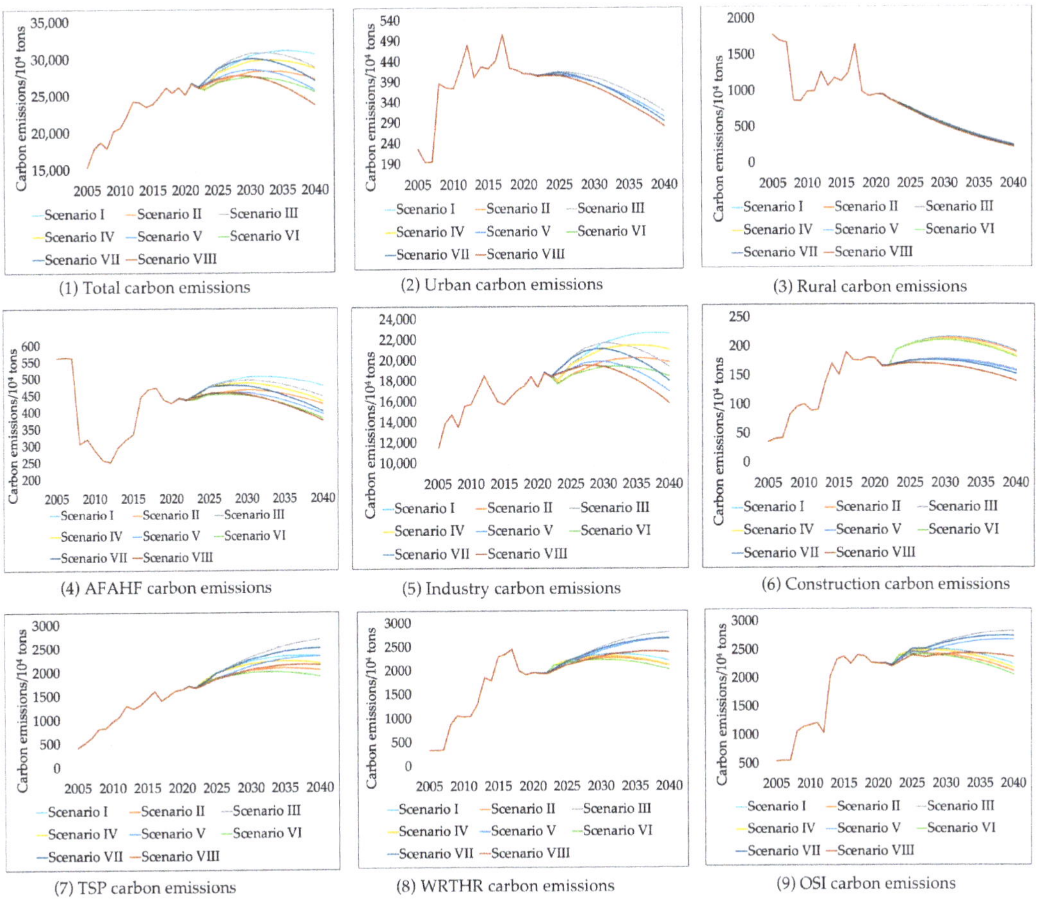

Figure 5. Trend prediction of carbon emission scenarios under eight scenarios.

3.3.1. Baseline Scenario

In this scenario, it was predicted that Guizhou Province would attain its highest level of carbon emissions among all eight scenarios, reaching a peak of 311.48 million tons in 2035. From the perspective of emission sectors, the industry remained the main contributor to carbon emissions. Industry, transport, storage, and post peaked in 2038, with 226.89 million tons and 23.66 million tons, respectively. Construction reached its peak of 2.16 million tons in 2031. Projections indicated that most sectors would not reach their peak emissions before 2030. Overall, under the scenario, this was disadvantageous to attaining Guizhou Province's carbon emission reduction goals, and carbon emission pressure was expected to continue increasing.

3.3.2. High-Quality Economic Development Scenario

The scenario reached a peak of 283.94 million tons in 2033. Compared with the baseline scenario, the peak was achieved two years in advance and reduced by 8.84 %. In terms of emission sectors, apart from the construction sector, the rest of the sectors had a significant influence in reducing carbon emissions, particularly the transport, storage, and post and industry sectors, which saw reductions of 8.25% and 7.74%, respectively, compared to the base scenario, and both reached their peaks in 2035. The peak for other service industries occurred in 2027, reaching 24.30 million tons, two years earlier than the baseline scenario. This indicated that carbon emissions in most sectors were more sensitive to adjusting the economic growth rate. Therefore, increased economic growth would result in higher carbon emissions, aligning with the LMDI model's decomposition analysis findings. Therefore, promoting high-quality economic development without sacrificing the environment was conducive to reducing carbon emissions.

3.3.3. Industrial Structure Optimization Scenario

Under this scenario, it was predicted that the peak carbon emissions of 30.88 million tons would be reached in 2032, years ahead of the baseline scenario, but with just a 0.85% reduction from the peak value. After reaching its peak, the rapid decrease in carbon emissions in Guizhou Province suggests that efforts to optimize the industrial structure could have helped the province achieve its carbon emission peak earlier. However, it is important to note that the initial decrease in emissions was relatively small. In terms of emission sectors, the industry was expected to peak in 2030, eight years ahead of the baseline scenario. In contrast, this measure led to an increase in carbon emissions from the tertiary sector, making it challenging for the industry to reach its peak by 2040. However, due to the relatively small contribution of the tertiary sector to carbon emissions, the general reduction in carbon emissions during the later period was more significant than that of the baseline scenario. Therefore, on the whole, industrial structure optimization helped achieve long-term carbon emission reduction goals.

3.3.4. Fossil Energy Intensity Adjustment Scenario

Under this scenario, it was predicted that carbon emissions would peak at 298.93 million tons in 2033, two years earlier than the baseline scenario, resulting in a reduction of 4.03% compared to the peak value. When examining the emissions sector, the decline in fossil energy intensity had an inhibitory influence on carbon emissions across all sectors, consistent with the conclusions drawn from the decomposition analysis of the LMDI model. Among them, the reductions in agriculture, forestry, animal husbandry, and fishery and urban residents showed more significant decreases, with an average decline of 4.28% and 3.07% during the forecast period compared to the baseline scenario. The industry and the transport, storage, and post peaked in 2035 at 215.05 million tons and 22.47 million tons, respectively. On the other hand, the wholesale, retail trade, hotel, and restaurants peaked in 2017 at 24.60 million tons, and other service industries peaked at 24.89 million tons in 2028.

3.3.5. Take Any Two Measures Scenario

Under Scenario V, Scenario VI, and Scenario VII, carbon emissions were projected to peak in 2030 at 286.23 million tons, 276.10 million tons, and 301.07 million tons, respectively. The respective reductions compared to the baseline scenario were 8.11%, 11.36%, and 3.34%, which indicated that the combined effect of two measures on emission reduction was more significant than that of a single measure. Among them, adjustments to the

GDP growth rate and the intensity of fossil energy had demonstrated a more significant influence on emission reduction. In terms of emission sectors, except for the tertiary sector, the impact of carbon emission reduction in other sectors was more significant than in single measures. In the combined scenario of industrial structure optimization and any other single measure, carbon emissions in the tertiary sector were projected to increase faster than in the baseline scenario but slower than in the industrial structure optimization scenario. This indicated that any other measure could offset the negative influence of industrial structure optimization on tertiary sector carbon emissions. Furthermore, it emphasizes the significance of implementing overlapping measures to achieve carbon emission reduction goals.

3.3.6. The Combined Scenarios

Under this scenario, carbon emissions were predicted to peak at 277.71 million tons in 2029, six years ahead of the baseline scenario, resulting in a 10.84% reduction in peak emissions. From the carbon emission sector perspective, all other sectors exhibited the lowest emission levels except for the tertiary sector. Furthermore, except for transport, storage, and post and other service industries, all other sectors were projected to reach their carbon emission peaks before 2030. Among them, the industry sector with the highest carbon emission contribution peaked in 2027, 11 years earlier than the baseline scenario, resulting in 195.45 million tons of carbon emissions and a reduction of 13.86%. In the combined scenario, although the carbon emissions of the tertiary sector were not at the lowest level, the overall decline rate of carbon emissions was still faster than in other scenarios due to the significant proportion of industrial carbon emissions, surpassing the tertiary sector. Therefore, overall, the combined scenario had the most notable impact on reducing carbon emissions, facilitating Guizhou Province to reach its peak emissions as soon as possible.

3.4. *Analysis of Decoupling States under Eight Scenarios*

The primary driver of carbon emissions in Guizhou Province is economic growth. A strong decoupling relationship between economic growth and carbon emissions has yet to be established. Therefore, early conversion to achieving strong decoupling becomes pivotal in attaining coordinated development between the economy and low-carbon. Consequently, conducting a comprehensive examination of the decoupling status between economic growth and carbon emissions under different scenarios can test measures to achieve strong decoupling as soon as possible.

The decoupling index results under eight scenarios are shown in Figure 6. When the GDP change is positive, if $0 < \varepsilon < 0.8$, it shows a weak decoupling state; if it is less than 0, it shows a strong decoupling state [44]. In the baseline scenario, a shift from weak to strong decoupling between economic growth and carbon emissions was projected to occur during 2036–2038, and the decoupling index would slowly decrease. In taking a single measure, the strong decoupling between economic growth and carbon emissions would be achieved between 2032 and 2034, indicating that any single measure could promote the early achievement of strong decoupling. Notably, industrial structure optimization exerted the most significant influence. In scenarios combining any two measures, three scenarios would achieve strong decoupling between 2030 and 2032, which was more evident than any single measure. Furthermore, implementing three emissions reduction measures led to a shift from weak to strong decoupling between 2028 and 2030, demonstrating the earliest achievement of strong decoupling and the fastest decline in the decoupling index. This indicated that the time required for strong decoupling between the variables progressively reduced with each additional emission reduction measure introduced.

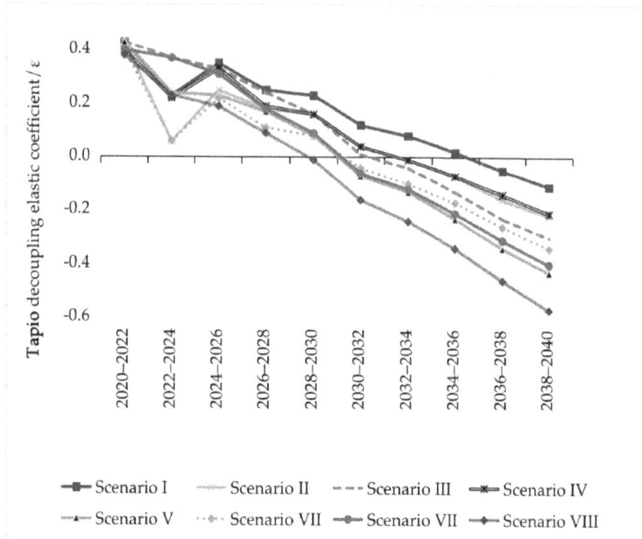

Figure 6. Prediction decoupling between economic growth and carbon emission.

When the above evidence is combined, it is discovered that each emission reduction measure can effectively reduce carbon emissions in Guizhou Province compared to the baseline scenario. Moreover, combining any two measures was more effective than a single measure. Nevertheless, it was difficult to achieve the peak before 2030, and the strong decoupling could only be achieved after 2030, which was not conducive to China's "3060" dual carbon goal. However, only in the combined scenario, where all three emission reduction measures were taken simultaneously, could Guizhou Province achieve the emission peak by 2030. A strong decoupling between economic growth and carbon emissions would be realized during the period spanning from 2028 to 2030. This scenario's carbon emission intensity and fossil energy consumption were all at low levels. In conclusion, the combination scenario with multiple emission reduction measures emerged as the most influential approach for curbing carbon emissions in Guizhou Province, providing an effective path towards achieving significant reduction.

4. Discussion and Policy Recommendations

4.1. Policy Recommendations

Based on the research, Guizhou Province can effectively achieve its carbon reduction goals by simultaneously implementing measures of high-quality economic development, industrial structure optimization, and energy intensity reduction. First, under the high-quality economic development scenario, carbon emissions peaked two years earlier than the baseline scenario, with a peak decline of 8.84%. Therefore, abandoning the unsustainable traditional growth model, transforming the economic growth model, and promoting high-quality economic development are crucial measures to curb carbon emissions in Guizhou Province. Second, under the industrial structure optimization scenario, carbon emissions peaked three years earlier than the baseline scenario. This indicates that substantial reductions in carbon emissions can be achieved by promoting industrial upgrading. Finally, accelerating the decline of fossil energy intensity had an emission reduction influence on all sectors. Consequently, enhancing the efficiency of fossil energy consumption, reducing the intensity of fossil energy, fostering the development of high-tech industries, and accelerating efforts to build an energy-saving society will aid in the earlier attainment of the carbon emission reduction goal.

4.2. Limitations

The following three aspects can be used to widen the scope of the study: First, the study only considers two emission sources. However, more emission sources, such as agriculture and waste disposal, can be considered when data is available. At the same time, carbon emission absorption can also be considered. Second, the model construction is constructed primarily on assumptions. Because of the different boundary settings, the prediction results will vary. As a result, it is difficult to compute an accurate value; more is to test the influence of carbon emission reduction measures. In the future, with more time, we can increase the model variables and expand the model boundaries to simulate reality more accurately and assist policymakers in making more reliable policy recommendations. Third, this research only constructs the model using provincial regions as the research object. In the future, the model has the potential to be applied to other areas of carbon emission reduction research.

5. Conclusions

This study takes Guizhou province, a less developed region in southwest China, as a representative case study to conduct a more comprehensive analysis of the carbon emission reduction path at the provincial level. The goal is to tackle the provincial carbon emission reduction challenge by merging three models of the LMDI, SD, and Tapio decoupling model. The LMDI model assists policymakers in understanding the key factors influencing carbon emissions. Based on this, a system dynamics model is constructed, and different scenarios are set to assess the future carbon emission trends under various emission reduction measures. Additionally, the Tapio decoupling index tests which scenario can achieve a strong decoupling relationship between economic growth and carbon emissions as early as possible to support policymakers in making decisions. The findings demonstrate that, firstly, carbon emissions in Guizhou province increased from 2005 to 2020, with the industrial sector contributing the most carbon emissions. Although the decoupling index is declining, it is still in a weak decoupling state. Secondly, economic scale and industrial fossil energy intensity are the primary driving and hindering factors in Guizhou Province's carbon emissions, respectively. Finally, only a combination of three measures simultaneously can Guizhou Province reach the peak carbon emission of 277.71 million tons in 2029, and a strong decoupling of economic growth and carbon emissions can be achieved early. This research framework also applies to studying other regions' carbon reduction issues.

Author Contributions: Conceptualization, H.W.; project administration, H.W.; funding acquisition, H.W.; investigation, W.X.; resources, W.X.; methodology, W.X.; writing—original draft, W.X.; validation, W.X.; writing—review & editing, W.X.; data curation, W.X.; visualization W.X.; Supervision, Y.Z. All authors have read and agreed to the published version of the manuscript.

Funding: This research was funded by the Humanities and Social Sciences of Ministry of Education Planning Fund of China, grant number 22YJA630096.

Institutional Review Board Statement: Not applicable.

Informed Consent Statement: Not applicable.

Data Availability Statement: Publicly available datasets were analyzed in this study. These data can be found here: http://stjj.guizhou.gov.cn/, https://data.stats.gov.cn/, https://data.cnki.net/yearBook (accessed on 1 September 2022).

Acknowledgments: Thanks to the anonymous reviewers and all the editors in the process of revision.

Conflicts of Interest: The authors declare no conflict of interest.

References

1. Jiang, J.; Ye, B.; Xie, D.; Tang, J. Provincial-level carbon emission drivers and emission reduction strategies in China: Combining multi-layer LMDI decomposition with hierarchical clustering. *J. Clean. Prod.* **2017**, *169*, 178–190. [CrossRef]
2. Wang, S.; Fang, C.; Wang, Y. Spatiotemporal variations of energy-related CO_2 emissions in China and its influencing factors: An empirical analysis based on provincial panel data. *Renew. Sustain. Energy Rev.* **2016**, *55*, 505–515. [CrossRef]
3. Li, G.; Chen, X.; You, X.-Y. System dynamics prediction and development path optimization of regional carbon emissions: A case study of Tianjin. *Renew. Sustain. Energy Rev.* **2023**, *184*, 113579. [CrossRef]
4. Huang, R.; Zhang, S.F.; Wang, P. Key areas and pathways for carbon emissions reduction in Beijing for the "Dual Carbon" targets. *Energy Policy* **2022**, *164*, 19. [CrossRef]
5. Huo, T.F.; Ma, Y.L.; Xu, L.B.; Feng, W.; Cai, W.G. Carbon emissions in China's urban residential building sector through 2060: A dynamic scenario simulation. *Energy* **2022**, *254*, 13. [CrossRef]
6. Li, B.; Han, S.W.; Wang, Y.F.; Li, J.Y.; Wang, Y. Feasibility assessment of the carbon emissions peak in China's construction industry: Factor decomposition and peak forecast. *Sci. Total Environ.* **2020**, *706*, 13. [CrossRef]
7. Chen, X.; Shuai, C.Y.; Wu, Y.; Zhang, Y. Analysis on the carbon emission peaks of China's industrial, building, transport, and agricultural sectors. *Sci. Total Environ.* **2020**, *709*, 9. [CrossRef]
8. Wang, F.; Ge, X. Inter-provincial responsibility allocation of carbon emission in China to coordinate regional development. *Environ. Sci. Pollut. Res.* **2022**, *29*, 7025–7041. [CrossRef]
9. Zhang, M.W.; Gao, F.F.; Huang, B.; Yin, B. Provincial Carbon Emission Allocation and Efficiency in China Based on Carbon Peak Targets. *Energies* **2022**, *15*, 9181. [CrossRef]
10. Cheng, X.J.; Ouyang, S.Q.; Quan, C.G.; Zhu, G.J. Regional allocation of carbon emission quotas in China under the total control target. *Environ. Sci. Pollut. Res.* **2023**, *13*, 66683–66695. [CrossRef]
11. Yang, F.; Lee, H.Y.S. An innovative provincial CO_2 emission quota allocation scheme for Chinese low-carbon transition. *Technol. Forecast. Soc. Chang.* **2022**, *182*, 10. [CrossRef]
12. Guo, F.; Zhang, L.; Wang, Z.; Ji, S. Research on Determining the Critical Influencing Factors of Carbon Emission Integrating GRA with an Improved STIRPAT Model: Taking the Yangtze River Delta as an Example. *Int. J. Environ. Res. Public Health* **2022**, *19*, 8791. [CrossRef] [PubMed]
13. Huang, Y.; Liu, J.; Shi, M. Analysis of influencing factors and prediction of carbon emissions of typical urban agglomerations in China: A case study of Beijing-Tianjin-Hebei region. *Environ. Sci. Pollut. Res.* **2023**, *30*, 52658–52678. [CrossRef] [PubMed]
14. Yue, H.; Bu, L. Prediction of CO_2 emissions in China by generalized regression neural network optimized with fruit fly optimization algorithm. *Environ. Sci. Pollut. Res.* **2023**, *30*, 80676–80692. [CrossRef] [PubMed]
15. Kim, Y.G.; Yoo, J.; Oh, W. Driving forces of rapid CO_2 emissions growth: A case of Korea. *Energy Policy* **2015**, *82*, 144–155. [CrossRef]
16. Liu, M.Z.; Zhang, X.X.; Zhang, M.Y.; Feng, Y.Q.; Liu, Y.J.; Wen, J.X.; Liu, L.Y. Influencing factors of carbon emissions in transportation industry based on C-D function and LMDI decomposition model: China as an example. *Environ. Impact Assess. Rev.* **2021**, *90*, 106623. [CrossRef]
17. Wang, Z.H.; Yang, Y.T. Features and influencing factors of carbon emissions indicators in the perspective of residential consumption: Evidence from Beijing, China. *Ecol. Indic.* **2016**, *61*, 634–645. [CrossRef]
18. Huang, Y.S.; Shen, L.; Liu, H. Grey relational analysis, principal component analysis and forecasting of carbon emissions based on long short-term memory in China. *J. Clean. Prod.* **2019**, *209*, 415–423. [CrossRef]
19. Wang, C.J.; Wang, F.; Zhang, X.L.; Yang, Y.; Su, Y.X.; Ye, Y.Y.; Zhang, H.G. Examining the driving factors of energy related carbon emissions using the extended STIRPAT model based on IPAT identity in Xinjiang. *Renew. Sustain. Energy Rev.* **2017**, *67*, 51–61. [CrossRef]
20. Yu, S.W.; Zhang, Q.; Li Hao, J.; Ma, W.T.; Sun, Y.; Wang, X.C.; Song, Y. Development of an extended STIRPAT model to assess the driving factors of household carbon dioxide emissions in China. *J. Environ. Manag.* **2023**, *325*, 11. [CrossRef]
21. Sun, L.; Yu, H.; Liu, Q.; Li, Y.; Li, L.; Dong, H.; Adenutsi, C.D. Identifying the Key Driving Factors of Carbon Emissions in 'Belt and Road Initiative' Countries. *Sustainability* **2022**, *14*, 9104. [CrossRef]
22. Ang, B.W. Decomposition analysis for policymaking in energy: Which is the preferred method? *Energy Policy* **2004**, *32*, 1131–1139. [CrossRef]
23. Xu, S.C.; He, Z.X.; Long, R.Y. Factors that influence carbon emissions due to energy consumption in China: Decomposition analysis using LMDI. *Appl. Energy* **2014**, *127*, 182–193. [CrossRef]
24. Yang, J.; Cai, W.; Ma, M.D.; Li, L.; Liu, C.H.; Ma, X.; Li, L.L.; Chen, X.Z. Driving forces of China's CO_2 emissions from energy consumption based on Kaya-LMDI methods. *Sci. Total Environ.* **2020**, *711*, 15. [CrossRef] [PubMed]
25. Cai, L.Y.; Luo, J.; Wang, M.H.; Guo, J.F.; Duan, J.L.; Li, J.T.; Li, S.; Liu, L.T.; Ren, D.P. Pathways for municipalities to achieve carbon emission peak and carbon neutrality: A study based on the LEAP model. *Energy* **2023**, *262*, 16. [CrossRef]
26. Xu, G.Y.; Schwarz, P.; Yang, H.L. Determining China's CO_2 emissions peak with a dynamic nonlinear artificial neural network approach and scenario analysis. *Energy Policy* **2019**, *128*, 752–762. [CrossRef]
27. Fang, K.; Tang, Y.Q.; Zhang, Q.F.; Song, J.N.; Wen, Q.; Sun, H.P.; Ji, C.Y.; Xu, A.Q. Will China peak its energy-related carbon emissions by 2030? Lessons from 30 Chinese provinces. *Appl. Energy* **2019**, *255*, 12. [CrossRef]

28. Sun, L.L.; Cui, H.J.; Ge, Q.S. Will China achieve its 2060 carbon neutral commitment from the provincial perspective? *Adv. Clim. Chang. Res.* **2022**, *13*, 169–178. [CrossRef]
29. Zhang, L.; Jiang, Z.; Liu, R.; Tang, M.; Wu, F. Can China Achieve its CO_2 Emission Mitigation Target in 2030: A System Dynamics Perspective. *Pol. J. Environ. Stud.* **2018**, *27*, 2861–2871. [CrossRef]
30. Gao, J.W.; Pan, L.Y. A System Dynamic Analysis of Urban Development Paths under Carbon Peaking and Carbon Neutrality Targets: A Case Study of Shanghai. *Sustainability* **2022**, *14*, 15045. [CrossRef]
31. Tan, X.C.; Lai, H.P.; Gu, B.H.; Zeng, Y.; Li, H. Carbon emission and abatement potential outlook in China's building sector through 2050. *Energy Policy* **2018**, *118*, 429–439. [CrossRef]
32. Zhao, L.T.; Zhao, T.; Yuan, R. Scenario simulations for the peak of provincial household CO_2 emissions in China based on the STIRPAT model. *Sci. Total Environ.* **2022**, *809*, 10. [CrossRef] [PubMed]
33. Yang, H.H.; Li, X.; Ma, L.W.; Li, Z. Using system dynamics to analyse key factors influencing China's energy-related CO_2 emissions and emission reduction scenarios. *J. Clean. Prod.* **2021**, *320*, 16. [CrossRef]
34. Du, L.L.; Li, X.Z.; Zhao, H.J.; Ma, W.C.; Jiang, P. System dynamic modeling of urban carbon emissions based on the regional National Economy and Social Development Plan: A case study of Shanghai city. *J. Clean. Prod.* **2018**, *172*, 1501–1513. [CrossRef]
35. Gu, S.; Fu, B.T.; Thriveni, T.; Fujita, T.; Ahn, J.W. Coupled LMDI and system dynamics model for estimating urban CO_2 emission mitigation potential in Shanghai, China. *J. Clean. Prod.* **2019**, *240*, 14. [CrossRef]
36. Kong, H.; Shi, L.; Da, D.; Li, Z.; Tang, D.; Xing, W. Simulation of China's Carbon Emission based on Influencing Factors. *Energies* **2022**, *15*, 3272. [CrossRef]
37. Fennell, P.; Driver, J.; Bataille, C.; Davis, S.J. Cement and steel—Nine steps to net zero. *Nature* **2022**, *603*, 574–577. [CrossRef]
38. Eggleston, H.; Buendia, L.; Miwa, K.; Ngara, T.; Tanabe, K. (Eds.) *2006 IPCC Guidelines for National Greenhouse Gas Inventories*; Institute for Global Environmental Strategies (IGES): Hayama, Japan, 2006.
39. Shan, Y.; Guan, D.; Zheng, H.; Ou, J.; Li, Y.; Meng, J.; Mi, Z.; Liu, Z.; Zhang, Q. China CO_2 emission accounts 1997–2015. *Sci. Data* **2018**, *5*, 170201. [CrossRef]
40. Kaya, Y. *Impact of Carbon Dioxide Emission Control on GNP Growth: Interpretation of Proposed Scenarios*; IPCC Energy and Industry Subgroup; Response Strategies Working Group: Paris, France, 1990.
41. Ang, B.W. LMDI decomposition approach: A guide for implementation. *Energy Policy* **2015**, *86*, 233–238. [CrossRef]
42. Ding, S.; Liu, Y.X. Adsorption of CO_2 from flue gas by novel seaweed-based KOH-activated porous biochars. *Fuel* **2020**, *260*, 10. [CrossRef]
43. Gu, C.L.; Guan, W.H.; Liu, H.L. Chinese urbanization 2050: SD modeling and process simulation. *Sci. China-Earth Sci.* **2017**, *60*, 1067–1082. [CrossRef]
44. Song, C.; Zhao, T.; Xiao, Y. Temporal dynamics and spatial differences of household carbon emissions per capita of China's provinces during 2000–2019. *Environ. Sci. Pollut. Res.* **2022**, *29*, 31198–31216. [CrossRef] [PubMed]

sustainability

MDPI

Article

Performance Study of Booster-Driven Hybrid Cooling Units for Free Cooling in Data Centers

Rong Zhuang [1,2], Feng Zhou [3,*], Xuwen Tian [3], Buqing Xu [3], Shaocong Li [3] and Guoyuan Ma [3]

1 State Key Laboratory of Air-Conditioning Equipment and System Energy Conservation, Zhuhai 519070, China
2 Gree Electric Appliances, Inc. of Zhuhai, Zhuhai 519000, China
3 Department of Refrigeration and Cryogenics Engineering, Beijing University of Technology, Beijing 100124, China
* Correspondence: zhoufeng@bjut.edu.cn

Abstract: In the data center, using ambient energy cooling technology can effectively reduce the average power use efficiency, and the heat pipe as an effective use of ambient energy device has attracted much attention. For the dynamic heat pipe, reducing the power consumption of the pump effectively is the key to improving the efficiency. In this paper, the rotary booster is selected as the gas phase booster device of the heat pipe unit, the standard unit of the rotary booster is improved, and three types of boosters are obtained, including two improved boosters and one standard unit. Comparative test studies are conducted on three different types of boosters, and the power of the booster shows a downward trend with the increase in indoor and outdoor temperature differences (outdoor temperature decreases). With the increase in indoor and outdoor temperature differences, the cooling capacity increases first and then decreases. When the indoor and outdoor temperature difference is greater than 20 °C, the suction pressure of the booster is greater than the saturated condensing pressure force under outdoor ambient temperature, and the work of the booster decreases. Among the three types of boosters, the medium pressure ratio booster energy efficiency ratio (EER) is the largest. After throttling the standard unit, results show that its cooling capacity unit increases, but the booster power also increases, and the EER is still smaller than that of the improved unit.

Keywords: data center; booster; free cooling; thermal performance

Citation: Zhuang, R.; Zhou, F.; Tian, X.; Xu, B.; Li, S.; Ma, G. Performance Study of Booster-Driven Hybrid Cooling Units for Free Cooling in Data Centers. *Sustainability* **2023**, *15*, 14558. https://doi.org/10.3390/su151914558

Academic Editors: Fuqiang Wang, Chao Shen, Dong Li and Zhonghao Rao

Received: 16 August 2023
Revised: 25 September 2023
Accepted: 27 September 2023
Published: 7 October 2023

1. Introduction

With the development of big data, data centers are widely used. The energy flow density of data centers is high, and the unit energy consumption increases rapidly, so reducing the energy consumption of the computer room is urgent [1]. Among the computer room equipment, the energy consumption of the air conditioning system accounts for a large proportion of about 30–50% [2], so reducing the air conditioning energy consumption is a technical route to save energy in the computer room. Using natural cooling sources to cool down the data centers is another way to reduce air conditioning energy consumption. The heat pipe as a device that can effectively utilize the natural cold source is widely studied and applied by scholars. Li et al. [3] studied the performance of heat pipe composite machine room air conditioning; when the outdoor ambient temperature is 7 °C, the separated heat pipe cooling capacity reaches a maximum value of 4575 W, and the energy efficiency ratio (EER) can reach 17.99. Shi et al. [4] combined the separated heat pipe unit with the vapor compression refrigeration unit to make a year-round machine room composite air conditioning system. Compared to the conventional base station air conditioning in the same conditions, it can save 30–45% energy in the experiments [5]. However, the separation of heat pipe as an auxiliary cooling equipment for computer room air conditioning occupies a large space in the computer room and has high requirements for unit installation. In the unit of the long pipeline or the machine room under the density of large heat flow, if the kinetic energy of the unit is low, a portion of the condenser may be in the "useless" state, the device increases the temperature difference, and the unit heat dissipation effect deteriorates [6]. Wang et al. [7] proposed a series of heat exchanger vapor compression and heat pipe composite system, heat pipe loop in the plate heat exchanger, and air-cooled heat exchanger. To overcome the flow resistance, the pump drive is added to the heat pipe loop [8]. The results showed that the addition of a pump to the liquid loop could solve the problems of underpowered split heat pipe units and excessive installation requirements.

Wang et al. [9] studied a shielded pump-driven loop heat pipe to cool the data room. The EER was 15.4 when the indoor and outdoor temperature difference was 20 °C, which is 3–5 times the EER than the ordinary vapor compression refrigeration. Zhou et al. [10] designed a shielded pump and magnetic pump-driven loop heat pipe unit based on the loop heat pipe, conducted experimental studies on start-up characteristics, heat transfer performance, and circulation characteristics, and applied the unit to a small data room in Beijing; the results showed that compared with the use of air-conditioning heat dissipation, 20.18% of electrical energy could be saved. The liquid pump vaporizes during operation, and the flow may be broken in some experiments and applications [11]. Therefore, Wei et al. [12] designed a gas-phase-driven heat pipe cycle using a sliding vane compressor and tested its cycle performance under different charging volumes, height differences, and temperature differences. The results showed that the experimental prototype reached an EER of 7.705 when the temperature difference was 30 °C. Wang et al. [13] investigated the gas- and liquid-powered separated heat pipes for the natural cooling needs of data centers and designed a 10 kW three-mode composite air-conditioning prototype. The performance tests were conducted in the standard enthalpy difference method laboratory. The results showed that when the indoor and outdoor temperature differences reached 20 °C, the gas-powered split heat pipe could be used to replace the vapor compression refrigeration system (under the same cooling capacity), which can save about 8–10% of energy. Shi et al. [14] used a DC speed-regulated compressor as a gas booster to drive the loop heat pipe, and the results showed that when the indoor and outdoor temperature differences were more than 20 °C, the EER of heat pipe circulation with gas pressurization separation reached 3.9, and the energy saving rate was about 8%. When the indoor and outdoor temperature differences were 30 °C, the EER reached 7.6, with an energy-saving rate of 70%.

As mentioned above, for the complex system with long pipelines or high heat flux dissipation needs, the conventional heat pipe assisted with gravity cannot work in many fields because of its essential height difference and weak driving force. The liquid pump-driven heat pipe can fix the insufficient driving force problem, but new challenges are emerging, including efficiency improvement under small temperature differences and system stability threats from cavitation. And for the booster-driven heat pipe (cooling unit), the performance deterioration or flow cutoff from cavitation of the liquid-driven heat pipe can be avoided. Moreover, the power consumption of vapor gets small under the same volume flow rate due to the compressibility. Then, the system performance can be enhanced because of the liquid mitigation in the evaporator and condenser.

Moreover, considering the indispensable vapor-compressed refrigeration in summer, the thermodynamic cycle is similar between booster-driven circulation and compressor circulation, which are both used to circulate vapor and present a good matching relationship. However, the key special component, booster, is still rare and is mostly the compressor working with low frequency. The development and improvement of the special booster and matching booster-driven system should be studied further to obtain good performance and high efficiency for good energy savings.

In this paper, to address these problems, the rotary booster is used as the power device of the booster-driven hybrid cooling unit to verify the working range as well as the safety and stability of the booster. The pressure difference between the inlet and outlet of the booster is reduced, the rotary booster is improved, and the power can be reduced while guaranteeing the cooling capacity to provide a reference for the design, research, and development of the booster-driven hybrid cooling unit.

2. Booster-Driven Hybrid Cooling Unit System

2.1. Booster Selection and Improvement

In this paper, the rotary booster is selected as the booster to drive the loop cooling unit. The rotary booster belongs to the volumetric booster, and R22 is selected as the refrigerant of the unit. The frequency of the booster is always 50 Hz, the speed of the booster is 2880 r/min, and the exhaust volume of the booster is 36 cm^3. The structure of the booster is shown in Figure 1.

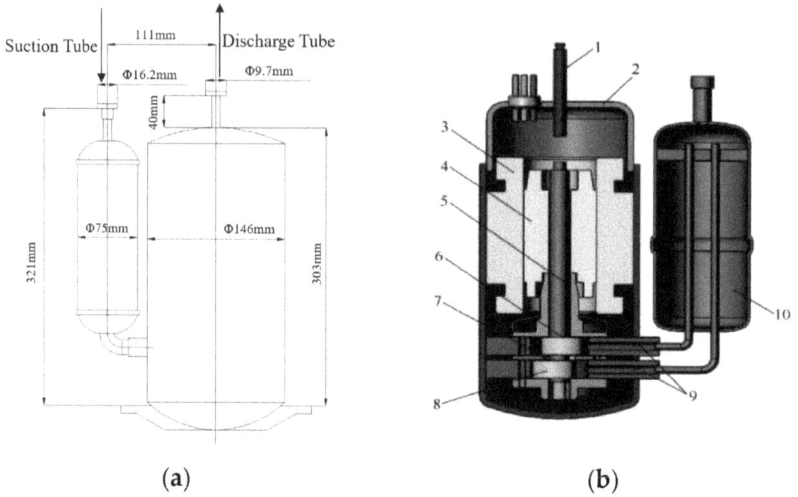

Figure 1. Schematic diagram of rotary booster structure. (**a**) Dimensions; (**b**) inner structure. 1-exhaust tube; 2-shell; 3-stator; 4-rotor; 5-crank axle; 6-cylinder 1; 7-baffle; 8-cylinder 2; 9-suction tube; 10-oil container.

For the booster-driven hybrid cooling units, the role of the rotary booster is to pressurize and supplement power. The booster-driven hybrid cooling unit can increase the temperature of the refrigerant to boost the pressure by using the evaporator heat transfer. So, the booster is not the only method to boost the pressure. In this context, the fixed pressure ratio of the booster is reduced, which can effectively reduce the power of the booster. Thus, the booster is improved to obtain two different improved units: rotary booster number 1 with a large torque and rotary booster number 2 with a small torque. The two kinds of improved units, as well as the unimproved standard unit, are used for comparison experiments in this paper.

2.2. Heat Exchangers and Piping

The heat exchanger of the unit is divided into an evaporative heat exchanger and a condensing heat exchanger. The two evaporative heat exchangers are connected in parallel in the unit. The condensing heat exchanger is also present in two parallel connections. The evaporation heat exchanger and the condensing heat exchanger use a copper tube–aluminum finned heat exchanger. These two heat exchangers have the same structural dimensions and are placed at the same height. The material of the copper tube is red copper, tube cluster arrangement for the positive triangle fork row, fins for the corrugated shape of the whole sheet of aluminum sets, and heat exchanger structural parameters, as shown in Table 1, and its design diagram is shown in Figure 2.

Table 1. Geometric parameters of the heat exchanger.

Parameter	Symbol	Value	Unit
Outer tube diameter	d_o	10	mm
Tube thickness	δ_t	0.5	mm
Inner tube diameter	d_i	9	mm
Tube length	l	500	mm
Number of tube rows on the windward side	n_x	24	/
Number of tube rows in the airflow direction	n_y	5	/
Center distance between tubes on the windward side	s_x	25	mm
Pipe center distance in the airflow direction	s_y	22	mm
Fin thickness	δ_f	0.2	mm
Fin spacing	s_f	2.3	mm

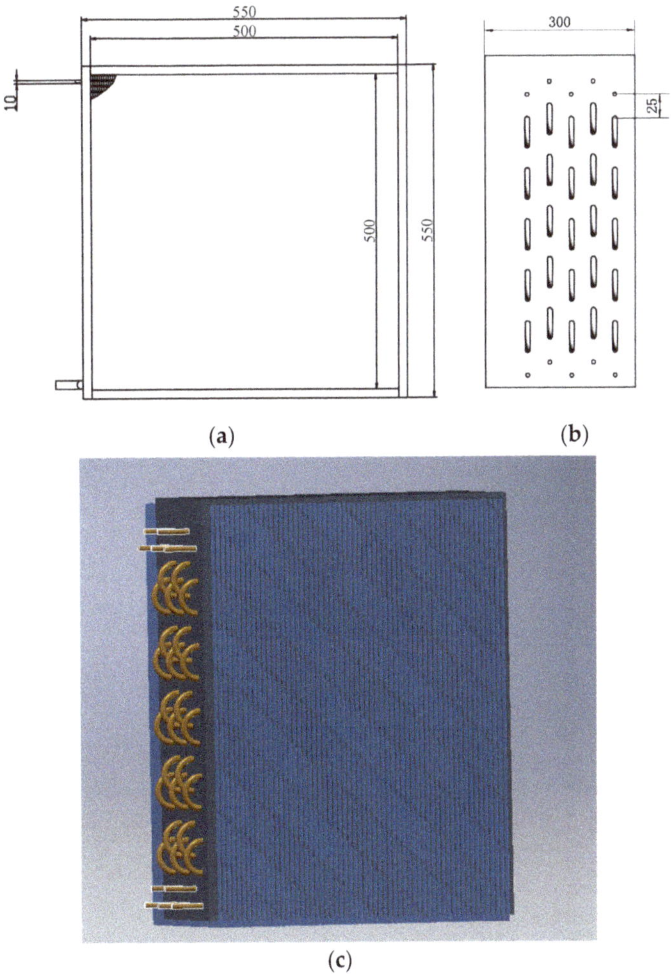

Figure 2. Design of the heat exchanger. (**a**) Main view; (**b**) left view; (**c**) three-dimensional figure of the heat exchanger.

2.3. Experimental Setup

The booster-driven hybrid cooling unit is built, and its experimental arrangement is shown in Figure 3. The experimental test platform and the unit are divided into four parts: control system, indoor unit system, outdoor unit system, and measurement system. The system mainly consists of an evaporator, condenser, rotary booster separator, rotary booster, shut-off valve, three-way valve, and connecting piping. The evaporator and other components are placed in the indoor test room and connected to the static pressure box through the air duct. The condenser, gas–liquid separator, and rotary booster are placed in the outdoor test room.

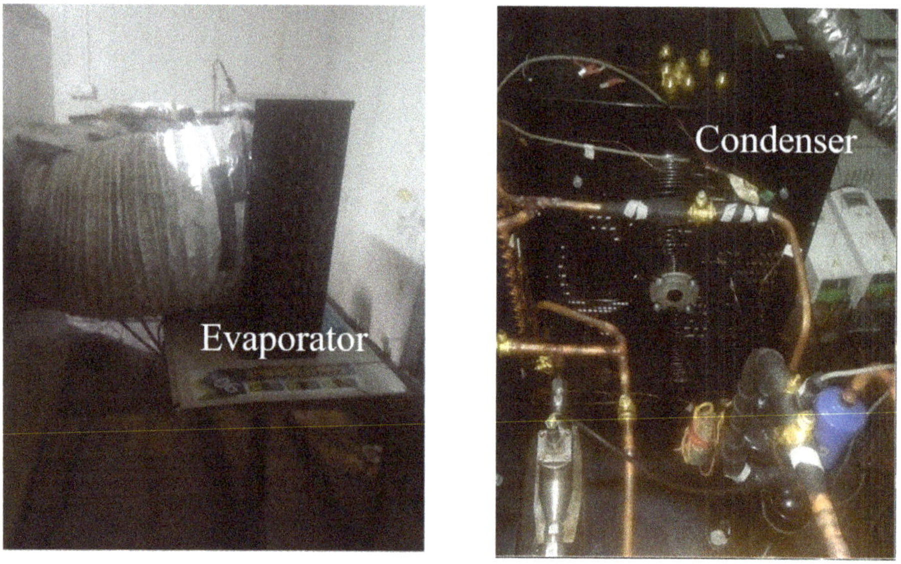

(a) (b)

Figure 3. Physical figure of indoor/outdoor units. (**a**) Indoor unit; (**b**) outdoor unit.

Figure 3 shows that for the heat exchange area as an influencing factor, the unit uses a double heat exchanger program, and each heat exchanger is equipped with a variable-frequency fan. The unit is arranged in a way that the evaporator and the condenser are on the same level. The main components outside the unit include the gas pump and the gas–liquid separator. The liquid side includes a flow meter. The condenser arrangement is placed in parallel, where shut-off valves 1 and 2 control the condensers 1 and 2 of the inlets, respectively, shut-off valves 9 and 12 control the condenser of the discharge, the shut-off valve opening and closing mode is open or closed. The unit indoor measurement evaporator is also placed in parallel. The shut-off valves 3 and 4 control evaporators 3 and 4 of the liquid inlets, respectively; the shut-off valves 10 and 11 control the exhaust of the evaporator. The shut-off valve opening and closing mode is open or closed.

2.4. Test Instrument

The data needed for the unit are condenser inlet and outlet air dry bulb temperature, evaporator inlet and outlet air dry and wet bulb temperature, rotary booster inlet and outlet pressure, condenser inlet and outlet pressure, evaporator inlet and outlet pressure, liquid side flow rate, pressure difference before and after the nozzle, actual area of each nozzle, dry and wet bulb temperature before the nozzle, power of condenser fan, power of evaporator fan, and power of air pump. The equipment used for the measurements of the rotary booster-driven hybrid cooling unit is shown in Table 2.

Table 2. Main parameters of instruments.

Instrumentation	Precision	Range	Brand/Model
Thermocouples	±0.2 °C	−30 °C–150 °C	Omega/TT-T
Pressure transducer	±0.2%	0–25 bar	Huba/YD512
Data logger	——	——	Agilent/34970A
Power monitor	±0.02%	——	YOKOGAWA/WT230
Pressure difference transducer	±0.25%	1000 pa	EJA-120/110
RTD platinum resistance	A class	0.00–60.00 °C	NRHS3/3wire
Mass flow meter	±0.2%	500 kg/h	BJSINCERITY/Ultrasound Mass flow meter

Figure 4 shows the arrangement of measurement points for the condenser inlet and outlet air dry bulb temperature measured by four T-type thermocouples. The evaporator inlet and outlet air dry bulb temperature are measured by four T-type thermocouples and an RTD platinum resistance. The evaporator inlet and outlet air wet bulb temperature are measured by a wet bulb thermometer. The pressures of the rotary booster, condenser, and evaporator are measured by a pressure sensor (where the outlet of the rotary booster and condenser inlet distance is short, so the pressure measurement point is one). The liquid side flow rate is measured by an ultrasonic mass flow meter. The pressure difference before and after the nozzle is measured by a pressure difference meter. The evaporator air dry and wet bulb temperatures represent the dry and wet bulb temperatures before the nozzle, and the nozzle area uses the calibration value. All data are collected and exported by a data acquisition instrument. The power of the rotary booster and fans of the two devices is measured by a power meter.

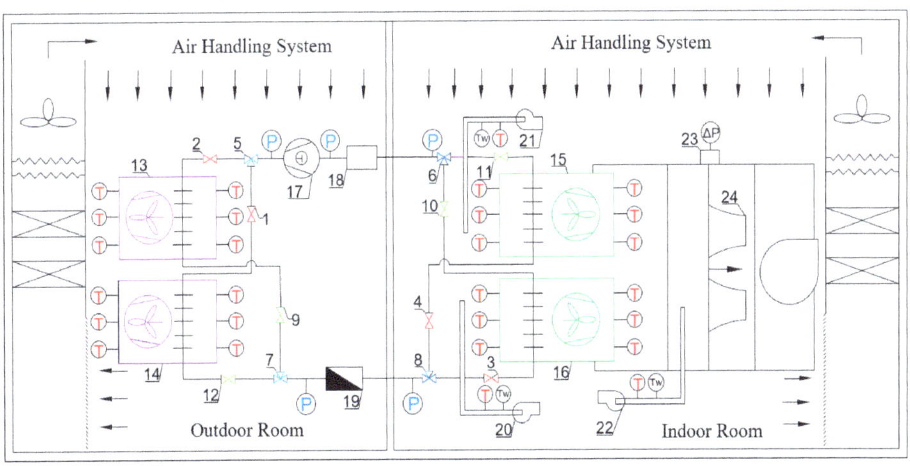

Figure 4. Experimental setup of booster-driven loop heat pipe energy recovery device. T—dry bulb temperature measurement point; Tw—wet bulb temperature measurement point; P—pressure measurement point; ΔP—pressure difference measurement point; 1, 2, 3, and 4—heat exchanger inlet shut-off valves; 5, 6, 7, and 8—three-way valves; 9, 10, 11, and 12—heat exchanger outlet shut-off valves; 13 and 14—condenser; 15 and 16—evaporator; 17—rotary booster; 18—gas/liquid separator; 19—mass flowmeter; 20, 21, and 22—sampling blower; 23—pressure difference measurement point; 24—nozzles.

Measurement errors are evaluated using the class B method of standard uncertainty [15], and the uncertainties of the experimental data are ±1.45% for the cooling capacity Q, ±0.02% for the electrical power P, and ±3.75% for the EER.

2.5. Performance Evaluation Indicators

In this paper, the performance of the rotary booster-driven hybrid cooling unit is mainly evaluated by three performance indicators: cooling capacity Q, power P, and EER.

1. Cooling capacity

The cooling capacity of the rotary booster-driven hybrid cooling unit is calculated by the product of the enthalpy difference between the inlet and outlet of the evaporator and the air volume. The enthalpy of the evaporator inlet and outlet and the enthalpy difference are determined from the dry bulb and wet bulb temperatures of the air inlet and outlet of the evaporator, and the air volume of the evaporator air can be calculated from the nozzle flow meter. The formula is as follows:

$$q_v = K \times C_d \times A_{nozzle} \times \sqrt{1000 \times \Delta p \times V_n}, \quad (1)$$

where q_v is the air volume at the measurement point inside the unit, m^3/s; K is the nozzle coefficient; C_d is the nozzle flow coefficient; A_{nozzle} is the nozzle area, m^2; Δ_p is the static pressure difference before and after the nozzle, Pa; and V_n is the specific volume of air at the nozzle inlet, m^3/kg.

A stronger cooling capacity means improved heat transfer performance of the unit. Conversely, the heat transfer performance of the unit is poor. In addition, the cooling capacity can be seen as the multiplication of the enthalpy difference between the evaporator inlet and outlet and the mass flow rate of the unit.

2. Power

The power consumption of the evaporator fan, the condenser fan, and the rotary booster can be measured by a digital power meter. The total power of the unit is obtained by adding the power of the rotary booster, $P_{booster}$, the power of the evaporator fan, $P_{fan,evap}$, and the power of the condenser fan, $P_{fan,cond}$. At the same frequency, the power of the rotary booster and the fan can be considered constant. A lower power of the rotary booster indicates that the rotary booster does lesser work and the rotary booster is more suitable for the rotary booster-drive loop cooling cycle. The formula is as follows:

$$P = P_{booster} + P_{fan,evap} + P_{fan,cond}, \quad (2)$$

where P is the total power of the unit, kW; $P_{booster}$ is the rotary booster power, kW; $P_{fan,cond}$ is the condenser fan power, kW; and $P_{fan,evap}$ is the evaporator fan power, kW.

If the power of the rotary booster is lower, it cannot ensure the unit cooling capacity is maintained above a certain level, so the energy efficiency ratio index is proposed.

3. Energy efficiency ratio

EER is the ratio of the cooling capacity to the effective input power when the unit operates under the rated working condition and specified conditions, and its unit is W/W. The higher the EER value, the less power is spent to achieve a higher cooling capacity and the better the performance of the unit.

$$EER = Q/P, \quad (3)$$

where EER is the energy efficiency ratio; Q is the cooling capacity, kW; and P is the total power of the unit, kW.

3. Results and Discussion

3.1. Operating Temperature Range of the Unit

3.1.1. Limit Outdoor Operating Temperature of Improved Unit Number 1

When improved unit number 1 is turned on and running, shutdown occurs when the outdoor ambient temperature increases from 30 °C to 35 °C and the indoor ambient temperature is 25 °C. The system parameters are shown in Figure 5. The outdoor temperature is stabilized at 35 °C, and the system pressure and temperature are relatively stable within 40 s. The suction pressure of the rotary booster is about 0.76 MPa, the suction temperature is 19 °C, the exhaust pressure is about 1.7 MPa, and the exhaust temperature is 61 °C. The horizontal coordinate after 50 s shows the system mass flow rate declines sharply during 50–60 s and is finally maintained at about 50 kg/h. At the same time, the system pressure and the temperature begin to change, and the rotary booster inlet and outlet pressure and temperature overlap. The current also exhibits irregular jumping. Regular analysis shows the rotary booster inlet and outlet pressures tend to coincide because the rotary booster inlet and outlet string gas or exhaust valve piece cannot be closed properly, which is the reason for the temperature change from the pressure change. The mass flow rate dropped twice; the reason for the first decline in the first 15 and 50–60 s is the outdoor ambient temperature from 30 °C to 35 °C, and the throttle valve opening is reduced, resulting in a decline in the mass flow rate. The second mass flow rate drop is accompanied by irregular jumps in the rotary booster current. Therefore, the unit cannot run stably in an environment above 35 °C.

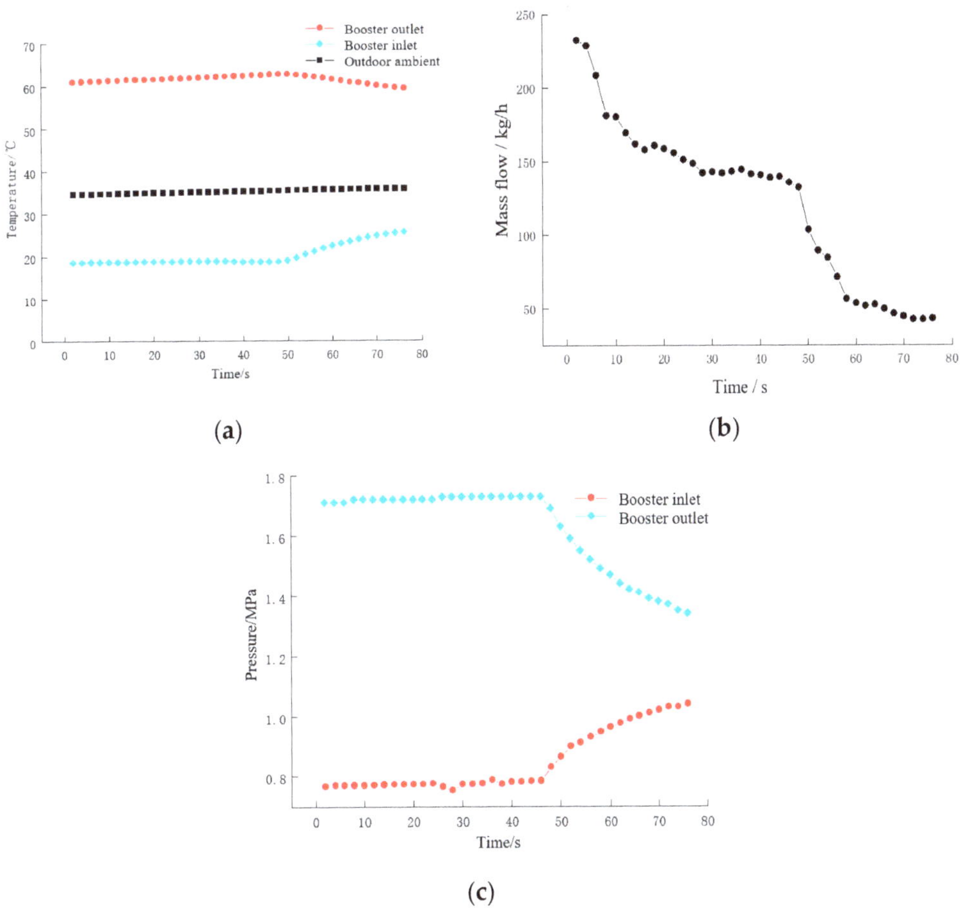

Figure 5. Operating parameters of number 1 improvement booster at 35 °C outdoor temperature. (**a**) Temperature of the unit measurement points; (**b**) mass flow rate; (**c**) pressure of the unit measurement points.

Thus, the outdoor environment control temperature is best set at 35 °C or below, or the indoor and outdoor temperature difference is less than 10 °C.

3.1.2. Limit Outdoor Operating Temperature of Improved Unit Number 1

For improved unit number 2, the outdoor ambient temperature is first kept at 40 °C. When the indoor ambient temperature is 25 °C, the unit is turned on, the unit runs for a period of time, and the same phenomenon of shutdown occurs. The specific parameters are shown in Figure 6.

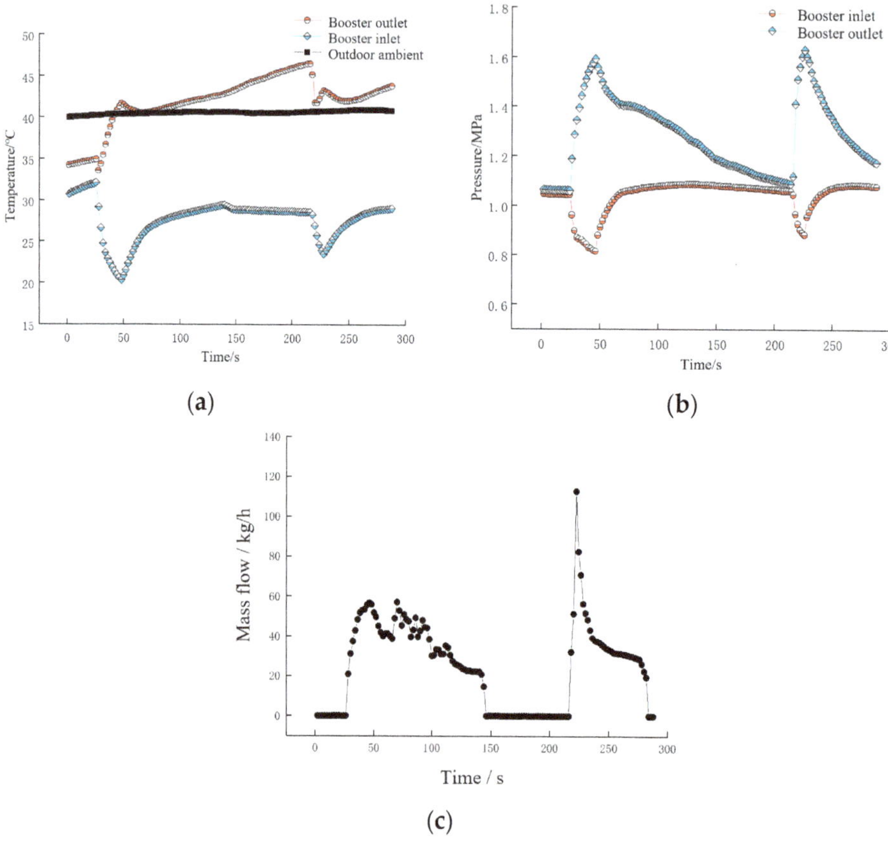

Figure 6. Operating parameters of number 2 improvement booster at 40 °C outdoor temperature. (**a**) Temperature of unit measurement point; (**b**) pressure of unit measurement point; (**c**) mass flow.

At 35 s in the horizontal coordinate, the unit is turned on, the mass flow rate rises, the suction pressure and temperature fall, and the exhaust temperature and pressure rise. However, the unit is suddenly shut down at 50 s, when the pressure and temperature of the rotary booster begin to change. However, the unit is restarted at 60 s and has been repeatedly shut down and started up. At 150 s, the unit is shut down for experimental safety reasons. At 220 s, the same problem still occurs when the unit is restarted. The reason for this phenomenon is that the rotary booster is under high temperature and high pressure. If the rotary booster is overheated or under high pressure, the unit starts overheating protection, and the improved rotary booster is not suitable for long-term operation in an outdoor high-temperature environment above 40 °C. Improved unit number 2 cannot complete the refrigeration work under the condition of 40 °C in an outdoor environment. Thus, the working range of the outdoor temperature is better below 40 °C.

Therefore, the experiment determines the subsequent operating temperature range, mainly for the data center in the winter operating conditions of the unit operation, that is, the outdoor temperature of −5 °C–20 °C with every 5 °C as a unit of change.

The experimental results show that the test operating temperature of the unit is controlled at outdoor temperatures of −5 °C–20 °C and indoor ambient temperature of 25 °C. In this paper, the analysis and research on the system pressure variation in different rotary booster-driven hybrid cooling units and the evaluation of the unit performance indexes in the range of indoor and outdoor temperature differences from 5 °C to 30 °C are discussed.

3.2. Pressure

An analytical study of the system pressure variation in different rotary booster-driven hybrid cooling units, as well as an evaluation of the unit performance indexes within the range of indoor and outdoor temperature differences from 5 °C to 30 °C, are performed.

3.2.1. Rotary Booster Pressure

Figure 7 shows the curves of the suction and exhaust pressure of the rotary booster and the pressure difference between the inlet and outlet of the rotary booster with indoor and outdoor temperature differences. Figure 7a,b show a curve of saturation pressure of refrigerant at outdoor ambient temperature.

Figure 7a reveals that the four exhaust pressure curves show a decreasing trend with the increase in indoor and outdoor temperature difference, the exhaust pressure curves of the standard unit in the throttled condition and the unthrottled condition are coincident, and the exhaust pressure curves of improved units numbers 1 and 2 coincide as well. The difference between the two curves is maintained at 0.04–0.05 MPa due to the exhaust pressure and the structure of the rotary booster. The condensing pressure of the curve gradually decreases with the increase in indoor and outdoor temperature differences, and the slope is the same as the four exhaust pressure lines. Thus, the difference between the two parts of the exhaust pressure curve reflects the difference between the improved unit and the standard unit.

Figure 7b shows the variation in the four suction pressure curves, with 20 °C as the boundary. When the between indoor and outdoor temperature difference is more than 20 °C, the suction pressure of the rotary booster decreases with increasing indoor and outdoor temperature differences, and the decreases are the same. When the indoor and outdoor temperature differences are less than 20 °C, all the slopes of the curves are less than the slopes when the indoor and outdoor temperature difference is more than 20 °C. The suction pressure at the throttle on the standard units is lower than the other curves. This result is consistent with the characteristics of throttling. Comparing the suction pressure curve with the saturated condensing pressure curve shows that when the indoor and outdoor temperature difference is more than 20 °C, all the curves of the suction pressure are more than the saturated condensing pressure, and the difference is more than 0.02 MPa. After the completion of the suction of the booster, the refrigerant does not need to be compressed and pressurized. It can directly complete the exhaust, and the pressure only needs to be greater than the spring force of the booster exhaust valve piece.

Figure 7c shows that the trend of the rotary booster pressure difference curve is generally decreasing with the increase in the indoor and outdoor temperature difference, but the decrease is slowing down. The rotary booster pressure difference in improved unit number 1 is always the smallest, from the highest point of 0.32 MPa to the lowest point of 0.07 MPa. Improved unit number 2 is the second one, from the highest point of 0.38 MPa to the lowest point of 0.12 MPa. When the indoor and outdoor temperature difference is more than 20 °C, no change is observed in the pressure difference in the rotary booster before and after throttling of the standard unit. When the indoor and outdoor temperature difference is less than 20 °C, the standard unit increases the pressure difference in the rotary booster after throttling, and it reaches the highest point of 0.62 MPa. The pressure difference curve of improved unit number 2 is always larger than that of improved unit number 1, which is smaller than the standard unit. The fixed pressure ratio relationship between the three follows the order of improved unit number 1 < improved unit number 2 < standard unit. Moreover, the pressure behavior with varying indoor and outdoor temperature differences indicates the working status in the cylinder of the developed booster, which provides changes in discharge and suction pressure ratio and indoor and outdoor temperature difference. Then, the pressure can be adjusted accordingly to match the temperature difference for high energy efficiency. When the temperature difference is large, the discharge and suction pressure ratio should be reduced to maintain the constant cooling capacity with small power consumption. Then, high efficiency can be obtained for meeting the same cooling load need. The relative statements have been marked in red in the manuscript.

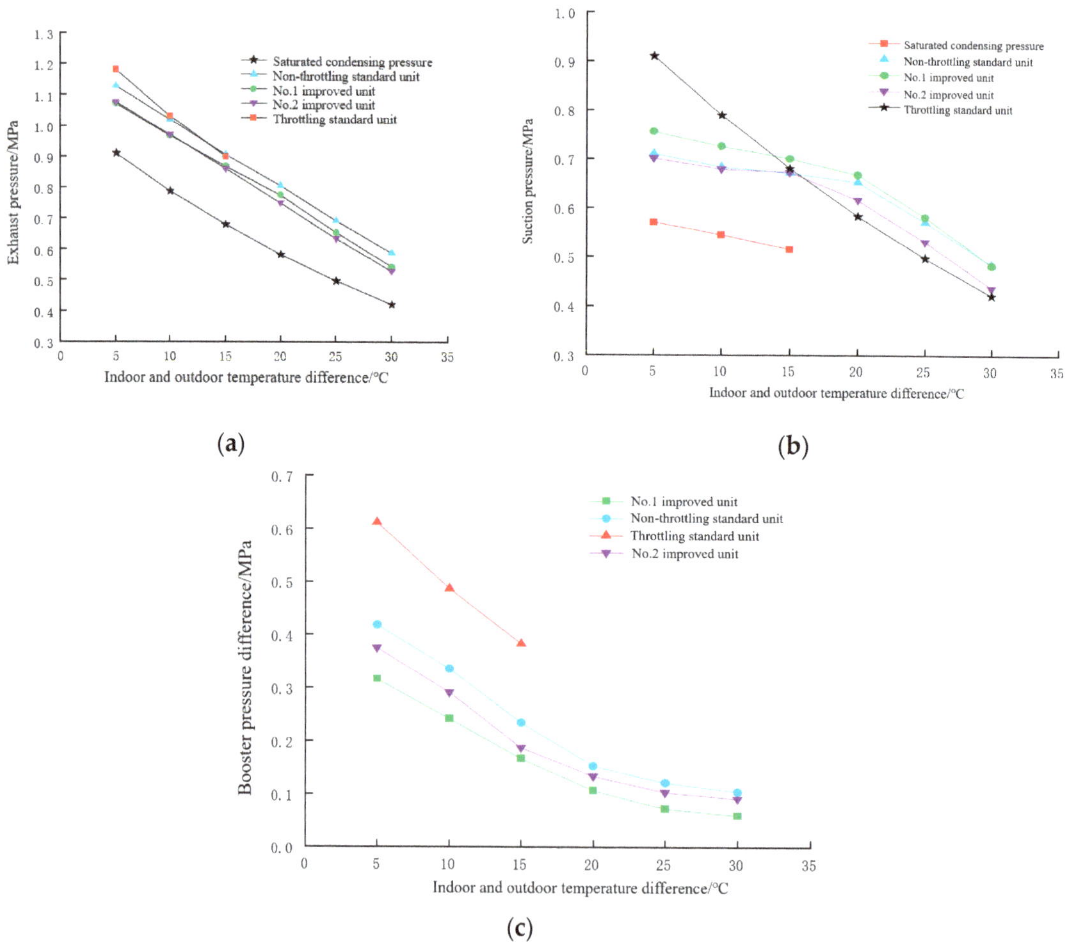

Figure 7. Variation in pressure with temperature difference in different boosters. (**a**) Exhaust pressure of system; (**b**) suction pressure of system; (**c**) pressure difference in booster.

3.2.2. Rotary Booster System Pressure

When different rotary boosters drive the cooling loop unit operation, the pressure changes in each position of the system are shown in Figures 8–12 below, including rotary booster inlet pressure, rotary booster outlet pressure, condenser inlet pressure, condenser outlet pressure, evaporator inlet pressure, and evaporator outlet pressure.

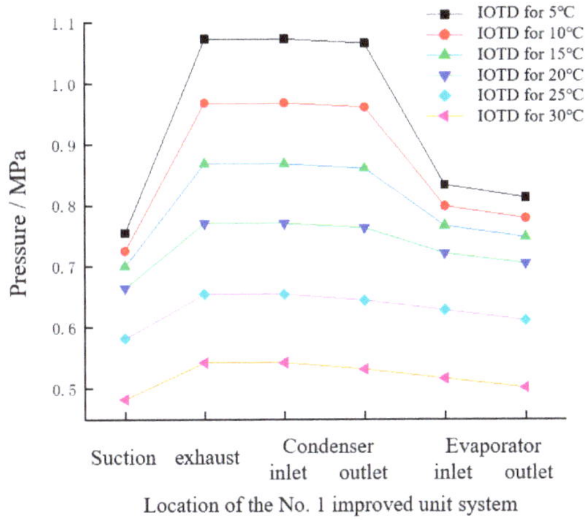

Figure 8. Pressure of number 1 improved unit system (indoor and outdoor temperature difference abbreviated as IOTD, the same below).

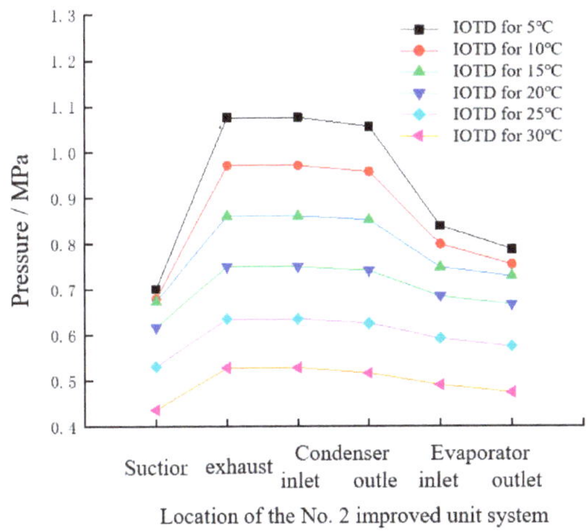

Figure 9. Pressure of number 2 improved unit system.

The graphs in Figures 8–10 depict the pressure distribution within the system at different temperature variances for diverse types of rotary booster operations. Four modes of operation are included for the three types of units: the unthrottled operation and throttled operation of improved unit number 1, improved unit number 2, and standard unit. Figure 8 includes the indoor and outdoor temperature differences from 5 °C to 30 °C with six solid lines in each panel. As for the standard unit after throttling, the test is conducted at indoor and outdoor temperature differences from 5 °C to 15 °C because of the very poor results at large temperature differences, as shown in the three dashed lines in Figure 9.

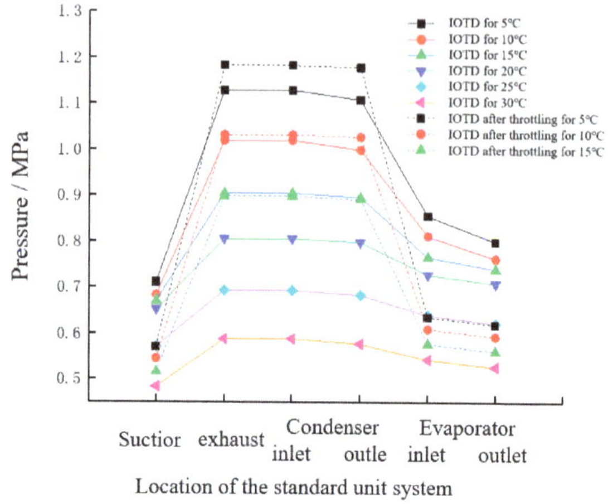

Figure 10. Pressure of standard unit system.

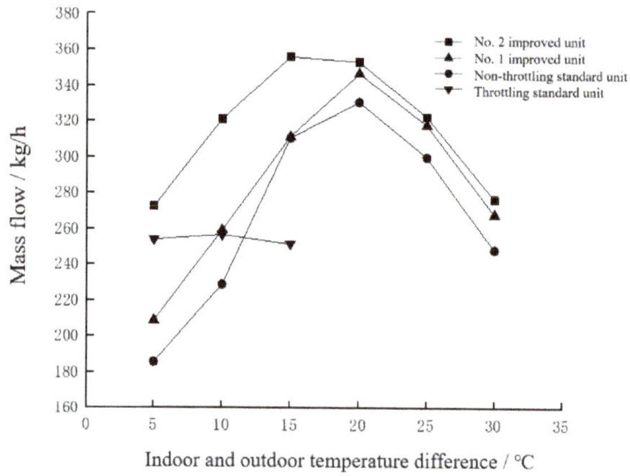

Figure 11. Variation in mass flow with temperature difference in different boosters.

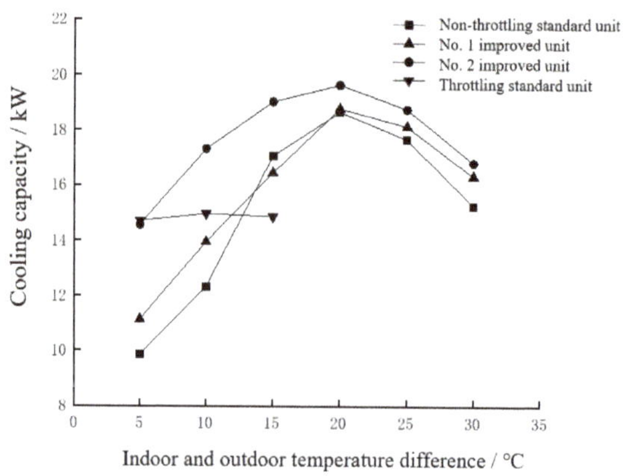

Figure 12. Variation in refrigeration capacity with temperature difference in different boosters.

Figures 8–10 show the same indoor and outdoor temperature differences between the curve of the variation is the same. From the suction to the exhaust, the pressure rises rapidly. From the exhaust to the condenser inlet and then to the condenser outlet, the pressure drops slightly. From the condenser outlet to the evaporator inlet, the pressure drops rapidly, but the magnitude of the drop is less than the magnitude between the suction and exhaust. From the evaporator inlet to the evaporator outlet, the pressure drops; the magnitude of the drop is less but greater than the condenser pressure drops. Finally, from the evaporator outlet to the inlet, the pressure drops. Taking the standard unit in the unthrottled condition, for example, the indoor and outdoor temperature difference of 5 °C from the suction pressure to the exhaust pressure rises mainly due to the rotary booster rotary pressurization, which is the only pressure rise in the system, that is, the source of the system power. From the exhaust to the condenser inlet and then to the condenser outlet, the pressure drops slightly mainly due to the manifold resistance from the rotary booster to the condenser inlet and the condenser heat transfer of the pressure drop generated in the process of the condenser and the condenser process resistance. From the condenser outlet to the evaporator inlet, the pressure drops mainly due to pipeline resistance, and the refrigerant is condensed even subcooled, which causes the pressure drop, as well as outdoor temperature changes; these make the liquid tube flow state changes and then causes the pressure loss. From the evaporator inlet to the evaporator outlet, the pressure drops mainly due to the evaporator along the pipeline process loss and the evaporator heat exchanger process of the pressure drops. From the evaporator outlet to the booster inlet, the pressure drops mainly due to the booster inlet along the pipeline process loss and the pressure loss of the gas–liquid separator.

Figure 9 shows that with the increase in the indoor temperature difference, the achievable maximum and minimum pressures of the unit decrease, and the difference between the maximum and minimum pressures also decreases. This result is mainly due to the decrease in outdoor temperature that leads to a decrease in saturated condensing pressure and a decrease in rotary booster exhaust pressure. The decrease in outdoor temperature makes the condensing temperature decrease, the condensing pressure decrease, and the evaporating pressure decrease, leading to a decrease in the suction pressure of the rotary booster.

After the standard unit throttling, the suction pressure and evaporator inlet and outlet pressure drop. When the indoor and outdoor temperature difference is 15 °C and 10 °C, the exhaust pressure is equal to that of the standard unit in the unthrottled condition, but when the indoor and outdoor temperature difference is 5 °C, the exhaust pressure is higher than that of the standard unit in the unthrottled condition. This result is mainly because after throttling, the evaporator inlet pressure plummets, resulting in a drop in the suction pressure of the rotary booster and an increase in the pressure difference.

3.3. *Analysis of System Performance Indicators*

3.3.1. Mass Flow

Figure 11 shows the mass flow rate of different rotary booster-driven loop units with the increase in indoor and outdoor temperature difference variation. For improved units numbers 1 and 2 and the standard unit unthrottled condition-driven loop unit, the mass flow rates with the indoor and outdoor temperature difference in the variation are the same. With the increase in the indoor and outdoor temperature difference, the mass flow rates all appear to increase and then decrease. The mass flow rate of each unit is minimized when the indoor and outdoor temperature difference is 5 °C. With the increase in the indoor and outdoor temperature difference (i.e., the decrease in the outdoor temperature), the mass flow rate starts to increase and reaches its respective peak when the indoor and outdoor temperature difference is 20 °C. After that, the indoor and outdoor temperature difference is further increased, but the mass flow rate starts to decrease. After the indoor and outdoor temperature differences exceed 20 °C, the mass flow rate starts to decrease because as the outdoor temperature decreases, the degree of subcooling starts to increase, resulting in a decrease in the suction pressure of the unit, at which time the specific volume of the rotary booster inlet increases and the mass flow rate decreases.

3.3.2. Cooling Capacity

Figure 12 shows the cooling capacity of different rotary booster-driven loop units with different indoor and outdoor temperature difference variations. The cooling capacity with the indoor and outdoor temperature difference is the same as the variation in the loop unit driven by improved unit number 1, improved unit number 2, and the standard unit in the unthrottled condition. The cooling capacity of all the units appears to increase and then decrease with the increasing indoor and outdoor temperature difference. The cooling capacity of each unit is the smallest when the indoor and outdoor temperature difference is 5 °C. The cooling capacities of improved unit number 1, improved unit number 2, and the standard unit in the unthrottled condition are 11.1, 14.6, and 9.87 kW, respectively. With the increase in the indoor and outdoor temperature difference (i.e., the drop of the outdoor temperature), the cooling capacity starts to increase and reaches its respective peaks at a temperature difference of 20 °C between indoor and outdoor temperatures. The cooling capacities of improved unit number 1, improved unit number 2, and the standard unit in unthrottled condition are 18.8, 19.6, and 18.6 kW, respectively. After that, the temperature indoor and outdoor temperature difference increases further, but the cooling capacity starts to decrease. When the indoor and outdoor temperature difference is 30 °C, the cooling capacities of improved unit number 2, improved unit number 1, and the standard unit are 16.8, 16.3, and 15.2 kW, respectively.

The unthrottled condition of the standard unit is compared with that of the improved unit. The cooling capacity of improved unit number 1 and improved unit number 2 is better than that of the standard unit over most of the range of indoor and outdoor temperature differences. Then, throttling the standard unit loop, the cooling capacity changes, and the cooling capacity stabilizes at 14.9 kW in the range of indoor and outdoor temperature differences from 5 °C to 15 °C. Compared with the standard unit in the unthrottled condition, the change rates of the cooling capacity are 49.1%, 21.5%, and −14.9%, which means the throttling of the standard unit enhances the cooling capacity in a certain range. However, when the indoor and outdoor temperature differences continue to increase, the cooling capacity of the standard unit after throttling is not as good as that of the unthrottled condition, so the experiments on the standard unit throttled with large temperature differences are not continued.

The cooling capacity curve of improved unit number 2 is always higher than the three other curves. The difference in cooling capacity between improved unit number 1 and improved unit number 2 is large at indoor and outdoor temperature differences of up to 20 °C. After the rotary booster of the unit is changed from improved unit number 1 to improved unit number 2, the growth rates of the cooling capacity when the indoor and outdoor temperature difference is from 5 °C to 15 °C are 30.7%, 24.2%, and 15.7%. However, when the indoor and outdoor temperature difference is up to 20 °C, the growth rate is below 5%. The difference between the standard unit and improved unit number 2 is even greater. The growth rate of the cooling capacity is above 40% for indoor and outdoor temperature differences of 5 °C and 10 °C. The growth rate of the cooling capacity for the rest of the working conditions is also around 10%.

3.3.3. Power

Figure 13 shows the variation in the power in different rotary booster-driven loop units with different indoor and outdoor temperature differences. For improved unit number 1, improved unit number 2, and the standard unit in the unthrottled condition-driven loop, the variations in power with temperature difference in different boosters are the same, which decreases with the increasing indoor and outdoor temperature difference.

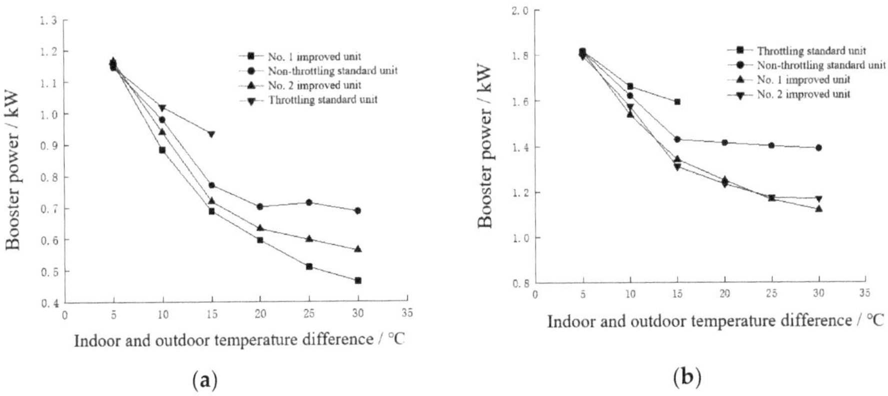

Figure 13. Variation in power with temperature differences in different boosters. (**a**) Power of booster; (**b**) power of unit.

Considering the rotary booster power, when the indoor and outdoor temperature difference is 5 °C, the rotary booster power of each unit is the largest. The rotary booster power of improved unit number 1, improved unit number 2, and the standard unit in the unthrottled condition are 1.17, 1.16, and 1.15 kW, respectively. With the increase in the indoor and outdoor temperature difference (i.e., the decrease in the outdoor temperature), the unit power begins to decline, and after the indoor and outdoor temperature difference of 15 °C, the decline of the booster power slows down in an indoor and outdoor temperature difference of 30 °C. At 30 °C, the booster power of improved unit number 1, improved unit number 2, and the standard unit are 0.46, 0.56, and 0.69 kW, respectively. The rotary booster power decreases mainly due to the decrease in the rotary booster pressure difference. At indoor and outdoor temperature differences from 5 °C to 15 °C, the mass flow rate increases, but it cannot compensate for the decrease in the pressure difference, so the power of the rotary booster decreases. At indoor and outdoor temperature differences from 20 °C to 30 °C, the mass flow rate and the rotary booster pressure difference decrease, so the rotary booster power also decreases. According to the relationship between suction pressure and saturation pressure at outdoor ambient temperature, the pressurization role of the rotary booster weakens.

The unthrottled case of the standard unit compared with the improved units: The rotary booster power of improved unit number 1 and improved unit number 2 are less than that of the standard unit in most of the indoor and outdoor temperature difference ranges. Then, after throttling the standard unit loop unit, the rotary booster power changes and becomes higher than the three other curves in the range of indoor and outdoor temperature differences from 5 °C to 15 °C. In the full range of indoor and outdoor temperature differences, the rotary booster power follows the order of standard unit throttled condition > standard unit unthrottled condition > improved unit number 2 > improved unit number 1.

3.3.4. EER

Figure 14 shows the variation in EER of different booster units with indoor and outdoor temperature differences. EER is the ratio of cooling capacity and unit power as Equation (3). The cooling capacity of improved unit number 2 is greater than that of improved unit number 1, the improved unit is greater than the standard unit unthrottled condition, the power of improved unit number 1 is less than that of improved unit number 2, and the improved unit is less than the standard unit for the throttled condition.

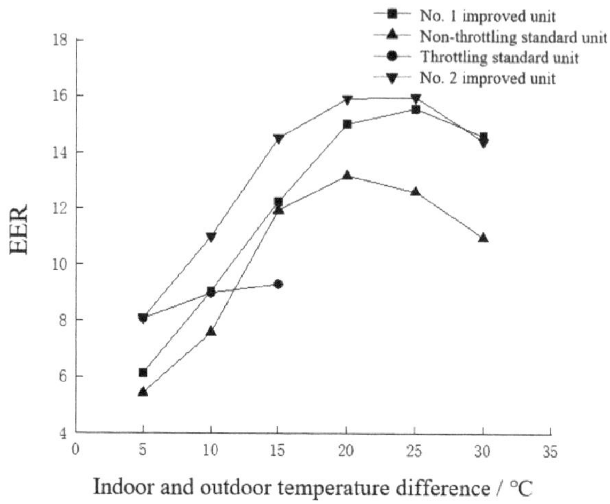

Figure 14. Variation in EER of different booster units with indoor and outdoor temperature differences.

Figure 14 shows the loop cooling units driven by improved unit number 1, improved unit number 2, and the unthrottled condition of the standard unit. The variations in EER with the indoor and outdoor temperature difference are the same. When the indoor and outdoor temperature difference is 5 °C, the EER of each unit is the smallest. The EER of improved unit number 1, improved unit number 2, and the standard unit in unthrottled condition are 6.14, 8.1, and 5.44, respectively. With the increase in the indoor and outdoor temperature difference (i.e., the drop in the outdoor temperature), the EER begins to increase. When the indoor and outdoor temperature difference is 20 °C, the unthrottled condition of the standard unit reaches the peak, and the unthrottled condition of the standard unit is 13.2. When the indoor and outdoor temperature difference is 25 °C, the EER of improved unit number 1 and improved unit number 2 reach the peak; the EER of improved unit number 1 is 15.6, and that of improved unit number 2 is 16.0. After that, the indoor and outdoor temperature differences increase further, but the EER starts to decline. At the indoor and outdoor temperature differences of 30 °C, the EERs of improved unit number 2, improved unit number 1, and the standard unit are 14.4, 14.6, and 11.0, respectively.

After throttling the standard unit, cooling capacity increases in the 5 °C–10 °C range. When the power of the unit rises as well, the EER does not exceed that of improved unit number 2, but it is better than that of improved unit number 1 and the unthrottled condition of the standard unit.

Finally, a comprehensive comparison of the EER curves reveals that improved unit number 2 owns the largest EER and the best performance, considering the higher suction pressure and cooling capacity due to the larger torque. Then, it is selected for the subsequent experiments.

4. Conclusions

In this paper, comparison experiments of different rotary booster-driven loop cooling systems are carried out without changing the system components, the outdoor temperature change is controlled to carry out the comparison experiment between different rotary booster units, and the limit working temperature of different air pumps and their operating performance under diverse indoor and outdoor temperature differences are studied. The following conclusions are drawn:

(1) When the standard unit is running, the temperature working range is from 40 °C to −5 °C in the outdoor environment for normal, stable operation; the temperature working range of improved unit number 1 rotary booster-driven hybrid cooling unit is

from 30 °C to −5 °C in the outdoor ambient temperature for stable operation; and the temperature working range of improved unit number 2 rotary booster-driven hybrid cooling unit is from 35 °C to −5 °C in the outdoor environment for normal, stable operation. The temperature below −5 °C is not in the research scope of this paper, so no experimental verification is performed to determine whether the operation is normal. Combined with the data center's year-round operating variations and characteristics, the outdoor temperature operating range is set between −5 °C and 20 °C and the indoor temperature is controlled at 25 °C.

(2) The three kinds of rotary booster system pressure variations are the same. From the suction port to the exhaust, the pressure rises rapidly. From the exhaust to the condenser inlet to the condenser outlet, the pressure drops slightly. From the condenser outlet to the evaporator inlet, the pressure drops rapidly, but the magnitude of the drop is less than the magnitude between the suction and exhaust. From the evaporator inlet to the evaporator outlet, the pressure drops, and the magnitude of the pressure drop is small but greater than the pressure drop of the condenser. Finally, the pressure drops from the evaporator outlet to the suction port.

(3) The pressure curve of the rotary booster shows that with the increase in the indoor and outdoor temperature differences, the suction pressure, the exhaust pressure, as well as the pressure difference in the rotary booster show a decreasing trend. Comparison with the saturated condensation pressure curves at the outdoor ambient temperature reveals that the suction pressure exhibits a different variation with the indoor and outdoor temperature differences of 20 °C as the boundary. When the indoor and outdoor temperature difference is less than 20 °C (small temperature difference), the suction pressure is less than the saturated condensation pressure at the outdoor ambient temperature. When the indoor and outdoor temperature difference is more than 20 °C (large temperature difference), the suction pressure is more than the saturated condensation pressure at the outdoor ambient temperature. This outcome indicates that the pressurizing effect of the rotary booster weakens under a large temperature difference.

(4) In terms of the system performance evaluation index, the cooling capacity in the indoor and outdoor temperature difference of 5 °C–15 °C of improved unit number 2 is better than that of improved unit number 1, and the growth rates are 30.7%, 24.2%, and 15.7%. However, after the indoor and outdoor temperature difference increases up to 20 °C, the growth rate is below 5%. The cooling capacity of improved unit number 2 is better than that of improved unit number 1, and the growth rate of the cooling capacity is more than 40% when the indoor and outdoor temperature difference is 5 °C and 10 °C. The growth rate of the cooling capacity for the rest of the working conditions is also about 10%.

(5) The system performance evaluation index, in all indoor and outdoor temperature difference ranges, shows the rotary booster power follows the order of standard unit in the throttled condition > standard unit in the unthrottled condition > improved unit number 2 > improved unit number 1. With the increase in the indoor and outdoor temperature difference, the rotary booster power shows a decreasing trend, and the decreasing trend stabilizes after the indoor and outdoor temperature difference of up to 20 °C. The reason is related to the suction pressure of the rotary booster and the saturated condensing pressure in ambient temperature.

(6) The EER curves of improved unit number 2 are better than those of the standard unit and improved unit number 1 under most of the indoor and outdoor temperature differences. For the development of a special booster, the one with a large torque owns the priority. The performance variation rules and matching relationship of the hybrid unit can be adopted for the system design and efficiency improvement.

Author Contributions: Conceptualization, F.Z. and G.M.; methodology, F.Z., G.M. and S.L.; validation, S.L., B.X. and X.T.; formal analysis, R.Z., S.L., B.X. and X.T.; investigation, R.Z., S.L., F.Z. and G.M.; data curation, R.Z., S.L., B.X. and X.T.; writing—original draft preparation, R.Z., S.L., B.X. and X.T.; writing—review and editing, F.Z., G.M., R.Z., S.L., B.X. and X.T.; visualization, R.Z., S.L. and B.X.; supervision, F.Z. and G.M.; project administration, R.Z., F.Z. and G.M.; funding acquisition, F.Z. and G.M. All authors have read and agreed to the published version of the manuscript.

Funding: This research was supported by the State Key Laboratory of Air-Conditioning Equipment and System Energy Conservation (ACSKL2021KT03), the General Project of Science and Technology Program of Beijing Municipal Education Commission (KM201910005017), and the Science and Technology Innovation Foundation for Urban Carbon Neutralization of Beijing University of Technology (No. 049000514122607).

Institutional Review Board Statement: Not applicable.

Informed Consent Statement: Not applicable.

Conflicts of Interest: The authors declare no conflict of interest.

References

1. Salim, M.; Tozer, R. Data centers' energy auditing and bench marking-progress update. *Ashrae Trans.* **2010**, *116*, 109–117.
2. Ebrahimi, K.; Jones, G.F.; Fleischer, A.S. A review of data center cooling technology, operating conditions and the corresponding low-grade waste heat recovery opportunities. *Renew. Sustain. Energy Rev.* **2014**, *31*, 622–638. [CrossRef]
3. Li, G.; Shi, Z.; Zhang, Z.; Jin, T.; Shao, S.; Tian, C. Experimental investigation on energy saving of the combined air conditioner in data center by separate heat pipe and vapor compressor. *Cryog. Supercond.* **2016**, *10*, 67–71.
4. Shi, W.; Han, L.; Wang, B. Principle of combined air conditioner by heat pipe and vapor compression and its application analysis in high heat density space. *Refrig. Air Cond.* **2011**, *11*, 30–36.
5. Lee, S.; Kang, H.; Kim, Y. Performance optimization of a hybrid cooler combining vapor compression and natural circulation cycles. *Int. J. Refrig.* **2008**, *32*, 800–808. [CrossRef]
6. Zhang, H.; Shao, S.; Xu, H.; Zou, H.; Tian, C. Free cooling of data centers: A review. *Renew. Sustain. Energy Rev.* **2014**, *35*, 171–182. [CrossRef]
7. Wang, F. Design and Test of 30 kW Dynamic Separate Type Heat Pipe. Ph.D. Thesis, Hefei University of Technology, Hefei, China, 2014.
8. Liu, J. Investigations on Running Characteristics of the Mechanically Pumped Cooling Loop for Space Applications. Ph.D. Thesis, Shanghai Jiao Tong University, Shanghai, China, 2008.
9. Wang, T.; Wang, F.; Li, H.; Wang, J.; Zhao, S.; Liu, H.; Lv, J. Design and experimental study of dynamic separate type heat pipe. *Refrig. Air Cond.* **2014**, *14*, 40–43.
10. Zhou, F.; Wei, C.; Ma, G. Development and analysis of a pump—Driven loop heat pipe unit for cooling a small data center. *Appl. Therm. Eng.* **2017**, *124*, 1169–1175. [CrossRef]
11. Zhang, P.; Wang, B.; Shi, W.; Li, X. Experimental investigation on two-phase thermosyphon loop with partially liquid-filled downcomer. *Appl. Energy* **2015**, *160*, 10–17. [CrossRef]
12. Wei, C.; Ma, G.; Xu, S.; Zhang, S.; Zhou, F. Design and experimental study on heat exchanger unit with a gas pump—Driven loop heat pipe for communication base stations. In *Proceedings of the 12th National Conference on Refrigerators (Cabinet), Air Conditioners and Compressors*; Small Refrigerator Low Temperature Biomedical Committee of Chinese Association of Refrigeration, Home Appliance Industry Technology Innovation Strategic Alliance: Hefei, China, 2014.
13. Wang, F.; Huang, D.; Shi, Z. Experimental study on two kinds of dynamic separate type heat pipe system. *Refrig. Air Cond.* **2017**, *17*, 53–57.
14. Shi, W.; Wang, F.; Huang, D.; Shi, Z. Development of gas pressurized composite air-conditioning unit and annual operation energy efficiency analysis. *Refrig. Air Cond.* **2017**, *17*, 11–16.
15. Fei, Y. *Error Theory and Data Processing*; Mechanical Industry Press: Beijing, China, 2010; pp. 82–86.

energies

MDPI

Article

Experimental Research on the Gas-Solid Flow Characteristics in Large-Scale Dual Fluidized Bed Reactor

Yubin Lin [1], Qinhui Wang [1,2,*], Chao Ye [3], Yao Zhu [1] and Haojie Fan [4]

1 State Key Laboratory of Clean Energy Utilization, Zhejiang University, Hangzhou 310027, China
2 Institute for Thermal Power Engineering, Zhejiang University (Yuquan Campus), 38 Zheda Road, Hangzhou 310027, China
3 School of Mechanical and Energy Engineering, Zhejiang University of Science and Technology, Hangzhou 310023, China
4 School of Mechanical Engineering, Shanghai Jiao Tong University, Shanghai 200240, China
* Correspondence: qhwang@zju.edu.cn; Tel.: +86-571-87952802

Abstract: A dual fluidized bed (DFB) reactor is the main operating system of various energy-efficient and clean utilization technologies. The gas-solid flow characteristics of the DFB reactor greatly affect the efficiency of various technologies. A large-scale DFB reactor with a maximum height of 21.6 m was built and relevant cold mode tests were carried out in this study. The effects of the superficial gas velocity of both beds, static bed height and particle size on the distribution of both pressure and solid suspension density, solid circulation rate, solid inventory distribution ratio and other characteristics were studied. For 282 μm-particles, the solid suspension density in the dense phase zone of the two beds was 100–400 and 400–800 kg/m^3, respectively, when the static bed height was 0.65 m; the solid circulation rate was about 0.87–1.75, 1.04–3.04 and 1.13–3.69 kg/(m^{2}s) when the static bed height was 0.65, 0.95 and 1.25 m, respectively. The solid circulation rate was positively correlated with the static bed height and the superficial gas velocity of both beds, yet negatively correlated with the particle size. Additionally, the empirical equation of solid circulation rate and the empirical equation of solid inventory distribution ratio were proposed, respectively. The material control method of the DFB reactor is put forward.

Keywords: dual fluidized bed reactor; distribution of pressure; solid circulation rate; solid inventory distribution ratio

Citation: Lin, Y.; Wang, Q.; Ye, C.; Zhu, Y.; Fan, H. Experimental Research on the Gas-Solid Flow Characteristics in Large-Scale Dual Fluidized Bed Reactor. *Energies* **2023**, *16*, 7239. https://doi.org/10.3390/en16217239

Academic Editors: Artur Blaszczuk, Fuqiang Wang, Chao Shen, Dong Li and Zhonghao Rao

Received: 4 September 2023
Revised: 17 October 2023
Accepted: 20 October 2023
Published: 25 October 2023

1. Introduction

A dual fluidized bed (DFB) reactor refers to a reactor system where two fluidized beds are coupled together [1]. In the reactor, material particles are fluidized and circulated between two fluidized beds, which facilitate the heat and mass transfer within the system. In the context of carbon neutrality, as the main operating system of various energy-efficient and clean utilization technologies, including polymerization processes [2], chemical looping combustion technology [3], carbon-capture utilization and storage technology [4] and biomass gasification technology [5], the DFB reactor is of self-evident importance. The specific roles of the two fluidized beds within the DFB reactor used in various technologies are shown in Table 1 [3–5]. The gas-solid flow characteristics of the DFB reactor, including the particle circulation characteristics, greatly affect the heat and mass transfer of various technologies, thus affecting their performance and efficiency. Therefore, it is of great necessity to carry out research on the DFB reactor to understand its gas-solid flow characteristics.

Experimental research on the DFB reactor has been reported by many researchers. Tobias Proll et al. built a 120 kW DFB reactor and carried out cold mode tests, and found that the reactor had good circulation performance [6]. Chunbao Zhou et al. conducted a pilot study of interconnected pyrolysis and gasification in a 50 kg/h DFB reactor designed by themselves, and preliminarily verified the feasibility of the reactor in the field of biomass utilization and carbon capture [7]. A. Charitos et al. conducted tests on a hydrodynamically-scaled cold model of the 10 kW_{th} calcium-looping DFB facility and identified a stable operating region bordered by two unstable regions [8]. Saurabh Gupta et al. investigated the hydrodynamics of a DFB gasifier designed for high-ash coal and found that secondary aeration is more effective in maintaining a proper balance between the solid holdup and the solid circulation rate when the bottom riser bed is operated in the fast-fluidized bed regime for in-bed solids discharge [9].

Table 1. The specific roles of DFB reactors in various technologies.

Technology	1# Fluidized Bed	2# Fluidized Bed
Chemical-Looping Combustion	Air reactor	Fuel reactor
Carbon-Capture Utilization and Storage	Carbonator	Regenerator
Biomass Gasification	Combustor	Gasifier

There are also many researchers carrying out numerical simulation studies on the DFB reactor. Yang Liu et al. analyzed gas-solid fluidization and coal-gasification reactions in a DFB unit by using the full-loop numerical simulation method, which provides a basis for industrial application [10]. Yangjun Wei et al. adopted an analysis approach considering gas-solid hydrodynamics, reaction kinetics and reacting species nonuniformity together in a dual-reactor system, and drew a conclusion that energy balance has a close relationship with the mass transfer in the DFB reactor system [11]. Peter Ohlemuller et al. and Asad H. Sahir et al. used Aspen to simulate the chemical-looping combustion process based on a DFB reactor under scales of 0.1, 1, 10 and 100 MW_{th} [12–14].

The types and scales of the above research are summarized in Table 2. As can be seen from Table 2, existing research on DFB reactors are divided into experimental research and numerical simulation research. On the one hand, most experimental studies are based on small DFB reactor systems, which are far from the actual production scale. This will lead to complex, unforeseen changes during the process of scaling up the device from the laboratory scale to the actual production scale, thus making the results of the research unable to provide theoretical experience for the design and operation of actual production. On the other hand, the existing numerical simulation research can be divided into Aspen simulations and CFD simulations. Aspen simulations focus on the economic evaluation of the efficiency and yield of various processes at the macro level but neglect various gas-solid flow characteristics inside the reactor. CFD simulations focus on various gas-solid flow characteristics inside the reactor, however, due to the lack of support from experimental data, the research results cannot be directly used to guide the actual production.

Table 2. The types and scales of the existing researches.

Researchers	Type	Scale
Tobias Pröll et al. [6]	Experiment	1.80 m
Chunbao Zhou et al. [7]	Experiment	4.00 m
A. Charitos et al. [8]	Experiment	5.30 m
Saurabh Gupta et al. [9]	Experiment	2.25 m
Yang Liu et al. [10]	CFD simulation	60.0 m
Yangjun Wei et al. [11,15]	Experiment & CFD simulation	15.5 m & 11.2 m
Peter Ohlemuller et al. [12,13]	Aspen simulation	0.1 & 1 MW_{th}
Asad H. Sahir et al. [14]	Aspen simulation	1, 10 & 100 MW_{th}

The distribution of pressure and solid suspension density in the furnace are the most basic gas-solid flow characteristics of a DFB reactor. These two characteristics have a great influence on the heat and mass transfer inside the reactor. Chengliang Han et al. studied the effect of bed material size on the distribution of pressure and the solid suspension density in a CFB (Circulating Fluidized Bed) reactor at low solid recirculation rates [16]. The solid circulation rate is another important gas-solid flow characteristic of the DFB system, which represents the degree of solid circulation between two fluidized beds, and is usually expressed by G_s [17,18]. It not only affects the gas-solid flow in the furnace but also has a significant impact on the furnace temperature, gas-solid heat transfer efficiency, solid fuel pyrolysis efficiency and component wear. Hu et al. studied the solid circulation rate of a pressurized CFB and an internal CFB [17,19] while Atipong Armatsombat et al. and Mona Mary Varghese et al. tried to investigate or predict the solid circulation rate of different kinds of CFB [20,21].

Apparently, it is of great significance to master the influencing factors and changing rules of the distribution of both pressure and solid suspension density and the solid circulation rate in the DFB system. However, the subjects of the above research were all single fluidized beds. In a DFB reactor system, two fluidized beds are coupled to each other and, consequently, the flow field of two fluidized beds will influence each other, which may bring unknown changes to the gas-solid flow characteristics of the DFB reactor. This means that the gas-solid flow characteristics of a DFB reactor cannot be described simply by that of a single fluidized bed.

All in all, the types and scales of the existing studies on DFB reactors cannot provide effective theoretical experience for actual production. At the same time, the coupling of two fluidized beds will make the existing research on gas-solid flow characteristics based on single beds fail to serve as a reference point. In view of this, a large-scale DFB reactor with a maximum height of 21.6 m was built and relevant cold mode tests were carried out in this study. The effects of superficial gas velocity of both beds, static bed height and particle size on the distribution of both pressure and solid suspension density, solid circulation rate, solid inventory distribution ratio and other characteristics were studied. Additionally, the empirical equation of solid circulation rate and the empirical equation of solid inventory distribution ratio were proposed. The material control method of a DFB reactor is put forward, and the research on DFB reactors is improved, so as to provide guidance and demonstration for industrial production.

2. Experiment

2.1. Experiment System

In this study, a large-scale DFB reactor cold mode test system was built with polymethyl methacrylate as the main material. As shown in Figure 1, the experimental system is mainly divided into 1# fluidized bed (1#FB for short) and 2# fluidized bed (2#FB for short), with two risers, four cyclones and four loop-seals. The whole system utilizes remote control through a distant control system.

The size and structure design of each part of the system are shown in Table 3.

Table 3. Size and structure design of main parts.

Item	Value
The height of 1#FB	21.60 m
The cross-sectional area of 1#FB	0.30 m × 0.40 m
The height of 2#FB	14.40 m
The cross-sectional area of 2#FB	0.25 m × 0.40 m
The relative height difference of the inlets	7.20 m
The design of riser inlet	Tapered

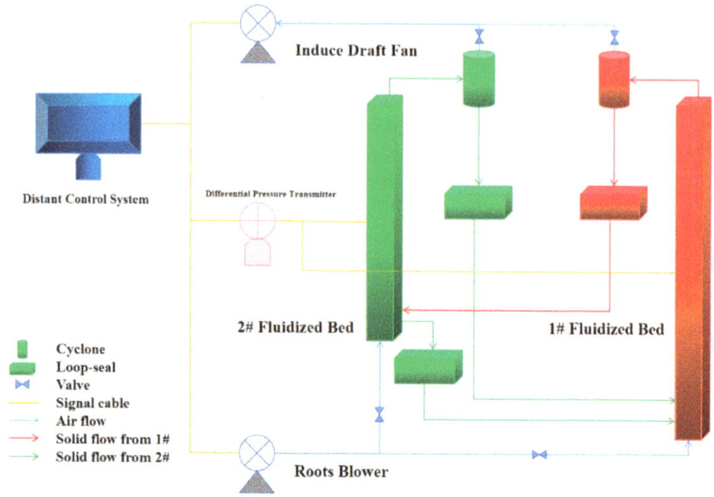

Figure 1. Schematic of the experiment system.

The fluidizing air of the riser, the return air and the loose air of the loop-seal are provided by a Roots blower, and an induced draft fan is connected to the exit of the cyclone at the top of the riser to maintain the internal pressure balance. A digital flowmeter is connected to the inlet of 1#FB and the inlet of 2#FB to obtain the superficial gas velocity of air at the inlet of the two risers. The two risers are divided into 12 and 18 sections with 1.2 m as the unit, and there are measuring points in the middle of each section for measuring the pressure at specific positions in the riser.

There are two grades of cyclones on the top of 2#FB, and two groups of cyclones are symmetrically arranged on the top of 1#FB. The outlet of the two cyclones on 1#FB and the secondary cyclone on 2#FB is connected to the induced draft fan.

Loop-seals are widely used in CFB systems to achieve a closed solid circulation loop [22]. Its biggest advantage is that it can not only realize solid circulation loop but also prevent or minimize airflow interference between different areas within the system. The system has four loop-seals. One of them is located at the bottom of 2#FB, and the other three are connected with the secondary cyclone of 2#FB and the two cyclones of 1#FB, respectively.

2.2. *Material Circulation and Typical Operation State*

In the operation process, quartz sand is used as the material in the experiment, and the internal material balance is achieved by adjusting the air volume. Part of the quartz sand in 2#FB is transported from the outlet to the upper loop-seal of 2#FB, through the two-grade cyclone, and then is sent back to 1#FB; the other part directly exits from the overflow port at the bottom of 2#FB to the bottom loop-seal of 2#FB, and then is sent back to 1#FB. The quartz sand in 1#FB reaches the corresponding loop-seal through the cyclone arranged symmetrically at the top and then is sent back to 2#FB. Figure 2 shows the typical operating state of each major component of the system during normal operation.

2.3. *Test Condition and Data Processing*

2.3.1. Test Condition

The experimental parameters are shown in Table 4. According to the Geldart particle classification method [23], the particles used in the experiment are all class B particles.

Figure 2. Typical operating state: (**a**) 2#FB; (**b**) 1#FB; (**c**) Loop-seal.

Table 4. Test condition.

Item	Value	Unit
Superficial gas velocity of 1#FB	2.5/3/3.5/4/4.5	m/s
Superficial gas velocity of 2#FB	2/2.5/3/3.5	m/s
Static bed height of 1#FB	0.65/0.95/1.25	m
Average particle size	282/641	μm

The static bed height of 1#FB refers to the height of the initial bed inventory in the 1#FB. The particle size distribution of the two kinds of sand used in the experiment is shown in Figure 3.

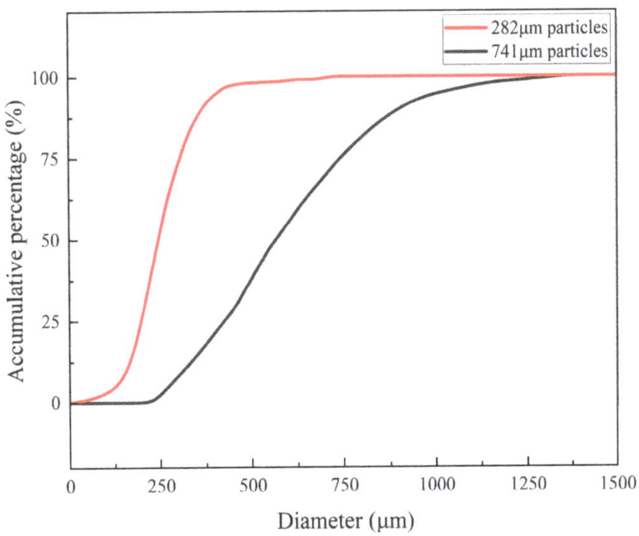

Figure 3. Particle size distribution.

2.3.2. Pressure Distribution

This system uses a differential pressure transmitter to measure the pressure of a preset testing point in real-time. As the pressure fluctuates continuously during the experiment, a computer (distant control system) is used to record the pressure measured by the differential pressure transmitter, and the mean value is taken after reaching a stable state. In order to maintain normal operation of the system, a slight negative pressure is maintained at the top outlet of the riser during the test. However, due to the frequent fluctuation of pressure, the negative pressure cannot be set to a definite value accurately. Therefore, in actual operation, the power of the induced draft fan and blower is adjusted to maintain the pressure at the top outlet of the riser between −50 Pa and −200 Pa. When comparing the pressure distribution under different working conditions, this range error will significantly affect the comparability of different series of data. In order to enhance the comparability between the data of different working conditions, the pressure value of each height of the riser is different from the outlet pressure under the same working condition. By comparing this relative value, the influence of different experimental parameters on the system pressure balance is explored.

2.3.3. Solid Suspension Density

Solid suspension density is calculated by the following process.

According to the overall pressure drop formula of the riser:

$$\Delta P = \frac{mg}{A} \tag{1}$$

That is:

$$m = \frac{\Delta P \cdot A}{g} \tag{2}$$

Therefore, the solid suspension density is:

$$\rho = \frac{m}{V} = \frac{\Delta P}{g \cdot \Delta h} \tag{3}$$

where, ΔP is the overall pressure drop in the riser, in Pa; m is the amount of material in the riser, in kg; A is the cross-sectional area of the riser, in m^2; and Δh is the height difference between two test points.

2.3.4. Solid Circulation Rate

The importance of solid circulation rate is discussed in the introduction. The solid circulation rate can be calculated as follows:

$$G_s = \frac{\Delta m}{t * A} \tag{4}$$

where, G_s is the solid circulation rate, in $kg/(m^2 s)$; Δm is the accumulated mass of solid in the standpipe within a period of time, obtained by multiplying the change of stack height in the standpipe within a period of time by the standpipe area and bulk density of solid, in kg; t is time, in s; and A is the cross-sectional area of the riser, in m^2.

2.3.5. Solid Inventory Distribution Ratio

The solid inventory distribution ratio is the ratio of materials in the two risers and is used to describe the difference in the distribution of solids between two reactors during steady-state operation. Its formula is given by the following procedure.

According to Equation (2), the solid inventory distribution ratio is:

$$I = \frac{m_1}{m_2} = \frac{\Delta P_1 \cdot A_1}{\Delta P_2 \cdot A_2} \tag{5}$$

where, ΔP is the overall pressure drop in the riser, in Pa; m is the amount of material in the riser, in kg; A is the cross-sectional area of the riser, in m^2; I is the solid inventory distribution ratio; and the subscripts 1 and 2 represent the 1#FB and 2#FB, respectively.

3. Results and Discussion

3.1. The Effects of Different Parameters on the Distribution of Pressure along the Riser Height

Figures 4 and 5 show the distribution of pressure drop along the riser height relative to the outlet in each riser when 282-μm particles are running in a steady state. In the figure, the abscissa is the relative height, and the ordinate is the pressure drop. Generally speaking, the pressure in 1#FB and 2#FB reaches a relatively balanced state. Both the superficial gas velocity and the static bed height have a comprehensive influence on the pressure balance of the two risers. The pressure drop distribution in both risers presents the same rule. The pressure drop is very high at the bottom due to the mass accumulation of the materials. The pressure drop decreases sharply upon entering the dilute phase zone and subsequently decreases slowly with the increase in the riser height.

Within the dense phase zone, the superficial gas velocity of 1#FB has a positive effect on the pressure drop of 2#FB and yet has a negative effect on that of 1#FB. Figure 4 shows the effect of the superficial gas velocity of 1#FB on the distribution of the pressure drop along the riser height. In 1#FB, the pressure drop in the bottom dense phase zone is between 3000 and 5000 Pa, while that in the dilute phase zone is between 0 and 600 Pa. In 2#FB, the pressure drop in the bottom dense phase zone is between 500 and 3000 Pa, while that in the dilute phase zone is between 0 and 300 Pa. As can be seen in the figure, when the superficial gas velocity of 2#FB remains unchanged, the pressure drop in the bottom dense phase zone of 1#FB decreases with the increase in the superficial gas velocity of 1#FB. This is because the increase in the superficial gas velocity of 1#FB enhances its transporting capacity, which will transport a large number of materials gathered at the bottom to the dilute phase zone of 1#FB and whereafter reach 2#FB through circulation, resulting in the decrease of the pressure drop at the bottom dense phase zone of 1#FB. On the contrary, the pressure drop in the bottom dense phase zone of 2#FB increases with the increase in the superficial gas velocity of 1#FB, which is because of the accumulation of materials at the bottom of 2#FB. The relationship between the pressure drop in the dense phase zone of the two risers presents a reverse equilibrium. At the same time, when the superficial gas velocity of 1#FB increases, the pressure drop in the dilute phase zone of both beds increases slightly.

The static bed height has a positive effect on the pressure drop, which is more reflected in the dense phase region. Figure 5 shows the effect of the static bed height on the distribution of the pressure drop along the riser height when the superficial gas velocity of 1#FB and 2#FB is 4.5 m/s and 2.5 m/s, respectively. In 1#FB, the pressure drop in the bottom dense phase zone is between 3000 and 5500 Pa, while that in the dilute phase zone is between 0 and 1200 Pa. In 2#FB, the pressure drop in the bottom dense phase zone is between 2500–5500 Pa, while that in the dilute phase zone is between 0 and 300 Pa. As can be seen in the figure, when other conditions remain unchanged, the pressure in the two risers increases as a whole with the increase in the static bed height, which is caused by the gradually increasing number of materials in the riser. Additionally, it can be clearly seen in the figure that when the static bed height increases, the pressure drop at the bottom of 1#FB rises significantly, which implies that the height of the dense phase zone also increases. At the same time, the pressure drop at the bottom of 2#FB increased. However, the pressure drop in the dilute phase zone of 2#FB almost did not change. This can be explained by the low superficial gas velocity of 2#FB, which only transports a limited quantity of material to the dilute phase zone, leaving most of the material gathered in the bottom dense phase zone.

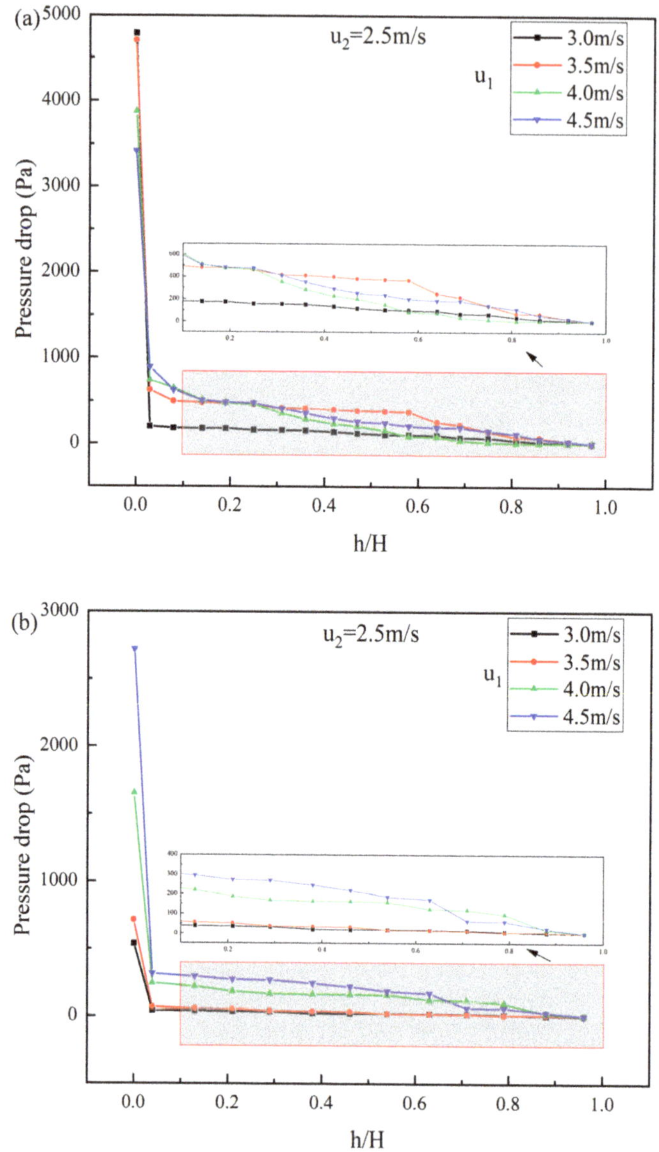

Figure 4. The effects of superficial gas velocity of 1#FB on the distribution of pressure drop along the riser height of (**a**) 1#FB and (**b**) 2#FB.

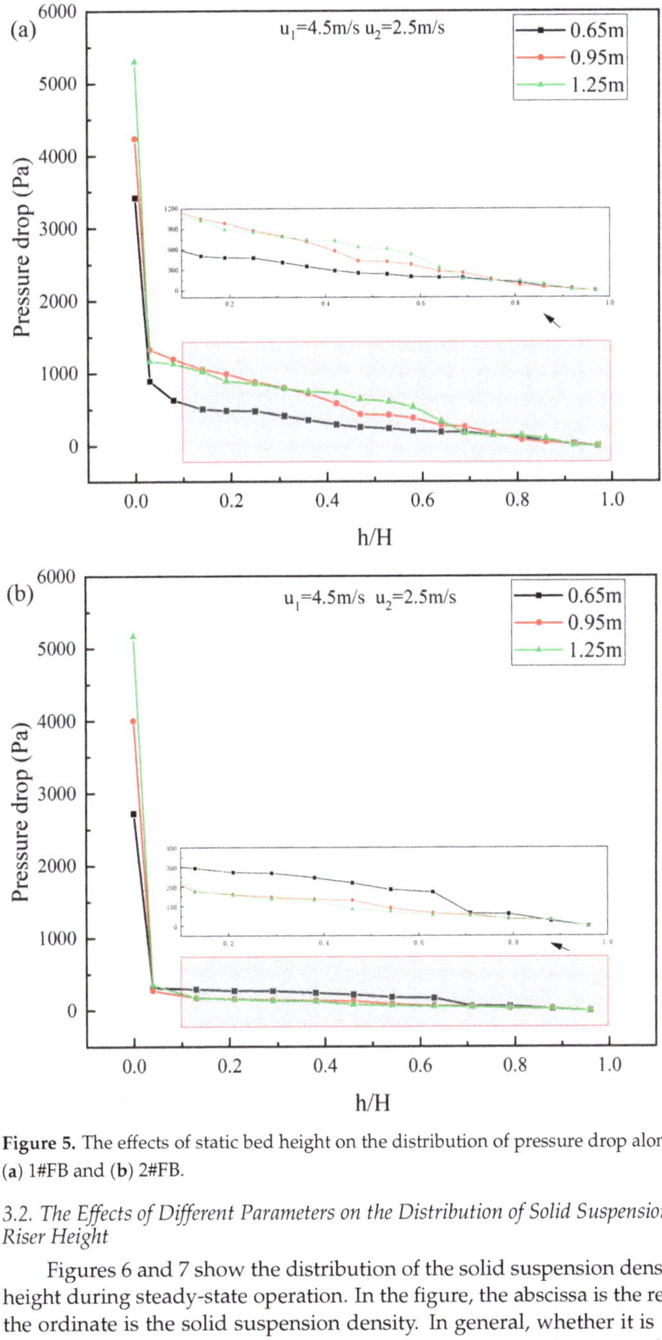

Figure 5. The effects of static bed height on the distribution of pressure drop along the riser height of (**a**) 1#FB and (**b**) 2#FB.

3.2. The Effects of Different Parameters on the Distribution of Solid Suspension Density along the Riser Height

Figures 6 and 7 show the distribution of the solid suspension density along the riser height during steady-state operation. In the figure, the abscissa is the relative height, and the ordinate is the solid suspension density. In general, whether it is 1#FB or 2#FB, the distribution of solid suspension density in the riser shows the same rule. Specifically, the solid suspension density at the bottom is very high, and the height of the dense phase zone is about one-tenth of the total height of the riser (h/H = 0.1). Upon entering the dilute phase zone, the solid suspension density decreases rapidly and remains relatively stable along the riser height. At the same time, there are some differences between the two risers. In terms of spatial arrangement, the bottom of 1#FB is 7.2 m lower than that of 2#FB. Therefore, when the superficial gas velocity of the two beds is close, the solid suspension density at the bottom of 1#FB is much larger than that of 2#FB. This difference can be reduced by increasing the superficial gas velocity of 1#FB or the static bed height.

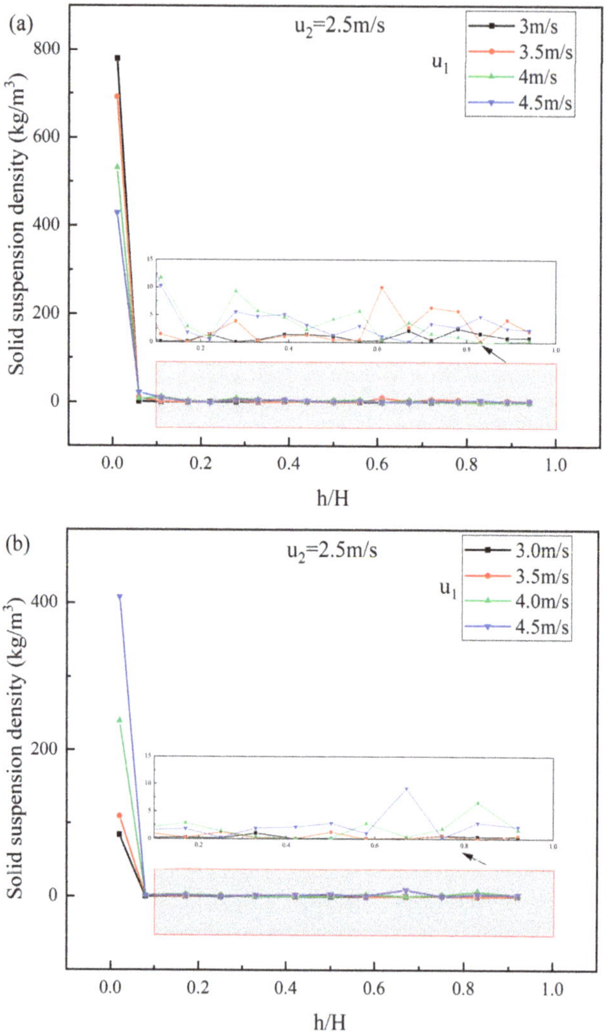

Figure 6. The effects of superficial gas velocity of 1#FB on the distribution of solid suspension density along the riser height of (**a**) 1#FB and (**b**) 2#FB.

Within the dense phase zone, the superficial gas velocity of 1#FB has a positive effect on the solid suspension density of 2#FB and yet has a negative effect on that of 1#FB. Figure 6 shows the effects of the superficial gas velocity of 1#FB on the distribution of solid suspension density along the riser height. In 1#FB, the solid suspension density in the bottom dense phase zone is roughly between 400–800 kg/m^3, while that in the dilute phase zone is between 0–15 kg/m^3. In 1#FB, with the increase in the superficial gas velocity of 1#FB, the solid suspension density in the dense phase zone decreases while the solid suspension density in the dilute phase zone increases slightly on the whole, and vice versa.

This is because the distribution of solids between the two risers reaches a certain degree of balance in steady operation. However, the increase in the superficial gas velocity of 1#FB breaks this balance, leading to the transfer of solids from 1#FB side to 2#FB side, and the decrease of solid suspension density in the bottom dense phase zone of 1#FB. Meanwhile, the increase in the superficial gas velocity improves the entraining effect on the solid, resulting in a slight increase in the solid suspension density in the dilute phase zone on the whole. In 2#FB, the solid suspension density at the bottom is roughly between 100 and 400 kg/m^3, while that in the dilute phase zone is between 0 and 20 kg/m^3. With the increase in the superficial gas velocity of 1#FB, the solid suspension density at the bottom increases and the solid suspension density in the dilute phase zone increases slightly on the whole, and vice versa. The reasons have been explained above.

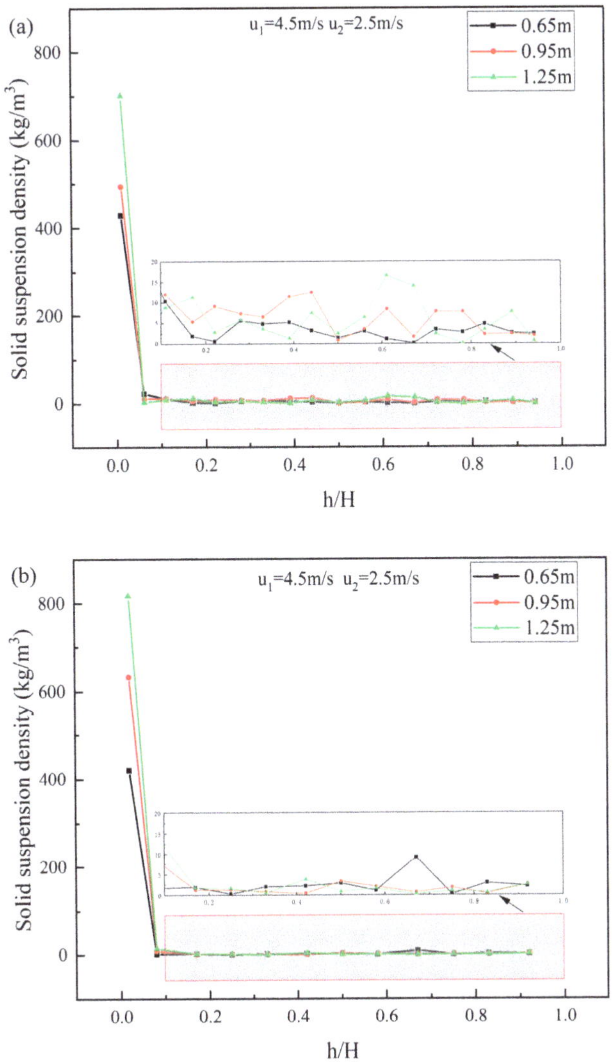

Figure 7. The effects of static bed height on the distribution of solid suspension density along the riser height of (**a**) 1#FB and (**b**) 2#FB.

Increasing the superficial gas velocity of 1#FB will cause the material balance to shift to the 2#FB side, which will enhance the heat and mass transfer of the 2#FB side. Meanwhile, the solid suspension density in the dilute phase zone of both risers is simultaneously increased and thus enhances the heat and mass transfer in the dilute phase zone. The effect of the superficial gas velocity of 2#FB on the material balance of two risers is opposite to that of 1#FB.

The static bed height has a positive effect on the solid suspension density, which is more reflected in the dense phase region. Figure 7 shows the effects of static bed height on the distribution of solid suspension density along the riser height when the superficial gas velocity of 1#FB and 2#FB is 4.5 m/s and 2.5 m/s, respectively. In 1#FB, the solid suspension density in the bottom dense phase zone is roughly between 400 and 800 kg/m^3, while that in the dilute phase zone is between 0 and 15 kg/m^3. In 2#FB, the solid suspension density in the bottom dense phase zone is roughly between 400 and 800 kg/m^3, while that in the dilute phase zone is between 0 and 20 kg/m^3. When the static bed height increases, more solids participate in the solid circulation loop in the reactor, which leads to the increase in the solid suspension density of the dense phase zone of the two risers, and thus intensifies the heat and mass transfer in this region. The influence of static bed height on the solid suspension density of the dilute phase zone is limited.

3.3. The Effects of Different Parameters on the Solid Circulation Rate

The solid circulation rate is positively correlated with the superficial gas velocity of both beds and yet negatively correlated with the particle size. Figure 8a shows the effects of the superficial gas velocity on the solid circulation rate when the static bed height is 0.65 m for 282-μm particles. In the figure, the abscissa is the superficial gas velocity of 1#FB, and the ordinate is the solid circulation rate. The two curves are the results when the superficial gas velocity of 2#FB is equal to 2 m/s and 2.5 m/s, respectively. As can be seen in the figure, for 282-μm particles, the solid circulation rate during steady operation is about 0.75–1.75 kg/(m^{2}s), which is positively correlated with the superficial gas velocity of both two beds. Therefore, increasing the superficial gas velocity of two risers is conducive to the solid circulation between the two risers, ensuring the heat and mass transfer between the two risers and improve the efficiency of the system. It is worth mentioning that when the superficial gas velocity of 1#FB and 2#FB is 2.5 m/s and 2 m/s, respectively, the solid circulation rate deviates significantly from the normal range, indicating that the solid circulation loop in the reactor is not normal at this time. In conclusion, for 282-μm particles, the system can operate normally only when the superficial gas velocity of 1#FB and 2#FB is greater than 2.5 m/s and 2 m/s, respectively.

Figure 8b shows the effects of the superficial gas velocity on the solid circulation rate when the static bed height is 0.65 m for 641-μm particles. The four curves in the figure are the results when the superficial gas velocity of 2#FB is equal to 2, 2.5, 3 and 3.5 m/s respectively. It can be seen in the figure that for 641-μm particles, the solid circulation rate during steady-state operation is about 0.02–0.16 kg/(m^{2}s), and its correlation with the superficial gas velocity is the same as that of 282-μm particles. When the superficial gas velocity of 1#FB and 2#FB is 3 m/s and 2 m/s respectively, it is hard to obtain the solid circulation rate during the experiment, which indicates that the reactor is not in a normal working state at this time. Apparently, the minimum superficial gas velocity required for normal operation is higher than 282-μm particles (2.5 m/s in 1#FB and 2 m/s in 2#FB).

Comparing the solid circulation rate of two types of particles in steady-state operation, it can be seen in Figure 8c that the increase in particle size greatly improves the difficulty of airflow carrying solid particles, which makes the solid circulation rate decrease by one order of magnitude, which seriously affects the solid circulation between the two risers. Therefore, the particle size should be reasonably low to enhance the efficiency of the system.

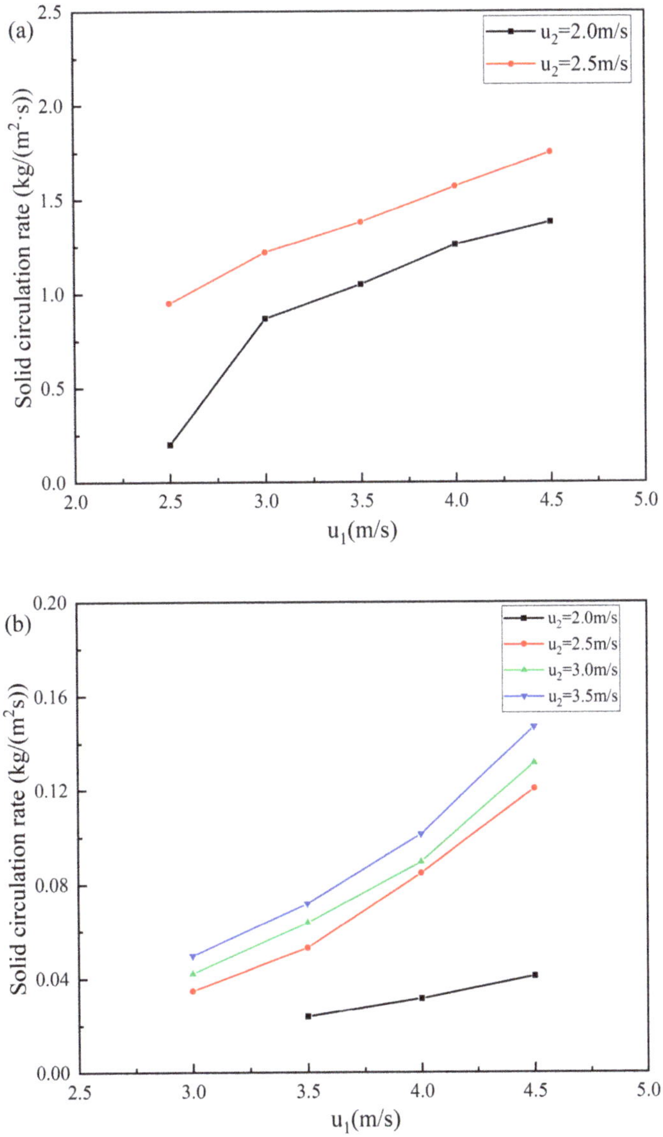

Figure 8. *Cont.*

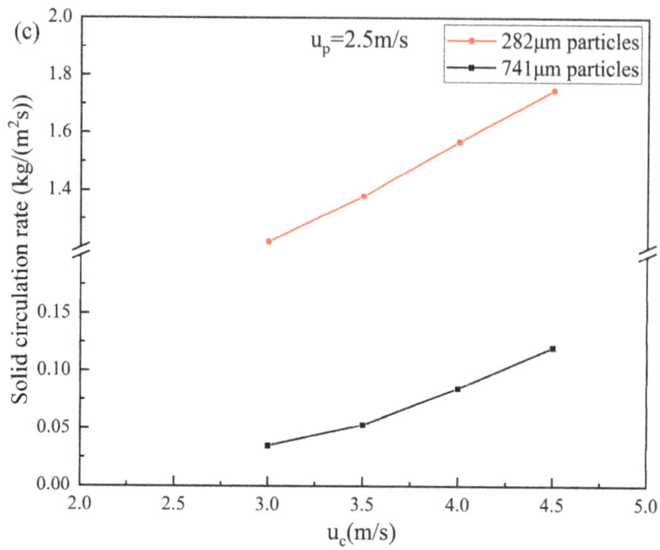

Figure 8. The effects of the superficial gas velocity on the solid circulation rate: (**a**) 282-μm particles; (**b**) 641-μm particles; (**c**) Comparison.

The static bed height has a positive influence on the solid circulation rate. The solid circulation rate is about 0.87–1.75, 1.04–3.04, and 1.13–3.69 kg/(m^2s) when the static bed height is 0.65, 0.95, and 1.25 m, respectively. Figure 9 shows the effects of the static bed height on the solid circulation rate under different superficial gas velocity settings for 282-μm particles. In the figure, the horizontal coordinate is the static bed height, and the vertical coordinate is the solid circulation rate. The three curves are the results of three different superficial gas velocity settings. As can be seen in the figure, the solid circulation rate under the three steady-state operating conditions is about 1–4 kg/(m^2s) and is positively correlated with the static bed height. Two more results can be observed in the figure. Firstly, when the static bed height is equal to 0.65 m, the effect of the increase in the superficial gas velocity on the solid circulation rate of the two risers is much lower than when the static bed height is equal to 0.95 m or 1.25 m. This shows that the influence of the superficial gas velocity has a marginally decreasing effect on the solid circulation rate. When the superficial gas velocity reaches a certain degree, the static bed height becomes the main factor restricting the solid circulation rate. Secondly, when the superficial gas velocity of 1#FB and 2#FB is equal to 3 m/s and 2 m/s, respectively, the effect of the increase in the static bed height on the solid circulation rate is much lower than that of the other two superficial gas velocity settings. This indicates that the static bed height also has a marginally decreasing effect. When the static bed height continuously increases until the amount of circulating solid in the reactor reaches the maximum transporting capacity of the given superficial velocity, the solid circulating rate of the reactor reaches its limit. Therefore, during the actual operation in production, it is necessary to fully consider the relationship between the static bed height (reflecting the total amount of materials in the reactor) and the superficial gas velocity of the two risers (reflecting the transporting capacity), so as to effectively increase the solid circulation rate of the reactor, consequently improving the heat and mass transfer effect and in the end improve the efficiency of the reactor.

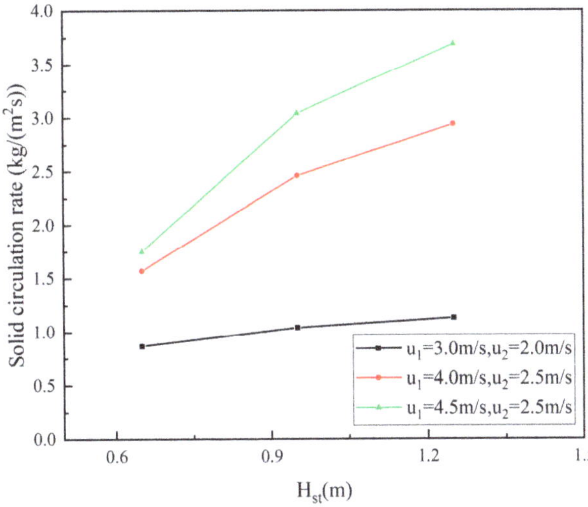

Figure 9. The effects of the static bed height on the solid circulation rate.

Figure 10 shows the fitting between the solid circulation rate of 282-μm particles and the superficial gas velocity of 1#FB, the superficial gas velocity of 2#FB, and the static bed height. During fitting, the parameters in the fitting formula were nondimensionalized based on the situation that the static bed height is 0.65 m, the superficial gas velocity of 1#FB is 2.5 m/s and the superficial gas velocity of 2#FB is 2 m/s when the solid circulation rate is 0.2 kg/(m^{2}s). Since the three parameters comprehensively affect the solid circulation rate and their effects all have marginally diminishing effect, the following form is adopted for fitting:

$$\frac{G_s}{0.2} = a\left(\ln\left(\frac{H_{st}}{0.65}\right)+1\right)\left(\ln\left(\frac{u_1}{2.5}\right)+\ln\left(\frac{u_2}{2}\right)\right)+1 \tag{6}$$

where, G_s is the solid circulation rate, in kg/(m^{2}s); a is a constant defined after fitting; H_{st} is the static bed height in m; u_1 is the superficial gas velocity of 1#FB, in m/s; and u_2 is the superficial gas velocity of 2#FB, in m/s.

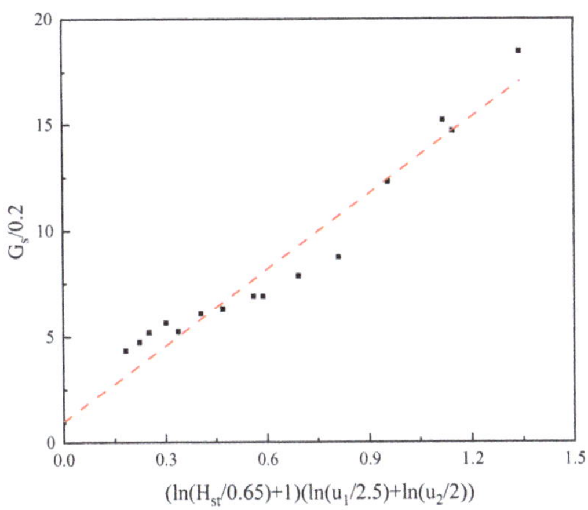

Figure 10. The solid circulation rate versus operating parameters.

Table 5 shows the specific values of the solid circulation rate under different parameters and the processing results according to the above formula. After fitting using the above data, a = 11.81 is obtained, and the empirical equation is as follows:

$$G_s = 2.362\left(\ln\left(\frac{H_{st}}{0.65}\right) + 1\right)\left(\ln\left(\frac{u_1}{2.5}\right) + \ln\left(\frac{u_2}{2}\right)\right) + 0.2,\ R^2 = 0.9816 \quad (7)$$

Table 5. The specific values of the solid circulation rate under different parameters and the processing results.

H_{st} (m)	u_1 (m/s)	u_2 (m/s)	G_s (kg/m^{2}s)	ln (H_{st}/0.65) + 1	ln (u_1/2.5)	ln (u_2/2)	G_s/0.2
0.65	2.50	2.00	0.20	1.00	0.00	0.00	1.00
0.65	3.00	2.00	0.87	1.00	0.18	0.18	4.35
0.65	3.50	2.00	1.05	1.00	0.34	0.34	5.25
0.65	4.00	2.00	1.26	1.00	0.47	0.47	6.30
0.65	4.50	2.00	1.38	1.00	0.59	0.59	6.90
0.65	2.50	2.50	0.95	1.00	0.22	0.22	4.75
0.65	3.00	2.50	1.22	1.00	0.41	0.41	6.10
0.65	3.50	2.50	1.38	1.00	0.56	0.56	6.90
0.65	4.00	2.50	1.57	1.00	0.69	0.69	7.85
0.65	4.50	2.50	1.75	1.00	0.81	0.81	8.75
0.95	3.00	2.00	1.04	1.38	0.25	0.18	5.20
0.95	4.00	2.50	2.46	1.38	0.96	0.69	12.30
0.95	4.50	2.50	3.04	1.38	1.12	0.81	15.20
1.25	3.00	2.00	1.13	1.65	0.30	0.18	5.65
1.25	4.00	2.50	2.94	1.65	1.15	0.69	14.70
1.25	4.50	2.50	3.69	1.65	1.34	0.81	18.45

In this work, the solid circulation rate is about 0.87–1.75, 1.04–3.04, and 1.13–3.69 kg/(m^{2}s) when the static bed height is 0.65, 0.95, and 1.25 m respectively. The results of different works are compared in Table 6. It can be easily seen that both the particle diameter and the maximum height of the riser have a significant influence on the solid circulation rate.

Table 6. Results of different work.

Researchers	Maximum Height of Riser (m)	Particle Diameter (μm)	G_s (kg/m^{2}s)
Tobias Pröll et al. [24]	1.80	54/161	40–60
Saurabh Gupta et al. [9]	2.25	322	6–24
This work	21.60	282	0.87–3.69

3.4. *The Effects of Different Parameters on the Solid Inventory Distribution Ratio*

Figure 11 shows the fitting between the solid inventory distribution ratio of 282-μm particles and the superficial gas velocity of 1#FB, the superficial gas velocity of 2#FB, and the static bed height. The following form is used for fitting [15]:

$$I = a\left(\frac{u_2}{u_1}\right)^{\frac{bH_{st}}{H}} \quad (8)$$

where, I is the solid inventory distribution ratio; u_1 is the superficial gas velocity of 1#FB, in m/s; u_2 is the superficial gas velocity of 2#FB, in m/s; H_{st} is the static bed height, in m; H is the height of 1#FB, which is 21.6 m; and a and b are constants defined after fitting.

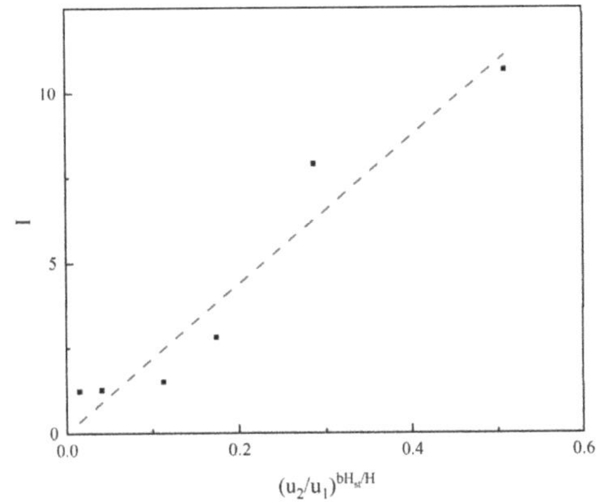

Figure 11. The solid inventory distribution ratio I versus operating parameters.

Table 7 shows the specific values of the solid inventory distribution ratio under different parameters and the processing results according to the above formula. After fitting using the above data, a = 21.85 and b = 123.38 are obtained by fitting, and the R^2 of the empirical equation was 0.9698.

Table 7. The specific values of the solid inventory distribution ratio under different parameters and the processing results.

u_1 (m/s)	u_2 (m/s)	H_{st} (m)	u_2/u_1	H_{st}/H	I
3.00	2.50	0.65	0.83	0.030	10.69
3.50	2.50	0.65	0.71	0.030	7.90
4.00	2.50	0.65	0.63	0.030	2.81
4.50	2.50	0.65	0.56	0.030	1.51
4.50	2.50	0.95	0.56	0.044	1.27
4.50	2.50	1.25	0.56	0.058	1.23

It can be concluded from this empirical equation that the higher the superficial gas velocity of 1#FB or the higher the static bed height, the smaller the ratio (the closer it is to one). This can be explained in the following two ways. On the one hand, as mentioned above, when other conditions remain unchanged, the increase in the superficial gas velocity of 1#FB makes the material in the 1#FB side transfer to the 2#FB side, which apparently makes the ratio smaller. On the other hand, when other conditions remain unchanged, the higher the static bed height is, with the absolute difference of material amount between the two risers almost unchanged, the smaller the relative difference is, which makes the ratio smaller. Therefore, the material balance between the two risers can be adjusted by reasonably setting the superficial gas velocity of the two risers and the static bed height.

4. Conclusions

In this paper, a DFB cold mode test system with a maximum height of 21.6 m was built independently and relevant experiments were carried out. The results show that the system can operate normally and stably. The research results in this paper can provide a reference for the design and operation of large-scale DFB reactor systems. Relevant conclusions are summarized as follows:

1. The pressure in the two risers reaches a relatively balanced state, with the superficial gas velocity and the static bed height having a comprehensive influence on the pressure balance of the two risers. The increase in the superficial gas velocity of 1#FB

will decrease the bottom pressure of 1#FB and increase that of 2#FB, and also have a positive effect on the pressure of the dilute phase zone of the two risers. The increase in the static bed height can significantly increase the pressure in the risers.
2. Increasing the superficial gas velocity on one side will cause the material balance to shift to the other side and enhance the heat and mass transfer in the other side. Meanwhile, the solid suspension density in the dilute phase zone of the two risers is increased to enhance the heat and mass transfer in the dilute phase zone. With the increase in the static bed height, the solid suspension density of the dense phase zone increases, which intensifies the heat and mass transfer in this zone. The influence of static bed height on the solid suspension density of the dilute phase zone is limited.
3. Increasing the superficial gas velocity of the two risers or the static bed height is helpful to promote the material circulation between the two risers, so as to ensure heat and mass transfer between the two risers and improve the efficiency of the system and yet the effects of both have diminishing, marginal effect. The particle size has a significant effect on the solid circulation rate. Different particle sizes require different minimum operating conditions. The empirical equation of solid circulation rate proposed in this paper can provide a reference for production operations.
4. The empirical equation of the solid inventory distribution ratio proposed in this paper shows that the material balance between two risers can be adjusted by reasonably setting the superficial gas velocity of the two risers and the static bed height.

Author Contributions: Conceptualization, Q.W. and H.F.; Methodology, Y.L., Q.W. and C.Y.; Investigation, Y.L. and C.Y.; Writing—original draft, Y.L.; Writing—review & editing, Y.Z. All authors have read and agreed to the published version of the manuscript.

Funding: This work was supporting by the Chinese National Natural Science Fund [No.52076138] and the Fundamental Research Funds for the Central Universities [2022ZFJH004].

Data Availability Statement: The data presented in this study are available on request from the corresponding author. The data are not publicly available due to privacy.

Conflicts of Interest: The authors declare that they have no known competing financial interests or personal relationships that could have appeared to influence the work reported in this paper.

Nomenclature

Symbols

Δm	accumulated mass of solid in the standpipe within a period of time (kg)
ΔP	overall pressure drop in the riser (Pa)
A	cross-sectional area of the riser (m^2)
G_s	solid circulation rate (kg/m^2s)
g	gravity constant (m/s^2)
H_{st}	static bed height (m)
I	solid inventory distribution ratio (-)
m	solid mass in riser (kg)
t	time (s)
u	superficial gas velocity (m/s)
ρ	solid suspension density (kg/m^3)

Subscripts

1	1#FB
2	2#FB

Abbreviations

1#FB	1# fluidized bed
2#FB	2# fluidized bed

References

1. Hanchate, N.; Ramani, S.; Mathpati, C.S.; Dalvi, V.H. Biomass gasification using dual fluidized bed gasification systems: A review. *J. Clean. Prod.* **2021**, *280*, 123148. [CrossRef]
2. Ghasem, N.M.; Ang, W.L.; Hussain, M.A. Dynamics and stability of ethylene polymerization in multizone circulating reactors. *Korean J. Chem. Eng.* **2009**, *26*, 603–611. [CrossRef]
3. Daneshmand-Jahromi, S.; Sedghkerdar, M.H.; Mahinpey, N. A review of chemical looping combustion technology: Fundamentals, and development of natural, industrial waste, and synthetic oxygen carriers. *Fuel* **2023**, *341*, 127626. [CrossRef]
4. McLaughlin, H.; Littlefield, A.A.; Menefee, M.; Kinzer, A.; Hull, T.; Sovacool, B.K.; Bazilian, M.D.; Kim, J.; Griffiths, S. Carbon capture utilization and storage in review: Sociotechnical implications for a carbon reliant world. *Renew. Sustain. Energy Rev.* **2023**, *177*, 113215. [CrossRef]
5. Schmid, J.C.; Benedikt, F.; Fuchs, J.; Mauerhofer, A.M.; Müller, S.; Hofbauer, H. Syngas for biorefineries from thermochemical gasification of lignocellulosic fuels and residues—5 years' experience with an advanced dual fluidized bed gasifier design. *Biomass Convers. Biorefinery* **2021**, *11*, 2405–2442. [CrossRef]
6. Pröll, T.; Rupanovits, K.; Kolbitsch, P.; Bolhàr-Nordenkampf, J.; Hofbauer, H. Cold Flow Model Study on a Dual Circulating Fluidized Bed (DCFB) System for Chemical Looping Processes. *Chem. Eng. Technol.* **2009**, *32*, 418–424. [CrossRef]
7. Zhou, C.; Chen, L.; Liu, C.; Wang, J.; Xing, X.; Liu, Y.; Chen, Y.; Chao, L.; Dai, J.; Zhang, Y.; et al. Interconnected pyrolysis and gasification of typical biomass in a novel dual fluidized bed. *Energy Convers. Manag.* **2022**, *271*, 116323. [CrossRef]
8. Charitos, A.; Hawthorne, C.; Bidwe, A.R.; Korovesis, L.; Schuster, A.; Scheffknecht, G. Hydrodynamic analysis of a 10 kWth Calcium Looping Dual Fluidized Bed for post-combustion CO_2 capture. *Powder Technol.* **2010**, *200*, 117–127. [CrossRef]
9. Gupta, S.; De, S. Investigation of cold flow hydrodynamics in a dual fluidized bed for gasification of high-ash coal. *Powder Technol.* **2021**, *384*, 564–574. [CrossRef]
10. Liu, Y.; Huo, P.; Li, X.; Qi, H. Numerical study of coal gasification in a dual-CFB Plant based on the generalized drag model QC-EMMS. *Fuel Process. Technol.* **2020**, *203*, 106363. [CrossRef]
11. Wei, Y.; Cheng, L.; Li, L. An analysis approach of mass and energy balance in a dual-reactor circulating fluidized bed system. *Chin. J. Chem. Eng.* **2021**, *40*, 18–26. [CrossRef]
12. Ohlemüller, P.; Alobaid, F.; Gunnarsson, A.; Ströhle, J.; Epple, B. Development of a process model for coal chemical looping combustion and validation against 100 kWth tests. *Appl. Energy* **2015**, *157*, 433–448. [CrossRef]
13. Ohlemüller, P.; Alobaid, F.; Abad, A.; Adanez, J.; Ströhle, J.; Epple, B. Development and validation of a 1D process model with autothermal operation of a 1 MW th chemical looping pilot plant. *Int. J. Greenh. Gas Control* **2018**, *73*, 29–41. [CrossRef]
14. Sahir, A.H.; Dansie, J.K.; Cadore, A.L.; Lighty, J.S. A comparative process study of chemical-looping combustion (CLC) and chemical-looping with oxygen uncoupling (CLOU) for solid fuels. *Int. J. Greenh. Gas Control* **2014**, *22*, 237–243. [CrossRef]
15. Wei, Y.; Cheng, L.; Wu, E.; Li, L. Experimental research on steady-state operation characteristics of gas–solid flow in a 15.5 m dual circulating fluidized bed system. *Chin. J. Chem. Eng.* **2021**, *32*, 70–76. [CrossRef]
16. Han, C.; Hu, L.; Song, T.; Zhang, Y.; Lyu, J.; Zhang, H.; Liu, Q.; Ma, S. Effect of bed material size on gas-solid flow characteristics in a CFB at low solid recirculation rates. *Fuel* **2023**, *333*, 126354. [CrossRef]
17. Hu, J.; Liu, D.; Li, H.; Liang, C.; Chen, X. Experimental study of the solid circulation rate in a pressurized circulating fluidized bed. *Particuology* **2021**, *56*, 207–214. [CrossRef]
18. Rahman, M.H.; Bi, X.T.; Grace, J.R.; Lim, C.J. Comparison of techniques for measuring CFB solids circulation rates at low and high temperatures. *Powder Technol.* **2020**, *360*, 43–54. [CrossRef]
19. Hu, J.; Liu, D.; Liang, C.; Ma, J.; Chen, X.; Zhang, T. Solids flow characteristics and circulation rate in an internally circulating fluidized bed. *Particuology* **2021**, *54*, 69–77. [CrossRef]
20. Armatsombat, A. An investigation of the aeration proportion effect at the supply and recycle chambers on the optimal operation point of the U-type loop seal and the maximum solid circulation rate in a fluidized bed reactor. *Adv. Powder Technol.* **2023**, *34*, 104196. [CrossRef]
21. Varghese, M.M.; Vakamalla, T.R.; Gujjula, R.; Mangadoddy, N. Prediction of solid circulation rate in an internal circulating fluidized bed: An empirical and ANN approach. *Flow Meas. Instrum.* **2022**, *88*, 102274. [CrossRef]
22. Kim, D.; Won, Y.; Hwang, B.W.; Kim, J.Y.; Kim, H.; Choi, Y.; Lee, Y.-R.; Lee, S.-Y.; Jo, S.-H.; Park, Y.C.; et al. Loop-seal flow characteristics of a circulating fluidized bed for 3 MWth scale chemical looping combustion system. *Energy* **2023**, *274*, 127271. [CrossRef]
23. Geldart, D. Types of gas fluidization. *Powder Technol.* **1973**, *7*, 285–292. [CrossRef]
24. Karl, J.; Pröll, T. Steam gasification of biomass in dual fluidized bed gasifiers: A review. *Renew. Sustain. Energy Rev.* **2018**, *98*, 64–78. [CrossRef]

energies

Article

Analysis of the Influence Factors of the Crude Oil Temperature Maintenance System of Solar Sewage Heat Pumps in Cold Regions

Shangwen Zhu [1], Xiaohua Gu [1,2,*], Mengyi Duan [1], Feiyang Yu [1], Danyi Zhao [1], Siwen Liu [3] and Yan Liu [1,*]

[1] School of Energy and Building Environment, Guilin University of Aerospace Technology, Guilin 541004, China; glzsw2337899@163.com (S.Z.); 15517977333@163.com (M.D.); 17773084776@163.com (F.Y.); zdy20020326@163.com (D.Z.)

[2] State Key Laboratory for Modification of Chemical Fibers and Polymer Materials, College of Materials Science and Engineering, Donghua University, Shanghai 200051, China

[3] College of Innovative Material & Energy, Hubei University, Wuhan 430062, China; 202121113012770@stu.hubu.edu.cn

* Correspondence: 2022043@guat.edu.cn (X.G.); 2019009@guat.edu.cn (Y.L.); Tel.: +86-18121138868 (X.G.); +86-18078343019 (Y.L.)

Abstract: Traditional crude oil heating methods that use fossil fuels or electricity have the disadvantages of high consumption of nonrenewable resources, low energy utilization, and high carbon emissions. Therefore, it is urgent to develop green and sustainable crude oil heating technologies. In this paper, a solar synergistic sewage heat pump (SSHS) dual heat source crude oil temperature maintenance system is proposed. The system utilizes clean and sustainable solar energy to heat crude oil while combining sensible heat storage technology and the waste heat utilization technology of a sewage source heat pump to solve the unstable fluctuation of the solar heating problem. A simulation and analysis model is established to analyze the influencing factors of the SSHS, and the optimal operation scheme is provided. The results show that the efficiency of the solar collector decreases and the proportion of crude oil heating increases with an increase in the solar energy guarantee rate, while the unit flow rate of the pump has a large impact on the performance of the sewage source heat pump. In order to avoid energy waste, it is more appropriate to adopt a 30% guarantee rate and an A3 pump unit flow rate, under which the solar collector efficiency is 50.18%, the proportion of solar heating of crude oil is 47.16%, the average temperature of crude oil is 42.59 °C, and the COP of the sewage source heat pump is 4.65. Further increases in the COP of the wastewater source heat pump can be realized by increasing the temperature of the wastewater supply. The results of this study provide a valuable reference for the optimization of crude oil storage heating systems.

Keywords: solar energy; sewage source heat pump; crude oil; TRNSYS; system operating characteristics

Citation: Zhu, S.; Gu, X.; Duan, M.; Yu, F.; Zhao, D.; Liu, S.; Liu, Y. Analysis of the Influence Factors of the Crude Oil Temperature Maintenance System of Solar Sewage Heat Pumps in Cold Regions. *Energies* **2023**, *16*, 8124. https://doi.org/10.3390/en16248124

Academic Editor: Philippe Leclère

Received: 13 November 2023
Revised: 13 December 2023
Accepted: 17 December 2023
Published: 18 December 2023

1. Introduction

With the rapid development of the economy and technology, the demand for crude oil as an essential building block for development has been increasing annually [1]. This sustained growth in demand for crude oil has prompted, for safety and security, the control of oil reserves to become a concern for global economic development. Large floating roof tanks have become important examples of oil storage equipment due to their structural and economic advantages [2]. At the same time, the global crude oil production regions are widespread and span a large area, with some oil fields being located in severely cold regions [3].

In cold regions, the temperature of the crude oil stored in large floating roof oil tanks gradually decreases through the tank roof, tank wall, and tank bottom soil to the outside world and via heat dissipation [4,5]. When the temperature of crude oil falls to freezing point, the crude oil undergoes the gelation phenomenon and gelatinized crude oil is attached to the edge of the tank body of the large floating roof tanks, thus forming a certain thickness of a solid condensate layer [6,7]. This results in increased flow resistance and even the accumulation of condensate, which blocks the import and export of the oil, thus affecting the delivery and receipt of the oil and other turnover operations.

Crude oil heating [8,9] is a necessary measure through which to prevent accidents during the storage and transportation of crude oil in large floating roof tanks. However, traditional boiler heating and electric heating have problems such as high energy consumption and carbon emissions, and air source heat pumps [10] and ground source heat pump [11] heating have limitations in the use of the environment; as such, the development of a new type of heat source supply is imminent. Solar energy [12,13] is a green and low-carbon energy source, and the utilization of solar energy as a heat source for maintaining the temperature of crude oil in large floating roof tanks can reduce primary energy consumption, greenhouse gas emissions, and environmental pollution. Notably, in China's oil-rich Heilongjiang and Neimenggu (which, respectively, belong to cold regions), there are abundant solar energy resources for the development of floating roof oil tanks, solar phase-change heat storage, and thermoregulation technology. Previous research on solar collectors has focused on their application in industrial production processes, and relatively little research has been carried out on the potential of solar energy in heating crude oil. However, its methodology can provide theoretical guidance for this study.

In the field of solar energy industrial applications, previous research efforts [14,15] have begun to explore its potential applications in depth. Mohammadi et al. [16] conducted a comprehensive technical and economic analysis for a solar industrial process heat (SIPH) plant in Salt Lake City, Utah. The results showed that the use of parabolic trough collectors can significantly reduce operating costs and greenhouse gas emissions compared to conventional natural, gas-fired power plants. On the other hand, certain scholars [17,18] have optimized the design of a solar industrial heating system to significantly reduce the annual lifecycle cost in order to generate a low-to-medium process heat so as to meet the heat demand of the industrial process while minimizing the dependence on fossil fuels.

As far as the solar heating of crude oil is concerned, researchers such as Wang [19] assessed the potential for applying solar energy in the global oil business via covering a wide range of segments such as extraction, transportation, and refining. It was found that the potential demand for solar photovoltaic (PV) and solar thermal power in this sector ranges from 17 to 95 gigawatts (GW) and 21 to 95 GW, respectively. These estimates clearly reveal the huge future demand for solar energy in the oil and gas industry. The demand for solar energy from the oil and gas industry is expected to continue to rise over the next two decades. It is predicted that solar energy will contribute about 2 quintillion joules (PJ) to the industry by 2035, which is about 5% of the overall industry's energy demand [20]. Altayib et al. [21] used a solar heating system to replace 10% of the heat required for preheating crude oil before processing. The overall energy efficiency and power generation efficiency of the system reached 60.94% and 19.34%, respectively. It was also found that the solar unit and heat exchanger were the major components contributing to the energy consumption in the system. Naseer Ahmad Khan et al. [22] simulated the solar synergistic TES heating system for the heating of crude oil through transient simulation, and the results of the study showed that it could provide a cost saving of USD 21.046 million over the maximum lifecycle and 116,944 MWh of heat per year; additionally, GHG emissions could be reduced by approximately 34.045 tons of CO_2 equivalent per year.

In the field of crude oil storage, the application of solar energy is different from other industrial sectors. This is due to the fact that the demand for thermal energy was continuous during the period of static storages of crude oil. Therefore, solar heating systems have the potential to meet thermal energy demand during crude oil storage under conditions of continuous and sufficient solar radiation. However, the intensity of solar radiation is characterized by periodic fluctuations due to natural factors such as the climate, geography, weather, and diurnal variations [21]. This leads to the possibility of overheating or underheating of the heating temperature of the solar heating system.

In order to solve this problem, the main methods include the use of solar photovoltaic technology combined with thermal storage [23], or the utilization of other heating technologies [24,25]. By these means, the temperature of crude oil can be stabilized over a longer period of time, thus compensating for the heating temperature instability that may be caused by fluctuations in the intensity of solar radiation. This innovative approach is expected to provide a more reliable solution for the application of solar energy in crude oil storage. The solar thermal storage technologies of sensible heat storage [26], latent heat storage (phase change storage) [27,28], and thermochemical energy storage [29] are three important methods for solar thermal utilization and peak shifting [30,31]. Among them, sensible heat storage is the most widely used and has a strong application basis [26,32], but the unit energy storage density of sensible heat storage is slightly insufficient compared with that of phase-change energy storage [33,34]. As such, combination with other heat sources is necessary [35,36] for auxiliary heat supply. At the same time, considering the large amount of low and medium temperature wastewater to be treated within the oil field operation area, the sewage source heat pump is of great significance as an auxiliary heat source for temperature-controlled solar systems of crude oil static storage. This auxiliary heat source can be used to realize multistage recycling of waste heat. This technology not only reduces transportation and handling costs, but also effectively reduces heat loss. There are also a few scholars [37] who introduced the possibility of using a solar- sewage heat pump system for heating, but they did not provide a detailed analysis of its application and influencing factors.

However, it is worth noting that the aforementioned studies mainly focused on the extraction, transportation, and refining processes of crude oil. During static storage of crude oil, conventional natural gas heating or electric heating is still widely used to assist solar energy in maintaining crude oil temperature. Currently, there is relatively little research on using a solar sewage source heat pump system to maintain static crude oil storage temperature; as such, a further exploration and in-depth investigation of its influencing factors are needed. Such research will help reveal the potential for the wider applications of solar energy in the petroleum industry, thereby leading to the advancement of energy efficiency and environmental sustainability.

In this study, a new crude oil temperature maintenance system—the maintenance system of crude oil temperature using dual heat sources with solar synergistic sewage heat pumps (SSHS)—is proposed; in addition, a system simulation model is established based on the TRNSYS platform, and its accuracy is verified through experiments. The annual operating characteristics of the SSHS for large floating top tanks were analyzed with the COP of the sewage source heat pump, the heating demand of the crude oil storage tank, the proportion of crude oil heated using solar energy, the solar collector efficiency, and the crude oil temperature as evaluation indexes. The main objective of the research results is to provide a reference basis for the application of solar thermal utilization technology combined with sewage source heat pump technology in the field of crude oil heating and storage, as well as to provide new ideas for industrial heating, which is of great significance in promoting the green and sustainable development of industrial heating technology.

2. Model Description and Numerical Simulation

To clarify the operation mechanism of the SSHS on crude oil heating, a simulation and modeling platform for the SSHS was established, and its physical and mathematical models are described here.

2.1. Physical Model

Based on a 10×10^4 m^3 floating roof oil tank in Daqing, China, a solar phase-change temperature maintenance system was established as an example. Figure 1 shows the physical model of the SSHS.

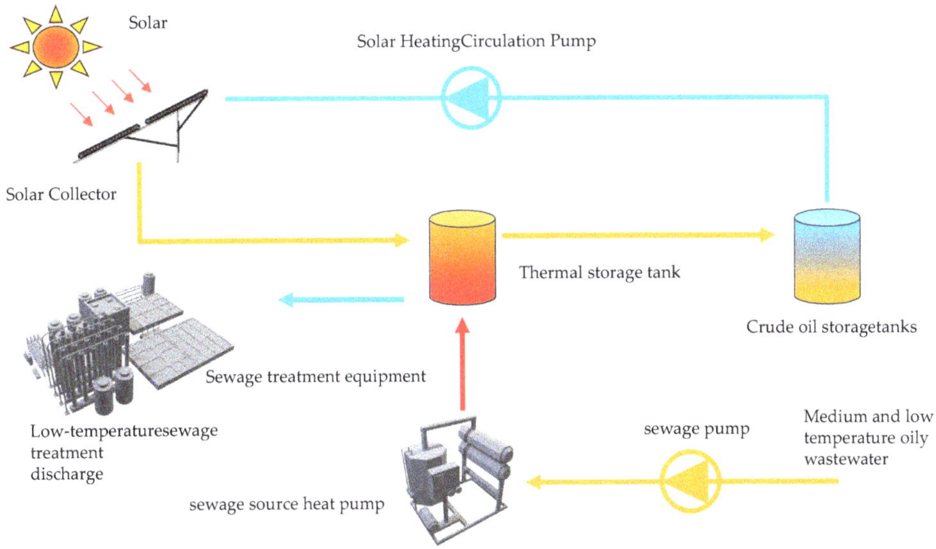

Figure 1. Physical model of the SSHS.

The SSHS designed in this study was divided into two main cycle sections: the solar temperature maintenance heating cycle and the sewage-assisted heating cycle. The purpose of the solar temperature maintenance heating cycle is to meet the heating demand of the floating roof tank to maintain the temperature of the static storage crude oil. The main components required for the cycle are solar collectors, sensible heat storage tanks, floating roof tanks, and solar heating circulation pumps. The specific process of the cycle is as follows: the solar radiation absorbed by the solar collector heats the heat transfer medium; the heated heat transfer medium flows into the sensible heat storage tank, where the control system judges whether or not to carry out the heat storage process; the floating top tank enters to heat the crude oil; and this then finally returns to the solar collector for the next cycle. The sewage-assisted heating cycle includes sewage pumps, sewage source heat pumps, and sewage treatment devices. When the heat generated using solar energy cannot meet the heating demand of the crude oil, the sewage source heat pump is turned on to utilize the waste heat of the medium- and low-temperature sewage to heat the crude oil in the floating top tank. This strategy maximizes the ability to ensure that the crude oil temperature is always above its freezing point temperature.

2.2. Mathematical Model

2.2.1. Design Parameter Calculation Model

To analyze the system operation characteristics of an SSHS, the following assumptions are proposed [38]: (1) Air is an ideal gas with a constant specific heat. (2) The change in kinetic and potential energy between different components is negligible. (3) Both the heat storage fluid and the heat transfer fluid are incompressible and isotropic. Therefore, based on the above modeling assumptions, the mathematical equation [39,40] of an SSHS for the static storage heating demand of crude oil can be expressed as follows:

$$Q_{oil} = \frac{(K_{wall}A_{wall} + K_{top}A_{top} + K_{bottom}A_{bottom})\Delta T}{1000} \tag{1}$$

where Q_{oil}, K_{wall}, K_{top}, K_{bottom}, A_{wall}, A_{top}, A_{bottom}, and ΔT are the crude oil static storage's thermal load (kW), thermal transfer coefficient of the wall, the top and bottom of the floating roof tank (W/(m·°C)), the area of the wall, the top and bottom of the floating roof tank (m^2), and the difference between the temperature of the environment and the maintenance temperature of the crude oil for static storage (°C), respectively.

The thermal transfer coefficient is calculated as follows:

$$K_{top} = \sum \frac{1}{\frac{1}{\alpha_{oil}} + \frac{\delta_i}{\lambda i} + \frac{1}{h_{top}}} \tag{2}$$

$$K_{wall} = \sum \frac{1}{\frac{1}{\alpha_{oil}} + \frac{\delta_i}{\lambda i} + \frac{1}{h_{wall}}} \tag{3}$$

$$K_{bottom} = \sum \frac{1}{\frac{1}{\alpha_{oil}} + \frac{\delta_i}{\lambda i} + \frac{\pi D}{8\lambda_s}} \tag{4}$$

where α_{oil}, λ_i, δ_i, h_{wall}, h_{top}, D, and λ_s are the heat transfer coefficient from the crude oil to the storage tanks (W/(m^2·°C)), the thermal conductivity of I (W/(m^2·°C)), the thickness of i (m), the convective heat transfer coefficients (W/(m^2·°C)) for tank walls and tops, the characteristic dimensions of the tank (m), and the thermal conductivity of the soil (W/(m^2·°C)), respectively.

The convective heat transfer coefficients for the tank walls and tops is calculated as follows:

$$h_{top} = \frac{\lambda_{steel}}{D}\left(0.664\text{Re}^{1/2}\text{Pr}^{1/3}\right) \tag{5}$$

$$h_{wall} = C\frac{\lambda_{baowen}}{D}\text{Re}^n \tag{6}$$

where λ_{steel}, Re, Pr, C, and λ_{baowen} are the thermal conductivity of steel (W/(m^2·°C)), the thickness of i (m), the Reynolds number, the Prandtl number, and the thermal conductivity of the insulation layer (W/(m^2·°C)), respectively.

The calculated convective heat transfer coefficients for crude oil storage tanks are 1.50 W/m^2·°C (h_{top}) and 4.57 W/m^2·°C (h_{wall}).

To ensure the safety of the operation of the large floating roof tanks, the maintenance temperature of crude oil in static storage should be higher than the freezing point temperature of 5–10 °C. After experimental testing, the freezing point temperature of crude oil in the Daqing area was found to be 30.2 °C. The dimensions and parameters, as shown in Table 1, were calculated from measurements and combined with references.

Table 1. The dimensions and physical parameters of large floating roof tanks.

Material	Value
Diameter of floating roof tank (m)	80
Height of floating roof tank (m)	21
Thermal conductivity of wall of floating roof tank (W/(m·°C))	0.475
Thermal conductivity of top of floating roof tank (W/(m·°C))	1.726
Thermal conductivity of bottom of floating roof tank (W/(m·°C))	0.11

The solar collector was a vacuum tube collector, and the main design parameter was the heat collection area, which was calculated as follows:

$$A_{solar} = \frac{86,400Q_{oil}f}{J_T\eta(1-\eta_L)} \tag{7}$$

where A_{solar}, J_T, f, η, and η_L are the collecting area of the vacuum tube collector (m^2), the annual average daily solar radiation on the receiving surface of the collector (kJ/m^2), the solar guarantee rate, the average collector efficiency, and the solar system average daily loss rate, respectively. The solar guarantee rate is the ratio of the effective heat gain from solar radiation to the required heating demand of a solar collector system during the system design phase.

The sensible heat storage tank is a key component of the SSHS that regulates the cyclical fluctuations of solar energy. The component relies primarily on temperature changes in the material to store and release heat. The volume of the thermal storage tank can be calculated using the following formula:

$$Q_H = \frac{c_{water} m_{water} \Delta t_{water}}{1000} \tag{8}$$

$$V_{water} = \frac{t Q_{oil}}{\rho_{water} Q_H} \tag{9}$$

where V_{water}, t_{water}, Q_H, ρ_{water}, m_{water}, and C_{water} are the volume of the sensible heat storage tank (m^3), the temperature rise of the sensible heat storage tank (°C), the heat storage of the sensible heat storage tank (kJ), the density of the heat storage material (kg/m^3), the mass of the heat storage material (kg), and the heat capacity of the heat storage material (J/kg), respectively. In this paper, the calculated volume size was about 300 m^3, radius was 4.37 m, and the height was 5 m for the cylindrical tanks with the same material and insulation as the crude oil storage tanks. The principle of regulating the temperature fluctuation of the sensible heat storage tank is mainly through the temperature change in the sensible heat material (e.g., water) inside the tank to realize the regulating effect on the crude oil heating temperature. When the intensity of the solar radiation was large, the excess heat of the collector was stored in the sensible heat storage tank in the form of a temperature rise; when the intensity of the solar radiation was small, the collector heating temperature could not meet the demand, and the sensible heat storage tank, in the form of a temperature drop, saw a release of heat to regulate the fluctuation in temperature.

The formula for calculating the heat loss from the pipe and sensible heat storage tanks is as follows:

$$Q_{storage,loss} = K_{storage} A_{storage} \Delta t_{water} \tag{10}$$

$$Q_{pipe,loss} = K_{pipe} A_{pipe} \Delta t_{HTF} \tag{11}$$

where $Q_{storage,loss}$, $Q_{pipe,loss}$ $K_{storage}$, K_{pipe}, $A_{storage}$, $A_{storage}$, Δt_{water}, and Δt_{HTFr} are the heat loss from the pipe and the sensible heat storage tanks (kW), the thermal transfer coefficient of the storage tank and pipe ($W/(m^2 \cdot °C)$), the area of the wall, the top and bottom of the floating roof tank (m^2), and the difference between the temperature of the environment and the temperature of the heat transfer fluid (°C), respectively.

The parameters of the crude oil, heat transfer fluid, and water used in this study are shown in Table 2.

Table 2. The parameters of the crude oil, heat transfer fluid, and water.

Material	Crude Oil	Heat Transfer Fluid	Water
Thermal conductivity (W/(m·°C))	0.1516	2.26	0.59
Specific thermal capacity (kJ/(kg·°C))	2	3.358	4.2
Density (kg/m³)	798	1064	1000

2.2.2. Heat Balance Model

The crude oil storage tanks are the ultimate component of the system. Their heat balance equation is the key to our comprehension of the thermal characteristics of the system. The heat balance equation for the tank boundary is as follows:

$$\lambda_{oil}\left(\frac{\partial}{\partial x}\left(\frac{\partial t(x,y)}{\partial x}\right)+\frac{\partial}{\partial y}\left(\frac{\partial t(x,y)}{\partial y}\right)\right)+K_i\frac{\partial(t_i - t(x,y))}{\partial x \partial y} = \rho_{oil} C_{oil}\frac{\partial t(x,y)}{\partial \tau} \tag{12}$$

where λ_{oil}, ρ_{oil}, C_{oil}, K_i, and t_i are the thermal conductivity (W/(m·°C)), the density (kg/m^3) and specific thermal capacity (kJ/(kg·°C)) of the crude oil, the heat transfer coefficient (W/(m^2·°C)) of i, and the temperature of the substance with which the crude oil heat is transferred (e.g., environmental temperature, the temperature of the heat transfer fluid, etc.) (°C), respectively.

The boundary conditions are expressed as follows:

$$\alpha_{oil}(t_{top} - t_{oil}) = K_{top}(t_0 - t_{oil}) \tag{13}$$

$$\alpha_{oil}(t_{wall} - t_{oil}) = K_{wall}(t_0 - t_{oil}) \tag{14}$$

$$\alpha_{oil}(t_{bottom} - t_{oil}) = K_{bottom}(t_{soil} - t_{oil}) \tag{15}$$

$$\alpha_{oil}(t_p - t_{oil}) = K_p(t_{HTF} - t_{oil}) \tag{16}$$

where t_p, t_{HTF}, and K_p are the temperature of (°C) the heat transfer coil, the temperature of the heat transfer fluids, and the heat transfer coefficient (W/(m^2·°C)) of the heat transfer coil, respectively.

The initial conditions are expressed as follows:

$$\tau = 0, t(x, y) = t_0 \tag{17}$$

where t_0 is the temperature (°C) of the environment.

2.3. System Simulation Model Construction

2.3.1. Transient Simulation Model

The SSHS is a heating method that utilizes vacuum tube collectors in conjunction with a sewage source heat pump to heat the HTF (heat transfer fluid) and transfer the heat to the crude oil storage tank. The system consists of a solar collector, a sewage source heat pump, a sensible heat storage tank, a heat pipe network, and a crude oil storage tank. The system simulation built using TRNSYS can further clarify the parameter relationships between the different internal devices. The SSHS simulation model designed in this study is shown in Figure 2.

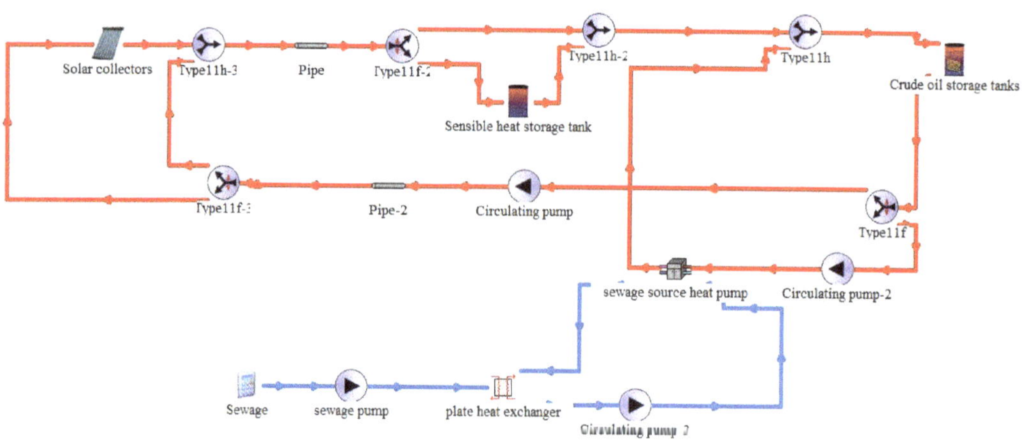

Figure 2. Transient simulation model of SSHS.

The SSHS simulation model was built based on TRNSYS 18.0 software. The main components of the SSHS simulation model are as follows: the meteorological parameter module (type 15-2), the vacuum tube collector (type 71), the auxiliary heating equipment (type 659), the crude oil storage tank (type 156), the storage tank (type 158), the circulation water pump (type 114), the plate heat exchanger (type 5), the controller (type 165), the diverter (type 11f), the mixer (type 11h), the sewage source heat pump (type 2256), the graphic output component (type 65c), and the data output component (type 28).

Among them, the solar collector (type 71), sewage source heat pump (type 2256), and circulation water pump (type 114) were found to be the main components of the system. The model and manufacturer of the RTC solar collector was Bailisheng New Energy Technology Co., Shaoxing, Zhejiang, China; the YEWS D-HP water source heat pump was manufactured by YORK Co., PA, USA.; and the water pumps were customized by Changsheng Technology Co., Daqing, China; these are the main parameters that are shown in Tables 3–5.

Table 3. RTC solar collector.

Parameters	Value
Combined efficiency (%)	45
Extreme Output Temperature (°C)	180
Length (mm)	3890
Width (mm)	2950
Height (mm)	1495
Mounting angle (°)	55

Table 4. YEWS D-HP water source heat pump.

Parameters	Value
Rated heat capacity (kW)	612
Input power (kW)	173
Evaporator water flow (L/s)	15.7
Evaporator water pressure drop (kPa)	16
Evaporator water flow (L/s)	29.5
Evaporator water pressure drop (kPa)	37

Table 5. Customized water pumps.

Parameters	Collector Heat Circulation Pumps	Condenser Side Water Pump	Evaporator Side Water Pump
Water flow (m^3/h)	306	106.2	56.52
Water pump lift (m)	30	20	20
Combined efficiency	0.75	0.75	0.75
Input power (kW)	35.54	8.21	4.11

2.3.2. Validation

Based on the experimental method, the accuracy of the TRNSYS simulation system was verified by heating the crude oil storage tank with a heat storage tank. Figure 3 shows the experimental setup and procedure, and Figure 4 shows the structure and thermocouple (Agilent 34450A, range −80−150 °C, and accuracy 0.002 °C) distribution (ABC(1,2,3), GHJ(1,2,3)) of the crude oil storage tank and heat storage tank. The main equipment used in this verification experiment (the experimental equipment in this study was customized by Changsheng Technology Co., Daqing, China) included a crude oil storage tank, heat storage tank, liquid storage tank, circulating water pump, and electric heating equipment, the dimensions of which are shown in Figure 4. The initial temperature of all three tanks was 10 °C, the power of the electric heating equipment was 6 kW, and the pump flow rate was 1.3 m^3/h. The verification experiment process was as follows: First, open the crude oil storage tank bypass valve, the electric heating equipment, and the circulating water pump.

Then, heat the heat storage tank to 57 °C and the make-up tank to 70 °C via the electric heating equipment. Subsequently, close the crude oil storage tank bypass valve and electric heating equipment, then use the heat storage tank and make-up tank to heat the crude oil storage tank. Lastly, record its internal temperature change using thermocouples.

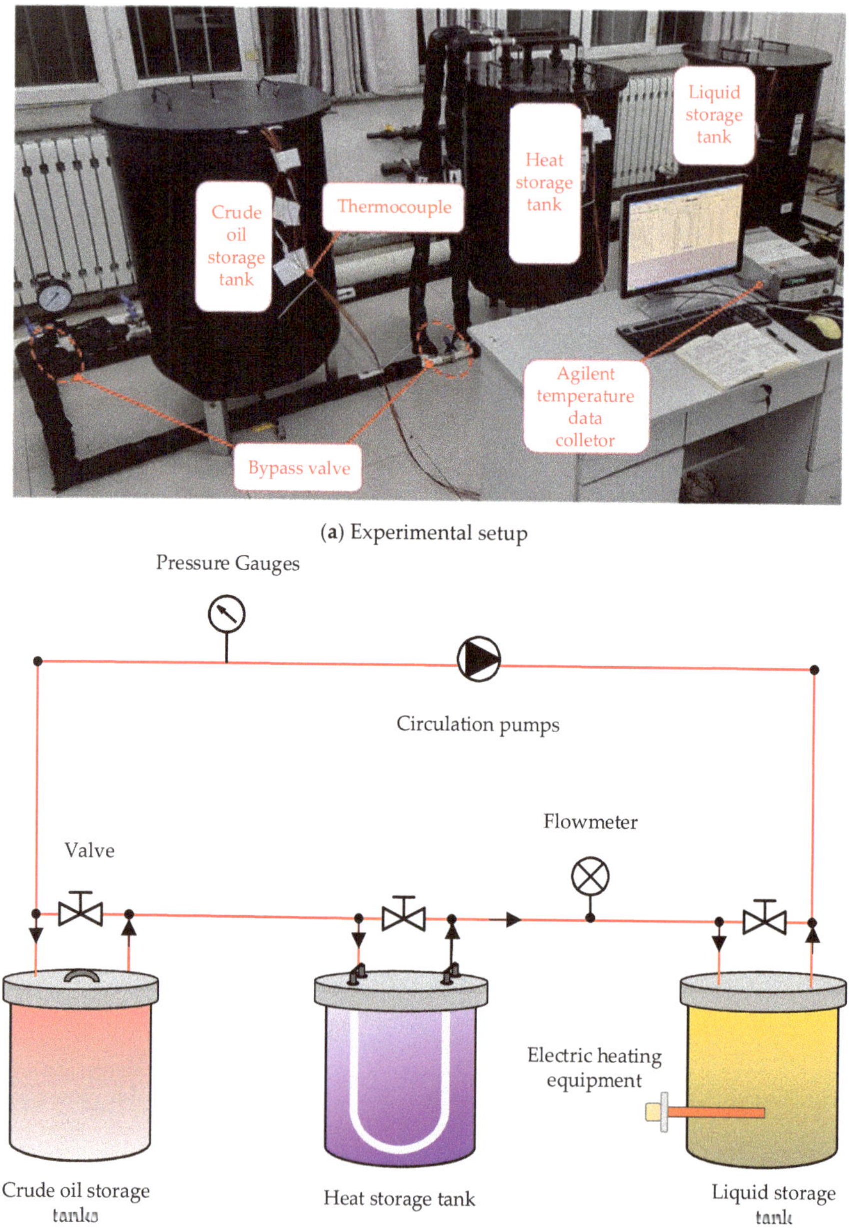

(**a**) Experimental setup

(**b**) Experimental procedure

Figure 3. Experimental setup and procedure.

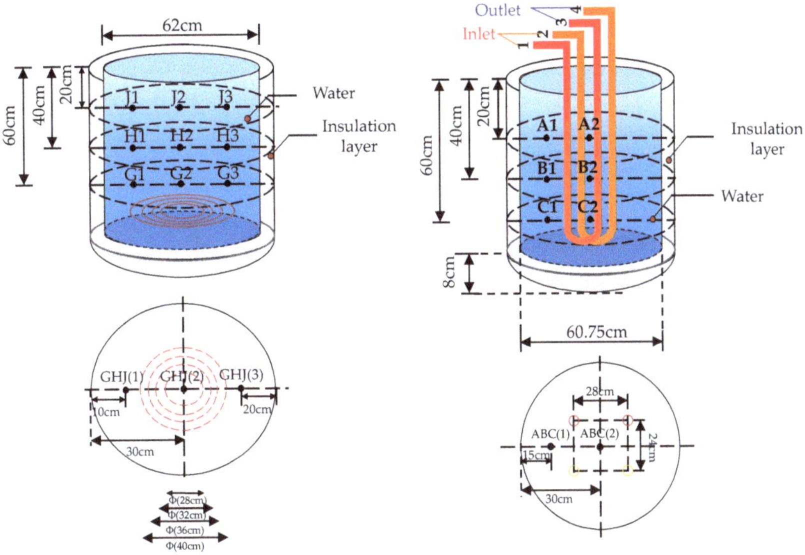

Figure 4. The structure and thermocouple distribution of the crude oil storage tank and heat storage tank.

Figure 5 illustrates the trend of the average temperature variation in the crude oil storage tank and heat storage tank. The results showed that the average error in the average temperature of the heat storage tank was 5.6%, and the average error in the average temperature of the floating top tank was 7.39%, which is within the error range [41]. The reason for the error was mainly caused by the accuracy gap between the experimental equipment and the simulation software, as well as the impurities produced by the heat transfer fluid during operation of the equipment. Therefore, the accuracy of the TRNSYS system simulation platform established in this study was acceptable.

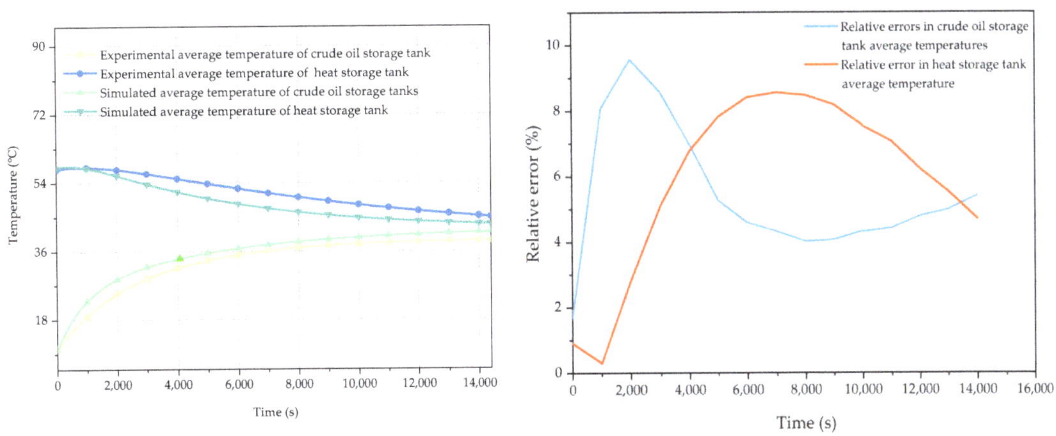

Figure 5. Average temperature variation in the crude oil storage tank and heat storage tank.

2.4. Performance Parameter Definitions

2.4.1. Proportion of the Crude Oil Heated Using Solar Energy

The proportion of the crude oil heated using solar energy is an important indicator for evaluating the thermal performance of a solar collector and is given by the following:

$$\phi_{solar} = \frac{Q_{solar,s}}{Q_{oil,r}} \times 100\% \tag{18}$$

where ϕ_{solar}, $Q_{solar,s}$, and $Q_{oil,r}$ are the proportion of crude oil heated by solar energy, the solar heat for crude oil (kJ), and the total heat of the crude oil being heated (kJ), respectively.

2.4.2. Heat Collection Efficiency of the Solar Collectors

The heat collection efficiency of the solar collectors is the ratio of the effective heat collection using the solar collectors to the total solar radiation projected onto the solar collectors, and this can be calculated by the following equation:

$$\eta_{\text{solar}} = \frac{Q_u}{I_u A_c} \tag{19}$$

where η_{solar}, Q_u, and I_u are the heat collection efficiency of the solar collectors, the effective heat collection using the solar collectors (MJ), and the total solar radiation on the inclined planet (MJ/m^2), respectively.

2.4.3. COP for Sewage Source Heat Pumps

The COP is the most important indicator for the evaluation of sewage source heat pumps, and its calculation formula is given as follows:

$$COP = \frac{Q_{sp}}{W_{sp}} \tag{20}$$

where COP, Q_{sp}, and W_{sp} are the sewage source heat pump coefficient of performance, the heat production (kJ), and the input energy (kJ) of the sewage source heat pumps, respectively.

2.5. Meteorological Parameters

The meteorological parameter reading module of the system simulation in this paper used a type15-2 component to read the meteorological data of a typical meteorological year in *.tm2 (the specific file format for weather parameter imports in TRNSYSY), which includes two main parameters: ambient temperature and solar radiation. Figure 6 shows the typical annual meteorological parameters for Daqing, Heilongjiang Province, China (data obtained from the Meteonorm database). From the figures, it can be concluded that the lowest temperature of the year occurs in January with a low of −26.4 °C, whereas the highest temperature occurs in July with a high of 29.4 °C. Solar radiation shows a cyclic variation throughout the year, reaching its maximum in July; the total monthly solar radiation was 300 kWh/m^2 for seven months of the year, with a total annual radiation of 3408 kWh/m^2.

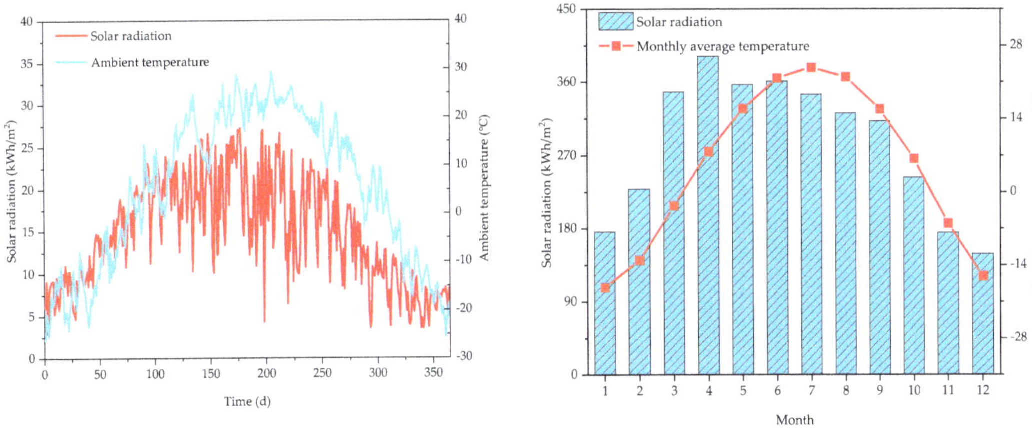

Figure 6. Meteorological parameters.

3. Results and Discussion

Based on the established simulation model of the SSHS system, the effects of different solar energy guarantee rates, pumping unit flow rates, and effluent supply temperatures on the operational characteristics of the SSHS system were analyzed. The simulation area was set in Daqing, China, and the time period was 1 year with a time step of 1 min.

3.1. Effect of Solar Guarantee Rate

The effect of the solar guarantee rate (f) on the thermodynamic properties of solar collectors, crude oil storage tanks, and the performance of sewage source heat pump in the SSHS was evaluated by testing different solar guarantee rates for the SSHS. Different solar guarantee values—namely, 10%, 20%, 30%, 40%, and 50%—were considered, and the results obtained are as follows:

Table 6 presents the calculated collector areas for the different solar guarantee rates based on Equations (1) and (2):

Table 6. The collector area at different solar guarantee rates.

f	Q_{oil} /kW	A_{solar} /m^2
10%	571.24	1226.32
20%	571.24	2452.64
30%	571.24	3678.95
40%	571.24	4905.27
50%	571.24	6131.59

To analyze the effect of the solar guarantee rate variation on the thermal collection efficiency of the solar collectors inside the SSHS, five different solar guarantee rates (10%, 20%, 30%, 40%, and 50%) were considered. The obtained solar collector efficiency, proportion of crude oil heated, average annual collector efficiency, and the average proportion of crude oil heated are shown in Figure 7. As expected, in Figure 7a–e, the solar collector's efficiency fluctuated less after the solar collector operation was stabilized. The proportion of the solar heat supply to crude oil rose gradually with the operation of the system. The main reason for this phenomenon is that, in the early stage of system operation, the heat collection capacity of the solar collector is small, and the collector outlet temperature is low, which cannot meet the demand of the static storage of crude oil. In addition, the proportion of heat supply to crude oil from the sewage source heat pump accounts for a larger proportion. With the operation of the system, the cumulative solar heat collection gradually increased, and the proportion of solar energy to crude oil heating also increased. It can also be found that, as the solar energy guarantee rate decreases, the phenomenon of decreasing the proportion of solar energy to crude oil heat supply occurs in the second half of the system operating time. Meanwhile, as shown in Figure 7f, the annual average collector efficiency gradually decreased, and the annual average proportion of the solar heat supply to crude oil also gradually increased, but the magnitude of the increase gradually decreased. This is because, with the increase in solar guarantee rate, the solar collector area increases; as such, the amount of heat collected gradually increases and can meet the demand of crude oil static storage. At the same time, the increase in collector area leads to an increase in collector heat loss, which reduces the collector's efficiency.

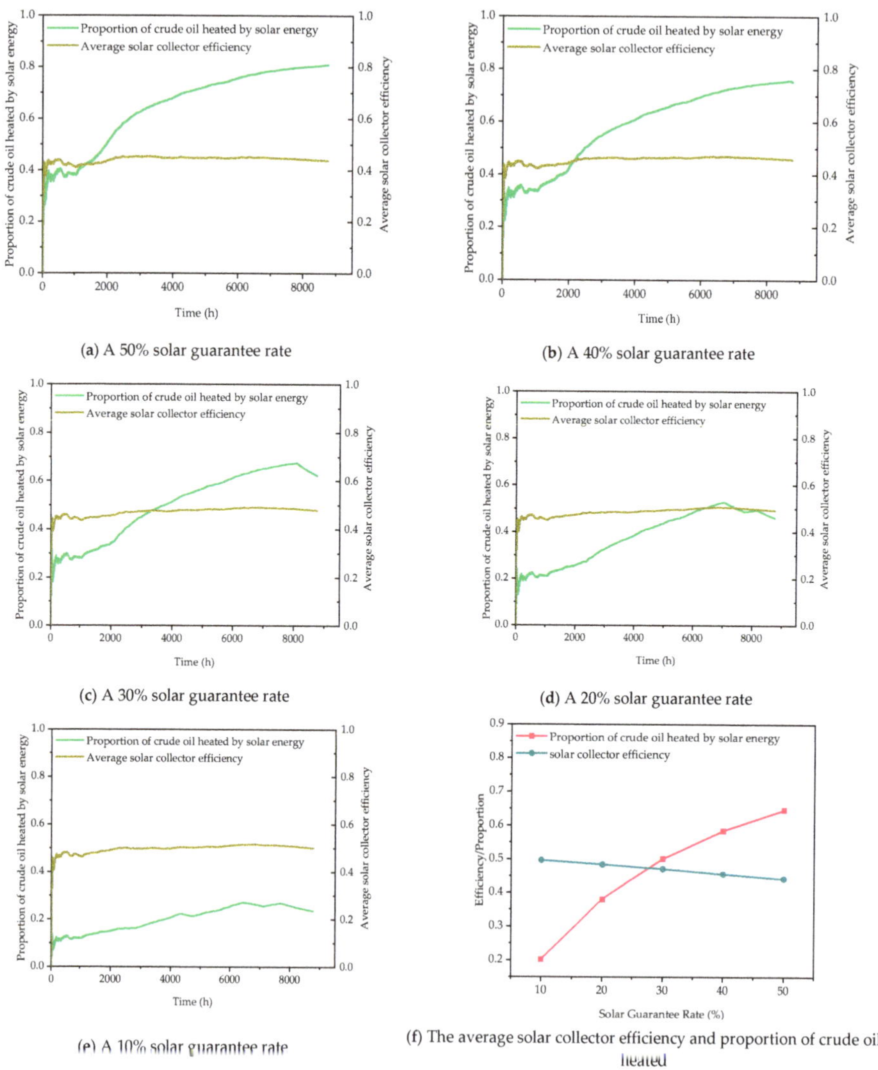

(a) A 50% solar guarantee rate

(b) A 40% solar guarantee rate

(c) A 30% solar guarantee rate

(d) A 20% solar guarantee rate

(e) A 10% solar guarantee rate

(f) The average solar collector efficiency and proportion of crude oil heated

Figure 7. Thermal characteristics of solar collectors with different solar guarantee rates.

In order to quantitatively analyze the effect of solar guarantee rates on the crude oil storage tanks, the heating demands of the crude oil storage tanks and the average annual temperature of the crude oil for all the considered scenarios were plotted, as shown in Figure 8a,b. As shown in Figure 8a, with the operation of the system, the heating demand of the crude oil storage tanks shows a trend of decreasing and then increasing, and the heating demand reached its maximum in December at 555,441.75 kWh, 514,226.38 kWh, 500,524.64 kWh, 499,047.9 kWh, and 506,489.14 kWh, respectively. Moreover, it reached its minimum in July at 263,940.78 kWh, 499,047.9 kWh, and 506,489.14 kWh, respectively, at 263,940.78 kWh, 230,099.16 kWh, 193,747.9 kWh, 172,799.63 kWh, and 160,613.90156 kWh (i.e., when the solar energy guarantee was 10%, 20%, 30%, 40%, and 50%), respectively. The main reason for this is that, when the system is working in the summer, the external ambient temperature is high and there is less heat exchange between the crude oil and the environment. However, when working in winter, the external ambient temperature is below zero; as such, the heat exchange between the crude oil and the environment is larger, leading to the phenomenon that the heating demand of the crude oil storage tank is large in winter and small in summer. Combined with Figure 8b, it was found that the average annual temperature of the crude oil increased with an increase in solar guarantee rates, leading to an increase in overall heating demand of the crude oil storage tanks.

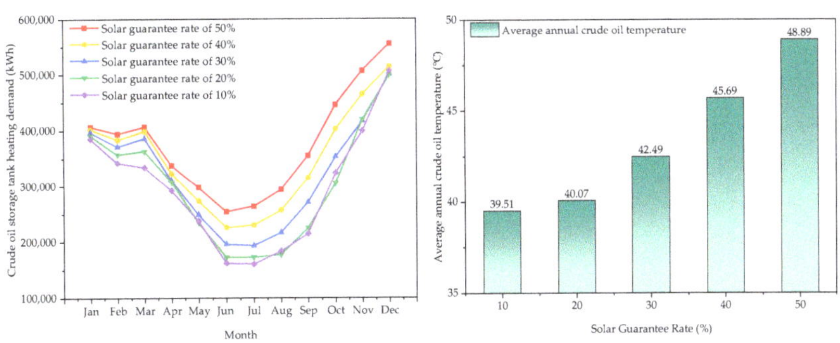

(**a**) Heating demand of crude oil storage tank (**b**) Average crude oil temperature

Figure 8. Thermal characteristics of the crude oil storage tank with different solar guarantee rates.

Figure 9a,b plots the hourly COP and average COP of the sewage source heat pump with different solar guarantee rates. As shown in Figure 9, the COP of the sewage source heat gradually decreased as the system operated, and the average COP of the sewage source heat also decreased gradually with the increase in solar guarantee rate. The main reason for this phenomenon was that, with the operation of the system, the cumulative heat collection of the solar collector increased, the temperature of the heat transfer fluid flowing through the condenser end of the sewage source heat pump in the circulation loop gradually increased, and the COP of the sewage source heat pump also decreased. When the solar collector collects heat to meet the crude oil heating demand, the sewage source heat pump is deactivated, at which time, the COP remains unchanged. When the sewage source heat pump is reactivated to meet the crude oil heating requirement (solar energy guarantee rate of 10%, the system operating time > 7000 h), the COP is reduced with the increase in the system temperature of the heat transfer fluid. By the same token, as the solar guarantee rate increased, the temperature of the system loop increased and the COP of the sewage source heat pump decreased. According to Figure 9b, the decrease value in the average COP of the sewage source heat showed a trend of decreasing, then increasing, and then decreasing as the solar guarantee rate changed from 10% to 20%, 20% to 30, 30% to 40%, and 40% to 50%, respectively. In addition, the average COP of the sewage source heat decreased by 0.64%, 0.43%, 1.53%, and 1.33%, respectively.

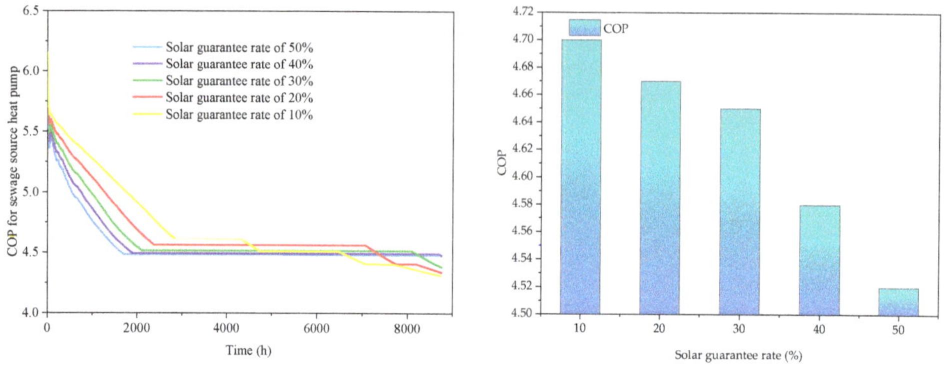

(**a**) COP (**b**) Average COP

Figure 9. The thermal characteristics of the sewage source heat pump with different solar guarantee rates.

In summary, the solar guarantee rate should not be too high in order to meet the demand for the heat stored in the crude oil when attempting to maximize solar collector efficiency; furthermore, a solar guarantee rate of 30% is the most appropriate.

3.2. Flow Rate of Pumping Units

As shown in Table 7, the flow rates of five different pump units (heat collecting water pump, sewage pump, and condenser-side water pump form a pump combination, and all three are increased at the same time, with the ratio of their increase being 50:20:25) were selected to study the operating characteristics of the main components of the SSHS when operated at different pump flow rates.

Table 7. Different pump flow rates.

Group	Heat Collecting Water Pump/ m^3/h	Sewage Pump/ m^3/h	Condenser Side Water Pump/ m^3/h
A1	80	10	25
A2	130	30	50
A3	180	50	100
A4	230	70	125
A5	280	90	150

Figure 10 details the solar collector efficiency, the proportion of crude oil heated, and the COP of the sewage source heat pump at different pump flow rates with a solar guarantee rate of 30%. As shown in Figure 10a, the solar collector efficiency and the proportion of crude oil heated showed a slight decreasing trend with an increase in the flow rate of pumping units. This is mainly due to the fact that an increase in the flow rate of the collector pump can take away the heat from the collector faster, while an increase in the flow rate of the sewage pump and the condenser-side pump can improve the heat transfer efficiency of the sewage source heat pump. However, when the flow rate of the collector pump is too high, solar radiation cannot be effectively converted into heat, while the flow rate of the sewage pump and condenser-side pump is too high, which also leads to insufficient heat exchange at the hot and cold ends of the sewage source heat pump, which in turn leads to a slight decrease in the amount of heat supplied by the crude oil. The analysis of Figure 10b shows that the COP of the sewage source heat pump demonstrated an increasing trend with the increase in the flow rate of pumping units. For the flow rate of pumping units from A1 to A3, the COP increase was larger, and, from A3 to A5, the COP increase was smaller. For example, when the pumping unit flow rate increased from A1 to A2 and A3, the COP increased by 9.29% and 8.24%, respectively. When the pumping unit flow rate increased from A3 to A4 and A5, the COP increased by 1.11% and 1.09%, respectively. Therefore, considering the economic and equipment constraints, the pumping unit flow rate of A3 was more appropriate.

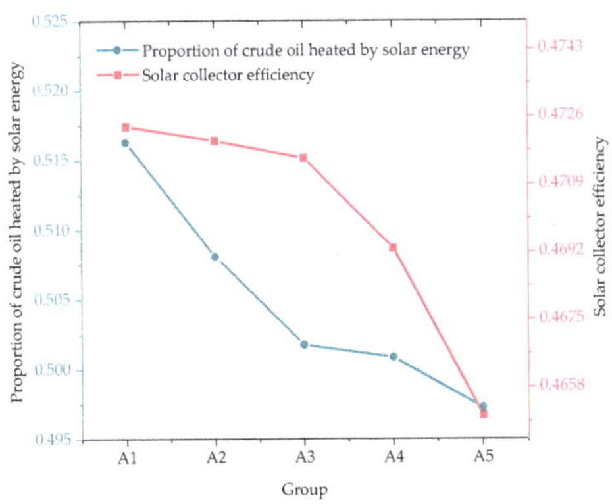

(**a**) Solar collector efficiency and proportion of crude oil heated

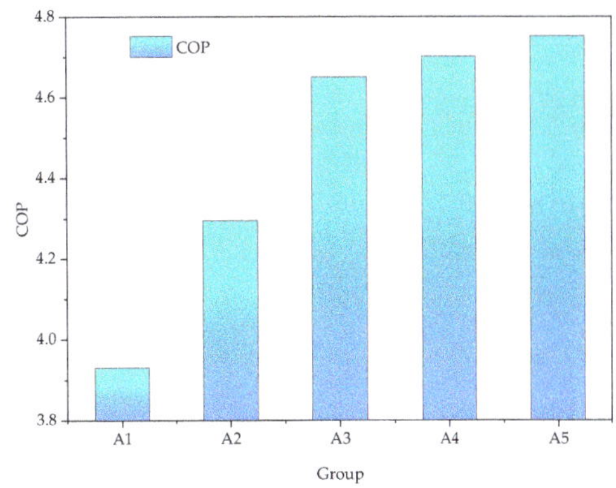

(**b**) COP

Figure 10. The solar collector efficiency, the proportion of crude oil heated and the COP of the sewage source heat pump at different pumping unit flow rates.

3.3. Sewage Supply Temperature

Five different sets of effluent supply temperatures (20 °C, 25 °C, 30 °C, 35 °C, and 40 °C) were selected for simulation to investigate the effect of effluent supply temperature on the operating characteristics of an SSHS under a solar energy guarantee of 30% and a pumping unit flow rate of A3.

The average crude oil temperature, the heating demand of crude oil storage tanks, and the COP of sewage source heat pumps at different sewage supply temperatures are shown in Figure 11. The results show that the average crude oil temperature, the heating demand of crude oil storage tanks, and the COP of sewage source heat pumps gradually increase with an increase in the sewage supply temperature. The main reason for this phenomenon is that when the sewage supply temperature increases, the power required for heating by the sewage source heat pump decreases, and the COP of the sewage source heat pump increases. As such, the temperature of the crude oil supplied by the sewage source heat pump also increases, which leads to an increase in the average temperature of the crude oil and an increase in the heating demand of the crude oil tanks. However, the effect of the sewage supply temperature on the increase in the heating demand of the crude oil tanks was found to be relatively slight; for example, if the temperature of the sewage supply temperature changed from 20 to 40 °C (an increase of 100%), the crude oil storage tank heating demand increase was only 1.43% and the sewage source heat pump COP increased by 9.03%. Therefore, the sewage supply temperature should be increased as much as possible when the conditions allow, which is favorable in terms of improving the COP of the sewage source heat pump.

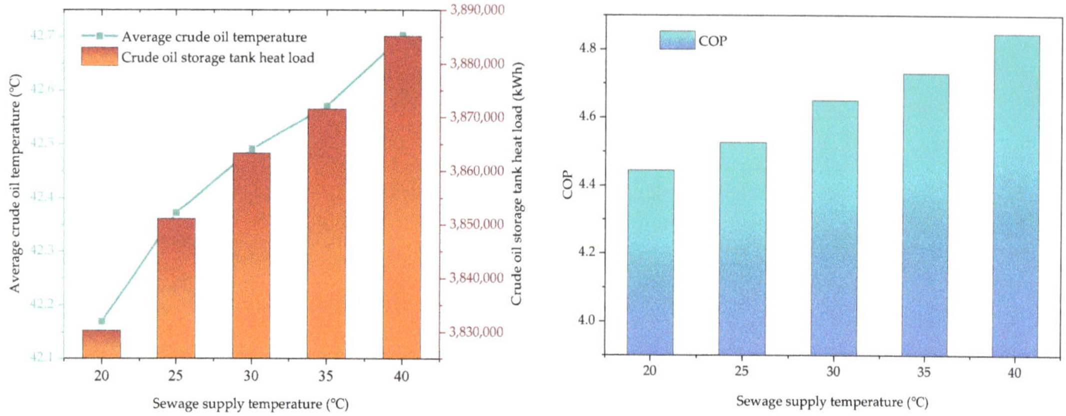

(**a**) Average crude oil temperature and heating demand of crude oil storage tank.

(**b**) COP

Figure 11. The average crude oil temperature, the heating demand of the crude oil storage tank, and the COP of the sewage source heat pump at different sewage supply temperatures.

4. Conclusions

Dual heat source supply solar sewage source heat pump (SSHS) technology for the static storage of crude oil was proposed to reduce the consumption of primary energy in the heating process of crude oil and to realize green and sustainable development in the field of crude oil temperature maintenance. An SSHS simulation model was developed and experimentally validated. In addition, the operational characteristics of different schemes were compared and analyzed. The relevant research contents and conclusions are as follows:

1. Solar guarantee rate is more important to SSHS; with the increase in solar guarantee rate, the annual average solar collector efficiency gradually decreases, and the annual average proportion of crude oil heated using solar gradually increases. In order to protect the unnecessary waste of heat, the solar guarantee rate should not be too high; in order to meet the demand for heat stored in the crude oil and at the same time maximize the protection of solar collector efficiency, a solar guarantee rate of 30% is most appropriate.
2. The pump unit flow rate has a relatively slight effect on the performance of the solar collector and a large effect on the performance of the sewage source heat pump. With the increase in pump set flow rate, the solar collector's collector efficiency and the proportion of crude oil heated slightly decreased, the COP of the sewage source heat pump showed an increasing trend. For the pump unit flow rate from A1 to A3, the COP increased by a large amount, and from A3 to A5, the COP increased by a small amount. Therefore, considering the economic and equipment limitations, it is more appropriate to use the pumping unit flow rate of A3. At this time, the solar energy guarantee rate is 30%, the pump flow rate is A3, the solar collector efficiency is 50.18%, the proportion of crude oil heated by solar is 47.16%, the average temperature of crude oil is 42.59 °C, and the COP of the wastewater source heat pump is 4.65.
3. The sewage supply temperature has a positive effect on the average crude oil temperature, the crude oil storage tank heating demand, and the COP of the sewage source heat pump. And since the effect of effluent supply temperature on the COP of the sewage source heat pump is much larger than that on the average crude oil temperature and the heating demand of the crude oil storage tank, the performance of the sewage source heat pump can be increased by increasing the sewage supply temperature.

In future work, considering the problems of sensible heat storage such as large volume, the density of heat storage can be enhanced by synergizing the phase change latent heat to further reduce system costs and carbon dioxide emissions and achieve green and sustainable development. Meanwhile, economic studies are needed to assess the feasibility of these practical applications.

Author Contributions: Conceptualization, Y.L.; Methodology, S.Z. and X.G.; Formal analysis, M.D.; Investigation, F.Y.; Resources, X.G.; Data curation, S.Z; Writing—original draft, S.Z. and S.L.; Writing review & editing, D.Z. and Y.L; Project administration, Y.L. All authors have read and agreed to the published version of the manuscript.

Funding: This research received no external funding.

Data Availability Statement: The data presented in this study are available upon request from the corresponding author. The data are not publicly available due to the laboratory interests involved in this data.

Conflicts of Interest: The authors declare no conflict of interest.

Abbreviations

Abbreviations and Mathematical Symbols	Defined and Described
SSHS	Maintenance system of crude oil temperature using dual heat sources with solar synergistic sewage heat pumps
SIPH	Solar industrial process heat
Q	Thermal load (kW)
K	Thermal conductivity (W/(m·°C))
A	Area (m^2)
T	Temperature (°C)
J	Annual average daily solar radiation (kJ/m^2)
f	Solar guarantee rate
η	Efficiency
η_L	Solar system average daily loss rate
V	Volume (m^3)
Δ	Difference in value
ρ	Density (kg/m^3)
m	Mass (kg)
C	Heat capacity (J/kg)
ϕ	Proportion
COP	Coefficient of performance
W	Input energy (kJ)
t	Temperature (°C)
Re	The Reynolds number
Pr	The Prandtl number

References

1. Chen, J. Oil Reserves: Definition, Calculation, Statistics, and Examples. *Investopedia* **2022**. Available online: https://www.investopedia.com/terms/o/oil-reserves.asp (accessed on 28 July 2022).
2. Ozsarac, V.; Brunesi, E.; Nascimbene, R. Earthquake-induced nonlinear sloshing response of above-ground steel tanks with damped or undamped floating roof. *Soil Dyn. Earthq. Eng.* **2021**, *144*, 106673. [CrossRef]
3. British Petroleum. The Statistical Review of World Energy. *Br. Pet.* **2022**. Available online: https://www.bp.com/de_de/germany/home/presse/nachrichten/statistical-review-of-world-energy-2022.html (accessed on 29 June 2022).
4. Zhao, J.; Dong, H.; Wang, X.; Fu, X. Research on heat transfer characteristic of crude oil during the tubular heating process in the floating roof tank. *Case Stud. Therm. Eng.* **2017**, *10*, 142–153. [CrossRef]
5. Kim, J.S.; Haines, R.J.; Lopez, J.P. A simplified dynamic approach for storage tank thermal inbreathing rates. *Hydrocarb. Process.* **2023**. Available online: https://www.hydrocarbonprocessing.com/magazine/2023/september-2023/tanks-terminals-and-storage/a-simplified-dynamic-approach-for-storage-tank-thermal-inbreathing-rates/#comments (accessed on 15 September 2023).

6. Zhao, J.; Liu, J.; Dong, H.; Zhao, W. Effect of physical properties on the heat transfer characteristics of waxy crude oil during its static cooling process. *Int. J. Heat Mass Transf.* **2019**, *137*, 242–262. [CrossRef]
7. Wang, M.; Zhang, X.; Shao, Q.; Li, J.; Yu, B. Temperature drop and gelatinization characteristics of waxy crude oil in 1000 m^3 single and double-plate floating roof oil tanks during storage. *Int. J. Heat Mass Transf.* **2019**, *136*, 457–469. [CrossRef]
8. Ji, S.; Cheng, Q.; Sun, W.; Qi, Y.; Wang, S. Study on optimal operation conditions in the heating process of a crude oil single-disk floating roof tank: Insights from exergy transfer analysis method. *Energy Rep.* **2023**, *9*, 3131–3147. [CrossRef]
9. Sun, W.; Liu, Y.; Li, M.; Cheng, Q.; Zhao, L. Study on heat flow transfer characteristics and main influencing factors of waxy crude oil tank during storage heating process under dynamic thermal conditions. *Energy* **2023**, *269*, 127001. [CrossRef]
10. Mary, E.; Brendan, D. Cold climate air source heat pumps: Industry progress and thermodynamic analysis of market-available residential units. *Renew. Sustain. Energy Rev.* **2023**, *188*, 113739. [CrossRef]
11. Wang, Z.; Shu, L.; Niu, S.; Su, X.; Zhang, S. Research on particle clogging during groundwater recharge of ground source heat pump system. *Geothermics* **2023**, *115*, 102810. [CrossRef]
12. Farjana, S.; Huda, N.; Mahmud, M.; Saidur, R. Solar process heat in industrial systems—A global review. *Renew. Sustain. Energy Rev.* **2018**, *82*, 2270–2286. [CrossRef]
13. Kumar, L.; Hasanuzzaman, M.; Rahim, N. Global advancement of solar thermal energy technologies for industrial process heat and its future prospects: A review. *Energy Convers. Manag.* **2019**, *195*, 885–908. [CrossRef]
14. Nathan, G.; Lee, L.; Ingenhoven, P.; Tian, Z.; Sun, Z.; Chinnici, A.; Jafarian, M.; Ashman, P.; Potter, D.; Saw, W. Pathways to the use of concentrated solar heat for high temperature industrial processes. *Sol. Compass* **2023**, *5*, 100036. [CrossRef]
15. Gil, J.; Topa, A.; Álvarez, J.; Torres, J.; Pérez, M. A review from design to control of solar systems for supplying heat in industrial process applications. *Renew. Sustain. Energy Rev.* **2022**, *163*, 112461. [CrossRef]
16. Mohammadi, K.; Khanmohammadi, S.; Immonen, J.; Powell, K. Techno-economic analysis and environmental benefits of solar industrial process heating based on parabolic trough collectors. *Sustain. Energy Technol. Assess.* **2021**, *47*, 101412. [CrossRef]
17. Hiben, Y.; Bayray, M.; Lauwaert, J. Optimizing solar-assisted industrial heating and cooling system for cost-effective installation. *Appl. Therm. Eng.* **2023**, *230*, 120778. [CrossRef]
18. Kumar, L.; Hasanuzzaman, M.; Rahim, N.; Islam, M. Modeling, simulation and outdoor experimental performance analysis of a solar-assisted process heating system for industrial process heat. *Renew. Energy* **2021**, *164*, 656–673. [CrossRef]
19. Wang, J.; O'Donnell, J.; Brandt, A. Potential solar energy use in the global petroleum sector. *Energy* **2017**, *118*, 884–892. [CrossRef]
20. Halabi, M.; Al-Qattan, A.; Al-Otaibi, A. Application of solar energy in the oil industry—Current status and future prospects. *Renew. Sustain. Energy Rev.* **2015**, *43*, 296–314. [CrossRef]
21. Altayib, K.; Dincer, I. Analysis and assessment of using an integrated solar energy based system in crude oil refinery. *Appl. Therm. Eng.* **2019**, *159*, 113799. [CrossRef]
22. Naseer, A.; Asif, H.; Naveed, A.; Fahid, R.; Mariam, M.; Majid, A.; Kalam, M.; Mujtaba, M. Solar-assisted hybrid oil heating system for heavy refinery products storage. *Case Stud. Therm. Eng.* **2023**, *249*, 103276. [CrossRef]
23. Pathak, S.; Tyagi, V.; Chopra, K.; Kalidasan, B.; Pandey, A.; Goel, V.; Saxena, A.; Ma, Z. Energy, exergy, economic and environmental analyses of solar air heating systems with and without thermal energy storage for sustainable development: A systematic review. *J. Energy Storage* **2023**, *59*, 106521. [CrossRef]
24. Jiang, M.; Feng, G.; Huang, K.; Liu, S.; Liang, D.; Li, H. Research and Analysis of Combined Operational Mode of Solar Energy Hot Water Heating System and Sewage Source Heat Pump Units. *Procedia Eng.* **2015**, *121*, 1544–1555. [CrossRef]
25. Zhang, X.; Wang, E.; Liu, L.; Qi, C.; Zhen, J.; Meng, Y. Analysis of the operation performance of a hybrid solar ground-source heat pump system. *Energy Build.* **2022**, *268*, 112218. [CrossRef]
26. Gautam, A.; Saini, R.P. A review on sensible heat based packed bed solar thermal energy storage system for low temperature applications. *Sol. Energy* **2020**, *207*, 937–956. [CrossRef]
27. Ryu, J.-Y. Alford A, Lewis G, Ding Y, Li Y, Ahmod HK, Park S-H, Park J-P, Branch S, Yu S, Ryu, C. A novel liquid air energy storage system using a combination of sensible and latent heat storage. *Appl. Therm. Eng.* **2022**, *203*, 117890. [CrossRef]
28. Crespo, A.; Barreneche, C.; Ibarra, M.; Platzer, W. Latent thermal energy storage for solar process heat applications at medium-high temperatures—A review. *Sol. Energy* **2019**, *192*, 3–34. [CrossRef]
29. Chen, X.; Wang, F.; Han, Y.; Yu, R.; Cheng, Z. Thermochemical storage analysis of the dry reforming of methane in foam solar reactor. *Energy Convers. Manag.* **2018**, *158*, 489–498. [CrossRef]
30. Chen, H.; Wang, Y.; Li, J.; Cai, B.; Zhang, F.; Lu, T.; Yang, J.; Jiang, L.; Zhang, Y.; Zhou, J. Experimental research on a solar air-source heat pump system with phase change energy storage. *Energy Build.* **2020**, *228*, 110451. [CrossRef]
31. Li, M.Y.; Li, B.; Liu, C.; Su, S.; Xiao, H.; Zhu, C. Design and experimental investigation of a phase change energy storage air-type solar heat pump heating system. *Appl. Therm. Eng.* **2020**, *179*, 115506. [CrossRef]
32. Liu, M.; Riahi, S.; Jacob, R.; Belusko, M.; Bruno, F. Design of sensible and latent heat thermal energy storage systems for concentrated solar power plants: Thermal performance analysis. *Renew. Energy* **2020**, *151*, 1286–1297. [CrossRef]
33. Punniakodi, B.; Senthil, R. Recent developments in nano-enhanced phase change materials for solar thermal storage. *Sol. Energy Mater. Sol. Cells* **2022**, *238*, 111629. [CrossRef]
34. Nazari, M.; Maleki, A.; Assad, M.; Rosen, M.; Haghighi, A.; Sharabaty, H.; Chen, L. A review of nanomaterial incorporated phase change materials for solar thermal energy storage. *Sol. Energy* **2021**, *228*, 725–743. [CrossRef]

35. Ma, Y.; Xi, J.; Cai, J.; Gu, Z. TRNSYS simulation study of the operational energy characteristics of a hot water supply system for the integrated design of solar coupled air source heat pumps. *Chemosphere* **2023**, *338*, 139453. [CrossRef] [PubMed]
36. FScarpa, L.A.; Tagliafico, G. T Integrated solar-assisted heat pumps for water heating coupled to gas burners; control criteria for dynamic operation. *Appl. Therm. Eng.* **2011**, *31*, 59–68. [CrossRef]
37. Qian, J.; Wang, Q. Development on solar—Assisted sewage source heat pump system for crude oil heating. *J. Harbin Univ. Commer. (Nat. Sci. Ed.)* **2017**, *33*, 477–481. [CrossRef]
38. Agrebi, S.; Chargui, R.; Tashtoush, B.; Guizani, A. Comparative performance analysis of a solar assisted heat pump for greenhouse heating in Tunisia. *Int. J. Refrig.* **2021**, *131*, 547–558. [CrossRef]
39. *GB 50495-2019*; Solar Heating Engineering Technical Standard. China Architecture Publishing House: Beijing, China, 2019.
40. *GB 50364-2018*; Civil Building Solar Water Heating System Application Technical Standard. China Architecture Publishing House: Beijing, China, 2018.
41. Li, X.; Qiu, H.; Wang, Z.; Li, J.; Yuan, G.; Guo, X.; Jin, L. Numerical Investigation of a Solar-Heating System with Solar-Tower Receiver and Seasonal Storage in Northern China: Dynamic Performance Assessment and Operation Strategy Analysis. *Energies* **2023**, *16*, 5505. [CrossRef]

energies

MDPI

Article

Numerical Simulation Study on the Gas–Solid Flow Characteristics of a Large-Scale Dual Fluidized Bed Reactor: Verification and Extension

Yubin Lin, Qinhui Wang *, Guilin Xie, Mengxiang Fang and Zhongyang Luo

State Key Laboratory of Clean Energy Utilization, Institute for Thermal Power Engineering, Zhejiang University, Yuquan Campus, 38 Zheda Road, Hangzhou 310027, China
* Correspondence: qhwang@zju.edu.cn; Tel.: +86-571-8795-2802

Abstract: Dual fluidized bed (DFB) reactor systems are widely used in gas–solid two-phase flow applications, whose gas–solid flow characteristics have a significant effect on the performance of many kinds of technologies. A numerical simulation model was established on the basis of a large-scale DFB reactor with a maximum height of 21.6 m, and numerical simulations focused on gas–solid flow characteristics were carried out. The effects of the superficial gas velocity of both beds and the static bed height and particle size on the distribution of the pressure and solid suspension density and the solid circulation rate were studied. The simulation results were in good agreement with the experimental data. With the strong support of the experimental data, the gas–solid flow characteristics of large-scale DFB reactors were innovatively evaluated in this numerical simulation study, which effectively makes up for the shortcomings of the current research. The results showed that the superficial gas velocity of both beds and the static bed height have different degrees of influence on the gas–solid flow characteristics. Specifically, for 282 μm particles, when the superficial gas velocity of both beds and the static bed height were 4.5 m/s, 2.5 m/s, and 0.65 m, respectively, under typical working conditions, the bottom pressure of the two furnaces was 3412.42 Pa and 2812.86 Pa, respectively, and the solid suspension density was 409.44 kg/m^3 and 427.89 kg/m^3, respectively. Based on the simulation results, the empirical formulas of the solid circulation rate were fitted according to different particle sizes. Under similar conditions, the solid circulation rates of particles with a particle size of 100 μm, 282 μm, 641 μm, and 1000 μm were 2.84–13.28, 0.73–4.91, 0.024–0.216, and 0.0026–0.0095 kg/(m^{2}s), respectively. It can be found that the influence of the particle size on the solid circulation rate is the most significant among all parameters.

Keywords: dual fluidized bed reactor; cold mode; gas–solid flow characteristic; solid circulation rate; numerical simulation

Citation: Lin, Y.; Wang, Q.; Xie, G.; Fang, M.; Luo, Z. Numerical Simulation Study on the Gas–Solid Flow Characteristics of a Large-Scale Dual Fluidized Bed Reactor: Verification and Extension. *Energies* **2024**, *17*, 1304. https://doi.org/10.3390/en17061304

Academic Editor: Artur Blaszczuk

Received: 19 January 2024
Revised: 22 February 2024
Accepted: 4 March 2024
Published: 8 March 2024

1. Introduction

In the context of carbon neutrality, the DFB (dual fluidized bed) reactor has attracted more and more attention [1]. A DFB reactor is a coupling of two fluidized beds, which is widely used in a variety of energy efficient and clean utilization technologies, including coal staged conversion technology [2], biomass gasification technology [3], chemical looping combustion technology [4], and calcium looping technology [5]. As shown in Figure 1, in different technologies, the two fluidized bed reactors cooperate with each other and play different roles in order to achieve the process flow required by the technologies.

The gas–solid flow characteristics such as pressure distribution, solid suspension density distribution, and solid circulation rate are the key and intuitive parameters to judge the normal operation of a fluidized bed reactor system. The same is true for a complex DFB reactor [1]. In particular, the normal operation of a DFB reactor depends on sufficient circulation of material and heat between the two fluidized beds. The solid circulation rate is a very intuitive representation of the level of material circulation between two fluidized beds, which makes this parameter more important. Therefore, it is of great necessity to carry out research on DFB reactors so as to understand their gas–solid flow characteristics, especially the solid circulation rate.

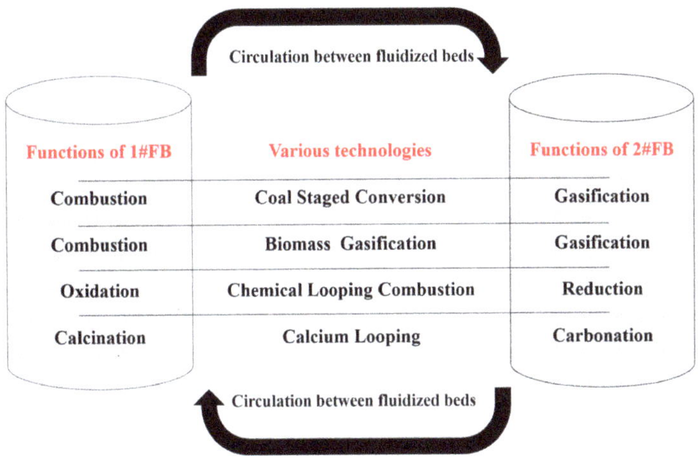

Figure 1. The specific roles of a DFB reactor in various technologies.

Various studies on the gas–solid flow characteristics of a fluidized bed have been reported by many researchers. Jie Xu et al. studied gas–solid flow characteristics at the riser wall of a CFB and proposed quantified empirical formulas for estimating key characteristics based on the experiment data [6]. Qitao Bian et al. used electrical capacitance volume tomography to explore particle flow characteristics in a CFB recirculation system and the experimental results showed that the branch had a great impact on the particle flow characteristics in the CFB riser [7]. Michael Stollhof et al. proposed an estimation method of the solid circulation rate in CFB systems after measuring the pressure drop and other parameters in experiments on four different fluidized bed units [8]. Pil-Sang Youn et al. gathered data from their own research and other references and successfully developed empirical formulae of the solids circulation rate and solids inventory in the riser of the CFB [9]. However, these studies were based on a single fluidized bed reactor. For a DFB reactor coupled by two fluidized beds, the internal flow field is simultaneously affected by the operating parameters of both reactors, which allows the gas–solid flow characteristics to change more unpredictably. For example, the solid circulation rate of a single fluidized bed reactor measures its own material circulation, while in a DFB reactor, the solid circulation rate measures the extent to which the material circulates between the two fluidized bed reactors. Therefore, the gas–solid flow characteristics of single fluidized bed reactors are not suitable for DFB reactors.

There are many researchers who have studied the DFB reactor in experimental way. Tobias Proll et al. conducted a series of cold flow model experiments on a 1.8 m scaled DFB reactor, whose results showed that the system ran with high solids circulations between the two reactors, even at low bed inventories [10]. Chunbao Zhou et al. designed and constructed a 4.0 m scaled DFB reactor and drew a conclusion that DFB is a promising system to achieve carbon negative economy, which deserves more research in the future [11]. Fangjun Wang et al. carried out investigations on biomass gasification of 4.8 m scaled compact-fast dual fluidized bed calcium looping and reported that an increase of the solid circulation rate can improve the concentration and the yield of hydrogen, and decrease the content of tar [12]. In general, the existing experimental studies on DFB reactors have not paid enough attention to the gas–solid flow characteristics of these reactors. Even if there are occasional studies focusing on the gas–solid flow characteristics, the research scale, limited by the experimental cost, is far from the actual production scale. Large gaps in scale are likely to lead to unpredictable changes, so the results of small-scale experiments have great limitations in guiding actual production.

Other researchers have carried out numerical simulations based on DFB reactors. Peter Ohlemuller et al. used a validated Aspen simulation model to study the effects of pressure drop, temperature, circulation characteristics, and other factors on the carbon capture performance of a DFB-based CLC pilot plant [13]. Yao Zhu et al. also used Aspen to simulate low-rank coal pyrolysis staged conversion polygeneration technology with a semi-coke heat carrier based on a dual-fluidized bed and found that this technology has absolute advantages in economy compared to the other two polygeneration methods [14]. The results of a numerical investigation reported by Liyan Sun et al. of a 1 MW full-loop chemical looping combustion unit with a DFB reactor showed that the system efficiency was positively correlated with the operating temperature and solid inventory [15]. Numerical simulation on a 1 MW_{th} pilot-scale chemical looping combustion system was carried out by Junjie Lin et al., showing that a higher coal feeding rate will raise the solid temperature, which has positive effects on product yields but negative effects on the transport of solid particles in the system [16]. In general, the existing numerical simulation research on DFB reactors is mainly divided into Aspen simulations and CFD simulations. Aspen simulations pay more attention to the theoretical efficiency and other economic parameters of the process, rather than the internal characteristics of the reactor. Although CFD simulations pay attention to various characteristics of the reactor, they tend to pay more attention to the reaction characteristics and pay less attention to the gas–solid flow characteristics, especially the solid circulation rate.

In summary, gas–solid flow characteristics, especially the solid circulation rate, are the key parameters used to judge the normal operation of a DFB reactor system. In the existing relevant research, the gas–solid flow characteristics studies based on single fluidized bed reactors have lost their reference significance because of the coupling of two fluidized beds. The scale of the experimental systems used in the experimental research based on DFB reactors has a large gap with the actual production, which leads to its reliability being greatly reduced when used to guide the actual production. The numerical simulation research based on DFB reactors pays little attention to the gas–solid flow characteristics, especially the solid circulation rate. Therefore, the existing research on the gas–solid flow characteristics of DFB reactors is still not sufficient.

In view of this, a numerical simulation model was established on the basis of a large-scale DFB reactor with a maximum height of 21.6 m, and a numerical simulation focused on gas–solid flow characteristics was carried out. In this paper, the effects of the superficial gas velocity, static bed height, and particle size on the gas–solid flow characteristics of the DFB reactor, including its pressure distribution, solid suspension density distribution, and solid circulation rate, were studied. The experimental data and simulation results were compared to verify the validity of the numerical simulation model. The validated model was used to simulate the extended working conditions, including a higher static bed height and different superficial gas velocities and particle sizes so as to complete and supplement the data curve. To focus on the solid circulation rate, an empirical formula for a solid circulation rate fitted based on the cold experimental results was firstly verified, and the empirical formula for the solid circulation rate applicable to different particle sizes was further proposed.

With the strong support of the experimental data, the gas–solid flow characteristics of large-scale DFB reactors were innovatively investigated in this numerical simulation study. In particular, a series of empirical formulas were proposed to pay attention to the solid circulation rate. These scientific and innovative results effectively make up for the shortcomings of the current research.

2. CFD Model

2.1. Simulation Objects

Based on the cold mode experiment, a 3D model was constructed according to an actual DFB reactor with a 1:1 ratio, and CFD numerical simulation was carried out using Ansys Fluent 19.0. This research is divided into two parts, verification and extension. Verification refers to using experimental data from previous work to verify the validity of a numerical simulation model. Extension refers to using the validated numerical simulation model to simulate the working conditions not studied in the experimental research, so as to extend and improve the data curve and improve the practicability of the research results.

2.1.1. Verification

Strictness was ensured through two ways in the verification part. On the one hand, the main preset parameters of the simulation, including superficial gas velocity, static bed height, and particle size, were the same as in the experimental study. On the other hand, the main simulation results were compared with the experimental data, including the pressure distribution, solid suspension density distribution, and the solid circulation rate.

2.1.2. Extension

In order to perfect the data curve and improve the reference significance of the research results for practical production applications, the working conditions not involved in the experimental research were simulated. In the first part, different superficial gas velocities and static bed height were used for the simulation. This part of the work improved the data curve of the effect of the superficial gas velocity and static bed height on the gas–solid flow characteristics, and it verified the empirical formula of the solid circulation rate summarized in the experimental study by using new conditions. In the second part, various particle sizes were used for simulation.

This part of the simulation results was sorted into different series according to different particle sizes. Combined with the form of the original empirical formula, the solid circulation rate was fitted with the superficial gas velocity and static bed height, so as to form a series of empirical formulas for practical production reference.

2.2. Model Structure

As shown in Figure 2, the reactor model is mainly divided into 1#FB and 2#FB, with two furnaces, four cyclones, and four loop-seals. The size and structural design of each part of the reactor are shown in Table 1.

Table 1. Size and structure design of main parts.

Item	Value
The height of 1#FB	21.60 m
The cross-sectional area of 1#FB	0.30 m × 0.40 m
The height of 2#FB	14.40 m
The cross-sectional area of 2#FB	0.25 m × 0.40 m
The relative height difference of the inlets	7.20 m
The design of furnace inlet	Tapered

Figure 2. The structure of the DFB system.

2.3. Simulation Considerations

Referring to the experiment, quartz sand and air were used as the circulating material and the fluidized gas, respectively, in the simulation. The quartz sand in 1#FB is returned to 2#FB by two loop-seals arranged symmetrically. Then, the quartz sand in 2#FB is returned to 1#FB by the upper loop-seal and the bottom loop-seal of 2#FB, respectively. In this way, material circulation between the two reactors is achieved.

Because cold mode simulation was developed, various chemical changes and heat transfer effects were not considered. An RNG (renormalization group) k-ε model was adopted as the turbulence model, which is widely used in numerical simulation. The RNG k-ε model, proposed by Yakhot and Orzag in 1986, is a mathematical model derived from renormalization group methods on instantaneous N-S equations [17]. The equations and coefficients are derived from analytical solutions, improving the ability to simulate high-strain flows in ε equations. Soo-Hwang Ahn et al. used a modified RNG k-ε model to propose an unsteady prediction of cavitating flow around a three dimensional hydrofoil and the results showed that strong turbulence is mainly produced in the condensation process, which delays the cavitation inception in the next cycle [18].

In the consideration of multiphase flow, due to the particularity of the materials used, the drag force between the air and sand must be emphasized. Some researchers consider interactions between particles [19,20]. Among the many drag models, the EMMS (energy minimization multi-scale) model, proposed by Jinghai Li, is widely recognized and adopted [21,22]. The EMMS model decomposes the local inhomogeneity of a gas–solid into the coexistence of three homogeneous subsystems, including a dense phase, a dilute phase, and a meso-scale interphase. The governing equations are listed for the three subsystems, and the energy consumption of transporting suspended particles in the system is taken as the stability condition to solve the problem so as to obtain the local flow parameters. Compared with other drag models, the EMMS can give more accurate drag values when the voidage is between 0.1 and 0.5. In Emad Ghadirian's work, the EMMS model performed better in several simulations compared to Syamlal O'Brien's model and Wen and Yu's model [23]. Timo Dymala et al. reported that simulations with the EMMS-based drag model show good agreement with the time-averaged axial solids concentration, circulation rate, and riser pressure drop [24]. Yuelin Yang et al. and Pedram Pakseresht et al. also carried out simulations using EMMS models [25,26].

The main conditions involved in the simulation are shown in Table 2.

Table 2. Numerical conditions.

Item	Settings
Density of sand	2600 kg/m^3
Diameter of sand	100, 282, 641, 1000 μm
Static bed height	0.65/0.95/1.25/1.55/1.85/2.15 m
Inlets	Velocity-inlet type, 3.0/3.5/4.0/4.5 m/s for 1#FB; 2.0/2.5/3.0/3.5 m/s for 2#FB
Outlets	Pressure-outlet type, −50 Pa for both
Elements of mesh	341,146
Average quality of mesh	0.83
Boundary condition of wall	No-slip
Restitution coefficient	0.95
Packing limit	0.6
Specularity coefficient	0.01
Time step size	0.001 s
Residual tolerance	0.001
Viscous model	RNG k-ε (C_μ = 0.0845, $C_{1\varepsilon}$ = 1.42, $C_{2\varepsilon}$ = 1.68, $C_{3\varepsilon}$ = 1.3, Pr = 0.75)
Turbulence multiphase model	Per phase
Solid shear viscosity	Syamlal-O'Brien
Granular bulk viscosity	Lun et al.
Solid pressure	Lun et al.

2.4. *Governing Equation*

2.4.1. Continuity Equation

For gas phase:

$$\frac{\partial}{\partial t}\left(\alpha_g\rho_g\right)+\nabla\cdot\left(\alpha_g\rho_g\vec{v}_g\right)=0 \tag{1}$$

For solid phase:

$$\frac{\partial}{\partial t}(\alpha_s\rho_s)+\nabla\cdot\left(\alpha_s\rho_s\vec{v}_s\right)=0 \tag{2}$$

2.4.2. Momentum Equation

For gas phase:

$$\frac{\partial}{\partial t}\left(\alpha_g\rho_g\vec{v}_g\right)+\nabla\cdot\left(\alpha_g\rho_g\vec{v}_g\vec{v}_g\right)=-\alpha_g\nabla p+\nabla\cdot\bar{\bar{\tau}}_g+\alpha_g\rho_g\vec{g}+K_{sg}(\vec{v}_s-\vec{v}_g) \tag{3}$$

where the gas stress tensor:

$$\overline{\overline{\tau}}_g = \alpha_g \mu_g \left(\nabla \vec{v}_g + \nabla \vec{v}_g^{\,T} \right) - \frac{2}{3} \alpha_g \mu_g \left(\nabla \cdot \vec{v}_g \right) \overline{\overline{I}} \tag{4}$$

For solid phase:

$$\frac{\partial}{\partial t} \left(\alpha_s \rho_s \vec{v}_s \right) + \nabla \cdot \left(\alpha_s \rho_s \vec{v}_s \vec{v}_s \right) = -\alpha_s \nabla p + \nabla \cdot \overline{\overline{\tau}}_s + \alpha_s \rho_s \vec{g} + K_{gs} (\vec{v}_g - \vec{v}_s) \tag{5}$$

where the solid stress tensor:

$$\overline{\overline{\tau}}_s = \alpha_s \mu_s \left(\nabla \vec{v}_s + \nabla \vec{v}_s^{\,T} \right) + \alpha_s \left(\lambda_s - \frac{2}{3} \mu_s \right)_s \left(\nabla \cdot \vec{v}_s \right) \overline{\overline{I}} \tag{6}$$

2.4.3. RNG k-ε Model

Equation of turbulent kinetic energy:

$$\frac{\partial(\rho k)}{\partial t} + \frac{\partial(\rho v_i k)}{\partial x_i} = \frac{\partial}{\partial x_i} \left[(\alpha_k \mu_{eff}) \frac{\partial k}{\partial x_i} \right] + G_k + G_b - \rho \epsilon - Y_M \tag{7}$$

Equation of dissipation rate of turbulent kinetic energy:

$$\frac{\partial(\rho \epsilon)}{\partial t} + \frac{\partial(\rho v_i \epsilon)}{\partial x_i} = \frac{\partial}{\partial x_i} \left[(\alpha_\epsilon \mu_{eff}) \frac{\partial \epsilon}{\partial x_i} \right] + C_{1\epsilon} \frac{\epsilon}{k} (G_k + C_{3\epsilon} G_b) - C_{2\epsilon} \rho \frac{\epsilon^2}{k} \tag{8}$$

2.4.4. EMMS Model

Momentum equation for dense phase:

$$\frac{3}{4} C_{Dc} \frac{f_c(1-\epsilon_c)}{d_s} \rho_g U_{slip-c}^2 + \frac{3}{4} C_{Di} \frac{f_c}{d_s} \rho_g U_{slip-i}^2 = f_c(1-\epsilon_c)(\rho_s - \rho_g)(g+a) \tag{9}$$

Momentum equation for dilute phase:

$$\frac{3}{4} C_{Df} \frac{(1-f_c)(1-\epsilon_c)}{d_s} \rho_g U_{slip-f}^2 = (1-f_c)(1-\epsilon_f)(\rho_s - \rho_g)(g+a) \tag{10}$$

Pressure drop balance equation:

$$C_{Df} \frac{(1-\epsilon_f)}{d_s} \rho_g U_{slip-f}^2 + C_{Di} \frac{f_c}{(1-f_c) d_{cl}} \rho_g U_{slip-i}^2 = C_{Dc} \frac{1-\epsilon_c}{d_s} \rho_g U_{slip-c}^2 \tag{11}$$

Mass conservation equation for gas phase:

$$U_g = f_c U_c + (1-f_c) U_f \tag{12}$$

Mass conservation equation for solid phase:

$$U_s = f_c U_{s-c} + (1-f_c) U_{s-f} \tag{13}$$

Diameter of cluster:

$$d_{cl} = \frac{d_s \left[\frac{U_s}{1-\epsilon_{max}} - \left(U_{mf} + \frac{U_s \epsilon_{mf}}{1-\epsilon_{mf}} \right) \right] g}{\frac{N_{st} \rho_p}{\rho_s - \rho_g} - \left(U_{mf} + \frac{U_s \epsilon_{mf}}{1-\epsilon_{mf}} \right) g} \tag{14}$$

Total voidage equation:

$$\alpha_g = \epsilon_c f_c + \epsilon_f (1-f_c) \tag{15}$$

Stability conditions:

$$\frac{N_{st}}{N_T} = \frac{U_g(1-\alpha_g) - fU_f(\epsilon_f - \alpha_g)(1-f_c)}{U_g(1-\alpha_g)} = \min \quad (16)$$

3. Results and Discussion

3.1. The Effect of Grid Resolution

It is necessary to determine the effect of grid resolution on the simulation results before proceeding to the formal simulation. In view of this, three simulation cases with a grid of 276,732, 341,146, and 421,208 cells are conducted. As shown in Figures 3 and 4, the pressure distribution along the furnace height of both furnaces was almost identical in all three cases when the superficial gas velocity of 1#FB and 2#FB was 3.5 m/s and 2.5 m/s and the static bed height was 0.65 m. This indicates that the grid resolution has very little effect on the simulation results. Therefore, considering both the accuracy of the results and the time cost, the formal simulations were carried out adopting the 341,146 cells grid.

3.2. Verification

In the verification parts, two series of simulations focusing on the superficial gas velocity of 1#FB and the static bed height were conducted using particles with a particle size of 282 μm as materials, and two series of results were obtained. The simulation results revealed the operation pattern of the DFB system preliminarily. After being compared with the experimental data, the simulation results effectively verified the validity of the numerical simulation model, which laid a solid foundation for the work in the expansion stage.

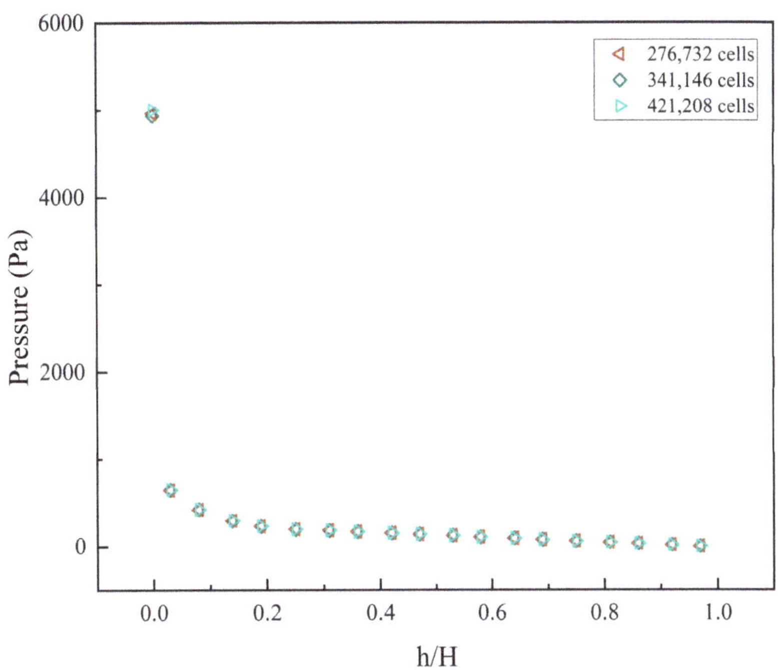

Figure 3. The pressure distribution along the height of 1#FB under different grid resolutions.

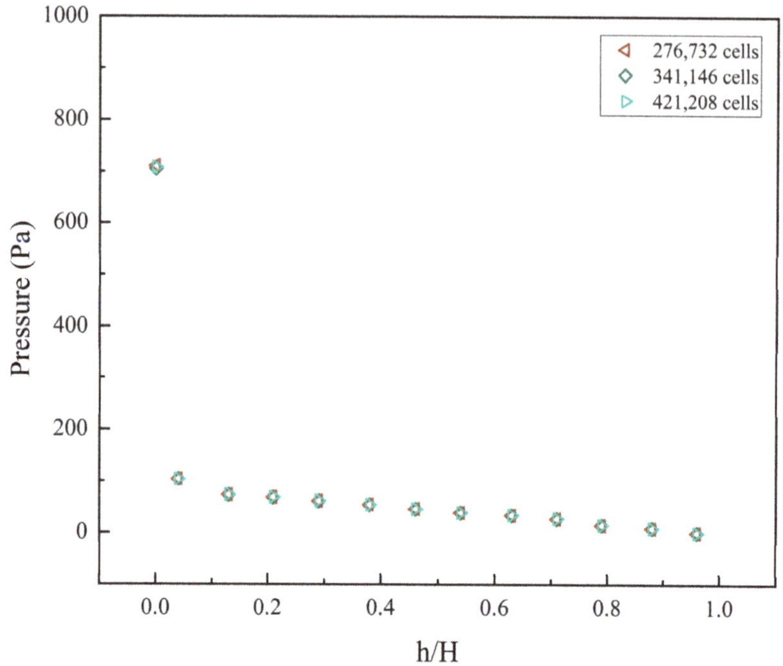

Figure 4. The pressure distribution along the height of 2#FB under different grid resolutions.

3.2.1. The Effects of Superficial Gas Velocity of 1#FB on Gas–Solid Flow Characteristics

Figures 5 and 6 show the effects of the superficial gas velocity of 1#FB on the distribution of the furnace pressure through the pressure contour and pressure distribution along the height of the furnace, respectively. In the verification parts, the particle size used was 282 μm. In this series, the static bed height and superficial gas velocity were fixed at 0.65 m and 2.5 m/s, respectively, while the superficial gas velocity of 1#FB was 3.0, 3.5, 4.0, and 4.5 m/s. In Figure 5, the superficial gas velocity of 1#FB corresponding to each pressure contour from left to right gradually increased. As can be seen from Figure 5, the pressure was balanced between important components such as the bottom of the two furnaces and four loop-seals. Specifically, the pressure at the bottom of the two furnaces was relatively large, and this bottom area was about 1/10 of the total furnace height. With the increase in height, especially when the critical point of 1/10 was broken, the pressure decreased rapidly. This is because in the material balance, the material is piled up at the bottom of the furnace, and is very thin in the upper part. As can be seen from the figure, with the increase of the superficial gas velocity of 1#FB, the pressure at the bottom of 1#FB obviously decreases. The pressure change at the bottom of 2#FB gradually increases slightly. Due to the scale range, the change in the bottom of 2#FB is not easily detected, which can be more intuitively observed in the pressure distribution diagram along the furnace height.

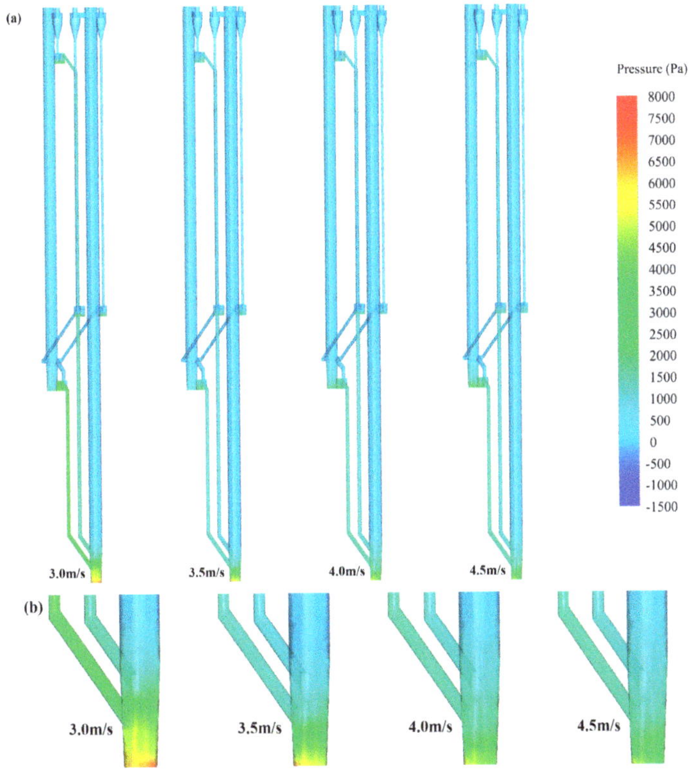

Figure 5. Pressure contour under different superficial gas velocities of 1#FB: (**a**) Full view; (**b**) partially enlarged view.

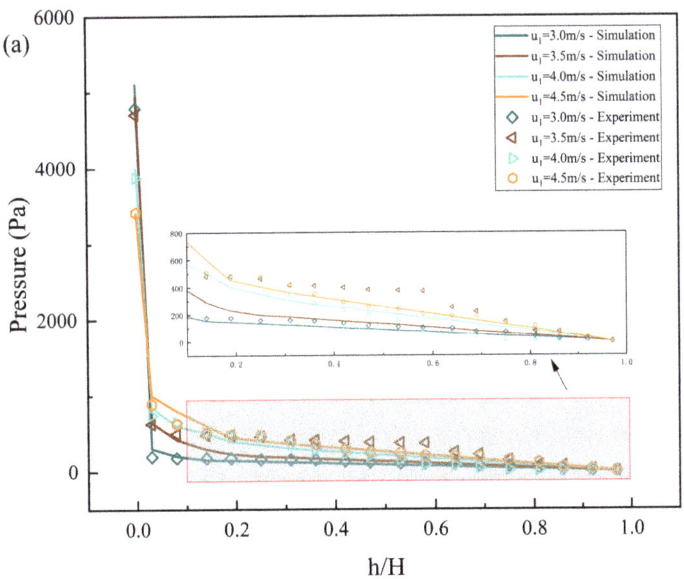

Figure 6. *Cont.*

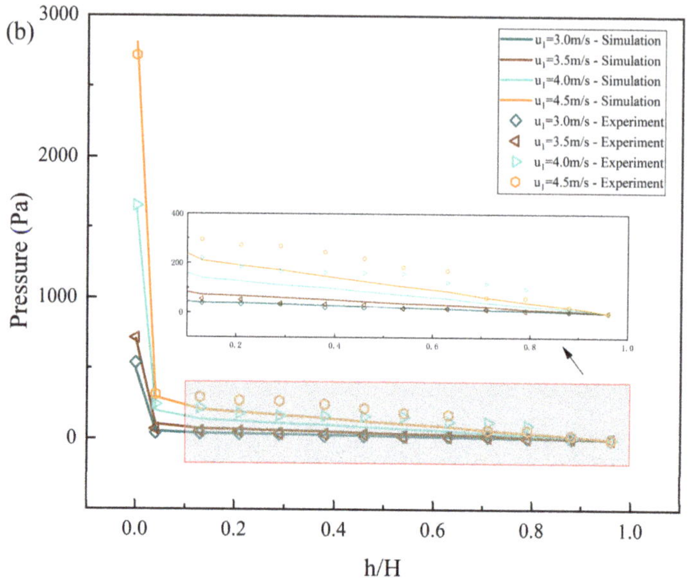

Figure 6. The effects of superficial gas velocity of 1#FB on the pressure distribution along the furnace height: (**a**) 1#FB; (**b**) 2#FB.

Figure 6a,b show the distribution of the pressure of 1#FB and 2#FB along the furnace height, respectively. Compared with the pressure contour, the graph shows more detailed changes. In addition, in the graph, the simulation results and experimental data are directly compared so as to verify the validity of the simulation results and the accuracy of the established model. The superficial gas velocity of 1#FB has a regular and clear effect on the pressure balance of the two furnaces. It can be clearly seen from Figure 6a that the furnace bottom pressure of 1#FB decreases with the increase of superficial gas velocity of 1#FB, while the pressure in the upper zone increases with the increase of the superficial gas velocity of 1#FB. Correspondingly, Figure 6b shows that the pressure at each height of 2#FB increases with the increase of superficial gas velocity of 1#FB. By comparing the simulation results with the experimental data, it can be found that the general trend of both series is consistent, and the data points in the two series match each other well. Specifically, the bottom data points fit well, which verifies the accuracy of the simulation results. In the middle and upper part, the pressure values obtained by the experiment fluctuate greatly but still have a certain law, while the simulation results completely show a relatively uniform law, which can be easily observed in the local magnification diagram. The reason can be considered that the experiment is subject to more external influences, and the error is relatively larger, which leads to large fluctuations of each data point under the constraint of a certain law, and causes great interference with the summary of the law. In contrast, simulations are carried out in idealized environments and settings, so that the results are easily uniform. Compared with the experiment, numerical simulation has some advantages in exploring and summarizing the trend and law.

Figures 7 and 8 show the effects of the superficial gas velocity of 1#FB on the solid suspension density distribution through the contour of the solid suspension density and the distribution of the solid suspension density along the furnace height, respectively. The setting of its working condition is consistent with that in Figures 5 and 6 above. In Figure 7, the superficial gas velocity of 1#FB corresponding to each solid suspension density contour from left to right gradually increases. For 1#FB, its dense phase zone height is about 1/10 of the total furnace height. With the increasing superficial gas velocity of 1#FB, the solid suspension density of the dense phase zone shows a decreasing trend. It is worth noting that its fluidization state also changes continuously. When the superficial gas velocity of 1#FB is 3.0 and 3.5 m/s, the fluidization state is roughly a turbulent bed. This state is conducive to adequate heat and mass transfer in the bottom region. When the superficial gas velocity of 1#FB is 4.0 and 4.5 m/s, the fluidization state is roughly in a fast bed state. This state facilitates the rapid transfer of materials throughout the furnace, thereby improving the level of material circulation between the two furnaces. For 2#FB, contrary to 1#FB, as the superficial gas velocity of 1#FB continues to increase, its solid suspension density in the bottom area shows an increasing trend. The change in the solid suspension density at the bottom of the two furnaces reflects the effects of the superficial gas velocity of 1#FB on the material balance. That is, the increase in the superficial gas velocity of 1#FB will cause the material balance to shift to 2#FB. By keeping the superficial gas velocity 2 constant at 2.5 m/s, the fluidization state of 2#FB is always in a bubbling bed state. This indicates that the residence time of the material in 2#FB is generally larger than in 1#FB.

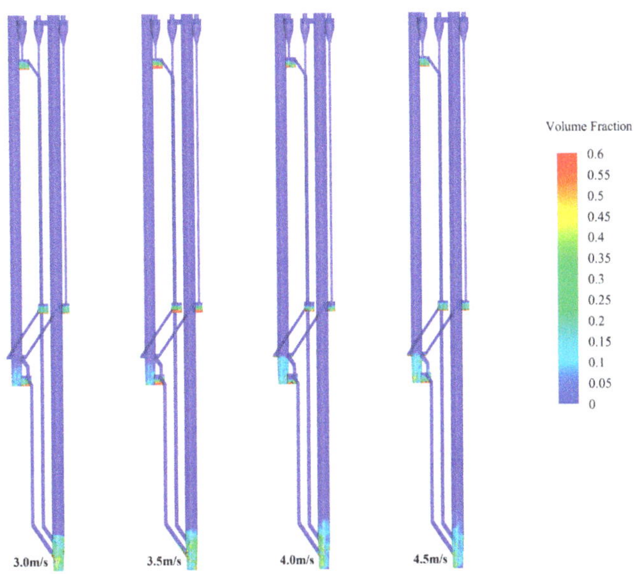

Figure 7. Volume fraction contour under different superficial gas velocities of 1#FB.

Figure 8a,b show the distribution of solid suspension density of 1#FB and 2#FB along the furnace height, respectively. Here, the simulation results are also compared with the experimental data. The superficial gas velocity of 1#FB also has a regular effect on the solid suspension density distribution of the two furnaces. It can be clearly seen from Figure 8a that the solid suspension density at the bottom of 1#FB decreases with the increase in the superficial gas velocity of 1#FB, and yet the solid suspension density in the upper part increases with the increase of the superficial gas velocity of 1#FB. Correspondingly, Figure 8b shows that the solid suspension density at each height of 2#FB increases with the increase of the superficial gas velocity of 1#FB. This is because the increase of the superficial gas velocity of 1#FB changes the material balance, resulting in the transfer of materials from 1#FB to 2#FB, leading to a decrease in the material density in the dense phase zone at the bottom of 1#FB. At the same time, the increase in the superficial gas velocity improves the conveying effect on the material, resulting in a slight increase in the density of 1#FB in the dilute phase zone on the whole.

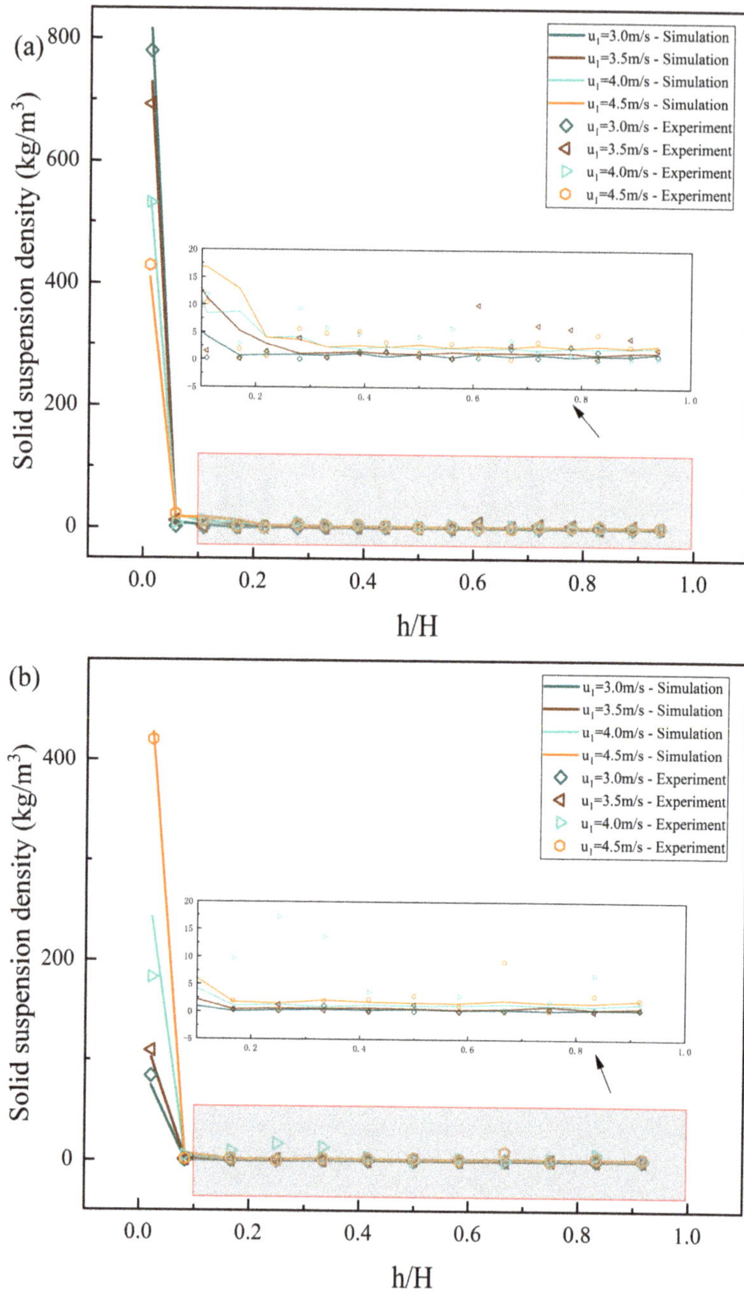

Figure 8. The effects of superficial gas velocity of 1#FB on the solid suspension density distribution along the furnace height. (**a**) 1#FB, (**b**) 2#FB.

In the comparison between the simulation results and experimental data, it can be seen from the figure that for most of the data points, the experimental values and simulation results are very close. Only when u = 4.0 m/s, in the dilute phase zone (h/H < 0.1), there is a large error in absolute value between the two. The reasons may be as follows. First of all, although the absolute error is large, the error is about 10, and the simulation results and experimental data of the bottom dense phase zone (h/H < 0.1) are about 200. The relative error is about 5%, which is acceptable. Secondly, because the other three series are well matched, and the computer used in the simulation is absolutely stable, the error when u = 4.0 m/s may come from the fluctuations and measurement errors of the experimental system during the experiment, which indicates that the data obtained from the experiment may not be completely reliable. Therefore, our simulation revised this part of the data to make the data more reliable. Finally, the consistency of the data in the bottom dense phase zone (h/H > 0.1) is more important because this region contains most of the material. In summary, a few absolute errors do not affect the consistency of the overall trend of the two, which once again verifies the accuracy of the simulation results.

Figure 9 shows the effects of the superficial gas velocity of 1#FB on the solid circulation rate and compares the experimental data with the simulation results. In the figure, the horizontal coordinate is the superficial gas velocity of 1#FB, and the vertical coordinate is the solid circulation rate. As can be seen from the figure, the trend of the experimental data and simulation results is the same, and the values are also very close in general. Specifically, the solid circulation rate is positively correlated with the superficial gas velocity of 1#FB. The solid circulation rate of the experimental data is about 0.20–1.75 kg/(m^{2}s), while the solid circulation rate of the simulation results is about 0.75–2.25 kg/(m^{2}s).

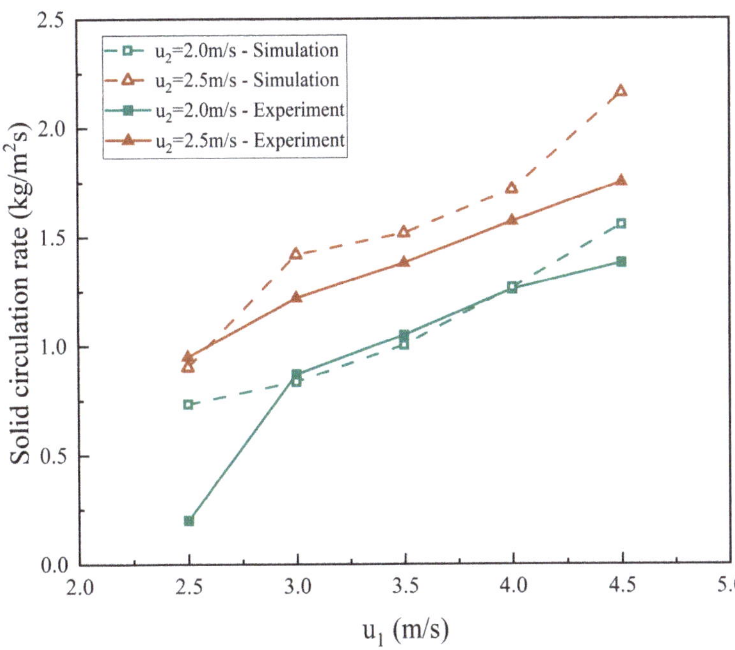

Figure 9. The effects of the superficial gas velocity of 1#FB on the solid circulation rate.

It is worth mentioning that when the superficial gas velocity of 1#FB and 2#FB are 2.5 m/s and 2.0 m/s, respectively, the experimental data significantly deviates from the normal range. This means that the superficial gas velocity is too low, which results in an abnormal circulating flow of the system material. What is significantly different is that the simulation results under this condition are in line with the normal trend; that is, the system is still in the normal circulation flow state at this time. The reason should be that the numerical simulation makes idealized assumptions, while the experiment has many practical limitations, which leads to the difference.

3.2.2. The Effects of Static Bed Height on Gas–Solid Flow Characteristics

Figures 10 and 11 show the effects of the static bed height on the pressure distribution through the pressure contour and pressure distribution along the furnace height. In this series, the superficial gas velocity of 1#FB and 2#FB were fixed at 4.5 m/s and 2.5 m/s, respectively, while the static bed heights were 0.65, 0.95, and 1.25 m, respectively. In Figure 10, the value of the static bed height corresponding to each pressure contour increases gradually from left to right. As can be seen from Figure 10, with the increase of static bed height, the pressure at the bottom of both furnaces increases significantly, which is an inevitable result of the gradual increase in the amount of materials at the bottom of the furnaces. Figure 11 shows more detailed data. The solid suspension density increased at all altitudes. Specifically, the dense phase zone in the bottom part is raised much more than the dilute phase zone in the upper part. This shows that limited by the superficial gas velocity conveying capacity in the furnace, the increase in bed materials caused by the increase in static bed height is mainly concentrated at the bottom of the furnace.

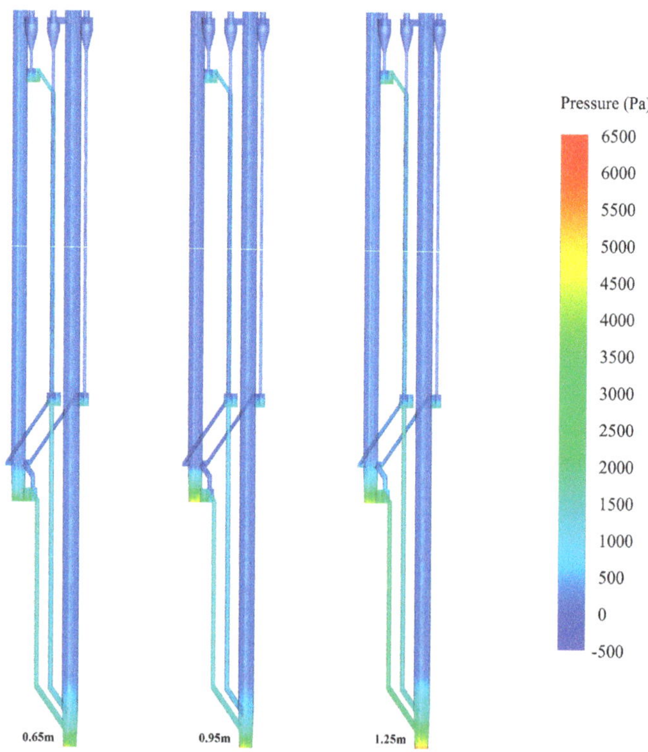

Figure 10. Pressure contour under different static bed heights.

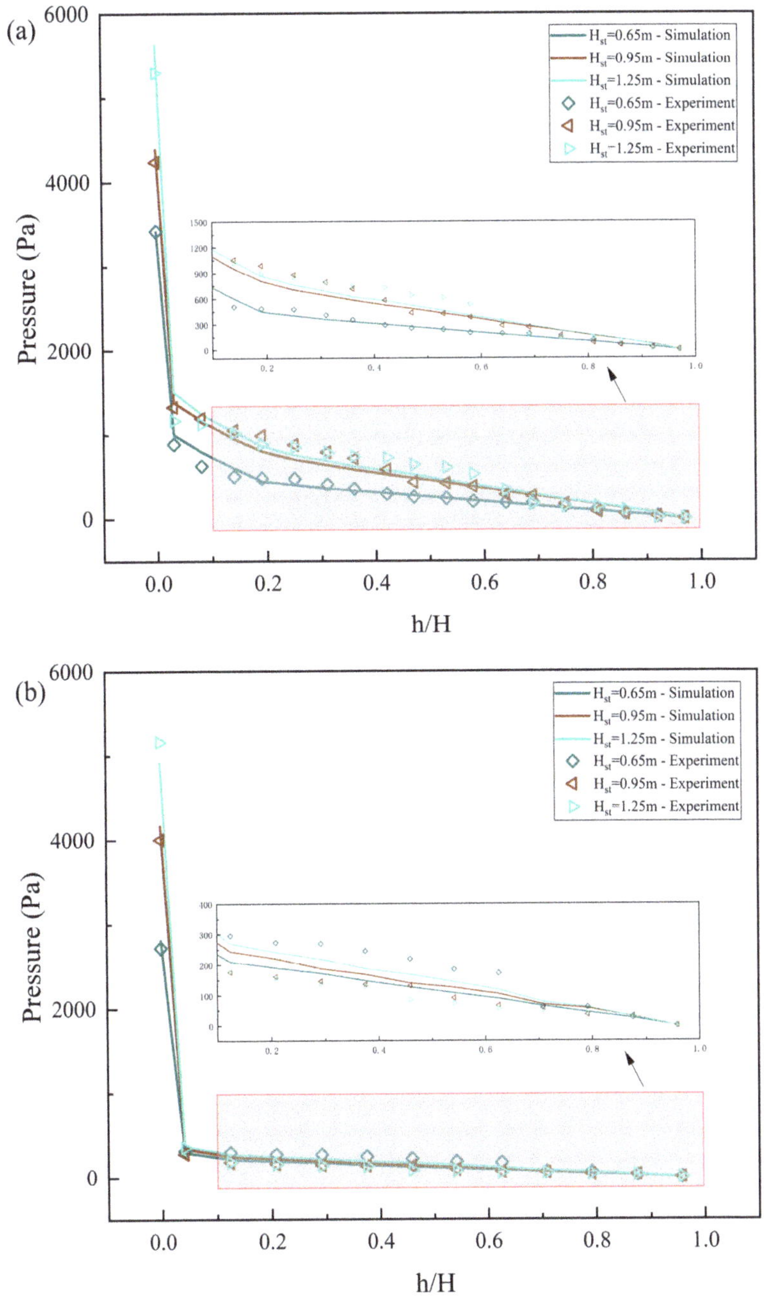

Figure 11. The effects of static bed height on the pressure distribution along the furnace height: (**a**) 1#FB; (**b**) 2#FB.

Figures 12 and 13 show the effects of the static bed height on the solid suspension density through the solid suspension density contour and the distribution graphs of the solid suspension density along the furnace height. The setting of its working condition is consistent with that in Figures 10 and 11 above. In Figure 12, the static bed height values corresponding to each solid suspension density contour gradually increase from left to right. As can be seen from the figure, the fluidization state of 1#FB is roughly in a fast bed state, while the fluidization state of 2#FB is always in a bubbling bed state. At the same time, with the gradual increase of static bed height, the height of the dense phase zone of the two furnaces also gradually increases. When the static bed height is 0.65, 0.95, and 1.25 m, the dense phase zone height of 1#FB is about 1.49, 1.86, and 2.23 m, while the height of the dense phase zone of 2#FB is about 1.12, 1.22, and 1.31 m, respectively. The reasons behind this are obvious. The increase of static bed height leads to the increase of the initial material in the bed, which also increases the solid suspension density of the material in the bed. In particular, since most of the materials are concentrated at the bottom of the furnace during operation, the increase in the bottom materials is particularly obvious, which correspondingly increases the height of the dense phase zone. Figure 13a,b show the distribution of solid suspension density along the furnace height of 1#FB and 2#FB, respectively. The increase in static bed height directly increases the solid suspension density of both furnaces at all heights. Among them, the increase in the bottom is significantly greater than that of the upper area. The reasons for this are as described above.

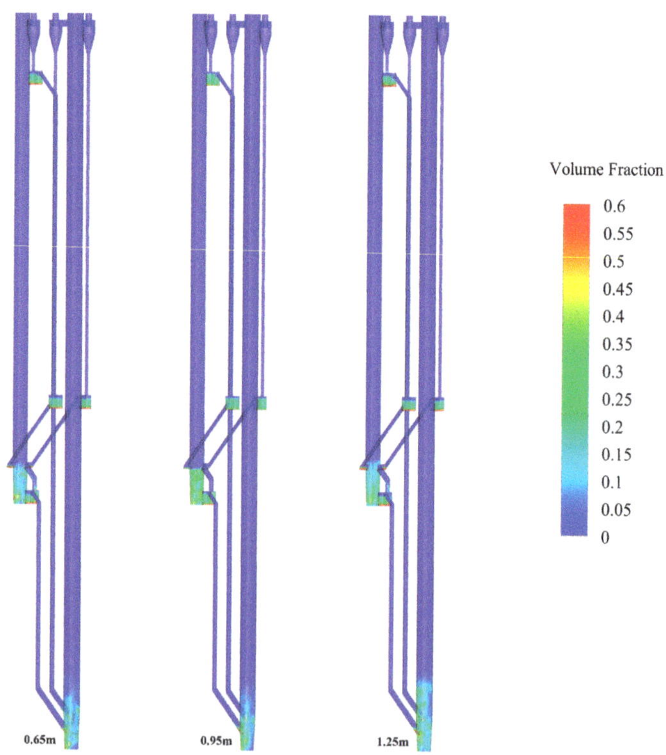

Figure 12. Volume fraction contour under different static bed heights.

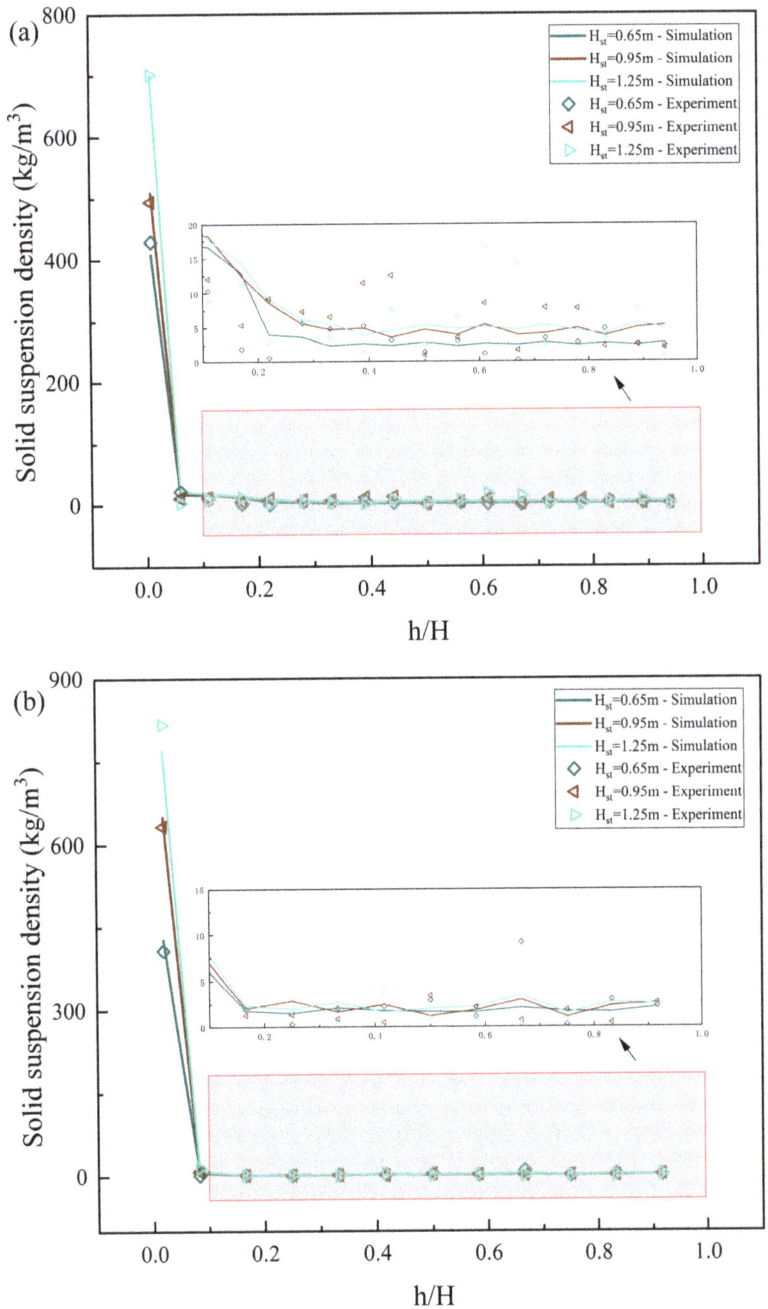

Figure 13. The effects of static bed height on the solid suspension density distribution along the furnace height: (**a**) 1#FB; (**b**) 2#FB.

Figure 14 shows the effects of the static bed height on the solid circulation rate and compares the experimental data with the simulation results. In the figure, the horizontal coordinate is the static bed height and the vertical coordinate is the solid circulation rate. As can be seen from the figure, the solid circulation rate is positively correlated with the static bed height. Specifically, when the three superficial gas velocity settings were used for steady operation, the solid circulation rate given by the experimental data and the simulation results was very close, which is about 0.8–3.6 kg/(m^{2}s). It is worth mentioning that in this series of data, a marginal decreasing phenomenon of the influence of the superficial gas velocity of 1#FB and the static bed height on the solid circulation rate is observed. When the static bed height is equal to 0.65 m, the impact of the superficial gas velocity on the solid circulation rate of the two furnaces is much lower than when the static bed height is equal to 0.95 m and 1.25 m. When the superficial gas velocity of 2#FB is equal to 2 m/s, and the superficial gas velocity of 1#FB is equal to 3 m/s, the effect of the static bed height on the solid circulation rate is much lower than when the other two superficial gas velocity settings are used.

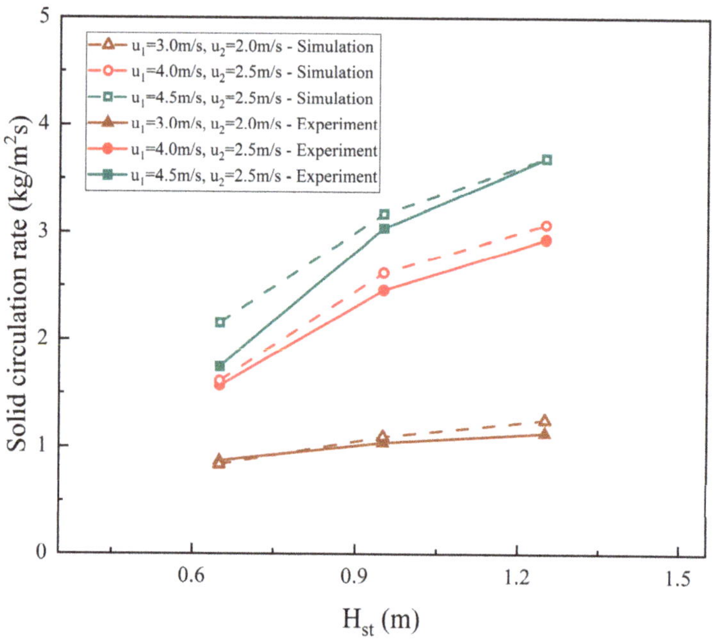

Figure 14. The effects of static bed height on the solid circulation rate.

3.3. *Extension*

In the extension stage, firstly, a particle with a particle size of 282 μm was used as the material in simulations focusing on the superficial gas velocity of 2#FB and the static bed height. The effects of the superficial gas velocity of 2#FB on the gas–solid flow characteristics were obtained. The effects of the static bed height on the gas–solid flow characteristics were expanded and deepened. In addition, aiming at the characteristics of the solid circulation rate, four particle sizes of 100 μm, 282 μm, 641 μm, and 1000 μm were used in the simulations, and a series of data were obtained and fitted to obtain empirical formulas. The effects of the particle size on the solid circulation rate were analyzed and summarized.

3.3.1. The Effects of Superficial Gas Velocity of 2#FB on Gas–Solid Flow Characteristics

Figures 15 and 16 are the pressure contour and solid suspension density contour, respectively, where the effects of the superficial gas velocity of 2#FB on the distribution of the furnace pressure and solid suspension density are presented. In this series, the particle size, superficial gas velocity of 1#FB, and static bed height were 282 μm, 4.5 m/s, and

0.65 m, respectively, while the superficial gas velocity of 2#FB was 2.0, 2.5, 3.0, and 3.5 m/s, respectively. In each group of graphs, the values of the superficial gas velocity of 2#FB corresponding to each contour from left to right gradually increases. When the superficial gas velocity of 2#FB increases, the solid suspension density at the bottom of 1#FB increases, while the solid suspension density at the bottom of 2#FB decreases. The same pattern applies to pressure. From the solid suspension density contour, it can be seen that with the gradual increase of the superficial gas velocity of 2#FB, the flow state of 2#FB continues to evolve. When the superficial gas velocity of 2#FB is 2.0 m/s, 2#FB is in a bubbling bed state. When the superficial gas velocity of 2#FB is 2.5 and 3.0 m/s, 2#FB gradually transforms into a turbulent bed, and the material balance gradually moves to 1#FB. When the superficial gas velocity of 2#FB is 3.5 m/s, the strong transport capacity of the superficial gas velocity of 2#FB superimposes the spatial position relationship, resulting in the material in 2#FB being unable to form a dense phase zone stably. Instead, it is quickly circulated to 1#FB. For 1#FB, with the continuous increase of superficial gas velocity 2, the material balance continues to push to 1#FB, and the materials in it continue to increase, thus raising the height of the dense phase zone.

Figure 17 shows the effect of the superficial gas velocity of 2#FB on the solid circulation rate when the superficial gas velocity of 1#FB is 4.5 m/s. The horizontal coordinate is the superficial gas velocity of 2#FB, and the vertical coordinate is the solid circulation rate. As can be seen from the figure, the solid circulation rate is positively correlated with the superficial gas velocity of 2#FB. In this series, the solid circulation rate was about 2.5–5.0 kg/(m^{2}s). As far as the growth rate of the solid circulation rate, the superficial gas velocity of 2#FB has the same effect as that of 1#FB.

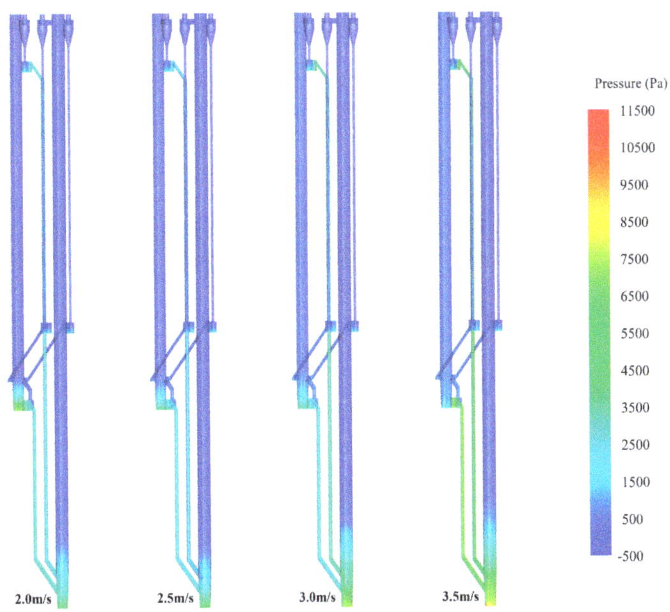

Figure 15. Pressure contour under different superficial gas velocities of 2#FB.

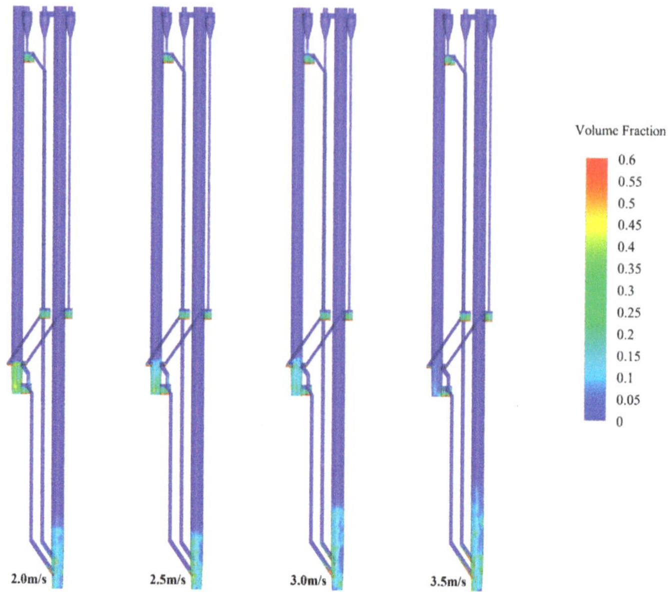

Figure 16. Volume fraction contour under different superficial gas velocities of 2#FB.

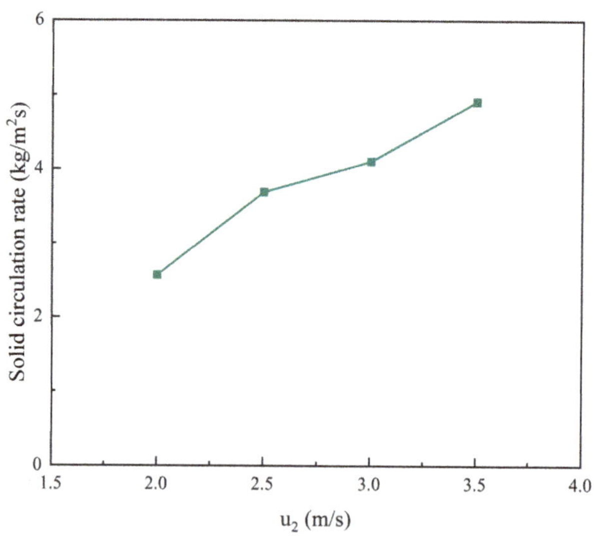

Figure 17. The effects of superficial gas velocity of 2#FB on the solid circulation rate.

3.3.2. The Effects of Static Bed Height on Gas–Solid Flow Characteristics

Figures 18 and 19 show the pressure contour and solid suspension density contour, respectively, showing the effects of the static bed height on the distribution of the pressure and solid suspension density, respectively. In this series, the particle size and superficial gas velocity of 1#FB and 2#FB were 282 μm for both, and 4.5 m/s and 2.5 m/s, respectively. The static bed height was 0.65, 0.95, 1.25, 1.55, 1.85, and 2.15 m, respectively. In each group of graphs, the value of the static bed height corresponding to each contour from left to right gradually increases. Compared with the verification phase, three new working conditions were added to obtain more simulation results and fully improve the data series to summarize deeper rules. When the static bed height increases, the amount of material in the bed increases, the overall solid suspension density in the bed increases, and the pressure also increases. Since the superficial gas velocity of the two furnaces remains constant, the flow states of 1#FB and 2#FB are always a turbulent bed and bubbling bed, respectively. At the same time, the height of the dense phase zone of the two furnaces continues to increase with the increase in the static bed height. When the static bed height is 1.55, 1.85, and 2.15 m, the height of the dense phase zone of 1#FB is 2.63, 2.82, and 3.01, while that of 2#FB is 1.52, 1.61, and 1.71 m, respectively. It can be seen that the height of the dense phase zone of 1#FB has a more obvious variation.

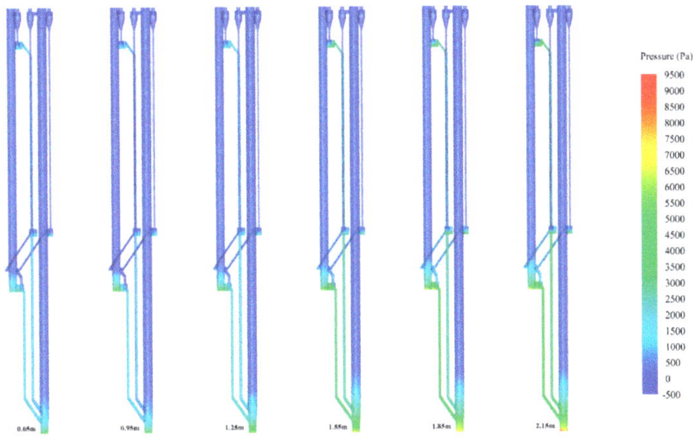

Figure 18. Pressure contour with new static bed height.

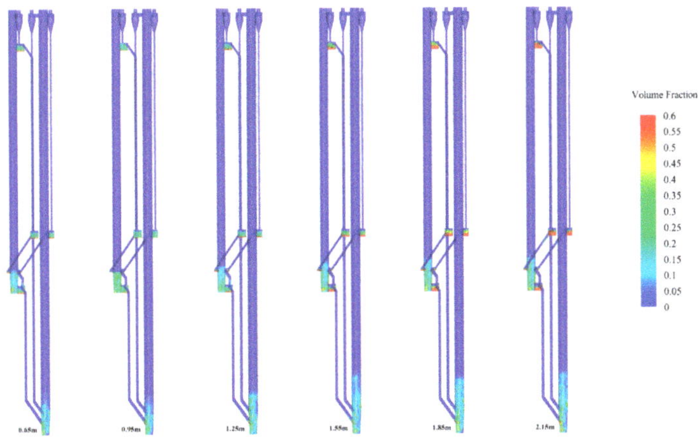

Figure 19. Volume fraction contour with new static bed height.

Figure 20 shows the effects of the static bed height on the solid circulation rate. In the figure, the horizontal coordinate is the static bed height and the vertical coordinate is the solid circulation rate. In this series, the solid circulation rate was about 1.9–4.5 kg/(m^{2}s). As can be seen from the figure, the solid circulation rate and static bed height are still positively correlated, but there is a marginal decreasing phenomenon. When the static bed height increases from 0.65 m to 0.95 m, the solid circulation rate jumps sharply from 1.9 kg/(m^{2}s) to 3.2 kg/(m^{2}s). Subsequently, with the continuous increase in the static bed height, the increase in the solid circulation rate is far less than before. This is because although the materials available for circulation in the bed are gradually increasing, the conveying capacity of the airflow of the two furnaces is limited, thus limiting the continuous increase of the solid circulation rate.

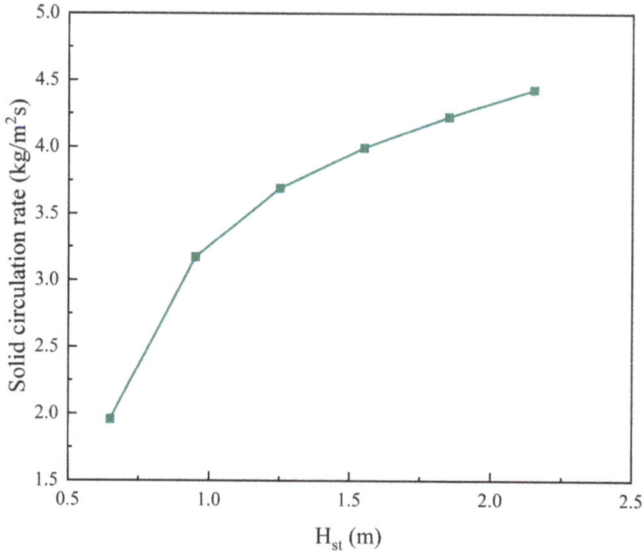

Figure 20. The effects of static bed height on the solid circulation rate (Extension).

3.3.3. Empirical Formulas of Solid Circulation Rate

Figure 21 shows the fitting of the solid circulation rate of the particles with a particle size of 282 μm with the superficial gas velocity of the two furnaces and the static bed height during steady operation. During fitting, the parameters in the fitting were nondimensionalized based on the simulation that the static bed height is 0.65 m, the superficial gas velocity of 1#FB is 2.5 m/s, the superficial gas velocity of 2#FB is 2 m/s, and the solid circulation rate is 0.2 kg/(m^{2}s). Since the three parameters comprehensively affect the solid circulation rate, and their effects all have a marginal diminishing effect, the following form of fitting is adopted:

$$\frac{G_s}{0.2} = a\left(\ln\left(\frac{H_{st}}{0.65}\right) + 1\right)\left(\ln\left(\frac{u_1}{2.5}\right) + \ln\left(\frac{u_2}{2}\right)\right) + 1 \tag{17}$$

where G_s is the solid circulation rate in kg/(m^{2}s); a is a constant defined after fitting; H_{st} is the static bed height in m; u_1 is the superficial gas velocity of 1#FB in m/s; and u_2 is the superficial gas velocity of 2#FB in m/s.

Here, three separate fittings were performed using different data sources. The first fitting was completely fitted using the experimental data as the basis for three fittings. The second fit took all of the simulation results and enriched the data points as a complement to the first fit. Due to the large deviation between the experimental data and the simulation when the superficial gas velocity of 1#FB was 2.5 m/s and the superficial gas velocity of 2#FB was 2 m/s, the intercept restriction was removed and the simulation results were fitted again for the third time.

a = 11.96 is obtained by fitting the experimental data, so the fitting result is as follows:

$$G_s = 2.392\left(\ln\left(\frac{H_{st}}{0.65}\right) + 1\right)\left(\ln\left(\frac{u_1}{2.5}\right) + \ln\left(\frac{u_2}{2}\right)\right) + 0.2 \tag{18}$$

a = 12.32 is obtained by fitting the simulation results, so the fitting result is as follows:

$$G_s = 2.464\left(\ln\left(\frac{H_{st}}{0.65}\right)+1\right)\left(\ln\left(\frac{u_1}{2.5}\right)+\ln\left(\frac{u_2}{2}\right)\right)+0.2 \tag{19}$$

After removing the intercept restriction, the simulation results were used and a = 11.59 and an intercept of 1.90 are obtained. Therefore, the fitting result is as follows:

$$G_s = 2.318\left(\ln\left(\frac{H_{st}}{0.65}\right)+1\right)\left(\ln\left(\frac{u_1}{2.5}\right)+\ln\left(\frac{u_2}{2}\right)\right)+0.38 \tag{20}$$

Combining the above equations and Figure 21, it can be seen that the fitting results obtained from the experimental data and the simulation results are very close.

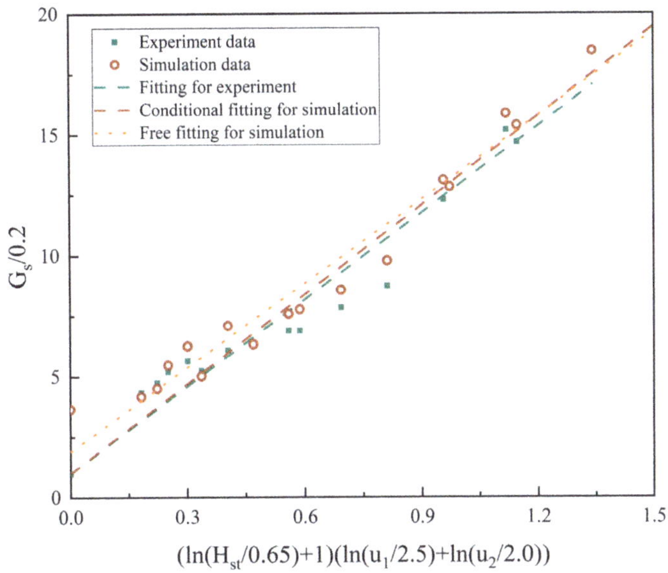

Figure 21. The solid circulation rate versus operating parameters—282 μm particles.

Figure 22 shows the fitting of the solid circulation rate with a particle size of 641 μm with the superficial gas velocity of the two furnaces and the static bed height during steady operation. During fitting, the parameters in the fitting were nondimensionalized based on the simulation that the static bed height was 0.65 m, the superficial gas velocity of 1#FB was 3.5 m/s, and the superficial gas velocity of 2#FB was 2 m/s when the solid circulation rate was 0.0238 kg/(m^{2}s). The fitting form is the same as for the 282 μm particles:

$$\frac{G_s}{0.0238} = a\left(\ln\left(\frac{H_{st}}{0.65}\right)+1\right)\left(\ln\left(\frac{u_1}{3.5}\right)+\ln\left(\frac{u_2}{2}\right)\right)+1 \tag{21}$$

Here, three separate fits were performed using different data sources. The first fit was performed entirely using experimental data. The second fit was completely fitted using the simulation results. The third fitting combined the experimental data and simulation results.

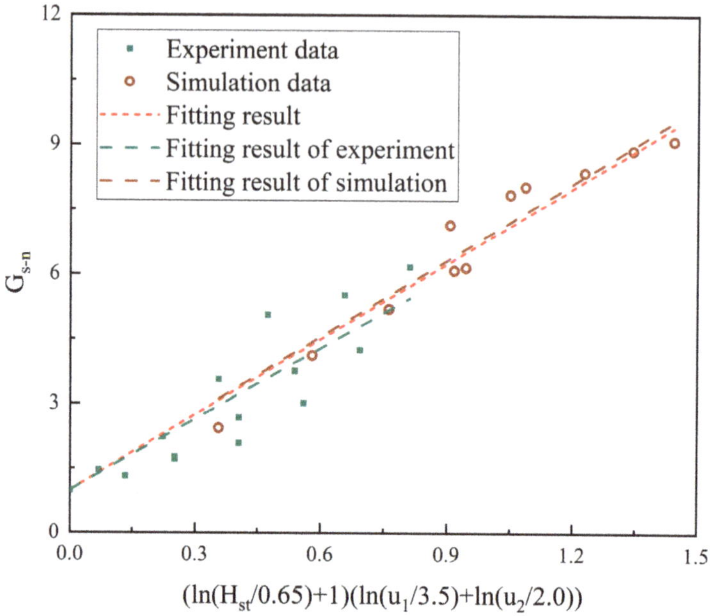

Figure 22. The solid circulation rate versus operating parameters—641 µm particles.

a = 5.48 is obtained by fitting experimental data, so the fitting result is as follows:

$$G_s = 0.1304\left(\ln\left(\frac{H_{st}}{0.65}\right)+1\right)\left(\ln\left(\frac{u_1}{3.5}\right)+\ln\left(\frac{u_2}{2}\right)\right)+0.0238 \quad (22)$$

a = 5.92 is obtained by fitting the simulation results, so the fitting result is as follows:

$$G_s = 0.1409\left(\ln\left(\frac{H_{st}}{0.65}\right)+1\right)\left(\ln\left(\frac{u_1}{3.5}\right)+\ln\left(\frac{u_2}{2}\right)\right)+0.0238 \quad (23)$$

a = 5.83 is obtained by fitting the experimental data and simulation results, so the fitting result is as follows:

$$G_s = 0.1388\left(\ln\left(\frac{H_{st}}{0.65}\right)+1\right)\left(\ln\left(\frac{u_1}{3.5}\right)+\ln\left(\frac{u_2}{2}\right)\right)+0.0238 \quad (24)$$

Combining the above equations and Figure 22, it can be seen that the fitting results obtained from the experimental data and the simulation results are also very close. Compared with 282-micron particles, the solid circulation rate of 641-micron particles during operation is generally an order of magnitude lower, which shows that particle size is the most critical factor affecting solid circulation rate.

Figure 23 shows the fitting of the solid circulation rate of 100 µm and 1000 µm particles with the superficial gas velocity of the two furnaces and static bed height during steady-state operation. For 100 µm particles, the parameters in the fitting were nondimensionalized based on the simulation that the static bed height was 0.65 m, the superficial gas velocity of 1#FB was 3 m/s, and the superficial gas velocity of 2#FB was 2 m/s when the solid circulation rate was 2.84 kg/(m^{2}s). The fitting form is as follows:

$$\frac{G_s}{2.84} = a\left(\ln\left(\frac{H_{st}}{0.65}\right)+1\right)\left(\ln\left(\frac{u_1}{3}\right)+\ln\left(\frac{u_2}{2}\right)\right)+1 \quad (25)$$

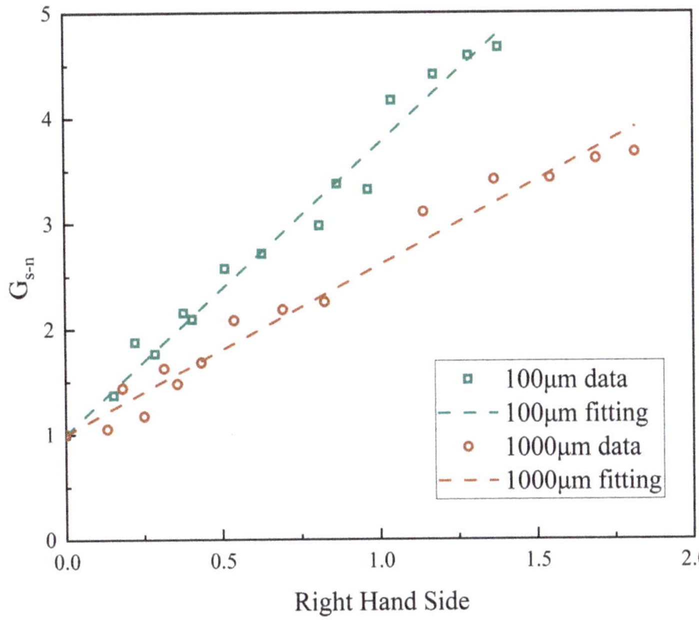

Figure 23. The solid circulation rate versus operating parameters—100 μm and 1000 μm particles.

For 1000 μm particles, the parameters in the fitting were nondimensionalized based on the simulation that the static bed height was 0.65 m, the superficial gas velocity of 1#FB was 3.5 m/s, and the superficial gas velocity of 2#FB was 2.5 m/s when the solid circulation rate was 0.00259 kg/(m^{2}s). The fitting form is as follows:

$$\frac{G_s}{0.00259} = a\left(\ln\left(\frac{H_{st}}{0.65}\right)+1\right)\left(\ln\left(\frac{u_1}{3.5}\right)+\ln\left(\frac{u_2}{2.5}\right)\right)+1 \tag{26}$$

a = 2.76 was obtained by fitting the simulation result of 100 μm, so the fitting result is as follows:

$$G_s = 7.84\left(\ln\left(\frac{H_{st}}{0.65}\right)+1\right)\left(\ln\left(\frac{u_1}{3}\right)+\ln\left(\frac{u_2}{2}\right)\right)+2.84 \tag{27}$$

a = 1.60 is obtained by fitting the 1000 μm simulation result, so the fitting result is as follows.

$$G_s = 0.00414\left(\ln\left(\frac{H_{st}}{0.65}\right)+1\right)\left(\ln\left(\frac{u_1}{3.5}\right)+\ln\left(\frac{u_2}{2.5}\right)\right)+0.00259 \tag{28}$$

Comparing a series of simulation results, it can be found that the solid circulation rate of 282 μm, 641 μm, and 1000 μm particles was three different orders of magnitude from large to small. The previous conclusion that particle size is the most important factor affecting solid circulation rate was verified again.

4. Conclusions

In this paper, a numerical simulation model was established on the basis of a large-scale DFB reactor with a maximum height of 21.6 m, and numerical simulations focused on gas–solid flow characteristics were carried out. The relevant conclusions are summarized as follows:

1. The simulation results are in good agreement with the experimental data in terms of the trend, and the specific values are generally similar. Therefore, the numerical model is scientific.

2. The superficial gas velocity of 1#FB, the superficial gas velocity of 2#FB, and the static bed height all have an effect on the gas–solid characteristics. The effects of the superficial gas velocity on the pressure and solid suspension density of the two furnaces are the opposite, while the static bed height has the same effect on the two furnaces. At the same time, the influences of different parameters on the gas–solid flow characteristics have marginal decreasing effects.
3. According to the simulation results, the empirical formula of the solid circulation rate was fitted according to the different particle sizes. It was found that the influence of particle size on the solid circulation rate could reach three levels of magnitude, and it is the most significant among all of the parameters.

Author Contributions: Conceptualization, Q.W., G.X., M.F. and Z.L.; Methodology, Y.L.; Investigation, Y.L.; Writing—original draft, Y.L.; Writing—review & editing, Q.W. All authors have read and agreed to the published version of the manuscript.

Funding: This work was supported by the Chinese National Natural Science Fund [No. 52076138] and the Fundamental Research Funds for the Central Universities [2022ZFJH004].

Data Availability Statement: Data is contained within the article.

Conflicts of Interest: The authors declare that they have no known competing financial interests or personal relationships that could have appeared to influence the work reported in this paper.

Abbreviations

Symbols

a	inertial term, m/s^2
C	constant in the turbulence model
C_D	drag coefficient
d_s	particle diameter, m
d_{cl}	cluster diameter, m
f	volume fraction
G_b	generation term of turbulent kinetic energy caused by buoyancy
G_k	generation term of turbulent kinetic energy caused by the mean velocity gradient
$\vec{g}$	gravitational acceleration, m/s^2
H	total height of the system, m
H_{st}	static bed height, m
$\bar{\bar{I}}$	unit tensor
K_{ij}	interphase momentum exchange coefficient from phase i to j
k	turbulent kinetic energy, m^2/s^2
N_{st}	mass-specific energy consumption for suspending and transporting particles, W/kg
N_T	mass-specific total energy consumption for particles, W/kg
p	pressure, Pa
t	time, s
u	superficial gas velocity, m/s
U_{slip}	superficial slip velocity, m/s
v or $\vec{v}$	velocity, m/s
Y_m	contribution of pulsation expansion in compressible turbulence
Greek Letters	
α	volume fraction
α_k	reciprocal of effective Prandtl number for turbulent kinetic energy
α_ε	reciprocal of effective Prandtl number for dissipation rate of turbulent kinetic energy
ε	dissipation rate of turbulent kinetic energy
ε_c	voidage of dense phase in the EMMS model
ε_f	voidage of dilute phase in the EMMS model
λ	bulk viscosity, Pa s
μ	viscosity, Pa s

μ_{eff}	effective viscosity, Pa s
ρ	density, kg/m^3
$\bar{\bar{\tau}}$	stress tensor, Pa
Subscripts	
1	1#FB
2	2#FB
c	dense phase in the EMMS model
f	dilute phase in the EMMS model
g	gas phase
i	meso-scale interphase in the EMMS model
mf	minimum fluidization
s	solid phase

References

1. Shrestha, S.; Ali, B.S.; Binti Hamid, M.D. Cold Flow Model of Dual Fluidized Bed: A Review. *Renew. Sustain. Energy Rev.* **2016**, *53*, 1529–1548. [CrossRef]
2. Gupta, S.; De, S. Investigation of Hydrodynamics and Segregation Characteristics in a Dual Fluidized Bed Using the Binary Mixture of Sand and High-Ash Coal. *Adv. Powder Technol.* **2021**, *32*, 2690–2702. [CrossRef]
3. Hanchate, N.; Ramani, S.; Mathpati, C.S.; Dalvi, V.H. Biomass Gasification Using Dual Fluidized Bed Gasification Systems: A Review. *J. Clean. Prod.* **2021**, *280*, 123148. [CrossRef]
4. Abuelgasim, S.; Wang, W.; Abdalazeez, A. A Brief Review for Chemical Looping Combustion as a Promising CO_2 Capture Technology: Fundamentals and Progress. *Sci. Total Environ.* **2021**, *764*, 142892. [CrossRef]
5. Chen, J.; Duan, L.; Ma, Y.; Jiang, Y.; Huang, A.; Zhu, H.; Jiao, H.; Li, M.; Hu, Y.; Zhou, H.; et al. Recent Progress in Calcium Looping Integrated with Chemical Looping Combustion (CaL-CLC) Using Bifunctional CaO/CuO Composites for CO_2 Capture: A State-of-the-Art Review. *Fuel* **2023**, *334*, 126630. [CrossRef]
6. Xu, J.; Lu, X.; Wang, Q.; Zhang, W.; Liu, C.; Xie, X.; Sun, S.; Fan, X.; Li, J. Visualization of Gas-Solid Flow Characteristics at the Wall of a 60-Meter-High Transparent CFB Riser. *Powder Technol.* **2018**, *336*, 180–190. [CrossRef]
7. Bian, Q.; Yang, D.; Wu, Z.; Liu, B.; Zhang, J.; Deng, R.; Pu, H. Investigation of Particle Flow Characteristics in a CFB Recirculation System Using Electrical Capacitance Volume Tomography. *Powder Technol.* **2022**, *408*, 117765. [CrossRef]
8. Stollhof, M.; Penthor, S.; Mayer, K.; Hofbauer, H. Estimation of the Solid Circulation Rate in Circulating Fluidized Bed Systems. *Powder Technol.* **2018**, *336*, 1–11. [CrossRef]
9. Youn, P.-S.; Lee, D.H.; Kim, D.; Won, Y.S.; Choi, J.-H.; Joo, J.B.; Ryu, H.-J. Solids Inventory and External Solids Circulation Rate in Risers of Circulating Fluidized Beds. *Adv. Powder Technol.* **2022**, *33*, 103810. [CrossRef]
10. Pröll, T.; Rupanovits, K.; Kolbitsch, P.; Bolhàr-Nordenkampf, J.; Hofbauer, H. Cold Flow Model Study on a Dual Circulating Fluidized Bed (DCFB) System for Chemical Looping Processes. *Chem. Eng. Technol.* **2009**, *32*, 418–424. [CrossRef]
11. Zhou, C.; Chen, L.; Liu, C.; Wang, J.; Xing, X.; Liu, Y.; Chen, Y.; Chao, L.; Dai, J.; Zhang, Y.; et al. Interconnected Pyrolysis and Gasification of Typical Biomass in a Novel Dual Fluidized Bed. *Energy Convers. Manag.* **2022**, *271*, 116323. [CrossRef]
12. Wang, F.; Zhou, W.; Chen, S.; Duan, L.; Xiang, W. Investigations on Biomass Gasification of Compact-Fast Dual Fluidized Bed Calcium Looping. *J. Clean. Prod.* **2023**, *405*, 137065. [CrossRef]
13. Ohlemüller, P.; Alobaid, F.; Abad, A.; Adanez, J.; Ströhle, J.; Epple, B. Development and Validation of a 1D Process Model with Autothermal Operation of a 1 MW Th Chemical Looping Pilot Plant. *Int. J. Greenh. Gas Control* **2018**, *73*, 29–41. [CrossRef]
14. Zhu, Y.; Li, K.; Wang, Q.; Cen, J.; Fang, M.; Luo, Z. Low-Rank Coal Pyrolysis Polygeneration Technology with Semi-Coke Heat Carrier Based on the Dual-Fluidized Bed to Co-Produce Electricity, Oil and Chemical Products: Process Simulation and Techno-Economic Evaluation. *Fuel Process. Technol.* **2022**, *230*, 107217. [CrossRef]
15. Sun, L.; Xiao, R.; Lin, J.; Kong, D.; Luo, K.; Fan, J. Numerical Investigation of a 1MW Full-Loop Chemical Looping Combustion Unit with Dual CFB Reactors. *Int. J. Greenh. Gas Control* **2023**, *123*, 103835. [CrossRef]
16. Lin, J.; Luo, K.; Hu, C.; Sun, L.; Fan, J. Full-Loop Simulation of a 1 MWth Pilot-Scale Chemical Looping Combustion System. *Chem. Eng. Sci.* **2022**, *249*, 117301. [CrossRef]
17. Yakhot, V.; Orszag, S.A. Renormalization-Group Analysis of Turbulence. *Phys. Rev. Lett.* **1986**, *57*, 1722–1724. [CrossRef] [PubMed]
18. Ahn, S.-H.; Xiao, Y.; Wang, Z.; Luo, Y.; Fan, H. Unsteady Prediction of Cavitating Flow around a Three Dimensional Hydrofoil by Using a Modified RNG K-ε Model. *Ocean Eng.* **2018**, *158*, 275–285. [CrossRef]
19. Al-Hanaya, A.; El-Sapa, S.; Ashmawy, E.A. Axisymmetric motion of an incompressible couple stress fluid between two eccentric rotating spheres. *J. Appl. Mech. Tech. Phy.* **2022**, *63*, 790–798. [CrossRef]
20. El-Sapa, S.; Albalawi, W. Interaction of Two Rigid Spheres Oscillating in an Infinite Liquid under the Control of a Magnetic Field. *J. Appl. Math.* **2023**, *2023*, 1146872. [CrossRef]
21. Li, J.; Tung, Y.; Kwauk, M. Method of energy minimization in multi-scale modeling of particle-fluid two-phase flow. In *Circulating Fluidized Bed Technology*; Elsevier: Amsterdam, The Netherlands, 1988; pp. 89–103, ISBN 978-0-08-036225-0.

22. Li, J.; Tung, Y.; Kwauk, M. Energy Transport and Regime Transition in Particle-Fluid Two-Phase Flow. In *Circulating Fluidized Bed Technology*; Elsevier: Amsterdam, The Netherlands, 1988; pp. 75–87, ISBN 978-0-08-036225-0.
23. Ghadirian, E.; Arastoopour, H. CFD Simulation of a Fluidized Bed Using the EMMS Approach for the Gas-Solid Drag Force. *Powder Technol.* **2016**, *288*, 35–44. [CrossRef]
24. Dymala, T.; Wytrwat, T.; Heinrich, S. MP-PIC Simulation of Circulating Fluidized Beds Using an EMMS Based Drag Model for Geldart B Particles. *Particuology* **2021**, *59*, 76–90. [CrossRef]
25. Yang, Y.; Zhang, Q.; Pan, T.; Qu, Y.; Liu, X.; Li, X.; Cheng, Y.; Wang, L.; Liu, W. The Hydrodynamics of Low-Density Particles and Optimization of the EMMS Calculation Process. *Powder Technol.* **2022**, *405*, 117511. [CrossRef]
26. Pakseresht, P.; Yao, Y.; Fan, Y.; Theuerkauf, J.; Capecelatro, J. A Critical Assessment of the Energy Minimization Multi-Scale (EMMS) Model. *Powder Technol.* **2023**, *425*, 118569. [CrossRef]

MDPI
St. Alban-Anlage 66
4052 Basel
Switzerland
www.mdpi.com

MDPI Books Editorial Office
E-mail: books@mdpi.com
www.mdpi.com/books

www.ingramcontent.com/pod-product-compliance
Lightning Source LLC
Chambersburg PA
CBHW061547220926
51850CB00005B/277
* 9 7 8 3 7 2 5 8 1 3 4 7 6 *